HEALTHY AGING AND LONGEVITY
Third International Conference

ANNALS OF THE NEW YORK ACADEMY OF SCIENCES
Volume 1114

HEALTHY AGING AND LONGEVITY

Third International Conference

Edited by Noah J. Weller and Suresh I. S. Rattan

Published by Blackwell Publishing on behalf of the New York Academy of Sciences
Boston, Massachusetts
2007

Library of Congress Cataloging-in-Publication Data

International Conference on Healthy Ageing and Longevity (3rd: 2006: Melbourne, Vic.)
Healthy aging and longevity: Third International Conference/editor Noah J. Weller, and Suresh I.S. Rattan
p. ; cm. – (Annals of the New York Academy of Sciences, ISSN 0077-8923 ; v. 1114)
Includes bibliographical references.
ISBN-13: 978-1-57331-680-4 (paper : alk. paper)
ISBN-10: 1-57331-680-6 (paper : alk. paper)
1. Aging–Congresses. 2. Longevity–Congresses. I. Weller, Noah J. II. Rattan, Suresh I.S. III. New York Academy of Sciences. IV. Title. V. Series.
[DNLM: 1. Aging–physiology. 2. Age Factors. 3. Disease Susceptibility–physiopathology. 4. Health. 5. Life Expectancy. W1 AN626YL v.1114 2007/WT 104 I5925h 2007]

RA776.75.I58 2007
613–dc22

2007034448

The *Annals of the New York Academy of Sciences* (ISSN: 0077-8923 [print]; ISSN: 1749-6632 [online]) is published 28 times a year on behalf of the New York Academy of Sciences by Blackwell Publishing with offices at 350 Main St., Malden, MA 02148 USA; 9600 Garsington Road, Oxford, OX4 2ZG UK; and 600 North Bridge Rd, #05-01 Parkview Square, 18878 Singapore.

Information for subscribers: For new orders, renewals, sample copy requests, claims, changes of address and all other subscription correspondence please contact the Journals Department at your nearest Blackwell office (address details listed above). UK office phone: +44 (0)1865 778315, fax +44 (0)1865 471775; US office phone: 1-800-835-6770 (toll free US) or 1-781-388-8599; fax: 1-781-388-8232; Asia office phone: +65 6511 8000, fax; +44 (0)1865 471775, Email: customerservices@blackwellpublishing.com

Subscription rates:
Institutional Premium The Americas: $4043 Rest of World: £2246
The Premium institutional price also includes online access to full-text articles from 1997 to present, where available. For other pricing options or more information about online access to Blackwell Publishing journals, including access information and terms and conditions, please visit www.blackwellpublishing. com/nyas
*Customers in Canada should add 6% GST or provide evidence of entitlement to exemption.
**Customer in the UK or EU: add the appropriate rate for VAT EC for non-registered customers in countries where this is applicable. If you are registered for VAT please supply your registration number.

Mailing: The *Annals of the New York Academy of Sciences* is mailed Standard Rate. Mailing to rest of world by DHL Smart & Global Mail. Canadian mail is sent by Canadian publications mail agreement number 40573520. **Postmaster:** Send all address changes to *Annals of the New York Academy of Sciences*, Blackwell Publishing Inc., Journals Subscription Department, 350 Main St., Malden, MA 02148-5020.

Membership information: Members may order copies of *Annals* volumes directly from the Academy by visiting www.nyas.org/annals, emailing membership@nyas.org, faxing 212-298-3650, or calling 800-843-6927 (US only), or 212-298-8640 (International). For more information on becoming a member of the New York Academy of Sciences, please visit www.nyas.org/membership. Claims and inquiries on member orders should be directed to the Academy at email: membership@nyas.org or Tel: 212-298-8640 (International) or 800-843-6927 (US only).

Printed in the USA. Printed on acid-free paper.

Annals are available to subscribers online at the New York Academy of Sciences and also at Blackwell Synergy. Visit www.blackwell-synergy.com or www.annalsnyas.org to search the articles and register for table of contents e-mail alerts. Access to full text and PDF downloads of *Annals* articles are available to nonmembers and subscribers on a pay-per-view basis at www.blackwell-synergy.com and www.annalsnyas.org.

The paper used in this publication meets the minimum requirements of the National Standard for Information Sciences Permanence of Paper for Printed Library Materials, ANSI Z39.48_1984.

ISSN: 0077-8923 (print); 1749-6632 (online)
ISBN-10: 1-57331-680-6 (alk. paper); ISBN-13: 978-1-57331-680-4 (alk. paper)

A catalogue record for this title is available from the British Library.

ANNALS OF THE NEW YORK ACADEMY OF SCIENCES

Volume 1114
October 2007

HEALTHY AGING AND LONGEVITY
Third International Conference

Editors
NOAH J. WELLER AND SURESH I. S. RATTAN

This volume is the result of a conference entitled **3rd International Conference on Healthy Ageing and Longevity**, held at the Melbourne Exhibition and Convention Centre, Australia, on October 13–15, 2006.

CONTENTS

Part II. Clinical Perspective

Part III. Socio-cultural Perspective

Part IV. Interventions

Preface

The International Research Centre for Healthy Ageing and Longevity and The International Conferences on Healthy Ageing and Longevity

As the "baby boomers," the largest generation in history, become elders, the unprecedented rise in the number of people aged over 60 years will dramatically expand market demand for medical and lifestyle solutions that will enable them to live healthier, longer, and more active lives.

Now, more than ever, a critical need exists for the development of appropriate policies so that aging is seen as a resource and not as an isolating and segregating experience. Solid research elucidating the processes of aging must be translated into strategies for clinical practice in order to respond to the needs of an aging population.

This volume is the proceedings of the Third International Conference on Healthy Ageing and Longevity. The mission of this series of meetings is to promote healthy aging and longevity through interdisciplinary collaboration amongst the world's leading experts on the science of health, aging, and longevity, and to disseminate evidence-based knowledge throughout the nations of the developed and developing world for the betterment of humankind.

Five basic domains that cover the key research areas of healthy aging and longevity have been identified:

1. Biological mechanisms of healthy aging and longevity,
2. Healthy aging and longevity determinants,
3. Interventions in aging and age-associated disease,
4. Care and support of the aging and elderly,
5. Policies and strategies for an aging society.

At every level, the various issues and domains listed above pose clear challenges for individuals, families, communities, and societies that require informed and evidence-based policy responses. Although the domains are relatively self-contained, particular issues may overlap several domains. This highlights the importance of maintaining a strong emphasis on a holistic overview.

For this reason, in organizing this conference series, we have endeavored to seek out and disseminate information on specialist activities while also

Ann. N.Y. Acad. Sci. 1114: xi–xv (2007). © 2007 New York Academy of Sciences.
doi: 10.1196/annals.1396.051

fostering awareness of how seemingly disparate areas of knowledge can come together in a total perspective on healthy aging and longevity.

The full spectrum of proven and potential aging interventions was explored at this international meeting, which successfully brought together over 100 high-profile speakers representing 20 countries, with the addition of 50 poster presenters who also made an important contribution to the program. The topics covered in the program included the following:

1) Basic Science and Mechanisms of Aging and Longevity
2) Health and Wellbeing in Older Age
3) Challenges for Aging Individuals, Families, Communities and Societies that require Informed and Evidence-Based Policy Responses
4) Depression and Dementia—Current Knowledge
5) Biological Predictors of Healthy Aging and Longevity
6) Changing Trends in the Quest for Healthy Aging and Longevity
7) Promoting Balance and Preventing Falls in an Aging Population
8) Hormone and Metabolic Interventions in Aging
9) Sustaining Optimal Aging: Inner Strength and Mutual Support
10) The New Caring: Financial and asset management, and substitute decision making by and for older people
11) Population Aging in Developing Countries
12) Baby-Boomer Work Force Participation
13) Natural Approaches to Age-Associated Disorders
14) Frontiers of Knowledge in Biogerontology
15) Wellbeing, Retirement Planning, and Expectations of the Baby-Boomer Generation
16) Quality of Care and Quality of Life for the Elderly
17) Evidence-Based Medicine in Older People
18) Nutritional Interventions in Aging and Age-Associated Disease: Diet and Supplements
19) Psychosocial Predictors of Healthy Aging and Longevity—Lessons learned from Longitudinal Studies
20) Respecting the Elders in our Care
21) Behavioral and Social Interventions for Healthy Aging and Longevity
22) Quality of Care and Quality of Life for the Elderly

The concluding session of the conference was an interactive panel discussion involving a diverse panel of international experts, which resulted in a general consensus on The Top 10 Tips for Healthy Ageing and Longevity based upon the main findings uncovered during the event.

The International Conferences on Healthy Ageing and Longevity have achieved recognition as significant world forums that bring together key opinion leaders from a diverse range of fields to share their ideas and exchange knowledge in an open and harmonious environment. The sponsor, the International Research Centre for Healthy Ageing and Longevity (IRCHAL), has conducted two previous conferences, in Sydney (2004) and Brisbane (2005), and is committed to facilitating a further 10 such world forums. Further information can be found online at www.longevity-international.com and www.irchal.org.

The conference programs are designed to stimulate discussion, debate, collaboration, and the exchange of ideas—creating fertile ground for forward thinking and decision making as the baby-boom generation population inevitably moves through the age structure.

The conferences explore the full spectrum of proven and potential aging interventions, including pharmaceutical, nutritional, clinical, educative, policy, complementary, preventive, and restorative means and the scientific underpinnings supporting them. At the same time the conferences recognize the diversity of aging individuals, cultures, and methods within all communities and across the globe.

All three conferences have been successful in attracting a wide cross-section of delegates, including scientists, academics, educators, policy makers, doctors and allied health professionals, students, and the general public.

In recognition of the quality of this series of conferences, the second and third conferences were co-sponsored by the World Health Organization (WHO), consistent with its view that they were recognized as an important world forum on healthy aging and longevity. The second and third conferences were also endorsed by the United Nations Focal Point on Ageing, the Prime Minister of Australia and the Australian Federal Government, and the Australian State Governments of Victoria and Queensland. This support is gratefully acknowledged.

We would also like to acknowledge the following members of IRCHAL's International Advisory Council for their excellent contribution to the planning and success of the Third International Conference on Healthy Ageing and Longevity:

1. **Prof. Suresh Rattan**, PhD, DSc, Research Professor, Danish Centre for Molecular Gerontology, Department of Molecular Biology, University of Aarhus, Aarhus, Denmark
2. **Prof. S. Jay Olshansky**, PhD, Professor, School of Public Health, University of Illinois, Chicago, IL, USA
3. **Prof. Helen Bartlett**, PhD, MSc, BA, Foundation Director, Australasian Centre on Ageing, The University of Queensland, Brisbane, QLD, Australia
4. **Prof. Leonard Hayflick**, PhD, Professor of Anatomy, School of Medicine, University of California, San Francisco, CA, USA
5. **Dr. Reubin Andres**, PhD, Clinical Director, National Institute on Aging, National Institutes of Health, Baltimore, MD, USA
6. **Prof. Hal Kendig**, AB (Calif), MPL, PhD (SCalif), FASSA, National Convener, ARC-NHMRC Research Network on Ageing Well. Research Professor of Ageing and Health, University of Sydney, NSW, Australia

7. **Dr. Donald K. Ingram**, PhD, Guest Researcher, Laboratory of Experimental Gerontology, National Institute on Aging, National Institutes of Health, Baltimore, MD, USA, Professor and Head, Nutritional Neuroscience and Aging Laboratory, Pennington Biomedical Research Center, Louisiana State University System, Baton Rouge, LA, USA
8. **Dr. George S. Roth**, PhD, CEO, GeroScience Inc, Pylesville, MD, Senior Guest Scientist, National Institute on Aging, National Institutes of Health, Baltimore, MD, USA
9. **Dr. James Joseph**, PhD, Lead Scientist, Laboratory Chief, Laboratory of Neuroscience, Jean-Mayer USDA Human Nutrition Research Center on Aging, Tufts University, Boston, MA, USA
10. **Dr. Bruce A. Carnes**, PhD, Associate Professor, Reynolds Department of Geriatric Medicine, University of Oklahoma College of Medicine, Oklahoma, OK, USA
11. **Prof. Colette Browning**, PhD, Professor of Healthy Ageing, Monash University, Melbourne, VIC, Australia. Convenor, Healthy Ageing Theme, ARC/NHMRC Research Network in Ageing Well
12. **Prof. John McCallum**, BEcon Hons Psych(Qld), MPhil, DPhil(Oxon), Deputy Vice-Chancellor, University of Victoria and Director of TAFE, Melbourne, VIC, Australia
13. **Dr. Bradley J. Willcox**, MD, MSc, Clinical Assistant Professor, Department of Geriatrics, John A. Burns School of Medicine, University of Hawaii. Investigator, Geriatrics, Pacific Health Research Institute, Honolulu, Hawaii, USA. Co-Principal Investigator, Okinawa Centenarian Study, Japan
14. **Dr. D. Craig Willcox**, PhD, Assistant Professor, Okinawa Prefectural University-College of Nursing, Naha City, Okinawa, Japan. Co-Principal Investigator, Okinawa Centenarian Study. Research Associate, Pacific Health Research Institute, Honolulu, Hawaii, USA
15. **Dr. S. Mitchell Harman**, MD, PhD, Director and President, Kronos Longevity Research Institute, Phoenix, AZ, USA
16. **Prof. Brian Morris**, PhD, DSc, FAHA, Professor, Basic and Clinical Genomics Laboratory, School of Medical Sciences, University of Sydney, NSW, Australia
17. **Prof. Rhonda Nay**, PhD, Director Gerontic Nursing Clinical School and ACEBAC, La Trobe University/ Bundoora Extended Care Centre, Bundoora, VIC, Australia
18. **Prof. David Le Couteur**, PhD, FRACP, Professor of Geriatric Medicine, Center for Education and Research on Ageing (CERA), University of Sydney, NSW, Australia
19. **Prof. Julie E. Byles**, BMed, PhD, Director, Research Centre for Gender, Health and Ageing (RCGHA), The University of Newcastle, NSW, Australia

20. **Prof. Brian F. C. Clark**, PhD, ScD, Leader, Danish Centre for Molecular Gerontology and Professor of Biostructural Chemistry, Institute of Molecular Biology, University of Aarhus, Denmark. Chairman, Task Group on International Relations (TGIR), European Federation of Biotechnology (EFB)
21. **Prof. Graham D. Burrows**, AO, KCSJ, MD ChB, BSc, DPM, FRANZCP, FRCPsych, MRACMA, Dip.M.Hlth.Sc (Clinical Hypnosis), FAChAM, Chairman, Mental Health Foundation, Australia. Professor of Psychiatry, Austin Health, Austin Hospital, Heidelberg, VIC, Australia

Special appreciation is due to Prof. Suresh Rattan of the University of Aarhus, Denmark, for his assistance in producing this volume, as well as the team at the *Annals of the New York Academy of Sciences* for their great support.

Sincere thanks are also due to Prof. Helen Bartlett of the Australasian Centre for Ageing, University of Queensland, and Prof. Hal Kendig of the Australian Research Council-National Health and Medical Research Council (ARC-NHMRC) Research Network on Ageing Well for their help in editorial handling of manuscripts, and Renae Nicolaou for her excellent work and unconditional support over the years.

It is our hope that this volume will be one of many to disseminate evidence-based knowledge to nations of the developed and developing world in order to create a healthier aging population.

—Noah J. Weller, John D. Weller, and Christian J. Weller

International Research Centre for Healthy Ageing and Longevity (IRCHAL), Melbourne, Australia

The Science of Healthy Aging

Genes, Milieu, and Chance

SURESH I. S. RATTAN

Laboratory of Cellular Aging, Department of Molecular Biology, University of Aarhus, Aarhus, Denmark

ABSTRACT: Healthy aging and longevity depend on successful and dynamic interactions among biological, psychological, and environmental factors. Biological aging occurs mainly during the period of survival beyond the evolutionarily required *essential lifespan* (ELS). Natural selection processes for survival and successful reproduction have selected for a range of genetically determined ELS-assuring maintenance and repair systems (MRSs). The progressive failure of MRSs, and the consequent accumulation of molecular heterogeneity and damage, underlie the biological basis of aging, age-related diseases, and eventual death. However, the genetic processes of MRSs operate in a complex hierarchy of factors which range from intracellular molecular factors to physiological, psychological, environmental, and other stochastic factors, including chance. This view also facilitates setting up a framework for understanding, researching, and developing effective and realistic strategies for aging intervention, prevention, and therapies. Manipulating genes and the milieu in which genes and gene products operate opens up novel possibilities of aging intervention and prevention. Gene therapy, stem cells, and modulation through functional foods, nutriceuticals, cosmeceuticals and lifestyle alterations, including mild stress-induced hormesis, are examples of such strategies at various levels of development and practice.

KEYWORDS: lifespan; survival; longevity; stress; hormesis; homeostasis; homeodynamics

INTRODUCTION

The doubling of human life expectancy during the last 100 years has occurred primarily because of a reduction in birth-related maternal and infant deaths. Some further improvements in the chances of survival of the elderly after the age of 80 years and living beyond 100 years are also noted, which

Address for correspondence: Suresh I. S. Rattan, Department of Molecular Biology, University of Aarhus, Gustav Wieds Vej 10C, DK8000 Aarhus - C, Denmark. Voice: +45 8942 5034; fax: +45 8612 3178.
rattan@mb.au.dk

Ann. N.Y. Acad. Sci. 1114: 1–10 (2007).
doi: 10.1196/annals.1396.044

TABLE 1. Three principles of aging and longevity

1. ***Evolutionary life history principle:*** Aging is an emergent phenomenon seen primarily in protected environments which allow survival beyond the natural lifespan of a species, termed *essential lifespan* (ELS).
2. ***Non-genetic principle:*** Although genes are involved in determining the ELS of a species, aging and longevity of an individual is not programmed in specific gerontogenes.
3. ***Mechanistic principle:*** Accumulation of molecular damage and increased molecular heterogeneity is the cause of age-related failure of homeodynamics.

are also due to improved biomedical and nutritional conditions. However, this extension in lifespan has not been matched so far by equivalent improvements in the health-span in old age. The main reason for this failure lies in the basic biology of our bodies, which has been shaped through millions of years of evolution, and which may not be amenable to simple interventions. Healthy aging and longevity depend on successful and dynamic interactions among biological, psychological, and environmental factors, and a combination of genes, environment, and chance determine the course and consequences of aging. This view also facilitates setting up a framework for understanding, researching, and developing effective and realistic strategies for aging intervention, prevention, and therapies. Gene therapy, stem cells, and modulation through functional foods, nutriceuticals, cosmeceuticals, and lifestyle alterations are examples of such strategies at various levels of development and practice. The aim of this article is to provide a brief overview of the biological basis of aging, which can provide the bases for the evaluation and development of potential interventions and therapeutic approaches for achieving healthy aging and longevity.

AGING: FROM DESCRIPTION TO PRINCIPLES

Biogerontology, the study of the biological basis of aging, has so far gathered a large body of descriptive data about the age-related changes in organisms, organs, tissues, cells, and macromolecules. These data clearly show that the progression and rate of aging are highly variable in different species, in organisms within a species, in organs and tissues within an organism, in cell types within a tissue, in subcellular compartments within a cell type, and in macromolecules within a cell. Although these data imply that there are no universally applicable markers of aging, three general principles of biological aging can be derived (TABLE 1).

First, the evolutionary *life history principle* describes aging as an emergent phenomenon that takes place primarily in protected environments which allow survival beyond the natural lifespan in the wild. The natural lifespan of a species, termed *essential lifespan* (ELS),[1,2] or the *warranty period*,[3] is the time required to fulfill the Darwinian purpose of life in terms of successful reproduction for the continuation of generations. Species that undergo fast maturation and have an early onset of reproduction with large reproductive potential generally have a short ELS, whereas slow maturation, late onset of reproduction, and small reproductive potential of a species is concurrent with its long ELS. For example, the ELS of *Drosophila* is less than a week as compared with that of about 50 years of *Homo sapiens*, even though in protected environments (laboratories and modern societies, respectively), a large proportion of populations of both species can and do live for much longer than that. Therefore, from an evolutionary point of view, ELS is the canvas against which the genetic selection and functional optimization unfold. Detailed arguments about evolutionary explanations for the origin and occurrence of aging can be accessed in extensive writings of several authors.[4,5]

Second, the *non-genetic principle of aging* specifies that aging and longevity of an individual are not programmed in specific gerontogenes. Unlike development, which is a highly programmed and well-coordinated genetic process in the evolutionary life history of an organism, there is no genetic program that determines the exact duration of survival of an organism. Furthermore, studies on establishing an association between genes and longevity have reported that the genetic heritability of variance in lifespan is less than 35%.[6,7] But a lack of specific gerontogenes with the sole purpose of causing aging and terminating the lifespan of an individual does not imply that genes do not or cannot influence survival, longevity, and the rate of aging. There is ample evidence from studies performed on yeast and other fungi, nematodes, insects, rodents, and humans that mutations in various genes can affect the lifespan, and some of these mutations are also the cause of premature aging syndromes in human beings.[8,9] Several studies have reported an association between human longevity and single nucleotide polymorphisms in a variety of genes in various biological pathways, including heat-shock response, mitochondrial functions, immune response, cholesterol metabolism, and others.[7,10–13]

An analysis of the various functions of the genes associated with aging and longevity shows that these genes cover a wide range of biochemical pathways, such as energy metabolism, kinases, kinase receptors, transcription factors, DNA helicases, membrane glucosidases, GTP-binding protein–coupled receptors, chaperones, and cell cycle checkpoint pathways. What is clear from the identification of the genes influencing aging and longevity is that whatever their normal function and mechanism of action may be, these gerontogenes did not evolve to cause and accumulate molecular damage, to cause functional disorders, and to terminate the life of the organism. Most of these genes have well-defined roles in normal metabolism, in intra- and intercellular signaling, and in maintenance and repair functions including stress response. It is the damage-induced changes in the regulation, structure, and/or activity of their gene products that result in their altered biological role with age. Therefore, such genes have been termed *virtual gerontogenes*.[14,15] Furthermore, a lack of evolutionary selection of virtual gerontogenes has given rise to the notion of post-genetics or "post-reproductive genetics" as an explanation for different biological roles played at different ages by the same genetic variants.[16]

Third, the *mechanistic principle* characterizes aging as a progressive increase in molecular heterogeneity and functional impairment, which is mainly due to the accumulation of molecular damage in nucleic acids, proteins, and lipids. Since homeostasis or homeodynamic ability of a living system is primarily due to its maintenance and repair systems (MRSs), it is the progressive failure of the MRSs that is the biochemical basis of aging and age-related diseases. The evidence that MRSs are crucial determinants of natural survival and longevity comes from experiments performed to retard aging and to increase the lifespan of organisms. For example, anti-aging and life-prolonging effects of caloric restriction (CR) are seen to be accompanied by the stimulation of various maintenance mechanisms. These include increased efficiency of DNA repair, increased fidelity of genetic information transfer, more efficient protein synthesis, more efficient protein degradation, more effective cell replacement and regeneration, improved cellular responsiveness, fortification of the immune system, and enhanced protection from free-radical- and oxidation-induced damage. Genetic selection of various organisms for longer lifespan also appears to work mainly through an increase in the efficiency of maintenance mechanisms, such as antioxidation potential. The identification of long-lived mutants of the nematode *Caenorhabditis elegans*, involving various genes, provides other examples that increased lifespan is accompanied by an increased resistance to oxidative damage, an increase in the activities of antioxidant enzymes, and an increase in thermotolerance. A comparative analysis of the ability of cells isolated from a variety of animals to resist oxidative stress also showed that species lifespan was directly related to the MRSs.

AGING INTERVENTIONS: THERAPY OR PREVENTION?

As a biomedical issue, the biological process of aging underlies all major human diseases, such as atherosclerosis, cancer, cardiovascular defects, cataracts, diabetes, dementia, macular degeneration, neurodegeneration, osteoporosis, and sarcopenia. Although the optimal treatment of each and every disease, irrespective of age, is a social and moral necessity, preventing the onset of age-related diseases by intervening in the basic process of aging is the best solution for improving the quality of human life in old age.

Unlike some other fields of research, it is central to biogerontology that effective means of intervention are found, developed, and applied to modulating human aging in order to prevent the onset of age-related diseases and improving the quality of life in old age. According to the three principles of aging and longevity described above, having the bodies that we have developed after millions of years of evolution, occurrence of aging in the period beyond ELS, and the onset of one or more diseases before eventual death appear to be the normal sequence of events. This viewpoint makes modulation of aging different from the treatment of one or more specific diseases.[17] In the case of a disease, such as a cancer of any specific kind, its therapy will, ideally, mean the removal and elimination of the cancer cells and restoration of the affected organ/tissue to its original disease-free state. What will be the "treatment" of aging and to what original "age-free" stage one would hope to be restored—to day 1?, year 1, 10, 30, 50? or what? Similarly, although piecemeal replacement of nonfunctional or half-functional body parts with natural or synthetic parts made of more durable material may provide a temporary solution to the problems of age-related impairments, it does not modulate the underlying aging process as such.

Scientific and rational anti-aging strategies aim to slow down aging, to prevent and/or delay physiological decline, and to regain lost functional abilities. However, the history of anti-aging research and therapy is replete with fraud, pseudoscience, and charlatanism, which has often given a bad name to the whole field. Claims for miraculous remedies and promises for an extremely long lifespan are prevalent even today. However, while not giving serious consideration to the claims made by charlatans, it cannot be ignored that several researchers are making genuine attempts to test and develop various means of intervention for the prevention and treatment of age-related diseases, for regaining functional abilities, and for prolonging the lifespan of experimental organisms.

$E = GMC^2$: A FORMULA FOR ETERNAL LIFE

A wide range of molecular, cellular, and physiological MRSs are well known, and these range from multiple pathways of nuclear and mitochondrial DNA repair to free radical–counteracting mechanisms, protein turnover and repair, detoxification mechanisms, and other processes including immune response and stress responses. All these processes involve numerous genes whose products and their interactions give rise to a "homeodynamic space" or the "buffering capacity," which is the ultimate determinant of an individual's chance and ability to survive and maintain a healthy state. A progressive shrinking of the homeodynamic space is the hallmark of aging and the cause of origin of age-related diseases. Therefore, in order to modulate aging and to achieve healthy old age and/or to extend lifespan indefinitely, three main conditions need to be fulfilled, as represented by the equation $E = GMC^2$, discussed below.

The first parameter G represents genes, which are the basis of survival, growth, development, and MRSs. If all the genes involved in MRSs, especially in the pathways of damage removal and repair, can be manipulated in such a way that their interactive accuracy and efficiency remains functional indefinitely, then the first requirement for eternal life is fulfilled. However, at present our knowledge about the numbers, variations, interactions, and means of manipulating genetic networks is too meager to even conceptualize what may be possible to do in this respect.

The second parameter *M* represent milieu—the environment in which living systems operate and survive. The milieu in which genes and gene products function ranges from the intracellular molecular and ionic milieu to all other orders of organization including the cellular, physiological, psychological, and societal. Almost all the ongoing work on aging modulation and intervention at present is aimed at modifying the milieu by either replenishing those enzymes, hormones, and other molecules, such as antioxidants and micronutrients, whose levels are reported to decrease during aging, or by increasing the levels of such biochemicals through nutritional supplemenation. Some of the main anti-aging approaches include supplementation with hormones including growth hormone, dehydroepiandrosterone, melatonin, and estrogen, and nutritional supplementation with synthetic and natural antioxidants in purified form or in extracts prepared from plant and animal sources, Although some of these approaches have been shown to have some clinical benefits in the treatment of some diseases in the elderly, none of these really modulate the aging process itself. Furthermore, claims for the benefits of intake of high doses of vitamins and various antioxidants and their supposed anti-aging and life-prolonging effects have very little scientific evidence to back them. In contrast to this, nutritional modulation through CR has been shown to be an effective anti-aging and longevity-extending approach in rodents and monkeys. But this is a highly debatable issue at present both in terms of the practicalities of defining CR and of applying CR in human beings in physiological and evolutionary contexts.[18]

Another promising area of milieu improvement for achieiving healthy aging and longevity is through mild stress-induced hormesis. As discussed above, a critical component of the homeodynamic property of living systems is their capacity to respond to stress. In this context, the term "stress" is defined as a signal generated by any physical, chemical, or biological factor (stressor) that, in a living system, initiates a series of events in order to counteract, adapt, and survive. While a successful and overcompensatory response to low doses of stressors improves the overall homeodynamics of cells and organisms, an incomplete or failed homeodynamic response leads to the damaging and harmful effects of stress, including death.

Several meta-analyses performed on innumerable research papers published in the fields of toxicology, pharmacology, and radiation biology have led to the conclusion that the most fundamental shape of the dose–response is neither threshold nor linear, but is U- or inverted U-shaped, depending on the endpoint being measured. This phenomenon of biphasic dose–response was termed as *hormesis* in 1940s (for the historical development of the term, see Calabrese[19]). The key conceptual features of hormesis are the disruption of homeodynamics, the modest overcompensation, the re-establishment of homeodynamics, and the adaptive nature of the process. A wide variety of physical, chemical, and biological agents exhibit hormetic dose–response, and these include heavy metals, pesticides, antibiotics, chemotherapeutic agents, ethanol, aldehyde, chloroform, pro-oxidants, vitamins, trace elements, and ionizing radiation.

The paradigm for considering the applicability of hormesis in aging intervention is the well-documented beneficial effect of exercise, which, at a biochemical level, results in the production of potentially harmful substances, such as free radicals, acids, and aldehydes.[20] Thus it is hypothesized that if biological systems are deliberately exposed to mild stress, so that their homeodynamic pathways of maintenance and repair are challenged and activated, this should lead to achieving beneficial hormetic effects, including health- and longevity-promoting effects. Hormesis in aging is, therefore, defined as the life-supporting beneficial effect resulting from the cellular responses to single or multiple rounds of mild stress.[20] Since the harmful effects of severe stress have long overshadowed the beneficial hormetic effects of low-level stress, applying hormesis in aging research and therapy is a relatively recent development.

Mild stresses that have been reported to delay aging and prolong longevity in various systems include temperature shock, irradiation (UV-, gamma- and X-rays), heavy metals, pro-oxidants, acetaldehyde, alcohols, hypergravity, exercise, and food restriction.[17,20–22] Hormesis-like beneficial effects of chronic but mild undernutrition have been reported for human beings. Intermittent fasting has been reported to have beneficial effects on glucose metabolism and neuronal resistance to injury. Although at present there are only a few studies performed that utilize mild stress as a modulator of aging and longevity, hormesis is a useful experimental approach in biogerontology.

Several dietary components, such as vitamins, antioxidants, trace elements, minerals, ethanol, and even herbicides and pesticides, have been shown to have typical hormetic dose–response. All such compounds (natural or synthetic), which bring about biologically beneficial effects by acting through one or more pathways of maintenance and repair and stress response, are termed *hormetins*.[20,23] The hormetic effects of various vitamins and macro- and microminerals, including iron, iodine, fluorine, selenium, and copper, have also been reported.[24] Additionally, the effects of zinc also show a typical hormetic dose–response, and its beneficial effects are considered to be achieved through stress response–induced alterations in gene expression of various maintenance and repair pathways.[25,26]

Dietary intake of moderate amounts of ethanol has been shown to have memory-enhancing beneficial effects in mice.[27] The cardioprotective, antioxidative, and other beneficial effects of wine are considered to be due to flavonoid and non-flavonoid components, such as resveratrol,[28] which also have hormetic dose–response. Since resveratrol's mode of action involves regulating various pathways of maintenance, repair, and metabolic rate, it qualifies to be called a hormetin.

Other potential hormetins are various antioxidants, including components of spices and other medicinal plants. Almost all antioxidants show hormetic dose–response and become pro-oxidants above certain doses. Furthermore, in some cases, such as alpha-lipoic acid and coenzyme Q10, it is their pro-oxidant activity in producing hydrogen peroxide, which induces defensive responses, which is the basis of their ultimately beneficial effects.[29] Components of various medicinal plants used frequently in the traditional Chinese medicine (TCM) and in the Indian Ayurvedic system of medicine are claimed to have anti-aging effects, which appear to be achieved through hormetic pathways. For example, celasterols and paeoniflorin, present in some medicinal herbs used in TCM, have cytoprotective effects and induce heat-shock proteins in human cells.[30,31] Similarly, curcumin, which is the active component in the commonly used yellow food spice *Haldi* and is derived from the roots of *Curcuma longa*, is a co-inducer of stress proteins and has wide-ranging biological effects, depending on its dosage.[23,32]

Hormesis may also be an explanation for the health-beneficial effects of numerous other foods and food components, such as garlic, gingko, and other fruits and vegetables.[24,33,34] Understanding the hormetic and interactive mode of action of natural and processed foods is a challenging field of research, and has great potential in developing nutritional and other lifestyle modifications for aging intervention and therapies. For example, it may be possible to develop multi-hormetin formulations as anti-aging drugs and nutriceuticals whose mode of action is through hormetic pathways by mild stress-induced stimulation of homeodynamic processes.

Finally, while the G and M components of the $E = GMC^2$ formula for eternal life are being taken care of by various experimental approaches, the third factor C represents chance, which is the probability of stochastic events leading to a cascade of error-catastrophe in complex interacting systems. Recent developments in our understanding of complex networks at all levels of organization from molecular to societal and global networks has highlighted the vulnerability of all strong and weak links, and has reasserted the significance of chance events, which are not amenable to regulation and manipulation. Therefore, whereas the science of healthy aging and longevity is based in solid footing, the dreams of eternal youth are best left to thrive in the domain of the unscientific.

ACKNOWLEDGMENTS

Research in the Laboratory of Cellular Aging is supported by grants from the Danish Medical Research council FSS, from shared cost action under the EU-Biomed & Health Programme.

REFERENCES

1. RATTAN, S.I.S. 2000. Biogerontology: the next step. Ann. N. Y. Acad. Sci. **908:** 282–290.
2. RATTAN, S.I.S. & B.F.C. CLARK. 2005. Understanding and modulating aging. IUBMB Life **57:** 297–304.
3. CARNES, B.A., S.J. OLSHANSKY & D. GRAHN. 2003. Biological evidence for limits to the duration of life. Biogerontology **4:** 31–45.
4. KIRKWOOD, T.B.L. 2002. Evolution of aging. Mech. Aging Dev. **123:** 737–745.
5. ROSE, M.R. 2004. Will human aging be postponed? Sci. Amer. **14:** 24–29.
6. RATTAN, S.I.S. 2006. Theories of biological aging: genes, proteins and free radicals. Free Rad. Res. **40:** 1230–1238.
7. SINGH, R., S. KØLVRAA & S.I.S. RATTAN. 2007. Genetics of longevity with emphasis on the relevance of HSP70 genes. Frontiers Biol. **12:** 4504–4513.
8. MARTIN, G.M. 2005. Genetic modulation of senescent phenotypes in *Homo sapiens*. Cell **120:** 523–532.
9. CHRISTENSEN, K., T.E. JOHNSON & J.W. VAUPEL. 2006. The quest for genetic determinants of human longevity: challenges and insights. Nat. Rev. Genet. **7:** 436–448.
10. ALTOMARE, K. *et al.* 2003. The allele (A)-110 in the promoter region of the HSP70-1 gene is unfavourable to longevity in women. Biogerontology **4:** 215–220.
11. TAN, Q. *et al.* 2001. Measuring the genetic influence in modulating the human life span: gene-environment interaction and the sex-specific genetic effect. Biogerontology **2:** 141–153.
12. SINGH, R. *et al.* 2004. Association between low self-rated health and heterozygosity for -110A-C polymorphism in the promoter region of HSP70-1 in aged Danish twins. Biogerontology **5:** 169–176.
13. BESSENYEI, B. *et al.* 2004. Single nucleotide polymorphisms: aging and diseases. Biogerontology **5:** 291–300.
14. RATTAN, S.I.S. 1995. Gerontogenes: real or virtual? FASEB J. **9:** 284–286.
15. RATTAN, S.I.S. 1998. The nature of gerontogenes and vitagenes: antiaging effects of repeated heat shock on human fibroblasts. Ann. N. Y. Acad. Sci. **854:** 54–60.
16. FRANCESCHI, C. *et al.* 2005. Genes involved in immune response/inflammation, IGF/insulin pathway and response to oxidative stress play a major role in the genetics of human longevity: the lesson of centenarians. Mech. Age. Dev. **126:** 351–361.
17. RATTAN, S.I.S. 2005. Anti-aging strategies: prevention or therapy? EMBO Rep. **6:** S25–S29.
18. LE BOURG, E. & S.I.S. RATTAN. 2006. Can dietary restriction increase longevity in all species, particularly in human beings? Introduction to a debate among experts. Biogerontology **7:** 123–125.
19. CALABRESE, E.J. 2002. Hormesis: changing view of the dose-response, a personal account of the history and current status. Mutat. Res. **511:** 181–189.
20. RATTAN, S.I.S. 2007. Hormesis in aging. Aging Res. Rev. In press.
21. RATTAN, S.I.S. 2004. Aging intervention, prevention, and therapy through hormesis. J. Gerontol. Biol. Sci. **59A:** 705–709.
22. MASORO, E.J. 2007. The role of hormesis in life extension by dietary restriction. Interdiscip. Top. Gerontol. **35:** 1–17.

23. ALI, R.E. & S.I.S. RATTAN. 2006. Curcumin's biphasic hormetic response on proteasome activity and heat shock protein synthesis in human keratinocytes. Ann. N. Y. Acad. Sci. **1067:** 394–399.
24. HAYES, D.P. 2007. Nutritional hormesis. Eur. J. Clin. Nutr. **61:** 147–159.
25. MOCCHEGIANI, E., M. MUZZIOLI & R. GIACCONI. 2000. Zinc, metallothioneins, immune responses, survival and aging. Biogerontology **1:** 133–143.
26. MOCCHEGIANI, E., L. RINK & M.A. BLASCO. 2006. Zinc and aging (ZINCAGE project). Biogerontology **7:** 305–306.
27. RITZMANN, R.F., *et al.* 1994. The interaction of ethanol with the cognitive enhancers tacrine, physostigmine, and AIT-082. J. Gerontol. **49:** B51–B53.
28. CORDER, R. *et al.* 2006. Red wine procyanidins and vascular health. Nature **444:** 566.
29. LINNANE, A.W. & H. EASTWOOD. 2006. Cellular redox regulation and prooxidant signaling systems. A new persepctive on the free radical theory of aging. Ann. N. Y. Acad. Sci. **1067:** 47–55.
30. WESTERHEIDE, S.D. *et al.* 2004. Celastrols as inducers of the heat shock response and cytoprotection. J. Biol. Chem. **279:** 56053–56060.
31. YAN, D. *et al.* 2004. Paeoniflorin, a novel heat shock protein-inducing compound. Cell Stress Chaperones **9:** 378–389.
32. RATTAN, S.I.S. & R.E. ALI. 2007. Hormetic prevention of molecular damage during cellular aging of human fibroblasts and keratinocytes. Ann. N. Y. Acad. Sci. **1100:** 424–430.
33. EVERITT, A.V. *et al.* 2006. Dietary approaches that delay age-related diseases. Clin. Intervent. Aging **1:** 11–31.
34. FERRARI, C.K.B. 2004. Functional foods, herbs and neutraceuticals: towards biochemical mechanisms of healthy aging. Biogerontology **5:** 275–289.

Pursuing the Longevity Dividend

Scientific Goals for an Aging World

S. JAY OLSHANSKY,[a] DANIEL PERRY,[b] RICHARD A. MILLER,[c] AND ROBERT N. BUTLER[d]

[a]*University of Illinois at Chicago, Chicago, Illinois, USA*

[b]*Alliance for Aging Research, Washington, DC, USA*

[c]*University of Michigan, Ann Arbor, Michigan, USA*

[d]*International Longevity Center, New York, New York, USA*

ABSTRACT: **The aging of humanity is about to experience a radical change as the demographic transformation to an older world is approaching its final stage. In recent decades, scientists have learned enough about the biological aging processes that many believe it will become possible to slow aging in humans. We contend that the social, economic, and health benefits that would result from such advances may be thought of as "longevity dividends," and that they should be aggressively pursued as the new approach to health promotion and disease prevention in the 21st century. The time has arrived for governments and national and international healthcare organizations to make research into healthy aging a major research priority.**

KEYWORDS: **aging; interventions; mortality**

Aging for both populations and individuals is on the verge of a new era. Humans are approaching old age in unprecedented numbers as a result of large baby-boom cohorts born in the middle of the 20th century that are approaching traditional retirement ages. Increases in the prevalence of age-related disease, frailty, and disability are visible harbingers of the potential costs and social burdens arising from this historic demographic shift. Advances in the scientific knowledge of aging, however, have now created new opportunities that may allow us and those that follow to live healthier and longer lives than our predecessors. We have reached a historical moment as scientists learn enough about aging to allow us to postpone a wide range of fatal and disabling diseases expressed throughout the life span, the result of which would be health and economic benefits for current and all future generations.

Address for correspondence: S. Jay Olshansky, Ph.D., University of Illinois at Chicago, 1603 West Taylor Street, Room 885, Chicago, IL 60612. Voice: 1-312-355-4668; fax: 1-847-537-7188. sjayo@uic.edu

Ann. N.Y. Acad. Sci. 1114: 11–13 (2007).
doi: 10.1196/annals.1396.050

Why are we so optimistic now? The primary reason is that science has revealed that aging is not the immutable process it was once thought to be. Interventions at a variety of genetic, cellular, physiological, nutritional, and behavioral levels not only increase longevity in laboratory organisms, but also dramatically increase the duration of disease-free life. The realization that some humans retain their physical and mental functioning for more than a century suggests that genes associated with the extension of healthy life already exist within the human genome. Biogerontologists have now gone from merely describing cellular aging and cell death to manipulating the mechanisms responsible for these phenomena. Important strides have also been made toward understanding the effects of hormones on cellular pathways that influence the rate of aging. Since these pathways have similar effects in both humans and laboratory organisms, intervention strategies can be evaluated quickly, in short-lived animals, to find the ones most likely to work in humans. In short, we now believe that extending the duration of healthy life in humans by slowing down the processes of aging is a scientifically plausible goal, and adequate funding in this area might well lead to dramatic advances in preventive medicine and public health within the next few decades.

Even a minor deceleration in the rate of human aging could have profound benefits for individuals and societies. Because prolonged, chronic illness is a powerful driver of medical costs, enormous cost savings would be achieved if mortality and morbidity could be compressed within a shorter duration of time at the end of life. At least some of the manipulations that appear to slow aging in animal models do just this, maintaining excellent physical and cognitive function well beyond the usual ages at which illness and disability start to affect most untreated individuals. In fact, aging interventions have the potential to do what no surgical procedure, behavior modification, or cure for any one major fatal disease can do; namely, extend youthful vigor throughout the life span. Extending the duration of physical and mental capacity would permit people to remain in the labor force longer, amass more income and savings, and thereby lessen the effect of shifting demographics on age-based entitlement programs, with a net benefit to national economies. The combined social, economic, and health bonuses accruing from a slowing of the rate of aging is what we call the *longevity dividend*—benefits that might begin with those now alive, and then continue for all generations that follow.

We now have good reasons to think that slowing aging in humans is scientifically plausible and, given sufficient research investment, might prove to be within our technical grasp in the foreseeable future. There are a number of compelling reasons why this effort should now be aggressively pursued: (1) the costs, to individuals and to society, of debilitating late-life illnesses are already increasing in many countries as the population of elderly people mounts to unprecedented levels, leading to escalating healthcare costs; (2) compressing mortality and morbidity into a shorter duration of time at the end of life will pay substantial health and financial dividends for members of the first generation to which they can be applied, dividends that will be compounded as new generations benefit from existing and expanding technological advances; and (3) a modest deceleration in the rate of biological aging would produce the equivalent of simultaneous major breakthroughs against every single fatal and non-fatal disease and disorder associated with growing older.

The time has arrived for governments and national and international health-care organizations to make research into healthy aging a major research priority, exploiting new discoveries towards the goal of manipulating the aging rate to prevent or postpone multiple forms of late-life illnesses and disabilities. We look forward to developing a national and international strategy that will lead to the permanent extension of healthy life that would result from a successful effort to slow the rate of aging.

ACKNOWLEDGMENTS

An expanded version of this manuscript was published in *The Scientist*.[1]

REFERENCE

1. OLSHANSKY, S.J., D. PERRY, R.A. MILLER & R.N. BUTLER. 2006. In pursuit of the longevity dividend. The Scientist **20:** 28–36.

Senescence Viewed through the Lens of Comparative Biology

BRUCE A. CARNES

The University of Oklahoma Health Sciences Center, D.W. Reynolds Department of Geriatric Medicine, Oklahoma City, Oklahoma, USA

ABSTRACT: Although mortality and longevity are inherently biological phenomena, their study has historically been the purview of demography and the actuarial sciences. An infusion of biological thinking into these disciplines transforms demography into biodemography and provides expectations and coherency to observations on age-determined mortality that would not be explainable otherwise. Comparative biology teaches us that reproduction is life's solution to the inevitability of death in the hostile environments of Earth. That solution, however, places a higher priority on investing physiological resources into reproduction that could otherwise have been used to maintain the soma (body) longer. As such, aging is an inescapable but inadvertent byproduct of imperfect maintenance and its attendant surveillance and repair. Biology also reveals that while bodies are not designed to fail, neither are they designed for extended operation. In other words, bodies are subject to biological warranty periods for normal operation. For sexually reproducing species, that warranty period includes the time from conception to sexual maturity, the production and nurturing of offspring, and a period of grand-parenting in some species. Humans are the only species capable of exploiting the loophole in the biological contract of life (bodies that are not designed to fail). Human ingenuity (science, medicine, public health) has produced interventions that manufacture survival time by delaying death, and in so doing, has created a phenomenon never before seen in the history of life—population aging (and all the societal and health consequences that go with it).

KEYWORDS: senescence; comparative biology; biodemography

INTRODUCTION

The first impression that emerges from a study of life on Earth is awe over the enormous diversity that exists. Not only is there a diversity of species, but, as Darwin so keenly chronicled, a remarkable amount of diversity also exists

Address for correspondence: Bruce A. Carnes, Ph.D., The University of Oklahoma Health Sciences Center, D.W. Reynolds Department of Geriatric Medicine, 921 N.E. 13th Street (11G), Oklahoma City, OK. Voice: (405) 271-8550; fax: (405) 271-3887.
Bruce-Carnes@ouhsc.edu

Ann. N.Y. Acad. Sci. 1114: 14–22 (2007). © 2007 New York Academy of Sciences.
doi: 10.1196/annals.1396.045

within species. Given this overwhelming evidence of widespread differences within and between species, is there any reason to suspect the existence of biological similarities? The most obvious example of this is the species concept where biological similarity is a requirement for the production of viable offspring. Although our attention is drawn to how species differ, it is their similarities that often provide important insights into the fundamental biology of life. For this reason, comparative biology is one of the oldest and most honored traditions in biology. As such, a primary goal of this essay is to use comparative biology in order to learn lessons about the biology of aging.

Biology is a massive discipline, far larger than any individual can master. As a result, biology is partitioned into areas of study based on levels of biological organization ranging from the molecular and cellular levels to the tissue and organ levels to the individual and population level. Researchers who work with multicellular organisms are able to observe evidence of senescence at each and every one of these levels of biological organization. Although the literature is replete with theories of senescence within levels of biological organization, very few efforts have been made to integrate these theories across levels of organization. Further, since senescence appears to be everywhere (systemic), theories of aging based on specific mechanisms or molecular pathways cannot capture the full multi-dimensionality of senescence. My plan is to take the opposite tack. By asking really broad questions about aging rather than tightly focused ones, I hope to reveal some of the profound lessons that can be revealed by the comparative biology approach. I will then make a transition to a series of more specific questions on how comparative biology can be used to explore the mortality consequences of aging.

Question 1: Why do organisms die?

Aging and death are inherently biological phenomenon. Interestingly, death or at least the risk of death has more to do with defining the biology of organisms than any other factor. The logic behind this statement begins with the observation that the environments of Earth, both past and present, are exceedingly hostile.[1] Just about every life form other than photosynthetic plants must eat other life forms in order to survive. Thus, even if an organism could be designed for indefinite survival (immortality), it would die. As such, the inevitability of death posed a challenge that had to be solved if life on earth was going to persist. Of course that challenge was overcome, and the solution to it became a defining characteristic of living matter, namely, the ability to self-replicate or reproduce.

Question 2: What determines duration of life?

The inevitability of death creates a race between reproduction and death. For example, animals that are prey, like mice, are under intense mortality pressures and therefore must achieve sexual maturity quickly in order to ensure that the instructions needed to build replacement mice are passed onto the next generation. By contrast, species like humans that experience much lower external mortality pressures give birth to altricial young that take up to 15 years to achieve sexual maturity. The Galapagos tortoises with their protective shells take even longer; males do not produce viable sperm until they are 35 years old. In this scenario, duration of life is calibrated to the time needed for maturation, reproduction, and a period of nurturing for some species.

From a biological perspective, the race between reproduction and death is so crucial that it cannot be left up to chance. As a result, growth and development involve highly regulated genetic programs.[2] Given the importance of the race to sexual maturity, there is no question that the tempos of these growth and development programs must be controlled by extremely precise and highly coordinated biological clocks. Note that these clocks have nothing to do with aging, sickness, and decay. Quite the contrary, every gene involved in this exquisitely complex process is designed for health, vigor, and the production of enough offspring to offset the mortality pressures of the environment. Essentially, this timeframe can be likened to a biological warranty period that defines duration of life for a particular life form.[3]

Question 3: Why does aging exist?

The previous reasoning leads to an important biological insight: reproduction is the biological imperative that defines the biology of all organisms. As such, investing physiological capital into maintaining the soma (body) beyond the ages needed for reproduction, nurturing offspring, and contributing to the reproductive success of those offspring (grand-parenting) would be an unnecessary and unwise investment of precious physiological resources.[4] In other words, sacrificing reproductive output for longevity simply does not work in a hostile world. However, comparative biology clearly reveals that sacrificing the maintenance of the soma does work. In other words, the inevitability of death established reproduction as a higher priority than maintenance of the soma. Senescence is the inevitable consequence of a disposable soma and its origins must date back to the origins of life.

Question 4: Did evolution produce senescence?

You cannot talk about evolution without talking about natural selection. A basic tenet of evolutionary biology is that natural selection alters the genetic composition of a population through the differential reproductive success of the individuals in that population. The corollary of this tenet is that selection is most effective during the timeframe before individuals begin reproducing, diminishes as the cumulative reproductive potential of individuals is achieved, and becomes weak or nonexistent once reproduction has ceased.[5] For example, genetic diseases that affect children are often exceedingly rare because they either kill the children before they attain the age of sexual maturity or they diminish reproductive success if they are not lethal. This age-based gradient for the effectiveness of selection has an exceeding important consequence: it permits the life-course of (sexually reproducing) species to be partitioned into biologically meaningful time periods—the pre-reproductive, reproductive, and post-reproductive periods.

Senescence has its most obvious manifestations during the post-reproductive period of the life span. Selection, however, does not operate during this time-frame. This lack of selection, in turn, implies that senescence did not evolve, but instead arose as an unintended byproduct of biological factors that evolved for other reasons. In this sense, senescence arose from evolutionary neglect rather than evolutionary intent.

Question 5: What are the biological consequences of evolutionary neglect?

Evolutionary neglect refers to the absence of selection in the post-reproductive period of the life span. In a genetic sense, the post-reproductive period is like the vacuum of space. There are no forces acting on gene expression in this period; the competition between alleles (variants of a gene) for representation in the next generation no longer exists. This analogy also implies that there are no evolutionary forces acting to either accelerate or decelerate the age-specific failure probabilities of the soma. If bodies are likened to rockets, then their trajectories of health and longevity were determined prior to their entry into the post-reproductive period. As such, an individual's genetic constitution at conception is important, but so is how they led their lives (i.e., behavior) and whether they lived in a relatively benign environment (e.g., protected from infectious disease early in life).

Evolutionary neglect also means that bodies are not designed to fail. However, neither are they designed for extended operation. These two sequelae of evolutionary neglect have profound implications. They indicate the existence of loopholes in the biological contract of life that can be exploited. For example, human ingenuity has already devised countless ways to repair the soma so that people who would have died from disease or injury are able to live longer.[6] In fact, so much survival time has been "manufactured" that the post-reproductive segment of populations around the world are becoming a progressively larger fraction of the total population. In other words, population aging has become a global phenomenon.[7]

It is important to note that the biological consequences of evolutionary neglect are almost exclusively a human issue. Organisms in nature do not often live long enough to experience the effects of senescence. When these organisms (e.g., laboratory animals, pets, and zoo animals) are protected, however, they exhibit all the signs of senescence. There are also organisms that live to exceedingly old ages without appearing to age (e.g., American lobster, bowhead whale, quahog clam, rockfish, and Galapagos tortoise). These animals and others like them tend to have indeterminate growth, and it is unclear at this point whether they do not age or whether they simply age at a very slow rate.[8]

The previous two paragraphs illustrate two issues that complicate the insights on senescence that can be gained through comparative biology. The first complication is that humans routinely benefit from interventions that delay death. Pets and zoo animals receive veterinary care, but animals in nature and those brought into the laboratory are rarely the beneficiaries of such life-extending interventions. The second complication is that clock time is not the same as biological time. Clock time is the same for all species, while the tempo of biological clocks varies among and within species. A third complication also exists. Species, by definition, have aspects of their biology that differ from those of other species. One consequence of this difference is that while species die from many of the same causes, they also die from different causes. All three of these issues must be accounted for when evaluating whether the consequences of evolutionary neglect are universal or species-specific.

Question 6: How can you compare senescence among species with different life spans?

This question begins the transition from broad questions about aging to issues that arise when comparative biology is applied to the mortality consequences of aging. The most obvious obstacle is that mice, dogs, humans, and tortoises have vastly different life spans. As a result, an age that is old for a mouse would be exceedingly young for a tortoise. If you measure rate of aging with an actuarial tool like the Gompertz function, you would confirm the obvious by finding that these organisms are aging, and hence dying, at vastly different rates. This problem becomes far more complicated when mortality comparisons are being made between organisms as diverse as yeast, fruitflies, roundworms, and humans—all organisms used in the study of aging.

Single-celled organisms obviously lack the complexity of interactions that mammals require in order to maintain homeostasis and integration among their various tissues and organs. Others, like the roundworms, lack dividing cells, which rules out the cancers that are a major cause of death in mammals. Although a step up in complexity, fruitflies either lack key anatomic structures (e.g., lungs) or possess far simpler forms of the structures (e.g., brain, heart) found in mammals. Further, none of these organisms possesses the regulatory capacity to maintain homeothermy. Although the ultimate goal of performing research on these organisms is to better understand aging in humans, the remaining discussion will be restricted to mammals in order to eliminate some of the obstacles to interspecies comparison just mentioned.

By restricting the discussion to mammals, we are also restricting ourselves to sexually reproducing species. Further, the mammals used in laboratory studies (e.g., mice and rats) are iteroparous, that is, offspring are produced over the course of a protracted period of reproduction. As such, these organisms have a life-course that can be partitioned into the pre-reproductive, reproductive, and post-reproductive periods, where the effectiveness of natural selection behaves as previously described (see question 4). There are many ways to convert clock time to biological time, but all of these scaling approaches are intended to allow species comparisons to be made within biologically comparable partitions of the life-course. This three-way partition achieves that goal. Therefore, what seemed like the biggest obstacle to species comparisons of age-determined mortality and longevity turns out to be one of the easiest to resolve.

Question 7: How can you compare species when they die from different causes?

In the previous section, there was no explicit discussion of what exactly was being compared between species. That was done intentionally in that the comparison could involve any biomarker deemed relevant to senescence observed at any of the levels of biological organization (i.e., molecular, cellular, tissue, organ) as long as those comparisons are being made within the same partition of the life-course. The current question addresses something more specific, namely, a comparison of mortality between species.

The most obvious observation is that if one species does not die from cause X, and cause X is the event of interest, then no comparison can be made. The question, however, is whether this issue negates all interspecies comparisons of mortality. A couple of examples may provide a possible way around to this apparent impasse. Leukemia in humans is predominately myelogenous, while leukemia in the laboratory mouse is lymphocytic-lymphoblastic. Similarly, lung tumors in humans are predominately bronchial, while those in the mouse are alveolar. In each case both species are dying from causes with the same name, but the cell of origin differs. However, if the diseases are biological surrogates of each other, then there may be value in comparing these mortality patterns. This issue is not as abstract as it may seem. Agencies like the Department of Energy and NASA are confronted with these very issues as they attempt to establish safety standards for humans (e.g., nuclear power workers and astronauts) from the results of radiation studies conducted on laboratory animals.[9]

A second observation pertains to the opposite extreme; namely, comparing mortality between species without making any distinction between cause of death. In demographic circles, this endpoint for comparison is called all-cause mortality. Remember that our objective is to compare mortality associated with senescence. All-cause mortality includes a large number of deaths that are caused by such things as accidents, homicide, suicide, war, infectious diseases, and natural disasters. Although links between these causes and senescence can be made, they would have to be characterized as ranging from tenuous to unrelated.

There is, however, a middle ground between analyses of specific causes of death and analyses that ignore cause of death. Attempts to combine causes of death associated with senescence into an aggregate endpoint for quantitative analysis dates back to the famous actuary William Makeham,[10] one of the founding fathers of the actuarial sciences. He felt that analyses based on such an endpoint would provide a better mathematical description of the mortality consequences of human senescence than traditional analyses based on all-cause mortality. Makeham not only identified senescence-related causes of death in humans, but he also invented the mathematical method (i.e., competing risk theory) needed to perform such an analysis. Since then, progressive improvements based on expanding knowledge have been made to Makeham's partition of mortality.[11]

Similar efforts have been made to identify aggregate endpoints for quantitative analyses of senescent mortality in laboratory animals.[12] The specific composition of these endpoints will not only vary between similar animals like strains of laboratory mice (variants derived from the house mouse *Mus musculus*), but they will, of course, also differ from those of other mammals like dogs and humans. Like the earlier discussion of leukemia and lung cancer, the issue is whether "senescent mortality" in the aggregate behaves similarly between species. The rationale for why it might do so goes back to the effectiveness of natural selection and how it permits the life-course to be partitioned into three comparable and biologically meaningful time segments.

Competing risk theory that occupies such a prominent position in textbooks on the analysis of mortality data is still based on the mathematics developed by Makeham in his 1867 paper. The name of the theory provides an apt image of the concept. Death abhors a vacuum; there is no age where the risk of death is zero. Similarly, the competition does not produce a single cause of death that dominates the entire age spectrum. Instead, windows of time along the age axis are dominated by a cause or set of causes that is associated with that window. This mortality scenario produces a series of overlapping failure distributions.

The discussion of aggregate endpoints suggests that the causes of death participating in the competition can be divided into two categories—those that are imposed on individuals by the environment and those that arise from the biology of the species.[13] It is the latter category that is of most interest to the study of age-determined mortality. By their nature, biological causes of death have a non-zero genetic component. As such, the effectiveness curve of natural selection should influence the age-specific failure distribution of each cause individually as well as the age-specific failure distribution for all biological causes of death in the collective. Selection that is highly effective in the pre-reproductive period and negligible or nonexistent in the post-reproductive period applies to all iteroparous species. The selection gradient within the reproductive period must also be generally similar because effectiveness diminishes with successive parities for all iteroparous species. Thus, despite dying from some different causes of death, there is a biological rationale for expecting senescent causes of death in the aggregate to exhibit an interspecies similarity after adjustment for differences in life span.

CONCLUSION

The study of mortality has historically been the purview of demography and the actuarial sciences. However, as I have attempted to argue in this paper, mortality and longevity are inherently biological phenomenon. As with other biological phenomenon, evolutionary biology provides expectations and coherency to observations on age-determined mortality that would not be explainable otherwise. Thus, this conceptual framework transforms demography into biodemography.

Evolutionary theories of senescence provide the paradoxical insight that forces of mortality external to the biology of species played a key role in molding the internal biology of those species. If Gould and Eldredge[14] are correct in their assertion that evolution occurs in fits and starts accompanied by long periods of stagnation (i.e., punctuated equilibrium), then another insight of considerable import is gained. The biology of *Homo sapiens* was established some 200,000 years ago and we are the most recent exemplars of that biology. Thus, in the context of comparative biology, the explanations for the aging and longevity consequences of a species' biology are found in the evolutionary past, not the present. In that time period, surviving to reproduction was a real race against death, and longevity, as we think of it today, was more theoretical than achievable. Thus, in the case of humans (and other organisms protected by humans), the intriguing question is not how much longer can they live, but rather why do they live as long as they do?

Gould[15] wrote a book describing the evolution of life from its origins to the present. Dawkins[16] also wrote a book on evolution, but he started with modern humans and traced our evolutionary lineage back to the origins of life. Both books emphasize another exceedingly important biological reality. All life that exists or has ever existed can (in theory) trace its origins back to a common ancestor. The modern tools of biology (e.g., DNA hybridization) reinforce this truth as they perform molecular taxonomy in order to confirm or adjust the tree of life developed through anatomic taxonomy. Shared DNA means shared biology at many levels of biological organization (molecules, mechanisms, developmental pathways, organs and organ systems, physiology). This shared biology also implies that species share the ramifications for health, aging, disease, longevity, and death that arise from that biology. This body of knowledge derived from comparative biology provides the underlying premise for using organisms other than humans in scientific studies in order to better understand the biology of humans.

Demography is the scientific study of human populations.[17] Demography and its sister discipline, actuarial science, have provided the elegant quantitative tools that form the basis of population science. These disciplines, however, focus solely on human populations. This approach has led to great gains in knowledge about the structure of populations, the ways in which they can change, and how they do so. Biodemography depends on demography, but augments it as well. Biodemography provides the theoretical grounding for why interspecies comparisons of mortality can be conducted, how these comparisons can be made between organisms that obviously differ in life span, and why cause of death is such an important factor in using demographic methods to gain insights into the health and longevity consequences of aging.

Although biodemography has roots going back to the origins of demography and the actuarial sciences, it is in a formal sense still a very young discipline. Biodemographers have generated considerable debate in the population sciences over such important issues as limits on the longevity of individuals and the life expectancy of populations. These debates have introduced new energy into the population sciences. There is also every reason to be optimistic that biodemography, with its emphasis on comparative biology, will continue to provide important biological and demographic insights into the health, mortality, and longevity consequence of senescence.

REFERENCES

1. CARNES, B.A., & S.J. OLSHANSKY. 1993. Evolutionary perspectives on human senescence.. Popul. Devel. Rev. **19:** 793–806.
2. Balinsky, B.I. 1970. An Introduction to Embroyology, 3rd ed. W.B. Saunders Co. Philadelphia.
3. CARNES, B.A., S.J. OLSHANSKY & D. GRAHN. 2003. Biological evidence for limits to the duration of life. Biogerontology **4:** 31–45.
4. KIRKWOOD, T.B.L. & R. HOLLIDAY. 1979. The evolution of ageing and longevity. Proc. R. Soc. London Biol. **205:** 531–546.
5. MEDAWAR, P.B. 1952. An Unsolved Problem of Biology. Lewis. London.
6. OLSHANSKY, S.J., B.A. CARNES & D. GRAHN. 1998. Confronting the boundaries of human longevity. Am. Sci. **86:** 52–61.
7. OLSHANSKY, S.J., B.A. CARNES & C. CASSEL. 1993. The aging of the human species. Sci. Am. **268:** 46–52.
8. FINCH, C.E. & S.N. AUSTAD. 2001. History and prospects: symposium on organisms with slow aging. Exp. Gerontol. **36:** 593–597.
9. CARNES, B.A., D. GRAHN & D. HOEL. 2003. Mortality of atomic bomb survivors predicted from laboratory animals. Radiat. Res. **160:** 159–167.
10. MAKEHAM, W.M. 1867. On the law of mortality. J. Inst. Actuar. **13:** 325–358.
11. SHRYOCK, H.S. & J. SIEGEL. 1975. The Methods and Materials of Demography, Vol. 2, 3rd ed. Bureau of the Census. Washington, DC.
12. CARNES, B.A., S.J. OLSHANSKY & DOUGLAS GRAHN. 1996. Continuing the search for a law of mortality. Popul. Devel. Rev. **22:** 231–264.
13. CARNES, B.A. & S.J. OLSHANSKY. 1997. A biologically motivated partitioning of mortality. Exp. Gerontol. **32:** 615–631.
14. GOULD S.J. & N. ELDREDGE. 1977. Punctuated equilibria: the tempo and mode of evolution reconsidered. Paleobiology **3:** 115–151.
15. GOULD, S.J. 1990. Wonderful Life: the Burgess Shale and the Nature of History. W.W. Norton. New York.
16. DAWKINS, R. 2005. The Ancestor's Tale: a Pilgrimage to the Dawn of Evolution. Mariner Books. Boston.
17. SIEGEL, J.S. & D.A. SWANSON. 2004. The Methods and Materials of Demography, 2nd ed. Elsevier Academic Press. London.

Cytomegalovirus Infection

A Driving Force in Human T Cell Immunosenescence

SVEN KOCH,[a] ANIS LARBI,[b] DENNIS ÖZCELIK,[b] RAFAEL SOLANA,[c] CÉCILE GOUTTEFANGEAS,[a] SEBASTIAN ATTIG,[a] ANDERS WIKBY,[d] JAN STRINDHALL,[d] CLAUDIO FRANCESCHI,[e] AND GRAHAM PAWELEC[b]

[a]*Department of Immunology, University of Tübingen, Auf der Morgenstelle 15, D-72076, Tübingen, Germany*

[b]*Center for Medical Research (ZMF), University of Tübingen, Waldhörnlestr. 22, D-72072 Tübingen, Germany*

[c]*Department of Immunology, University of Córdoba, Hospital Reina Sofia, Ave. Menedez, Pidal, s/n, ES-14004 Córdoba, Spain*

[d]*Department of Natural Science and Biomedicine, School of Health Sciences, Jönköping University, Box 1026, 551 11 Jönköping, Sweden*

[e]*Department of Experimental Pathology, University of Bologna, via S. Giacomo 12, I-40126 Bologna, Italy*

ABSTRACT: **The human immune system evolved to defend the organism against pathogens, but is clearly less well able to do so in the elderly, resulting in greater morbidity and mortality due to infectious disease in old people, and higher healthcare costs. Many age-associated immune alterations have been reported over the years, of which probably the changes in T cell immunity, often manifested dramatically as large clonal expansions of cells of limited antigen specificity together with a marked shrinkage of the T cell antigen receptor repertoire, are the most notable. It has recently emerged that the common herpesvirus, cytomegalovirus (CMV), which establishes persistent, life-long infection, usually asymptomatically, may well be the driving force behind clonal expansions and altered phenotypes and functions of CD8 cells seen in most old people. In those few who are not CMV-infected, another even more common herpesvirus, the Epstein-Barr virus, appears to have the same effect. These virus-driven changes are less marked in "successfully aged" centenarians, but most marked in people whom longitudinal studies have shown to be at higher risk of death, that is, those possessing an "immune risk profile" (IRP) characterized by an inverted CD4:8 ratio (caused by the accumulation primarily of $CD8^+$ $CD28^-$ cells). These findings support the**

Address for correspondence: Graham Pawelec, University of Tübingen Center for Medical Research, Waldhörnlestr. 22, D-72072 Tübingen, Germany. Voice: +4970712982805; fax: 004970718884679. grahampawelec@aol.com; graham.pawelec@uni-tuebingen.de

Ann. N.Y. Acad. Sci. 1114: 23–35 (2007). © 2007 New York Academy of Sciences.
doi: 10.1196/annals.1396.043

hypothesis that persistent herpesviruses, especially CMV, act as chronic antigenic stressors and play a major causative role in immunosenescence and associated mortality.

KEYWORDS: aging; cell aging; cytology; cytomegalovirus; cytomegalovirus infections; humans; immune system; immunity; immunology; longitudinal studies; memory; mortality; phenotype; risk factors; T lymphocytes; virology

INTRODUCTION

Infection with the β-herpesvirus cytomegalovirus (CMV) is usually asymptomatic, even in neonates, but once established, its containment becomes a priority for the immune system, which is unable completely to eliminate it. There are potentially interesting correlates regarding the impact on the immune system of CMV and other persisting sources of antigen, such as immunogenic cancers or long-term-residing parasites, in terms of their detrimental effects predominantly on T cells. A prominent and effective immune response against viral infections is mediated by $CD8^+$ cytotoxic T lymphocytes, although it is clear that CMV-specific $CD4^+$ T cells may also mediate cytotoxicity and be involved in antiviral responses. Clinically, CMV infection is generally believed to be asymptomatic in normal hosts, with T cell memory rapidly established and maintained indefinitely, even in neonates. The crucial importance of maintaining a constant defense against such an asymptomatic infection is dramatically demonstrated in immunosuppressed persons, where viral escape may be fatal. However, even healthy immunocompetent people may display symptoms of CMV infection more often than previously appreciated,[1] sometimes even with serious consequences and with age implicated as a risk factor.[2] Nonetheless, even the very elderly are generally accepted as maintaining efficient CMV immunosurveillance without distressing symptoms, in contrast to the related virus Varicella zoster. This virus is responsible for chickenpox in early life and may reactivate, usually in later life, as the very painful "shingles" syndrome, defense against which decreases with age.

The consequences to the immune system of the necessity for maintaining this constant CMV vigilance may be grave. Recent reports on the very young and the very old document that CMV infection results in similar alterations to $CD8^+$ T cell subset surface phenotypes.[3,4] This gave rise to the idea that apparently age-associated changes could be secondary to age-associated increases in prevalence of CMV infection. This, coupled with the identification of CMV infection as the putative driving force behind the CD8 clonal expansions that had been reported earlier in middle-aged and elderly humans, suggested an initially surprising impact of CMV on the aging T cell immune system.[4] The characteristics of these clonally expanded CD8 cells are now being intensively analyzed by several investigators including ourselves in terms of their antigenic specificity,[5] their surface phenotype,[6] and function[7] by using tetramer technology[8] in combination with polychromatic flow cytometry. Such findings are not seen only in humans, although in mice they may not necessarily be due to CMV (as is also sometimes the case in humans). In mice, it is likely that the CD8 clonal expansions are caused by the other main source of chronic antigenic stress almost ubiquitous in old lab mice, and unfortunately often found in humans as well, namely cancer antigens.[9,10] In the following brief review, we will summarize the effects of persistent human cytomegalovirus (HCMV) infection on the T cell repertoire in the elderly, T cell homeostasis, and possible interventions to overcome these effects.

Biomarkers and CMV pp65 (495-503)–Specific CD8 Cells

We have studied one major subset of CMV-specific T cells in elderly populations followed longitudinally for mortality in studies aimed at assessing risk categories by means of identifying immune biomarkers. We targeted CD8 cells binding HLA-A2/pp65 (495–503) tetramers (fluorescent reagents staining only those T cells carrying antigen receptors specific for this one viral epitope). We then quantified these cells and assessed their functional status. Phenotypic analysis of these CMV pp65 (495–503)–specific CD8 cells demonstrated that the proportion of cells coexpressing CD27 and CD28 is greatly decreased in elderly compared with young individuals. Thus, in the latter, a significant proportion of CMV-specific CD8 cells is included in the naïve subpopulation ($CCR7^{+}CD45RA^{+}$), but in elderly donors CMV-specific CD8 T cells mostly possess a phenotype associated with effector-memory T cells ($CCR7^{null}$ $CD45RA^{low}$ or $CCR7^{null}$ $CD45RA^{+}$). These results indicate that the phenotypes of the CD8 cells with exactly the same CMV specificity are different in young and old individuals. There is an age-associated expansion of effector-memory CD8 cells that do not express CD27, CD28, CCR7, or CD45RA and of terminally differentiated effector cells with a $CD27^{null}$, $CD28^{null}$, $CCR7^{null}$ $CD45RA^{+}$ phenotype.[6,11–14] Recent studies have investigated the nature of such $CD8^{+}$ $CD28^{-}$ and $CD8^{+}$ $CD28^{+}$ cells using DNA microarray technology.[15,16] In addition to the differentiation markers and lack of co-stimulatory receptors, these cells also express the genes for a variety of NK-related molecules and NK receptors (FIG. 1). We and others have confirmed the presence of several of these at the protein level on the cell surface[17–19] (Koch *et al.*, unpublished data). The accumulation of these highly differentiated CD8 T cells in the elderly seems to be mostly restricted to CMV-specific T cells; this is consistent with findings that EBV-specific CD8 T cells maintain expression of CD28 and have a lower expression of CD45RA than is observed in CMV-specific CD8 T cells.[20,21]

Regarding the functional integrity of the clonally expanded CMV-specific cells, we found that on average the number of *functional* CMV-specific CD8 cells is quite similar in young and old individuals—most commonly, the number is actually slightly increased in the elderly, possibly reflecting increased viral

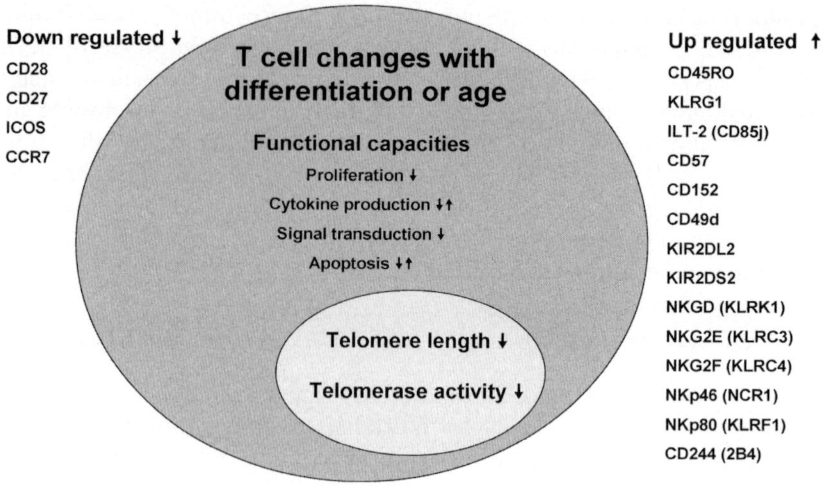

FIGURE 1. T cell changes with age. T cells from aged individuals or after several rounds of clonal expansion have a changed phenotype and altered functional capacity.

load.[19] This is consistent with suggestions that these cells may contribute to the pro-inflammatory status often observed in the elderly and suggested to contribute to frailty and mortality. In contrast, however, the number of *dysfunctional* CD8 cells carrying receptors for this CMV antigen is much more markedly increased in the elderly.[22,23] These cells can respond to mitogenic stimulation by secreting IFN-γ, but they do not respond to stimulation with specific antigen *ex vivo*, and are therefore considered to be anergic.[19,22,23] We have seen that some of the CMV-specific cells in the young express CD28, but very few do in the elderly; and whereas essentially all cells in the young and elderly are killer cell lectin-like receptor G-1 (KLRG1$^+$), KLRG1 expressed by memory cells (CD28$^-$) is a marker of end-stage differentiation[24] and apoptotic resistance.[25] It is interesting to note that our recent data suggest that in "successfully aged" populations of centenarians, it is exactly these parameters that more closely resemble the status of the young rather than the old.[12] As well as KLRG1, CD57 (HNK epitope 1) is also found to be expressed by effector cells and is upregulated in the elderly (data not shown).

The main difference, therefore, between CD8$^+$ CMV-specific cells in the young and old may be the relative proportions of KLRG1- and CD28-bearing cells. In the elderly, there is an accumulation of KLRG1$^+$, CD28$^-$ cells; these have short telomeres and no proliferative capacity, but are resistant to spontaneous or induced apoptosis and so persist.[25] Another marker which is present on terminally differentiated T cells with a replicatively senescent phenotype and upregulated in the elderly is CD57,[26] as mentioned above. We hypothesized that use of a combination of CD57 and KLRG1 as markers to identify dysfunctional CD8 T cells might allow their deletion and hence removal of

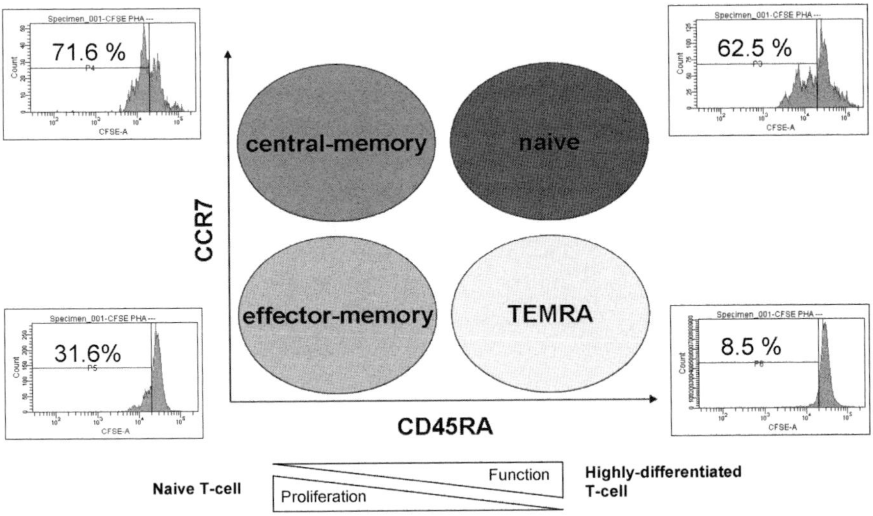

FIGURE 2. T cell subsets and proliferative capacity. Differentiation subsets identified by lymph node-homing chemokine receptor CCR7 and phosphatase isoform CD45RA,[11] $CCR7^+$ $CD45RA^+$ naïve, $CCR7^+$ $CD45RA^-$ central-memory, $CCR7^-$ $CD45RA^-$ effector-memory and late-differentiated effector-memory $CCR7^-$ CD45RA TEMRA T cells. Functions, such as cytokine production and cytotoxicity, increase during differentiation, whereas the proliferation capacity, depicted here as CFSE staining, decreases.

these dysfunctional, possibly negatively regulatory cells from the system. According to Ibegbu *et al.*[27] and our own data (FIGS. 2 and 3), $KLRG1^+CD57^+$ T cells were not able to proliferate, whereas the other subsets ($KLRG1^-CD57^-$, $KLRG1^-CD57^+$, and $KLRG1^+CD57^-$) were still able to proliferate after TCR-mediated stimulation. Thus, from the functional point of view, we found that this marker combination does identify cells that are *dysfunctional* in terms of proliferative capacity. This finding is consistent with the accumulation of $CD28^-$, $KLRG1^+CD57^+$ cells present preferentially in CMV-seropositive elderly donors belonging to the IRP-positive at-risk group in the longitudinal studies mentioned earlier.[28] Note that the number of $CD3^+$ $CD28^+$ (*functional*) T cells in elderly seropositive people was the same as in seronegatives (again consistent with the absolute requirement to maintain immunosurveillance against CMV)—but the number of $CD28^-$ cells was greatly increased in the former.[29] Thus, the overall picture (FIG. 4A) is that clonal expansions of apoptosis-resistant CMV-specific cells (or under certain ciurcumstances, those resulting from any other chronic antigenic stress), late-differentiation-stage CD8 cells with effector and effector-memory phenotypes accumulate in the elderly and may thus reduce the amount of "space" available for T cells of other types and specificities. From studies analyzing the numbers of different expanded clones in the samples from the longitudinal studies, Hadrup *et al.*[30] found that the variety of CMV-specific clones increases

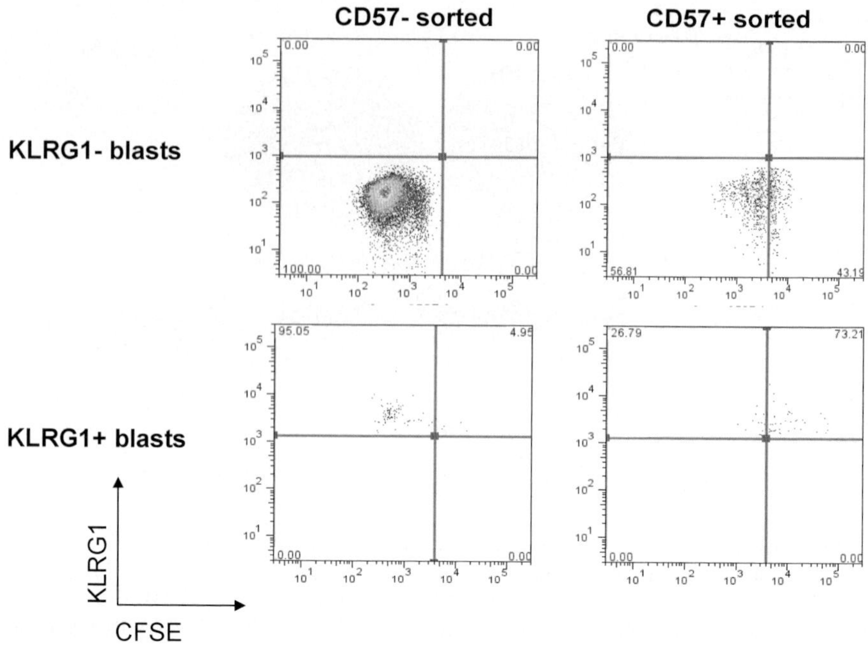

FIGURE 3. Proliferative capacity is reflected by the marker combination CD57 and KLRG1. $CD57^{-}$ and $CD57^{+}$ cells were sorted with MACS beads, then stimulated with CD3/CD28 Dynal beads. Proliferative capacity of $KLRG1^{-}$ and $KLRG1^{+}$ blasts was measured using polychromatic flow cytometry and CFSE staining. Only the $CD57^{+}$ $KLRG1^{+}$ double-positive cells failed to proliferate after TCR-mediated stimulation.

with age up to the terminal phase, but then goes into decline in the final years, especially in the IRP cohort, and that this number correlates with remaining survival time (FIG. 4B). In the most recent follow-up of this study, some rare exceptions were found which moved out of the IRP by a process of immune suppression (i.e., reduction of the number of CD8 cells), including increases in IL-6 and IL-10 and decreases in the number of $CD3^{+}CD8^{+}CD28^{-}$ cells.[31] These subjects had a survival advantage over those who were in the IRP group but did not exit it. In the meantime, the first subjects in these longitudinal studies have acquired centenarian status; these individuals had never entered the IRP and in fact, as alluded to above, like other centenarians, their immune system appears "young."[32]

It Is Not Only CMV pp65 (495-503)–Specific CD8 Cells That Are Expanded in the Elderly

It is remarkable that a significant proportion of the entire peripheral CD8 cell pool in the elderly may consist of T cells carrying receptors for one single

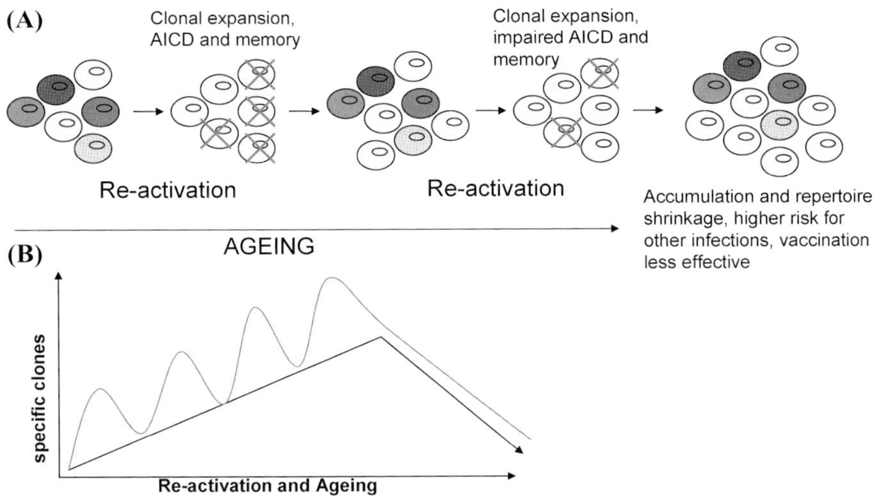

FIGURE 4. Model for disrupted homeostasis due to accumulation of CMV-specific T cells after several rounds of clonal expansion. (**A**) Chronic antigenic stress caused by persistent viruses such as HCMV leads to unremitting clonal expansion, activation-induced cell death (AICD) and memory generation. Over time, AICD and memory homeostasis is disrupted, leading to an accumulation of these cells in the elderly, which results in a reduction of diversity of the immunologic repertoire by reducing the space available for cells of other specificities. (**B**) Number of specific clones increases during persistent reactivation, but finally declines because of the loss of individual clones at their Hayflick limits in the final stage of life. (In color in Annals online.)

epitope of one single virus. This has now been confirmed with several different tetramers representing different combinations of HLA alleles and viral epitopes. Moreover, tetramer specificity has been repeatedly confirmed[8] (Britten *et al.*, manuscript submitted for publication). The high frequencies of CMV pp65 (495–503)–specific cells are likely to be a general phenomenon, not limited to this viral epitope in HLA-A2-positive donors only. Relatively little is known about responses to other epitopes on other genetic backgrounds, but it is clear that individual epitope selection by the immune system does occur, as with other viruses. Screening experiments have shown that a large variety of HCMV epitopes is indeed immunogenic, resulting in responses by up to 10% of all $CD4^+$ and 10% of all $CD8^+$ T cells, dominating the memory T cell compartment even in younger individuals.[33] Nonetheless, despite this huge commitment of resources to CMV immunosurveillance, this virus is not itself unique in causing clonal expansions of dysregulated CD8 cells; as mentioned above, the same phenomenon is seen with EBV, albeit usually to a quantitatively lesser extent.[34] In CMV/EBV co-infection, the effect of CMV is dominant, but in those rare cases of CMV-seronegativity studied, all of which were EBV-positive, the immune system was now fixated on EBV instead of

CMV.[35] This possibly quite general "dominance" of CMV has serious indirect clinical consequences because it is associated with decreased responsiveness to other viruses (e.g., in vaccination[36,37] and HIV progression).[38] It is also possible that in the absence of CMV or EBV or under certain circumstances even in their presence, other persistent antigens could cause a similar detrimental effect on immunity. One thinks especially of cancer, in which tumors stimulating immune response mediated by CD8 cells nonetheless persist and can induce anergy in specific T cells.[39]

The Impact of CMV Is Not Limited to CD8 Cells

CD8 cells and antigen-specific CD4 MHC class II-restricted "helper" cells function both independently and in concert to combat infection. There is less but increasing information on CMV-specific CD4 cells in humans, with some data emerging on age-related changes. CD4 cells also seem to recognize "immunodominant" epitopes from the pp65 protein preferentially,[40,41] and normal healthy subjects can possess very high levels of $CD4^+$ CMV-specific cells.[42] These manifest proliferative responses to pp65, but whether a fraction is anergic and whether this fraction increases with age, as with CD8 cells, is not known. The functional CD4 cells mediate peptide-specific cytotoxicity and secrete IFN-γ.[43] $CD4^+$ $CD28^-$ cells are more common in CMV-infected persons and also express biomarkers such as the KLRG1 and CD57 constellation that we identified on CD8 cells. They also contain high levels of granzyme B and perforin, and secrete IFN-γ, TNF-α and MIP-1β, as well as lacking molecules for secondary lymphoid organ homing, suggesting a direct role at sites of infection.[42,44–46] These data all suggest that $CD4^+$ T cells play a significant role in CMV infection. Some differences between CMV-specific CD8 and $CD4^+$ cells may be the greater loss of CD27 expression by the latter.[47] Modulation of CD70 (the ligand for CD27) can influence the size and the phenotype of CMV-specific T cells.[48] The CD4 cell–derived cytokine IL-21 is one factor that can block CD70. Thus, the CD27/CD70/IL-21 network may be of major importance in CMV infection. The frequencies of CMV-specific CD4 cells at different ages have not yet been clearly established. We hypothesize that they also increase with age and correlate with increased IRP. This is being tested.

Interventions to Prevent CMV Infection or Restore T Cell Immunity

What can be done to limit the damage resulting from persistent CMV infection (TABLE 1)? First and most obvious would be to prevent or delay early infection with CMV prenatally or during the breast-feeding period. A lesson from HIV may result in strategies to prevent mother-to-child transmission during birth.[49] Thereafter, the use of pasteurized breast milk to prevent postnatal

TABLE 1. Interventions to improve the T cell response

	Interventions to improve the T cell response
Prevention of early CMV infection	Antenatal and perinatal interventions,[49] pasteurized breast milk[50]
Antiviral treatments	Valganciclovir, ganciclovir, foscarnet, cidofovir[51]
CMV vaccination	Without T cell response
Thymic involvement	IL-7 treatment increases output of naïve cells[56,57]
Restoration of T cell function	PD-1[51,52] or IL-10R[53,55] blocking

infection may also be helpful.[50] In fact, at least in industrialized countries, the rate of CMV infection does now seem to be decreasing. Prophylactic vaccination would accelerate this tendency. However, for those already infected, reducing viral load with drugs is another option, and new approaches in AIDS and transplant patients are resulting in the development of improved antiviral agents with fewer side effects, such as ganciclovir, valganciclovir, foscarnet, or cidofovir.[51] These are now being used extensively, with a good safety record, which could suggest use in the elderly. Possibly, the elderly could be monitored at regular intervals and treated when their CMV-specific CD8 cell clonal expansions become excessive or when viral reactivation is demonstrated. Different treatment options could also be envisaged, again derived from experience with HIV, which may be able to restore function of the clonally expanded "exhausted" T cells to some extent. One strategem would be to apply antibodies to PD-1 ligand or PD-1, which has been reported to restore the function of "exhausted" HIV-specific T cells.[52,53] A second may be to use IL-10R blocking to restore T cell function *in vivo*.[54–56] These approaches to make "space" and "restore function," together with other methods directed, for example, at reinvigorating the thymus to produce more naïve T cells (e.g., using IL-7), might result in reconstituting immunity and extension of healthy longevity.[57,58]

ACKNOWLEDGMENTS

We thank Stefan Stefanovic and Dagmar Sigurdardottir for providing class-I tetramers, and Hans-Peter Pircher for KLRG1 antibody. The authors' collaborations were supported by the European Commission (Contracts QLK6-CT1999-02031 "ImAginE" and QLK6-CT2002-02283 "T-CIA"), and the Deutsche Forschungsgemeinschaft (SFB 685) (to G.P.) and FIS 01/0478 and 03/1383 (to R.S.).

REFERENCES

1. Wreghitt, T.G., E.L. Teare, O. Sule, *et al.* 2003. Cytomegalovirus infection in immunocompetent patients. Clin. Infect. Dis. **37:** 1603–1606.

2. GALIATSATOS, P., I. SHRIER, E. LAMOUREUX & A. SZILAGYI. 2005. Meta-analysis of outcome of cytomegalovirus colitis in immunocompetent hosts. Dig. Dis. Sci. **50:** 609–616.
3. KOMATSU, H., A. INUI, T. SOGO, *et al.* 2006. Large scale analysis of pediatric antiviral CD8$^+$ T cell populations reveals sustained, functional and mature responses. Immun. Ageing **3:** 11.
4. OLSSON, J., A. WIKBY, B. JOHANSSON, *et al.* 2000. Age-related change in peripheral blood T-lymphocyte subpopulations and cytomegalovirus infection in the very old: the Swedish longitudinal OCTO immune study Mech. Ageing Dev. **121:** 187–201.
5. KHAN, N., N. SHARIFF, M. COBBOLD, *et al.* 2002. Cytomegalovirus seropositivity drives the CD8 T cell repertoire toward greater clonality in healthy elderly individuals. J. Immunol. **169:** 1984–1992.
6. WILLS, M.R., G. OKECHA, M.P. WEEKES, *et al.* 2002. Identification of naive or antigen-experienced human CD8(+) T cells by expression of costimulation and chemokine receptors: analysis of the human cytomegalovirus-specific CD8(+) T cell response. J. Immunol. **168:** 5455–5464.
7. PITTET, M.J., A. ZIPPELIUS, D.E. SPEISER, *et al.* 2001. *Ex vivo* IFN-gamma secretion by circulating CD8 T lymphocytes: implications of a novel approach for T cell monitoring in infectious and malignant diseases. J. Immunol. **166:** 7634–7640.
8. ALTMAN, J.D., P.A. MOSS, P.J. GOULDER, *et al.* 1996. Phenotypic analysis of antigen-specific T lymphocytes. Science **274:** 94–96.
9. LI, F., D.A. YARILIN, J. VALIANDO, *et al.* 2002. Tumor antigen drives a persistent oligoclonal expansion of CD8$^+$ T cells in aged mice. Eur. J. Immunol. **32:** 1650–1658.
10. POSNETT, D.N., D. YARILIN, J.R. VALIANDO, *et al.* 2003. Oligoclonal expansions of antigen-specific CD8$^+$ T cells in aged mice. Ann. N. Y. Acad. Sci. **987:** 274–279.
11. SALLUSTO, F., J. GEGINAT & A. LANZAVECCHIA. 2004. Central memory and effector memory T cell subsets: function, generation, and maintenance. Annu. Rev. Immunol. **22:** 745–763.
12. KOCH, S., R. SOLANA, R.O. DELA & G. PAWELEC. 2006. Human cytomegalovirus infection and T cell immunosenescence: a mini review. Mech. Ageing Dev. **127:** 538–543.
13. RUFER, N., A. ZIPPELIUS, P. BATARD, *et al.* 2003. Ex vivo characterization of human CD8$^+$ T subsets with distinct replicative history and partial effector functions. Blood **102:** 1779–1787.
14. TOMIYAMA, H., T. MATSUDA & M. TAKIGUCHI. 2002. Differentiation of human CD8(+) T cells from a memory to memory/effector phenotype. J. Immunol. **168:** 5538–5550.
15. FANN, M., W.K. CHIU, W.H. WOOD, III, *et al.* 2005. Gene expression characteristics of CD28null memory phenotype CD8$^+$ T cells and its implication in T-cell aging. Immunol. Rev. **205:** 190–206.
16. WENG, N.P. 2006. Aging of the immune system: how much can the adaptive immune system adapt? Immunity **24:** 495–499.
17. BANDRES, E., J. MERINO, B. VAZQUEZ, *et al.* 2000. The increase of IFN-gamma production through aging correlates with the expanded CD8(+high)CD28(−)CD57(+) subpopulation. Clin. Immunol. **96:** 230–235.
18. NORTHFIELD, J., M. LUCAS, H. JONES, *et al.* 2005. Does memory improve with age? CD85j (ILT-2/LIR-1) expression on CD8 T cells correlates with 'memory inflation' in human cytomegalovirus infection Immunol. Cell Biol. **83:** 182–188.

19. OUYANG, Q., W.M. WAGNER, D. VOEHRINGER, *et al.* 2003. Age-associated accumulation of CMV-specific CD8$^+$ T cells expressing the inhibitory killer cell lectin-like receptor G1 (KLRG1). Exp. Gerontol. **38:** 911–920.
20. VESCOVINI, R., A. TELERA, F.F. FAGNONI, *et al.* 2004. Different contribution of EBV and CMV infections in very long-term carriers to age-related alterations of CD8$^+$ T cells. Exp. Gerontol. **39:** 1233–1243.
21. KHAN, N., A. HISLOP, N. GUDGEON, *et al.* 2004. Herpesvirus-specific CD8 T Cell immunity in old age: cytomegalovirus impairs the response to a coresident EBV infection. J. Immunol. **173:** 7481–7489.
22. OUYANG, Q., W.M. WAGNER, A. WIKBY, *et al.* 2003. Large numbers of dysfunctional CD8$^+$ T lymphocytes bearing receptors for a single dominant CMV epitope in the very old. J. Clin. Immunol. **23:** 247–257.
23. OUYANG, Q., W.M. WAGNER, W. ZHENG, *et al.* 2004. Dysfunctional CMV-specific CD8(+) T cells accumulate in the elderly. Exp. Gerontol. **39:** 607–613.
24. VOEHRINGER, D., M. KOSCHELLA & H. PIRCHER. 2002. Lack of proliferative capacity of human effector and memory T cells expressing killer cell lectinlike receptor G1 (KLRG1). Blood **100:** 3698–3702.
25. PAWELEC, G., S. KOCH, H. GRIESEMANN, *et al.* 2006. Immunosenescence, suppression and tumour progression. Cancer Immunol. Immunother. **55:** 981–986.
26. BRENCHLEY, J.M., N.J. KARANDIKAR, M.R. BETTS, *et al.* 2003. Expression of CD57 defines replicative senescence and antigen-induced apoptotic death of CD8$^+$ T cells. Blood **101:** 2711–2720.
27. IBEGBU, C.C., Y.X. XU, W. HARRIS, *et al.* 2005. Expression of killer cell lectin-like receptor G1 on antigen-specific human CD8$^+$ T lymphocytes during active, latent, and resolved infection and its relation with CD57. J. Immunol. **174:** 6088–6094.
28. WIKBY, A., B. JOHANSSON, J. OLSSON, *et al.* 2002. Expansions of peripheral blood CD8 T-lymphocyte subpopulations and an association with cytomegalovirus seropositivity in the elderly: the Swedish NONA immune study. Exp. Gerontol. **37:** 445–453.
29. PAWELEC, G., A. AKBAR, C. CARUSO, *et al.* 2005. Human immunosenescence: is it infectious? Immunol. Rev. **205:** 257–268.
30. HADRUP, S.R., J. STRINDHALL, T. KOLLGAARD, *et al.* 2006. Longitudinal studies of clonally expanded CD8 T cells reveal a repertoire shrinkage predicting mortality and an increased number of dysfunctional cytomegalovirus-specific T cells in the very elderly. J. Immunol. **176:** 2645–2653.
31. WIKBY, A., B.O. NILSSON, R. FORSEY, *et al.* 2006. The immune risk phenotype is associated with IL-6 in the terminal decline stage: findings from the Swedish NONA immune longitudinal study of very late life functioning. Mech. Ageing Dev. **127:** 695–704.
32. STRINDHALL, J., B.O. NILSSON, S. LOFGREN, *et al.* 2007. No immune risk profile among individuals who reach 100 years of age: findings from the Swedish NONA immune longitudinal study. Exp. Gerontol. May 21 [Epub ahead of print].
33. SYLWESTER, A.W., B.L. MITCHELL, J.B. EDGAR, *et al.* 2005. Broadly targeted human cytomegalovirus-specific CD4$^+$ and CD8$^+$ T cells dominate the memory compartments of exposed subjects. J. Exp. Med. **202:** 673–685.
34. OUYANG, Q., W.M. WAGNER, S. WALTER, *et al.* 2003. An age-related increase in the number of CD8$^+$ T cells carrying receptors for an immunodominant Epstein-Barr virus (EBV) epitope is counteracted by a decreased frequency of their antigen-specific responsiveness. Mech. Ageing Dev. **124:** 477–485.

35. WIKBY, A., F. FERGUSON, R. FORSEY, *et al.* 2005. An immune risk phenotype, cognitive impairment, and survival in very late life: impact of allostatic load in Swedish octogenarian and nonagenarian humans. J. Gerontol. A Biol. Sci. Med. Sci. **60:** 556–565.

36. BRYDAK, L.B., M. MACHALA, J. MYSLIWSKA, *et al.* 2003. Immune response to influenza vaccination in an elderly population. J. Clin. Immunol. **23:** 214–222.

37. TRZONKOWSKI, P., J. MYSLIWSKA, E. SZMIT, *et al.* 2003. Association between cytomegalovirus infection, enhanced proinflammatory response and low level of anti-hemagglutinins during the anti-influenza vaccination: an impact of immunosenescence. Vaccine **21:** 3826–3836.

38. SINICCO, A., R. RAITERI, M. SCIANDRA, *et al.* 1997. The influence of cytomegalovirus on the natural history of HIV infection: evidence of rapid course of HIV infection in HIV-positive patients infected with cytomegalovirus. Scand. J. Infect. Dis. **29:** 543–549.

39. WALTER, S., G. BIOLEY, H.J. BUHRING, *et al.* 2005. High frequencies of functionally impaired cytokeratin 18-specific CD8$^+$ T cells in healthy HLA-A2+ donors. Eur. J. Immunol. **35:** 2876–2885.

40. KERN, F., T. BUNDE, N. FAULHABER, *et al.* 2002. Cytomegalovirus (CMV) phosphoprotein 65 makes a large contribution to shaping the T cell repertoire in CMV-exposed individuals. J. Infect. Dis. **185:** 1709–1716.

41. LI, P.G., L. BOTTONE, F. IVALDI, *et al.* 2005. Generation of cytomegalovirus (CMV)-specific CD4 T cell lines devoid of alloreactivity, by use of a mixture of CMV-phosphoprotein 65 peptides for reconstitution of the T helper repertoire. J. Infect. Dis. **191:** 215–226.

42. SESTER, M., U. SESTER, B. GARTNER, *et al.* 2002. Sustained high frequencies of specific CD4 T cells restricted to a single persistent virus. J. Virol. **76:** 3748–3755.

43. WEEKES, M.P., M.R. WILLS, J.G. SISSONS & A.J. CARMICHAEL. 2004. Long-term stable expanded human CD4$^+$ T cell clones specific for human cytomegalovirus are distributed in both CD45RAhigh and CD45ROhigh populations. J. Immunol. **173:** 5843–5851.

44. APPAY, V., J.J. ZAUNDERS, L. PAPAGNO, *et al.* 2002. Characterization of CD4(+) CTLs ex vivo. J. Immunol. **168:** 5954–5958.

45. CASAZZA, J.P., M.R. BETTS, D.A. PRICE, *et al.* 2006. Acquisition of direct antiviral effector functions by CMV-specific CD4$^+$ T lymphocytes with cellular maturation. J. Exp. Med. **203:** 2865–2877.

46. VAN LEEUWEN, E.M., E.B. REMMERSWAAL, M.T. VOSSEN, *et al.* 2004. Emergence of a CD4$^+$CD28- granzyme B+, cytomegalovirus-specific T cell subset after recovery of primary cytomegalovirus infection. J. Immunol. **173:** 1834–1841.

47. FLETCHER, J.M., M. VUKMANOVIC-STEJIC, P.J. DUNNE, *et al.* 2005. Cytomegalovirus-specific CD4$^+$ T cells in healthy carriers are continuously driven to replicative exhaustion. J. Immunol. **175:** 8218–8225.

48. GAMADIA, L.E., E.M. VAN LEEUWEN, E.B. REMMERSWAAL, *et al.* 2004. The size and phenotype of virus-specific T cell populations is determined by repetitive antigenic stimulation and environmental cytokines. J. Immunol. **172:** 6107–6114.

49. NEWELL, M.L. 2003. Antenatal and perinatal strategies to prevent mother-to-child transmission of HIV infection. Trans. R. Soc. Trop. Med. Hyg. **97:** 22–24.

50. HAMPRECHT, K., J. MASCHMANN, D. MULLER, *et al.* 2004. Cytomegalovirus (CMV) inactivation in breast milk: reassessment of pasteurization and freeze-thawing. Pediatr. Res. **56:** 529–535.

51. Biron, K.K. 2006. Antiviral drugs for cytomegalovirus diseases. Antiviral Res. **71:** 154–163.
52. Day, C.L., D.E. Kaufmann, P. Kiepiela, *et al.* 2006. PD-1 expression on HIV-specific T cells is associated with T-cell exhaustion and disease progression. Nature **443:** 350–354.
53. Trautmann, L., L. Janbazian, N. Chomont, *et al.* 2006. Upregulation of PD-1 expression on HIV-specific $CD8^+$ T cells leads to reversible immune dysfunction. Nat. Med. **12:** 1198–1202.
54. Blackburn, S.D. & E.J. Wherry. 2007. IL-10, T cell exhaustion and viral persistence. Trends Microbiol. **15:** 143–146.
55. Brooks, D.G., M.J. Trifilo, K.H. Edelmann, *et al.* 2006. Interleukin-10 determines viral clearance or persistence *in vivo*. Nat. Med. **12:** 1301–1309.
56. Ejrnaes, M., C.M. Filippi, M.M. Martinic, *et al.* 2006. Resolution of a chronic viral infection after interleukin-10 receptor blockade. J. Exp. Med. **203:** 2461–2472.
57. Aspinall, R. 2006. T cell development, ageing and interleukin-7. Mech. Ageing Dev. **127:** 572–578.
58. Henson, S.M., R. Snelgrove, T. Hussell, *et al.* 2005. An IL-7 fusion protein that shows increased thymopoietic ability. J. Immunol. **175:** 4112–4118.

Transcriptional Regulation of Telomerase Activity

Roles of the the Ets Transcription Factor Family

JULIE DWYER, HE LI, DAKANG XU, AND JUN-PING LIU

Department of Immunology, Monash University, Melbourne, Australia

ABSTRACT: Telomerase maintains telomeres to preclude cell senescence. It remains elusive how telomerase activity is repressed in differentiated cells, but retained at high levels in stem cells and cancer. Recent studies suggest that the Ets transcription factor family, downstream of the mitogen signaling pathways of MAP kinase, regulates telomerase activity at the gene transcription level of human telomerase reverse transcriptase (hTERT). Several Ets transcription factors are involved in regulating hTERT gene expression, both directly and indirectly through the proto-oncogene c-myc. ER81 may mediate telomerase activation in telomerase-negative fibroblasts stimulated by oncogenes Her2/Neu, Ras, and Raf. Ets2 may also play an important role in regulating the hTERT gene; but further studies are required to decipher the mechanisms in the regulation of telomerase activity.

KEYWORDS: Ets; transcription factor; telomerase; hTERT; telomeres; tumorigenesis; cancer; senescence

INTRODUCTION

In most tissues and organs, cells exhibit limited lifespan; cell aging occurs inevitably at different paces.[1] Why cells bear different lifespans across different tissues is not fully understood. Dividing cells have a short lifespan of 20–40 population doublings, and after a finite number of divisions, the cells stop dividing, begin to age, and undergo apoptosis, whereas stem cells and cells that are required to continuously renew and expand can proliferate with extended lifespan or immortality. The difference between cells with short and long lifespan has been recognized to involve the mechanisms of telomeres (ends of chromosomes). If the maintenance of telomeres is carried out by the enzyme called telomerase, cells will have a long proliferative lifespan. Without

Address for correspondence: Dr. Jun-Ping Liu, Department of Immunology, AMREP, Monash Medical School, Commercial Road, Prahran, Victoria 3181, Australia. Voice: 61-3-99030715; fax: 61-3-99030120.

jun-ping.liu@med.monash.edu.au

**Ann. N.Y. Acad. Sci. 1114: 36–47 (2007). © 2007 New York Academy of Sciences.
doi: 10.1196/annals.1396.022**

telomerase, telomeres shed off ~200 base pairs (bp) in each round of chromosome replication, or shorten from ~15 to ~5 kb in a cell lifespan of ~50 population doublings. Characteristic of replicative senescence (aging), short telomeres trigger cells to exit the cell cycle permanently.[2–5]

Telomeres are specialized structures at the ends of each eukaryotic chromosome, providing protective mechanisms to prevent chromosomes from recombination, fusion, and degradation.[2] Human telomeres comprise arrays of a few thousand tandem repeats of TTAGGG complementary strands, and their binding proteins. Although the means by which the structure and function of telomeres are regulated during cell divisions and aging are yet to be fully elucidated, it is known that activation of telomerase stabilizes telomeres and enables cells to divide continuously.[2] Consistently, telomerase is undetectable in normal human somatic cells that subsequently undergo senescence,[6] but is at high levels in stem cells, expanding lymphocytes, and cancer cells. Expression of telomerase in telomerase-negative cells *in vitro* also immortalizes the cells without transformation[7,8] (reviewed in Refs. 9–11). Inhibition of telomerase activity in cancer arrests cancer cell proliferation and induces cancer cell senescence and death.[12–16] Thus, telomerase activity plays a vital role in extending cell lifespan and evading cell senescence.

THE MECHANISMS OF REGULATION OF TELOMERASE ACTIVITY

Regulation of telomerase activity takes place at both telomerase protein and gene levels.[9,17,18] Human telomerase contains the catalytic subunit human telomerase reverse transcriptase (hTERT), a polypeptide of 1132 amino acid residues,[19–23] and also the human telomerase RNA component (hTERC), which consists of 450 nucleotides.[24] Responsible for lengthening of telomeres by reverse transcription and for continuous proliferation of highly proliferative cells, hTERT is the rate-limiting catalyst. The structure of hTERT features four major functional regions: the N-terminal regulatory (R) domain, RNA-binding (RB) domain, reverse transcription (RT) domain, and C-terminal dimerization domain (C) (FIG. 1A).[25] Several motifs (GQ, T2, QFP, and T) in the R and RB domains of hTERT are conserved in different species. The motifs from A to E in the RT domain (FIG. 1A) are conserved in other reverse transcriptases in man. Studies show that telomerase activity requires molecular chaperones for appropriate assembly of the holoenzyme.[26,27] Telomerase activity is also regulated positively by phosphorylation of the hTERT by protein kinase C and negatively by dephosphorylation by protein phosphatase 2A.[28,29]

The hTERT gene from the human genome comprises more than 35 kb DNA and codes for an mRNA including 16 exons and 15 introns.[30–32] The coding of hTERT sequences in 16 exons separated by 15 introns is consistent with the presence of various hTERT isoforms and with the possibility of complex

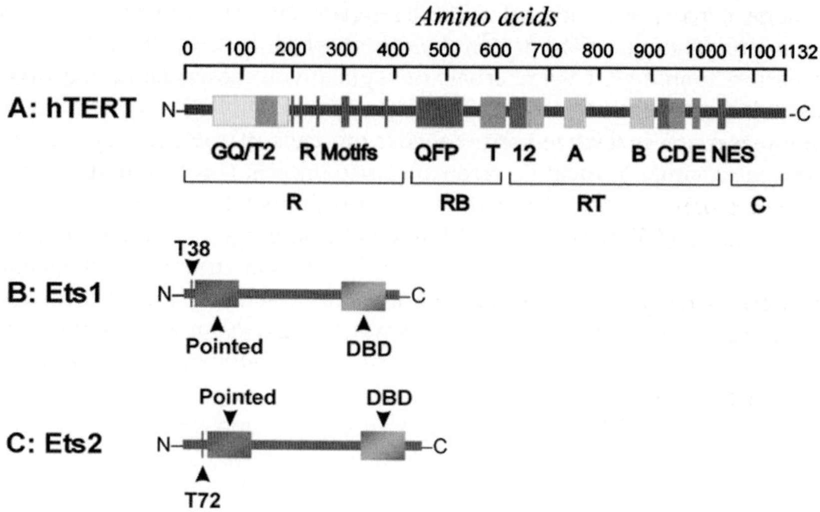

FIGURE 1. Domain structures of hTERT, Ets1, and Ets2. (**A**) hTERT has an N-terminal domain (N) including the conserved GQ block (hypomutable domain I), polyproline (PP) domain, RNA-binding (RB) domain containing conserved motifs CP, QFP and T (hypomutable domains II, III, and IV, respectively), the reverse transcriptase (RT) domain containing the seven conserved RT motifs, and a weakly conserved C-terminal extension. (**B** and **C**) The Ets protein is characterized by two highly conserved domains—the pointed domain and the Ets DNA binding domain (DBD). MAPK phosphorylation of Ets1 at position T38 and Ets2 at position T72 is suggested to activate Ets-mediated transcriptional regulation.

regulation of hTERT gene expression and function in mammals. The regulatory mechanisms underlying these alternative splicing events for various hTERT messages remain unknown. Neither is fully elucidated concerning the transcriptional control mechanisms of the hTERT gene in normal tissues and diseases.[10,33,34] The gene transcription of hTERT has been suggested to be a predominant step in the regulation of telomerase activity. To investigate the control mechanisms of hTERT gene transcription, researchers have investigated the structures of the hTERT gene promoter region by unidirectional deletion of the 5′ flanking region of the hTERT gene and expression of these hTERT promoter regions in a luciferase reporter assay. It has been found that the hTERT gene promoter lies within a region extending to 330 bp upstream of the translation start site (−330) to the second exon of the gene (+228), with a region of 59 bp (−208 to −150) required for maximal promoter activity.[30–32]

A number of binding sites for transcription factors and repressors have been identified in the hTERT promoter. These include two E boxes for the Myc oncogene, several GC boxes for Sp1, and the *cis*-regulatory elements that may recruit AP1, AP2, AP4, ATF, CREB, ER, PR, GC, IK2, MYOD, NF1, T3R and USF.[17] As a primary transcription factor of the hTERT gene, c-myc stimulates hTERT gene transcription by binding to the E box in the hTERT gene promoter. It has been shown that the hTERT transcription activity of c-myc is competitively controlled by the Mad-Max network.[35] More interestingly, estrogen may regulate the action of c-myc in human tumor cells.[36] Estrogen has also been shown to stimulate hTERT gene by direct binding to the estrogen-response element in ovarian epithelial cells and cancer cell lines.[10] In addition, epidermal growth factor (EGF) may activate hTERT gene through mitogen-activated protein (MAP) kinase and transforming growth factor-β (TGF-β) inhibits the hTERT transcription.[34] Signals transmitted through MAP kinase can increase and decrease hTERT transcription, depending on downstream mediators in response to various stimuli.[37–42] In the exquisite responses of MAP kinase signaling to the extracellular and intracellular environment, the Ets transcription factor ER81 mediates the protein oncogene Ras to stimulate hTERT gene transcription,[40] whereas the transcription factor activator protein 1 also binds to the hTERT gene promoter, inducing hTERT gene repression.[38]

THE ETS TRANSCRIPTION FACTOR FAMILY

The Ets (for E26 transformation–specific) transcription factor family consists of more than thirty evolutionarily conserved transcription factors.[43–45] This family was discovered from identification of the v-ets as part of a fusion oncogene in the E26 avian leukemia retrovirus in 1983. All Ets family proteins have conserved genes in phylogenetically divergent species from *Drosophila* and *C. elegans* to the human.[46] Ets transcription factors are the targets of several major signaling cascades elicited by activation of the cell membrane tyrosine kinase receptor for EGF and the closely related HER2/Neu pathway.[47,48] With a conserved winged helix-turn-helix DNA binding domain (ETS domain), Ets proteins recognize the core consensus DNA sequence GGAA/T of target genes. Many Ets proteins have been implicated in cell proliferation and tumor progression, especially the Ets1 and Ets2 transcription factors. Ets1 and Ets2 are similar transcription factors of 441 and 469 amino acid residues, respectively, containing a MAP kinase phosphorylation site, the pointed domain and the DNA-binding domain[49] (FIG. 1B and C). Importantly, a growing number of human malignancies are associated with high levels of Ets transcription factor, frequently Ets1 and Ets2.[50,51] Constitutive activation of Ets2 is often detected in human breast cancer specimens in patients with advanced disease, but not in normal breast epithelial cells.[50,52] Inhibition of the Ets factors reduces tumorigenicity.[53,54] In mice, overexpression of Ets1 and Ets2 increases cell proliferation, anchorage-independent growth, and tumorigenicity,[55] whereas a singly targeted Ets2 allele restricts development of mammary tumors.[56] Despite the upregulation of Ets2-targeted genes in cell cycle control and of anti-apoptosis factors, such as cyclin D1, Bcl-2, and Bcl-XL, as reported in prostate cancer,[54,57] how Ets2-mediates oncogenesis is yet to be fully understood.

DIFFERENTIAL REGULATIONS OF TELOMERASE ACTIVITY BY Ets

Acting on its binding elements in gene promoter regions, Ets transcription factors may regulate more than 200 genes for various cellular functions.[58] With the involvement of the Ets family in stem cell development, cell senescence and death, and tumorigenesis,[59] it has been thought that certain Ets family members participate in the regulations of multiple critical genes required for cell proliferation during development and oncogenic evolution. This hypothesis is particularly significant considering that certain Ets are up in tumors, and mediate the signaling of MAP kinase pathways that play a crucial role in the regulation of cell proliferation and neoplastic evolution. Several groups of researchers determined the roles of Ets in regulating telomerase activity that is required for extended cell proliferation.

Maida *et al.* reported that EGF stimulates telomerase activity that is abolished by MAP kinase inhibitor U0126.[39] In their search for the mechanisms of MAP kinase mediation of EGF regulation of the hTERT gene, the investigators found that mutation of a putative Ets binding element close to the hTERT gene transcription start codon inhibits EGF stimulation of telomerase activity, suggesting that Ets transcription factor is involved in mediating EGF- and MAP kinase-induced hTERT gene expression.[39] Consistently, overexpression of the Ets2 gene stimulates the hTERT gene promoter activity, whereas the dominant negative form of truncated Ets2 blocks EGF-induced hTERT gene promoter activity.[39] Although this report did not show a direct binding of the Ets2 protein to the hTERT gene, and whether or not other Ets proteins would produce a similar effect, it provides evidence clearly showing the opposing effects of wild-type and mutant Ets2 on the recombinant hTERT gene promoter activity in vulvar cancer A-431 cells and NIH3T3 cells.[39]

To investigate the roles of Ets in the underlying mechanisms of high and low telomerase activity in the human leukemic U937 cell lines, Xiao *et al.* carried out protein-DNA binding analysis and found that in addition to the bindings of c-myc, mad1, and Sp1 to the hTERT gene promoter, Ets1 and Ets2 also bind.[60] Even though the binding by Ets1 and Ets2 does not explain the varied telomerase activities in the U937 clones, the study has demonstrated a direct binding of Ets1 and Ets2 to the hTERT gene. In their further work, these investigators found that another Ets family member, Fli1, and Id2 (Ets binding partner) also bind to the hTERT gene promoter.[61] Surprisingly, however, these studies show that Ets1 or Ets2 is inhibitory to telomerase activity when overexpressed in human myeloid erythroleukemia line K562 cells.[61] The inhibitory effect of Ets1 and Ets2 is not associated with changes in c-myc or Sp1, suggesting a direct effect of the Ets on the hTERT gene without involving the transcription factors c-myc and Sp1. The specificity of the results has been demonstrated by the lack of effect of Fli1 and the opposite effect of Id2 when similarly overexpressed.[61] Thus, these data suggest an inhibitory regulation of telomerase activity by the Ets1 and Ets2 proteins,[61] which is in contrast to the study showing that Ets2 stimulates the hTERT gene promoter activity when overexpressed.[39]

Several possibilities may exist to explain the discrepancies of the functions of Ets2 in positive or negative regulation of the hTERT gene. First, Ets2 may stimulate the c-myc gene[62] to upregulate the hTERT gene, thereby indirectly stimulating the hTERT gene through c-myc. In the absence of increased c-myc,[61] increased Ets2 may interfere with the transcription effect of c-myc on the hTERT gene by binding directly to an Ets binding element juxtaposed to the E box, producing an inhibitory effect. This interfering effect of Ets2 on c-myc action on the hTERT gene promoter could be overturned competitively by increased c-myc. In addition, different levels of Ets2 in different cell types may produce different ratios of the Ets proteins involved in regulating the hTERT gene. Moreover, the binding of Ets2 to different Ets binding elements may lead to different outcomes of the hTERT gene promoter activity. Indeed, Xiao *et al.* have explored the potentially different functions of the multiple Ets binding elements on the hTERT gene promoter by site-directed mutagenesis.[61] The investigators designed a 408-bp hTERT promoter with each Ets binding site at a different location mutated, and they found that mutations of Ets binding elements result in both negative and positive regulation of the hTERT gene promoter activity, with the Ets binding site at −293 mediating an inhibitory effect and the Ets binding site at −36 mediating a stimulatory effect.[61]

To determine whether endogenous Ets1 or Ets2 might be involved in telomerase activity through mediating the hTERT gene transcription, we generated the small interference RNA (siRNA) expressing constructs to silence Ets1 and Ets2 genes specifically, using U6 polymerase III promoter. Having conducted both transient and stable gene expression of the Ets1 and Ets2 siRNA individually in cultured human breast cancer MCF-7 cells, we found that both Ets1 and Ets2 siRNA selectively silenced the Ets1 and Ets2 gene, respectively. Of interest, silencing Ets1 gene expression resulted in no change in the hTERT gene expression and telomerase activity, but silencing the Ets2 gene induced a significant reduction in the hTERT gene expression and telomerase activity (unpublished data). The data suggest that the endogenous levels of Ets2 gene expression in human breast cancer cells are essential for maintaining the high levels of hTERT gene expression for telomerase activity. Further studies are required to decipher the mechanisms underlying Ets2-positive regulation of the hTERT gene expression by direct and/or indirect mechanisms. Preliminary experiments showed that silencing Ets2 was also associated with a significant reduction of the c-myc gene expression, suggesting an involvement of c-myc in mediating Ets2 regulation of telomerase activity. These findings are consistent with the hypothesis that the c-myc gene is a transcription target of Ets2 transcription factor, so withdrawal of Ets2 switches off c-myc gene transcription and in turn c-myc transcriptional effect on the hTERT gene (FIG. 2). However,

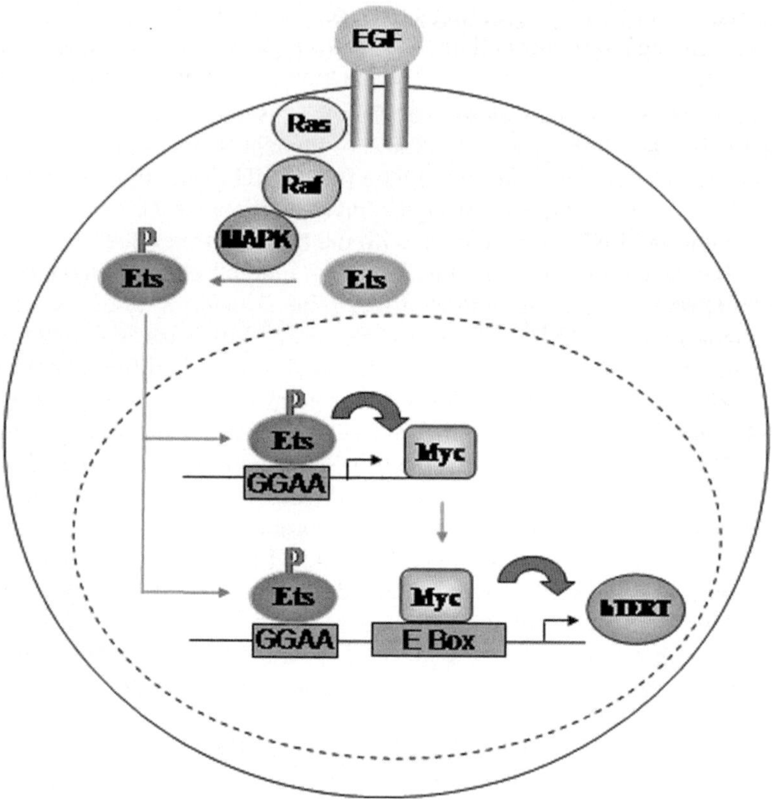

FIGURE 2. A working model of dual mechanisms of Ets regulation of the hTERT gene. Phosphorylated by the MAP kinase in response to growth factor, Ets migrates into the nucleus to bind the hTERT gene promoter to activate hTERT gene transcription. Ets also activates the c-myc gene that in turn stimulates the hTERT gene. Increased activity of the oncoprotein HER2, Ras, and Raf due to dominant mutations could hijack the MAP kinase to activate Ets transcriptional activation of the hTERT gene in cancer.

a direct action of the Ets2 transcription factor on the hTERT gene cannot be ruled out since several Ets binding consensus elements are present in the promoter region of the hTERT gene[39,61] and Ets2 has been shown to bind to the hTERT.[61]

In addition, recent studies have further identified the involvement of other Ets transcription factors in regulating the hTERT gene expression. Takahashi *et al.* show that in Ewing's sarcoma, an aggressive bone neoplasia in young adults and adolescents, the fusion proteins from the *EWS* fusion gene with one of the five members of the *ETS* gene family (*FLI1*, *ERG*, *ETV1/ER81*, *E1AF/PEA3*, and *FEV*) in the chromosome 22 are involved in stimulating hTERT gene expression.[63,64] Both EWS/FLI1 and EWS/E1AF activate telomerase activity by upregulating TERT gene transcription in NIH3T3 cells and in two Ewing's sarcoma cell lines.[63] However, EWS/FLI1 and EWS/E1AF stimulate the TERT gene promoter in an Ets binding site–independent manner; EWS/E1AF is found in the initiation complex of TERT transcription and to interact with CBP/p300 in activating the hTERT promoter, suggesting that in Ewing's sarcoma, EWS/ETS functions as a co-activator of TERT transcription.[63,64]

Goueli and Janknecht show that oncogenic Ras stimulates the hTERT gene expression by the Ets transcription factor ER81.[40] Three oncoproteins upstream of MAP kinase, HER2/Neu, Ras, and Raf, all stimulate hTERT promoter activity via ER81, and collaborate to enhance the stimulation of hTERT gene expression and telomerase activity in telomerase-negative BJ foreskin fibroblasts.[40] Dominant-negative mutant, mutation of the ER81 binding sites in the hTERT promoter, or inhibition of MAP kinase-dependent phosphorylation of ER81 render hTERT gene promoter unresponsive to HER2/Neu.[40] Therefore, ER81 may play a key role in mediating hTERT gene expression and telomerase activity in telomerase-negative cells stimulated by HER2/Neu, Ras, and Raf oncoproteins.

Consistent with direct transcriptional regulation of the hTERT gene by Ets family, Hsu *et al.* have found significant correlations between single nucleotide polymorphism of the Ets binding element, telomerase activity, and telomere length in non–small cell lung cancer.[65] The researchers sequenced the PCR products of a 716-bp DNA, upstream of the transcription starting code in the hTERT gene promoter, from 66 non–small cell lung cancers. They found that the thymine at the position −245 in the $^{-246}$TTCC^{-243} motif of the hTERT gene is mutated to cytosine in both alleles in about 10% of the patients, and single allele homozygote mutation has occurred to about 53% of the patients.[65] More interesting, while the wild-type Ets motif correlates with high levels of telomerase activity and long telomeres, the mutant ($^{-246}$T$\underline{\mathrm{C}}$CC^{-243}) correlates with low levels of telomerase activity and short telomeres.[65] These findings suggest that the Ets motif at the location of −246−243 mediates a positive regulation of telomerase activity in cancer patients.[65] Future studies are required to clarify the functions of other potential Ets binding elements in mediating Ets regulation of telomerase activity in other cancers. Establishment of the nucleotide polymorphisms in the Ets binding elements of the hTERT gene would shed light on the mechanisms of Ets regulation of the hTERT gene, and provide possible approaches for clinical diagnosis and molecular targets for molecular intervention of aging-related disorders such as cancer.

CONCLUSIONS

Ets transcription factors regulate hTERT gene expression. ER81 is a potent transcription factor of the hTERT gene in mediating growth factor–stimulated activation of telomerase. Ets2 may regulate the hTERT gene both positively and negatively, perhaps depending on the gene expression levels of c-myc. Stimulating the c-myc gene, Ets2 upregulates telomerase activity, but binding to the hTERT gene promoter, Ets2 may interfere with c-myc action on the hTERT gene, resulting in downregulation of telomerase activity. Of the Ets binding elements, a single nucleotide mutation from the thymine to cytosine at the position of −245 may be responsible for reduced telomerase activity and short telomeres in non–small cell lung cancers. Further studies are required to establish the molecular mechanisms of action of Ets2 and its relationships with other transcription factors in the regulation of telomerase activity in aging and cancer cells.

ACKNOWLEDGMENTS

This work was supported by grants from the National Health and Medical Research Council of Australia, Australia Research Council, and Cancer Council of Victoria, Australia. J.D. was a recipient of a scholarship from the Faculty of Medicine, Nursing and Health Sciences, Monash University.

REFERENCES

1. HERBIG, U. *et al.* 2006. Cellular senescence in aging primates. Science **311:** 1257.
2. BLACKBURN, E.H. 2001. Switching and signaling at the telomere. Cell **106:** 661–673.
3. BLACKBURN, E.H. 2005. Telomeres and telomerase: their mechanisms of action and the effects of altering their functions. FEBS Lett. **579:** 859–862.
4. DE LANGE, T. 2004. T-loops and the origin of telomeres. Nat. Rev. Mol. Cell Biol. **5:** 323–329.
5. SHAY, J.W. & W.E. WRIGHT. 2007. Hallmarks of telomeres in ageing research. J. Pathol. **211:** 114–123.
6. EPEL, E.S. *et al.* 2004. Accelerated telomere shortening in response to life stress. Proc. Natl. Acad. Sci. USA **101:** 17312–17315.
7. BODNAR, A.G. *et al.* 1998. Extension of life-span by introduction of telomerase into normal human cells. Science **279:** 349–352.
8. YANG, G. *et al.* 2007. Disruption of the retinoblastoma pathway by small interfering RNA and ectopic expression of the catalytic subunit of telomerase lead to immortalization of human ovarian surface epithelial cells. Oncogene **26:** 1492–1498.
9. LIU, J.P. 1999. Studies of the molecular mechanisms in the regulation of telomerase activity. FASEB J. **13:** 2091–2104.
10. BAYNE, S. & J.P. LIU. 2005. Hormones and growth factors regulate telomerase activity in ageing and cancer. Mol. Cell Endocrinol. **240:** 11–22.
11. LIU, J.P. *et al.* 2006. Mechanisms of cell immortalization mediated by EB viral activation of telomerase in nasopharyngeal carcinoma. Cell Res. **16:** 809–817.

12. SHAY, J.W. & W.E. WRIGHT. 2005. Mechanism-based combination telomerase inhibition therapy. Cancer Cell **7:** 1–2.
13. KELLAND, L.R. 2005. Overcoming the immortality of tumour cells by telomere and telomerase based cancer therapeutics: current status and future prospects. Eur. J. Cancer **41:** 971–979.
14. LI, S. *et al.* 2005. Cellular and gene expression responses involved in the rapid growth inhibition of human cancer cells by RNA interference-mediated depletion of telomerase RNA. J. Biol. Chem. **280:** 23709–23717.
15. DIKMEN, Z.G. *et al.* 2005. *In vivo* inhibition of lung cancer by GRN163L: a novel human telomerase inhibitor. Cancer Res. **65:** 7866–7873.
16. GELLERT, G.C. *et al.* 2006. Effects of a novel telomerase inhibitor, GRN163L, in human breast cancer. Breast Cancer Res. Treat. **96:** 73–81.
17. LIU, J.P. 2001. Molecular mechanisms regulating telomerase activity. *In* Telomerase, Aging and Disease. M.P. Mattson & T. Pandita, Eds.: 33–59. Elsevier Science. Amsterdam, the Netherlands.
18. LIU, J.P. *et al.* 2002. Regulation of telomerase activity. *In* Telomeres and Telomerases: Cancer and Biology. G. Krupp & R. Parwaresch, Eds.: 134–159. Landes Bioscience/Eurekah. Austin, TX.
19. NAKAMURA, T.M. *et al.* 1997. Telomerase catalytic subunit homologs from fission yeast and human. Science **277:** 955–959.
20. MEYERSON, M. *et al.* 1997. hEst2, the putative human telomerase catalytic subunit gene, is up-regulated in tumor cells and during immortalization. Cell **90:** 785–795.
21. NAKAYAMA, J. *et al.* 1997. TLP1: a gene encoding a protein component of mammalian telomerase is a novel member of WD repeats family. Cell **88:** 875–884.
22. HARRINGTON, L. *et al.* 1997. A mammalian telomerase-associated protein. Science **275:** 973–977.
23. KILIAN, A. *et al.* 1997. Isolation of a candidate human telomerase catalytic subunit gene, which reveals complex splicing patterns in different cell types. Hum. Mol. Genet. **6:** 2011–2019.
24. COLLINS, K. & J.R. MITCHELL. 2002. Telomerase in the human organism. Oncogene **21:** 564–579.
25. KELLEHER, C. *et al.* 2002. Telomerase: biochemical considerations for enzyme and substrate. Trends Biochem. Sci. **27:** 572–579.
26. HOLT, S.E. *et al.* 1999. Functional requirement of p23 and hsp90 in telomerase complexes. Genes Dev. **13:** 817–826.
27. KEPPLER, B.R. *et al.* 2006. The biochemical role of the heat shock protein 90 chaperone complex in establishing human telomerase activity. J. Biol. Chem. **281:** 19840–19848.
28. LI, H. *et al.* 1998. Telomerase is controlled by protein kinase Calpha in human breast cancer cells. J. Biol. Chem. **273:** 33436–33442.
29. LI, H. *et al.* 1997. Protein phosphatase 2A inhibits nuclear telomerase activity in human breast cancer cells. J. Biol. Chem. **272:** 16729–16732.
30. CONG, Y.S. *et al.* 1999. The human telomerase catalytic subunit hTERT: organization of the gene and characterization of the promoter [In Process Citation]. Hum. Mol. Genet. **8:** 137–142.
31. HORIKAWA, I. *et al.* 1999. Cloning and characterization of the promoter region of human telomerase reverse transcriptase gene [In Process Citation]. Cancer Res. **59:** 826–830.

32. Takakura, M. *et al.* 1999. Cloning of human telomerase catalytic subunit (hTERT) gene promoter and identification of proximal core promoter sequences essential for transcriptional activation in immortalized and cancer cells. Cancer Res. **59:** 551–557.
33. Li, H. *et al.* 2005. Telomerase down-regulation does not mediate PC12 pheochromocytoma cell differentiation induced by NGF, but requires MAP kinase signalling. J. Neurochem. **95:** 891–901.
34. Li, H. *et al.* 2006. Transforming growth factor beta suppresses human telomerase reverse transcriptase (hTERT) by smad3 interactions with c-myc and the hTERT gene. J. Biol. Chem. **281:** 25588–25600.
35. Xu, D. *et al.* 2001. Switch from Myc/Max to Mad1/Max binding and decrease in histone acetylation at the telomerase reverse transcriptase promoter during differentiation of HL60 cells. Proc. Natl. Acad. Sci. USA **98:** 3826–3831.
36. Sarkar, P. *et al.* 2006. Activation of telomerase in BeWo cells by estrogen and 2,3,7,8-tetrachlorodibenzo-p-dioxin in co-operation with c-Myc. Int. J. Oncol. **28:** 43–51.
37. Xu, D. *et al.* 2006. Inhibition of telomerase targeting MAP kinase signaling. *In* Methods in Molecular Biology: telomerase Inhibition: strategies and Protocols: L. G. Andrews & T. O. Tollefsbol, Eds.: in press. Humana Press Inc. Totowa, NJ.
38. Takakura, M. *et al.* 2005. Function of AP-1 in transcription of the telomerase reverse transcriptase gene (TERT) in human and mouse cells. Mol. Cell. Biol. **25:** 8037–8043.
39. Maida, Y. *et al.* 2002. Direct activation of telomerase by EGF through Ets-mediated transactivation of TERT via MAP kinase signaling pathway. Oncogene **21:** 4071–4079.
40. Goueli, B.S. & R. Janknecht. 2004. Upregulation of the catalytic telomerase subunit by the transcription factor ER81 and oncogenic HER2/Neu, Ras, or Raf. Mol. Cell Biol. **24:** 25–35.
41. Wang, Z. *et al.* 2000. Progesterone regulates human telomerase reverse transcriptase gene expression via activation of mitogen-activated protein kinase signaling pathway [In Process Citation]. Cancer Res. **60:** 5376–5381.
42. Alfonso-De Matte, M.Y. *et al.* 2002. Telomerase is regulated by c-Jun NH2-terminal kinase in ovarian surface epithelial cells. Cancer Res. **62:** 4575–4578.
43. Sharrocks, A.D. 2001. The ETS-domain transcription factor family. Nat. Rev. Mol. Cell Biol. **2:** 827–837.
44. Sementchenko, V.I. & D.K. Watson. 2000. Ets target genes: past, present and future. Oncogene **19:** 6533–6548.
45. Maroulakou, I.G. & D.B. Bowe. 2000. Expression and function of Ets transcription factors in mammalian development: a regulatory network. Oncogene **19:** 6432–6442.
46. Hsu, T. *et al.* 2004. Ets proteins in biological control and cancer. J. Cell Biochem. **91:** 896–903.
47. Svensson, S. *et al.* 2005. ERK phosphorylation is linked to VEGFR2 expression and Ets-2 phosphorylation in breast cancer and is associated with tamoxifen treatment resistance and small tumours with good prognosis. Oncogene **24:** 4370–4379.
48. Buggy, Y. *et al.* 2006. Ets2 transcription factor in normal and neoplastic human breast tissue. Eur. J. Cancer **42:** 485–491.

49. Sharrocks, A.D. 2001. The ETS-domain transcription factor family. Nat. Rev. Mol. Cell Biol. **2:** 827–837.
50. Myers, E. *et al.* 2005. Associations and interactions between Ets-1 and Ets-2 and coregulatory proteins, SRC-1, AIB1, and NCoR in breast cancer. Clin. Cancer Res. **11:** 2111–2122.
51. Seth, A. & D.K. Watson. 2005. ETS transcription factors and their emerging roles in human cancer. Eur. J. Cancer **41:** 2462–2478.
52. Sapi, E. *et al.* 1998. Ets-2 transdominant mutant abolishes anchorage-independent growth and macrophage colony-stimulating factor-stimulated invasion by BT20 breast carcinoma cells. Cancer Res. **58:** 1027–1033.
53. Foos, G. & C.A. Hauser. 2000. Altered Ets transcription factor activity in prostate tumor cells inhibits anchorage-independent growth, survival, and invasiveness. Oncogene **19:** 5507–5516.
54. Carbone, G.M. *et al.* 2004. Triplex DNA-mediated downregulation of Ets2 expression results in growth inhibition and apoptosis in human prostate cancer cells. Nucleic Acids Res. **32:** 4358–4367.
55. Hahne, J.C. *et al.* 2005. Ets-1 expression promotes epithelial cell transformation by inducing migration, invasion and anchorage-independent growth. Oncogene **24:** 5384–5388.
56. Neznanov, N. *et al.* 1999. A single targeted Ets2 allele restricts development of mammary tumors in transgenic mice. Cancer Res. **59:** 4242–4246.
57. Sementchenko, V.I. *et al.* 1998. ETS2 function is required to maintain the transformed state of human prostate cancer cells. Oncogene **17:** 2883–2888.
58. Oikawa, T. & T. Yamada. 2003. Molecular biology of the Ets family of transcription factors. Gene **303:** 11–34.
59. Xu, D. *et al.* 2002. Ets1 is required for p53 transcriptional activity in UV-induced apoptosis in embryonic stem cells. EMBO J. **21:** 4081–4093.
60. Xiao, X. *et al.* 2002. Identification and characterization of rapidly dividing U937 clones with differential telomerase activity and gene expression profiles: role of c-Myc/Mad1 and Id/Ets proteins. Leukemia **16:** 1877–1880.
61. Xiao, X. *et al.* 2003. Role of Ets/Id proteins for telomerase regulation in human cancer cells. Exp. Mol. Pathol. **75:** 238–247.
62. Roussel, M.F. *et al.* 1994. Dual control of myc expression through a single DNA binding site targeted by ets family proteins and E2F-1. Oncogene **9:** 405–415.
63. Takahashi, A. *et al.* 2003. EWS/ETS fusions activate telomerase in Ewing's tumors. Cancer Res. **63:** 8338–8344.
64. Shindoh, M. *et al.* 2004. E1AF, an ets-oncogene family transcription factor. Cancer Lett. **216:** 1–8.
65. Hsu, C.P. *et al.* 2006. Ets2 binding site single nucleotide polymorphism at the hTERT gene promoter-effect on telomerase expression and telomere length maintenance in non-small cell lung cancer. Eur. J. Cancer **42:** 1466–1474.

Potential Roles for Estrogen Regulation of Telomerase Activity in Aging

SHARYN BAYNE,[a] MARGARET E.E. JONES,[b] HE LI,[a] AND JUN-PING LIU[a]

[a]*Department of Immunology, Central and Eastern Clinical School, Monash University, Prahan, Victoria 3181, Australia*

[b]*Prince Henry's Institute of Medical Research, Melbourne, Australia*

ABSTRACT: **Estrogen regulates tissue development, function, and aging by regulating expression of critical genes. Recent studies suggest that estrogen promotes cell proliferation by stimulating telomerase activity, an enzyme that plays a key role in extending cell life span. In some estrogen-responsive tissues, estrogen appears important in regulating telomerase activity *in vivo*. Further characterization will provide a greater understanding of the link between estrogen, telomerase, and aging.**

KEYWORDS: **estrogen; telomerase; aging; adrenal**

INTRODUCTION

With an ever-increasing aging population, focus has intensified on understanding the mechanisms that regulate cell life span. Healthy aging and longevity of the organism depend on a finely tuned balance of gene expression profiles at cellular levels in proliferative potentials, cell renewal, and senescence. With time, these balances are altered by both environmental and pre-programmed genetic events, contributing to aging. For example, dietary restriction may reduce accumulation of metabolic damages to increase longevity in a number of model species through the Sirt1 gene[1] and reduced expression of insulin-like growth factors (IGFs), which in turn regulate a variety of genes associated with longevity.[2] In addition, a number of aging-related diseases, including cardiovascular disease, cancer, and type-2 diabetes, demonstrate a gender-bias: for example, men are at higher risk of developing coronary disease and cancer than women.[3,4] However, the risk levels change in women after the onset of menopause, so it is thought that some risk factors are related to the steroid hormone estrogen. While estrogen appears to have protective effects on

Address for correspondence: Dr. Jun-Ping Liu, Department of Immunology, AMREP, Monash Medical School, Commercial Road, Prahran, Victoria 3181, Australia. Voice: 61-3-99030715; fax: 61-3-99030120.
jun-ping.liu@med.monash.edu.au

Ann. N.Y. Acad. Sci. 1114: 48–55 (2007). © **2007 New York Academy of Sciences.**
doi: 10.1196/annals.1396.023

the cardiovascular[5] and neural[6] systems, prolonged exposure to estrogen has also been implicated in the development of cancers, such as breast,[7] endometrial,[8] and ovarian cancers.[9,10] This mini-review focuses on how estrogen is involved in tumorigenesis and the implications of estrogen-deficiency in aging through regulating the enzyme telomerase.

TELOMERASE AND TELOMERES

Telomerase is a large ribonucleoprotein complex, consisting of two core components, telomerase reverse transcriptase (TERT) and telomerase RNA template (TER). During embryonic development, telomerase activity is at high levels to support rapid and continuous cell divisions. As cells differentiate, telomerase is downregulated and becomes undetectable in most normal human somatic tissues after birth.[11–13] However, telomerase activity is detectable in highly proliferative cells, such as primary germ line cells[14] and activated lymphocytes,[15–17] and within reproductive tissues, including those of the endometrium,[18,19] ovary,[14] and testis.[20] Hematopoietic stem cells (HSCs) show detectable levels of telomerase activity, which is required to support the function of HSCs in replenishing the short-lived blood cells. In contrast, aberrant reactivation of telomerase activity occurs in approximately 90% of human cancers.[21]

Telomerase operates by maintaining telomeres—specialized structures at the termini of chromosomes. Telomeres serve to prevent loss of chromosomes[22,23] due to the inability of conventional DNA polymerases to replicate lagging strand DNA.[24] Therefore, telomeres progressively shorten by approximately 150 bp with each cell division.[22] Continuous attrition of telomeric DNA eventually triggers the onset of cell senescence. Maintenance of telomere length, however, ensures continuous cell proliferation and this is most commonly achieved through activation of telomerase. Telomerase synthesizes telomeric DNA to prevent telomere loss incurred during cell division.[25] In HSCs, when telomerase activity is insufficient to prevent cell division–dependent telomere loss, cells undergo age-dependent telomere attrition *in vivo.*[26–28] In contrast, clonally expanding B lymphocytes have upregulated telomerase activity to support increased cell division, but continued propagation is limited by telomerase inhibition and gradual loss of telomeres.[16,29]

ESTROGEN STIMULATION OF TELOMERASE ACTIVITY IN TUMORIGENESIS

Estrogen exerts stimulatory effects on tissue growth in various tissues and organs, including reproductive organs, bone, cardiovascular system, central nervous system, and the immune system. Mechanistic studies performed in cultured cells show that estrogen stimulates hTERT gene expression and telomerase activity in several human estrogen receptor (ER)–positive cell types, including breast cancer,[30] choriocarcinoma,[31] ovary epithelial,[32] vascular endothelial,[33,34] and endothelial progenitor cells.[35] The role of ERs in mediating the effect of estrogen on hTERT gene expression is also demonstrated by the finding that expression of ERα but not ERβ increases hTERT promoter activity in the presence of estrogen.[32] While telomerase activity is not detected in human ovary epithelial cell lines grown in the absence of estrogen, adding the hormone induces telomerase activity within 3 hours of treatment, suggesting that estrogen alone can activate telomerase in normal human ovarian surface epithelial cells that give rise to 90% of ovarian tumors.[32]

To stimulate hTERT gene expression, estrogen and its receptor complex may bind to a non-consensus estrogen-response element (ERE) located at −2677 of the hTERT promoter region, as deletion of the imperfect palindromic ERE dramatically impairs estrogen-induced stimulation of telomerase activity.[30,32] However, deletion of two c-myc binding sites in the hTERT promoter completely abrogates estrogen upregulation of telomerase activity.[30] Consistent with c-myc playing a predominant role in estrogen-induced hTERT gene expression are also the findings that estrogen alone fails to induce telomerase activity in c-myc null HO15.19 cells.[31] It is thus possible that the ER mediates estrogen regulation of the hTERT gene transcription via *cis* and *trans* mechanisms with both c-myc and ER being involved in initiating transcription of the hTERT gene. Further studies are required to investigate the relationship between ERE and E box that mediates c-myc binding in different estrogen-responsive cell types. Recent studies show that ER physically interacts with c-myc to stabilize the ER co-activator complex, resulting in chromatin remodelling and increased gene transcription.[36] It would be interesting to determine whether the ERE and E box mediate an interaction between the ER and c-myc protein–DNA complexes in estrogen-induced hTERT gene transcription.

In addition to stimulating hTERT gene transcription, studies using cultured human cells have suggested that estrogen induces the phosphorylation of Akt and hTERT. Inhibition of the phosphatidylinositol 3-kinase (PI3K) upstream of Akt by LY294002 attenuates the estrogen-induced increase of telomerase activity and hTERT promoter activity. Moreover, estrogen also stimulates phosphorylation of hTERT and its nuclear accumulation via the PI3K/Akt/NFkappaB pathway.[35,37,38] Thus, estrogen may activate Akt to stimulate hTERT gene expression and protein phosphorylation required for telomerase activity. Consistently, raloxifene, a selective ER modulator in the chemoprevention of breast cancer, like estrogen, induces telomerase activity and hTERT expression via ERα and ERβ, which is attenuated by LY294002.[34] In contrast to vascular endothelial cells, however, the researchers also show that raloxifene inhibits not only the estrogen-induced expression of hTERT, but also the estrogen-induced phosphorylation of both Akt and hTERT in human breast cancer cell line.[37] The mechanisms underlying the opposite effects of raloxifene with or without estrogen in different cell types on telomerase have not been determined. To investigate phosphorylation of hTERT, we showed previously that Akt does not phosphorylate hTERT and regulate telomerase activity because mutations of all putative Akt phosphorylation sites in hTERT have no effect on telomerase activity in telomerase-positive neuroendocrine PC12 cells and telomerase-negative osteosarcoma Saos-2 cells.[39] These data suggest that further studies are required to delineate post-translational regulation of hTERT by the Akt pathway and estrogen, and its role in regulating telomerase activity in various types of cells and tissues.

In human endometrium, glandular epithelial cells, but not stromal cells, display cyclical changes in telomerase activity. Telomerase activity levels peak during the proliferative stage and fall in the secretory and resting phases of the menstrual cycle.[40–43] Estrogen may thus be involved in stimulating telomerase activity to protect telomere length in the glandular epithelial cells to ensure continued reproductive potential. It is possible that the onset of menopause and withdrawal of estrogen result in telomerase inhibition leading to cell senescence, which contributes to cessation of the reproductive life span.[41] On the other hand, withdrawal of estrogen may also be an important mechanism to prevent the development of endometrial cancers by inhibiting telomerase, which prevents immortalization of neoplastic cells.

LOSS OF ESTROGEN CONTRIBUTES TO AGING

Loss of estrogen levels, due, for example, to menopause in women, has been associated with changes in a number of physiological processes, including loss of reproductive function and bone density[44] and increased risk of developing cardiovascular disease.[5] Consistently, inhibition of estrogen by a dominant negative mutant of ERα induces replicative cell senescence, abrogating tumorigenic potential in the presence of activated Ras oncogene.[45] Given that estrogen exerts its effects through various molecular mediators, it is possible that some response of cell senescence to estrogen deficiency may be associated with the withdrawal of stimulation of telomerase activity by estrogen.

To determine tissue-specific responses to estrogen deficiency *in vivo,* we have been examining telomerase activity in an estrogen-deficient mouse model. The aromatase knockout (ArKO) mouse is unable to convert androgens to estrogens on account of a targeted disruption in the *cyp19* gene that encodes the enzyme aromatase.[46] In contrast to humans, mice possess detectable levels of telomerase activity in a number of normal primary tissues. High levels of telomerase activity were observed in tissues with high levels of proliferative capacity, such as liver, although telomerase activity was not detected in the tissues with terminally differentiated cells, such as brain and heart[20,47,48] (data not shown). We hypothesized that telomerase activity in certain estrogen-responsive tissues with highly proliferative cells requires estrogen upregulation of TERT gene expression; withdrawal of estrogen prompts repression of the TERT gene and inhibition of telomerase activity in the certain tissues.

In our preliminary studies, we have screened a range of tissues for changes in weight, telomerase activity, and mTERT gene expression, and compared these parameters between wild-type and ArKO mice. Since estrogen is known to target specific tissues, it was not surprising to find that most tissues examined were unaffected by the lack of estrogen. Interestingly, some changes have been observed in the adrenal gland of ArKO mice and these changes are currently being characterized. We propose that estrogen may play a role in stimulating telomerase activity in the adrenal gland, where it could be involved in stimulating tissue growth and mediating cellular responses to stress. This idea is consistent with the changes observed in human adrenal function with age. In humans, the adrenal gland produces basal levels of hormones with age, but the ability to produce these hormones in response to stimuli decreases,[3] coinciding with decreased sex steroid hormone production. This contributes to a reduced capacity to respond to stress and may be affected by reduced expression of critical genes including TERT because of estrogen withdrawal. If we assume that estrogen is important for regulating telomerase in the adrenal gland, then a loss of estrogen production in women with age may contribute to reduced ability to respond to stress stimuli, and estrogen replacement therapy may therefore lessen stress responses by promoting cell proliferation and prohibiting cellular senescence.

Recent studies have also shown that in human endothelial progenitor cells isolated from peripheral blood, cell senescence is inhibited dose-dependently by estrogen *in vitro*.[35] Measurement of telomerase activity further shows that estrogen significantly increases telomerase activity by stimulating gene expression of TERT. Consistent with previous findings that TERT is regulated by PI3K pathway, the PI3K blockers, wortmannin or LY294002, inhibit estrogen-induced TERT gene expression. Furthermore, in addition to reducing cell senescence, estrogen increases mitogenic potential of the endothelial progenitor cells.[35] The estrogen deprivation–induced cell senescence should be reversible by estrogen replacement therapy, to increase cell proliferative potential by activating telomerase, thereby providing a mechanism for molecular intervention in cell aging.

PERSPECTIVES

In view of the significant upregulation of telomerase activity by estrogen in a number of cell types, through direct transcriptional activation and indirect action of the transcription factor c-myc on the TERT gene, estrogen may have important effects on cell proliferative lifespan. Indeed, recent studies have begun to unveil the effects of estrogen on cell proliferation and senescence in human cell cultures and certain tissues. Future *in vivo* studies in human tissues and animal models should reveal more specific actions of estrogen on telomere homeostasis in tissue growth, regeneration, and repair of certain organs and tissues. Analysis of estrogen withdrawal will show estrogen deficiency–induced, telomerase-dependent cell aging and tissue atrophy. In addition, more specific tissue progenitor cell types will be identified for estrogen to regulate cell proliferative capacity and the timing of cell differentiation. Although the molecular mechanisms of regulation of telomerase activity at both gene transcriptional and post-translational levels remain a challenge, targeting the estrogen–telomerase signaling axis, through estrogen agonist, antagonists, and modulators, will shed light on how to improve cellular functionality in a way that is important in cell repair and replacement therapy for healthy aging and longevity.

ACKNOWLEDGMENTS

This work was supported by grants from the National Health and Medical Research Council of Australia and Australian Research Council. S.B. is a recipient of an Australian Postgraduate Award.

REFERENCES

1. GUARENTE, L. & F. PICARD. 2005. Calorie restriction: the SIR2 connection. Cell **120:** 473–482.
2. TATAR, M. *et al.* 2003. The endocrine regulation of aging by insulin-like signals. Science **299:** 1346–1351.
3. PERRY, H.M., 3rd. 1999. The endocrinology of aging. Clin. Chem. **45:** 1369–1376.
4. SHAY, J.W. & W.E. WRIGHT. 2007. Hallmarks of telomeres in ageing research. J. Pathol. **211:** 114–123.
5. ROSANO, G.M. *et al.* 2007. Menopause and cardiovascular disease: the evidence. Climacteric **10**(Suppl 1): 19–24.
6. YAO, M. *et al.* 2007. Estrogen regulates Bcl-w and Bim expression: role in protection against beta-amyloid peptide-induced neuronal death. J. Neurosci. **27:** 1422–1433.
7. ALLAIN, J.E. *et al.* 2002. Immortalization of a primate bipotent epithelial liver stem cell. Proc. Natl. Acad. Sci. USA **99:** 3639–3644.
8. SAITO, T. *et al.* 1997. Proliferation-associated regulation of telomerase activity in human endometrium and its potential implication in early cancer diagnosis. Biochem. Biophys. Res. Commun. **231:** 610–614.
9. MURAKAMI, J. *et al.* 1997. Telomerase activity in ovarian tumors. Cancer **80:** 1085–1092.
10. LACEY, J.V., JR. *et al.* 2002. Menopausal hormone replacement therapy and risk of ovarian cancer. JAMA **288:** 334–341.
11. FORSYTHE, H.L. *et al.* 2001. Stable association of hsp90 and p23, but not hsp70, with active human telomerase. J. Biol. Chem. **276:** 15571–15574.

12. KIM, N. *et al.* 1994. Specific association of human telomerase activity with immortal cells and cancer. Science **266:** 2011–2015.
13. ULANER, G.A. *et al.* 1998. Telomerase activity in human development is regulated by human telomerase reverse transcriptase (hTERT) transcription and by alternate splicing of hTERT transcripts. Cancer Res. **58:** 4168–4172.
14. WRIGHT, W. *et al.* 1996. Telomerase activity in human germline and embryonic tissues and cells. Dev. Genet. **18:** 173–179.
15. HU, B.T. *et al.* 1997. Telomerase is up-regulated in human germinal center B cells *in vivo* and can be re-expressed in memory B cells activated in vitro. J. Immunol. **159:** 1068–1071.
16. NORRBACK, K.F. *et al.* 1996. Telomerase activation in normal B lymphocytes and non-Hodgkin's lymphomas. Blood **88:** 222–229.
17. WENG, N.P. *et al.* 1997. Telomere lengthening and telomerase activation during human B cell differentiation. Proc. Natl. Acad. Sci. USA **94:** 10827–10832.
18. LEHNERT, H. *et al.* 1993. Extrapituitary effects of corticotropin-releasing hormone and thyrotropin-releasing hormone. Neuropsychobiology **28:** 54–61.
19. TANAKA, M. *et al.* 1998. Expression of telomerase activity in human endometrium is localized to epithelial glandular cells and regulated in a menstrual phase-dependent manner correlated with cell proliferation. Am. J. Pathol. **153:** 1985–1991.
20. BURGER, A.M. *et al.* 1997. Telomerase activity in normal and malignant mammalian tissues: feasibility of telomerase as a target for cancer chemotherapy. Br. J. Cancer **75:** 516–522.
21. SHAY, J.W. & S. BACCHETTI. 1997. A survey of telomerase activity in human cancer. Eur. J. Cancer **33:** 787–791.
22. FULNECKOVA, J. & J. FAJKUS. 2000. Inhibition of plant telomerase by telomere-binding proteins from nuclei of telomerase-negative tissues. FEBS Lett. **467:** 305–310.
23. KARLSEDER, J. *et al.* 2002. Senescence induced by altered telomere state, not telomere loss. Science **295:** 2446–2449.
24. OLOVNIKOV, A.M. 1973. The incomplete copying of template margin in enzymic synthesis of polynucleotides and biological significance of the phenomenon. J. Theor. Biol. **41:** 181–190.
25. BLACKBURN, E. 2001. Switching and signaling at the telomere. Cell **106:** 661–673.
26. VAZIRI, H. *et al.* 1994. Evidence for a mitotic clock in human hematopoietic stem cells: loss of telomeric DNA with age. Proc. Natl. Acad. Sci. USA **91:** 9857–9860.
27. ZIMMERMANN, S. & U.M. MARTENS. 2005. Telomere dynamics in hematopoietic stem cells. Curr. Mol. Med. **5:** 179–185.
28. HIYAMA, K. *et al.* 1995. Activation of telomerase in human lymphocytes and hematopoietic progenitor cells. J. Immunol. **155:** 3711–3715.
29. LIU, K. *et al.* 1999. Constitutive and regulated expression of telomerase reverse transcriptase (hTERT) in human lymphocytes [In Process Citation]. Proc. Natl. Acad. Sci. USA **96:** 5147–5152.
30. KYO, S. *et al.* 1999. Estrogen activates telomerase. Cancer Res. **59:** 5917–5921.
31. SARKAR, P. *et al.* 2006. Activation of telomerase in BeWo cells by estrogen and 2,3,7,8-tetrachlorodibenzo-p-dioxin in co-operation with c-Myc. Int. J. Oncol. **28:** 43–51.

32. MISITI, S. *et al.* 2000. Induction of hTERT expression and telomerase activity by estrogens in human ovary epithelium cells [In Process Citation]. Mol. Cell Biol. **20:** 3764–3771.
33. LING, S. *et al.* 2006. Effects of 17beta-estradiol on growth and apoptosis in human vascular endothelial cells: Influence of mechanical strain and tumor necrosis factor-alpha. Steroids **71:** 799–808.
34. DOSHIDA, M. *et al.* 2006. Raloxifene increases proliferation and up-regulates telomerase activity in human umbilical vein endothelial cells. J. Biol. Chem. **281:** 24270–24278.
35. IMANISHI, T. *et al.* 2005. Estrogen reduces endothelial progenitor cell senescence through augmentation of telomerase activity. J. Hypertens. **23:** 1699–1706.
36. CHENG, A.S.L. *et al.* 2006. Combinatorial analysis of transcription factor partners reveals recruitment of c-MYC to estrogen receptor-[alpha] responsive promoters. Mol. Cell **21:** 393–404.
37. KAWAGOE, J. *et al.* 2003. Raloxifene inhibits estrogen-induced up-regulation of telomerase activity in a human breast cancer cell line. J. Biol. Chem. **278:** 43363–43372.
38. KIMURA, A. *et al.* 2004. Induction of hTERT expression and phosphorylation by estrogen via Akt cascade in human ovarian cancer cell lines. Oncogene **23:** 4505–4515.
39. LI, H. *et al.* 2005. Telomerase down-regulation does not mediate PC12 pheochromocytoma cell differentiation induced by NGF, but requires MAP kinase signalling. J. Neurochem. **95:** 891–901.
40. WONG, K.K. *et al.* 2003. Telomere dysfunction and Atm deficiency compromises organ homeostasis and accelerates ageing. Nature **421:** 643–648.
41. WILLIAMS, C.D. *et al.* 2001. A prospective, randomized study of endometrial telomerase during the menstrual cycle. J. Clin. Endocrinol. Metab. **86:** 3912–3917.
42. YOKOYAMA, Y. *et al.* 1998. Telomerase activity in the human endometrium throughout the menstrual cycle. Mol. Hum. Reprod. **4:** 173–177.
43. KYO, S. *et al.* 1997. Telomerase activity in human endometrium. Cancer Res. **57:** 610–614.
44. BUCKLER, H. 2005. The menopause transition: endocrine changes and clinical symptoms. J. Br. Menopause Soc. **11:** 61–65.
45. KATO, K. *et al.* 2002. Contribution of estrogen receptor alpha to oncogenic K-Ras-mediated NIH3T3 cell transformation and its implication for escape from senescence by modulating the p53 pathway. J. Biol. Chem. **277:** 11217–11224.
46. FISHER, C.R. *et al.* 1998. Characterization of mice deficient in aromatase (ArKO) because of targeted disruption of the cyp19 gene. Proc. Natl. Acad. Sci. USA **95:** 6965–6970.
47. CHADENEAU, C. *et al.* 1995. Telomerase activity in normal and malignant murine tissues. Oncogene **11:** 893–898.
48. PROWSE, K.R. & C.W. GREIDER. 1995. Developmental and tissue-specific regulation of mouse telomerase and telomere length. Proc. Natl. Acad. Sci. USA **92:** 4818–4822.

Mechanisms of Action of TGF-β in Cancer

Evidence for Smad3 as a Repressor of the hTERT Gene

HE LI AND JUN-PING LIU

Department of Immunology, Monash University, Melbourne, Australia

ABSTRACT: Transforming growth factor-β (TGF-β) induces cell differentiation and suppresses cell proliferation, but the mechanisms underlying the actions of TGF-β remain to be fully elucidated. Recent studies suggest that TGF-β suppresses neoplastic cell development by employing Smad3 protein to repress the gene of human telomerase reverse transcriptase (hTERT). In human breast cancer cells, TGF-β induces rapid phosphorylation and subsequent entry of Smad3 into the nucleus. In the nucleus, Smad3 binds to the hTERT gene promoter directly and inhibits hTERT gene transcription activity. By interacting with c-myc, Smad3 also represses the c-myc gene. Thus, TGF-β prevents continuous cell proliferation by switching off telomerase activity through Smad3 repression of the hTERT gene and the action of c-myc. Modulating the interface between Smad3 and the hTERT gene, and the potential feedback loop from telomeres to Smad3 via Smurf2, may represent a novel approach to regulate cell lifespan of proliferation.

KEYWORDS: TGF-beta; Smad3; telomerase; hTERT; telomeres; tumorigenesis; cancer; inhibition

INTRODUCTION

Normal human somatic cells have a finite lifespan of proliferation; cells enter senescence after a limited number of cell divisions. Cell senescence is a status of the cell's permanent exit from the cell cycle, characteristic of enlarged cell bodies and flattened cell morphology. Recent studies suggest that transforming growth factor-β (TGF-β) plays a role in cell senescence, regulating cell growth and differentiation.[1] TGF-β functions as an inhibitor as well as stimulator of cell proliferation through different intracellular signaling mechanisms.[2–4]

Address for correspondence: Drs. He Li or Jun-Ping Liu, Department of Immunology, AMREP, Monash Medical School, Commercial Road, Prahran, Victoria 3181, Australia. Voice: 61-3-99030715; fax: 61-3-99030120.
he.li@med.monash.edu.au
Jun-ping.liu@med.monash.edu.au

Ann. N.Y. Acad. Sci. 1114: 56–68 (2007). © 2007 New York Academy of Sciences.
doi: 10.1196/annals.1396.016

In normal development, TGF-β inhibits cell proliferation and promotes cell differentiation, whereas during tumorigenesis TGF-β secreted from the tumor cells often loses its inhibitory function in favor of its oncogenic activity.[5] TGF-β may promote tumor growth through induction of extracellular matrix, inhibition of immune system function, and instigation of angiogenesis. How TGF-β is involved in tumorigenesis by controlling specific genes coding for cellular effectors and regulatory factors remains to be fully elucidated. It is clear that TGF-β operates through its membrane receptors to trigger intracellular signaling molecules to regulate genes further downstream. Thus, TGF-β may play differential roles in cell differentiation and tumorigenesis by mechanisms depending on intracellular signaling and subsequent activation/inactivation of specific downstream genes.

TGF-β INTRACELLULAR SIGNALING AND SMAD3 FUNCTION IN CANCER

Understanding the functions of TGF-β requires insight into the intracellular mechanisms of TGF-β signaling during normal development and tumor formation. Two transmembrane receptor serine/threonine kinases form a heterotetrameric complex to mediate TGF-β signaling. The receptor complex consists of a ligand binding type II receptor (TRII), a signaling type I receptor (TRI), and Smad (small mother against decapentaplegic) proteins.[6] The Smad1, Smad2, Smad3, Smad5, and Smad8 proteins bind to TRI (as regulatory or R-Smads). Of these R-Smads, Smad3 and Smad2 mediate TGF-β action, whereas Smad1, Smad5, and Smad8 often mediate the actions of other members of the TGF-β superfamily. These Smads complex with the common Smad 4 (C-Smad) and are antagonized by Smad6 and -7 inhibitory Smads (I-Smads).[7] Upon phosphorylation by TRII, TRI phosphorylates R-Smads, and phosphorylated R-Smads then form signaling heterodimers with Smad4 to translocate into the nucleus to regulate gene expression.

In cancer, many defects of the TGF-β signaling pathway have been reported. The loss of expression of TRII in association with resistance to the growth-inhibitory effect of TGF-β has been reported in various types of cancer, including gastric, colon, breast, and prostate cancer.[8] TGF-β post-receptor signaling is also impaired by deletions or mutations in the Smad proteins. Smad4 has been identified as a tumor-suppressor gene that is lost in pancreatic tumors, deleted in pancreatic cancer locus 4 (DPC4).[9] Smad2 has also been shown to be mutated in colorectal and lung cancers, although at a lesser frequency than Smad4 genes.[10] Moreover, Smad3 has been reported to be undetectable in 37.5% of cases of gastric cancer; re-establishment of Smad3 reduces the tumorigenicity of gastric cancer.[11]

Smad3 is composed of 425 amino acid residues forming two conserved structural domains, the N-terminal MH1 (mad homology 1) domain and the

C-terminal MH2 domain. The SXS motif in the C-terminal tail region provides two serine phosphorylation sites, and phosphorylation of the SXS motif is required for Smad3 transcriptional activity. Using its MH1 domain to interact with specific DNA sequences containing the CAGA box, Smad3 mediates the tumor-suppressor activity of TGF-β by transcriptional repression of the proto-oncogene c-myc,[7] notwithstanding the inhibition that may occur in some cells circumstantially.[12] Deletion of Smad3 induces variable frequencies of spontaneous colon cancer in mice.[13,14] Deletion of Smad3 also impairs mucosal immunity and accelerated wound healing in mice.[15,16] Although loss of Smad3 appears insufficient to cause leukemia, further loss of the $p27^{Kip1}$ cyclin-dependent kinase inhibitor (which is frequently altered in human T cell acute lymphoblastic leukemia) renders leukemia in Smad3-deficient mice.[17] Furthermore, in children suffering from acute lymphoblastic leukemia, deficiency of Smad3 has been identified to be a specific feature.[17] Taken together, these studies suggest that Smad3 suppresses neoplasms by repressing critical genes that are activated during oncogenic evolutionary development.

TELOMERASE AND ITS REGULATORY MECHANISMS

Telomerase is activated in majority of cancers (85%).[18] The activation confers on cells the ability to avoid cellular senescence and thus to divide indefinitely. The primary mechanism of cell senescence is the progressive shortening of telomeres during normal diploid cell division.[19,20] Telomeres shorten approximately 150 base pairs from each cell division, and in the lifetime of human somatic cells, telomeres shorten from 15 to 5 kb until cell senescence.[19,20] However, during neoplastic development, telomere shortening is inhibited, so that the cells escape from the limitation on the proliferative potential imposed by cellular senescence. Telomerase activation to maintain telomeres by *de novo* synthesis of telomeric DNA is the most common mechanism to prevent telomere shortening.[21,22] By binding to telomeric DNA 3′ overhang, telomerase also provides a protective capping function shielding the 3′ end from cellular activities.[23,24] Withdrawal of telomerase—for example, through gene silencing—may induce a rapid telomere decapping-associated change followed by telomeric DNA shortening. Decapping of telomeres may subject telomeric DNA termini to be exposed as DNA breaks, triggering p53-dependent apoptotic responses.

Human telomerase contains the catalytic subunit—human telomerase reverse transcriptase (hTERT)—that contains a polypeptide of 1132 amino acid residues[25,26] as well as the human telomerase RNA component (hTERC) that consists of 450 nucleotides.[21] Consistently, inhibition of hTERT, by gene mutation or silencing, reduces telomerase activity and cancer cell proliferation *in vitro* and in animal models.[27–34] However, no specific inhibitor to telomerase ribonucleoprotein complex has hitherto been found. All inhibitors to the actions of telomerase have been targeting to hTERT gene expression, hTERC gene expression, the reverse transcriptase motifs of telomerase nonspecifically,[28,30] or to the telomeric DNA substrate structures.[31] However, a number of studies have shown that telomerase activity is shut down during terminal differentiation of many types of cells, including immortalized cell lines,[35,36] mouse F9 cells, and human HL-60 promyelocytic cells,[37] embryonic stem cells,[37–39] muscle cells,[33,40] mesenchymal stem cells and chondrocytes,[38] leukemia cells,[41] and neuroendocrine and epithelial cells.[33] These suggest collectively that agents inducing cell differentiation may be involved in the downregulation of telomerase activity.[33]

Transcriptional activation of the hTERT gene is a critical, initial rate-limiting step in hTERT function and telomerase activity.[42–44] Reflecting its multifactorial regulation, the hTERT gene promoter in man has multiple sites for transcriptional regulation. There are two typical E boxes and several GC boxes for the transcription factors c-Myc/max and Sp1, respectively.[45–48] Expression of N-Myc or c-Myc entrains a direct binding of Myc to the E box[46,49] and induction of hTERT gene transcription, followed by cell proliferation.[50,51] Another E box binding protein is upstream stimulatory factor (USF), which also upregulates hTERT promoter activity, with binding negatively regulated by the N-terminal truncated form USF2, as an inhibitory competitor whose levels are increased in telomerase-negative cells.[52] Little is known of the mechanisms whereby the hTERT gene is repressed during cell development and differentiation. Recent studies show that transcription factor activator protein 1 (AP-1) is involved in repressing hTERT gene transcription; combinations of c-Fos and c-Jun, or c-Fos and JunD, suppress hTERT gene activation by binding to two AP-1 sites in the hTERT gene promoter, suggesting a broad involvement of AP-1 in the regulation of telomerase in cell proliferation, differentiation, carcinogenesis, and apoptosis.[53]

SMAD3 REPRESSION OF THE hTERT GENE

Repression of the hTERT gene occurs during cell differentiation, and derepression takes place in tumorigenesis in most cancers by mechanisms that remain largely unexplored.[33,54–56] As a pleiotropic, autocrine, and paracrine cytokine in a variety of tissues, TGF-β stimulates cell differentiation and inhibits cell proliferation in epithelial, endothelial, and hematopoietic cell types. This TGF-β checkpoint has been implicated as playing an important role in mediating cell senescence.[57–59] Studies on the regulation of telomerase activity have now suggested a link between TGF-β and hTERT which forms a regulatory system to inhibit cell proliferative potential.[33,34,60–66] In addition to interactions with other growth-control genes, such as $p21^{WAF1/Cip1}$, $p15^{ink4b}$, Cdc25A, Cdk, MAP kinase, and Akt,[49,67–70] TGF-β downregulates hTERT gene expression in normal and cancerous cells in human and rat.[34,66]

Smad3 Is Required for TGF-β Repression of the hTERT Gene

TGF-β induces cell senescence in several cell types and models.[57–59] When human skin diploid fibroblasts are exposed to UV, the cells undergo premature senescence with enhanced TGF-β signaling; inhibition of TGF-β signaling using specific neutralizing antibodies against TGF-β receptors markedly reduces UV-induced cellular senescence.[59] The molecular basis of the connections between TGF-β and cell senescence has been further elucidated by recent studies on molecular regulation of telomerase activity.[33,34,66] In the cell cultures of rat normal smooth muscle cells, highly proliferative smooth muscle cells, lung fibroblasts, immortal kidney epithelial cells, and human breast cancer cells, TGF-β is shown to downregulate the TERT gene expression.[33,34,66] Consistent with the findings that inhibition of hTERT may mediate TGF-β-induced cell senescence, activation of endogenous telomerase or ectopic expression of hTERT induces resistance to TGF-β-induced senescence of human mammary epithelial cells.[71,72] Furthermore, the relationship between TGF-β and hTERT may be two-way signaling. While TGF-β suppresses the hTERT gene and telomerase maintenance of telomeres, thereby resulting in short telomeres, short telomeres may then activate the Smurf2 gene,[73] and the Smurf2 protein is an E3 ubiquitin ligase that may in turn inhibit TGF-β signaling by targeting Smad3 protein to degradation.[73] This telomere-Smurf2 mechanism may, on one hand, induce senescence of the cells with short telomeres,[73] and on another hand, allow telomeres to be lengthened by telomerase through removing the TGF-β suppressive signaling (FIG. 1). This forward and feedback loop from Smad3 to telomeres and from telomeres to Smurf2 and Smad3 may be important in maintaining telomere stability and cell proliferative status (FIG. 1). We hypothesize that the TGF-β and telomere signaling axis containing the components of Smad3, hTERT, and Smurf2 play a pivotal role in programming cell proliferative lifespan during cell development, aging, and tumorigenesis by mechanisms of telomere homeostasis. Short telomeres may trigger Smurf2 gene expression to target Smad3 degradation, thereby lifting the bar of Smad3 repression of hTERT in order to lengthen telomeres; short telomeres that cannot be lengthened may then trigger cell senescence through Smurf2-dependent and -independent mechanisms. Manipulation of the two-way signaling circuit would have an impact on the progress of cell aging and neoplastic evolution.

It is worth noting that inhibition of telomerase activity induced by TGF-β appears incomplete (about 60% of that in control) in human breast cancer cells and rat lung fibroblasts.[34,66] This may refect several possibilities. First, the TGF-β signaling pathway may be altered in cancer, with insufficient or aberrant transmission of the binding of TGF-β to its receptors. Supporting this possibility are the findings that epigenetic gene expression of Smad3 to bypass TGF-β receptor resulted in almost complete repression of the hTERT gene promoter.[34] Indeed, loss of TRII has been reported in various types of

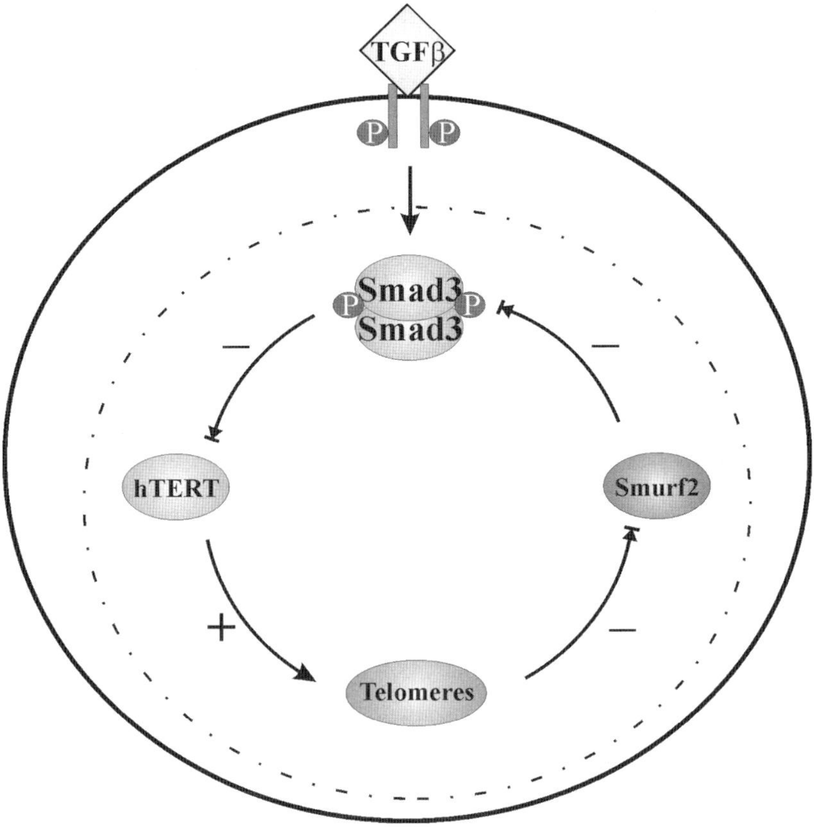

FIGURE 1. Schematic presentation on the feedback circuit of TGF-β-telomere–Smurf2 signaling. TGF-β signaling through Smad3 suppresses telomerase activity by inhibiting the hTERT gene. Inhibition of telomerase activity by Smad3 may entrain telomere shortening, which triggers Smurf2 gene expression. Smurf2, an E3 ubiquitin ligase, targets Smad3 to degradation to unleash Smad3 restraint of telomerase so as to lengthen telomeres. Thus, Smad3 signaling to telomerase may represent a mechanism to prevent telomere maintenance, whereas Smurf2 interaction with Smad3 may correspond to a mechanism to allow telomere repairs by telomerase before committing the cells to senescence.

cancer, including breast cancer,[8] and loss of endogenous Smad3 is also reported in gastric cancer cells.[11] Second, it cannot be ruled out that cancer cells have developed refractory mechanisms against TGF-β signaling. Although this could occur at several levels, it is interesting to note that inhibition of MAP kinase signaling by the specific MAP kinase inhibitor PD98059 abolishes TGF-β suppression of the hTERT gene expression, suggesting that MAP kinase signaling pathway plays a role in telomerase inhibition[33,74] (see also the article by Dwyer *et al.*[75] in this volume). In addition, recent studies have shown that short telomeres, as it is the case in cancer, trigger the gene expression of Smurf2 that inhibits TGF-β signaling by targeting Smad proteins to ubiquitinization pathway for degradation in the proteasomes.[73] Third, the reported work is done by using recombinant TGF-β1; given the large family of TGF-β, it is yet to be fully determined whether other TGF-β family members may have effects on the hTERT gene. We further hypothesize that other TGF-β superfamily members, such as the bone morphogenic protein family, are involved in the TGF-β–telomere signaling axis. This hypothesis is supported by the involvements of TGF-β family members in regulating embryonic stem cell development and the apparent lack of phenotypes of TGF-β inhibition of telomerase in the cell cultures.[34] Thus, future studies are required to focus on the potential roles of other TGF-β family members and their signaling mechanisms in regulating the hTERT gene in stem cells and transformed tumor cells.

As a central player in the TGF-β1 regulation of telomerase activity, Smad3 is shown for the first time not only to be required, but also to act alone predominantly to repress the hTERT gene when overexpressed in human and rat cells.[34] Smad3 responds to TGF-β stimulation by its phosphorylation, migration into breast cancer cell nucleus, binding to TERT gene promoter, and repression of TERT expression.[34,66] However, the repressive action of Smad3 appears to be completely overturned by Smad7 if the latter is overexpressed.[34] The findings that Smad7 prevents Smad3 phosphorylation and telomerase inhibition suggest that the negative regulation of the hTERT gene by Smad3 is reversible and balanced by intracellular antagonist Smad proteins. Consistent with Smad3's being the predominant signaling molecule mediating TGF-β regulation of the hTERT gene are the findings that Smad3 overexpression mimics, but is not synergistic for, the suppression of hTERT gene promoter activity induced by TGF-β.[34] What is not determined, though, includes the role of Smad4. This Smad4 co-Smad has been shown to be dispensable for Smad3 nuclear migration, but also to be required for transcriptional regulation carried by Smad3.[11] It would be important to determine the role of Smad4 in Smad3 repression of the hTERT gene. In addition, it is unclear whether other Smad proteins might be involved in the regulation of telomerase activity to program cell development, particularly considering their roles in embryonic development. Exploring the roles of other Smads, it is reminiscent to highlight the potential regulatory roles of TGF-β family members in telomerase activity as mentioned before.

How Does Smad3 Repress the hTERT Gene?

Studies show that Smad3 is capable of binding to the hTERT gene promoter directly.[34,66] The action of Smad3 on the hTERT gene promoter appears to be mediated by a direct binding of the protein to the promoter DNA as shown in both human breast cancer[34] and rat lung fibroblasts.[66] The relatively rapid suppression of the hTERT gene after 2–6 hours of TGF-β treatment reflects the need of Smad3 to act on the hTERT gene as a major downstream target in response to the action of TGFβ.[34] The direct binding between Smad3 and the hTERT gene promoter has been demonstrated by *in vitro* binding analysis and *in vivo* chromatin immunoprecipitation assays in human epithelial cells and rat fibroblasts.[34,66] However, the Smad3 binding element is suggested to be contained in the sequence of rTERT promoter DNA -748-729 (distal to the transcriptional start site of rTERT) in rat,[66] in contrast to that within the 330 base pairs of the hTERT promoter in human.[34] Gene deletion analysis suggests that the Smad3 binding site on the hTERT gene is the CAGA box at the -262-259 from the transcription start site. The discrepancy of the proximal and distal locations of Smad3 binding elements may reflect the largely heterologous structures of the TERT gene promoter between the two different species (~50% homology)[66] as well as different stringencies in the regulation of the hTERT gene transcription. In addition, we cannot rule out a second binding site on these genes. Moreover, the findings that mutation of Smad3 binding site on the hTERT gene results in an elevation of the basal transcriptional activity of the hTERT gene promoter may suggest an inhibitory role of Smad3 under basal conditions that may be lost in human cancer cells.[11] This possibility of Smad3's function as a housekeeping checkpoint to prevent hTERT gene expression in differentiated cells may be supported by the high levels of Smad3 gene expression compared to Smad1 and Smad2 in the absence of TGF-β challenge in rat lung fibroblasts.[66] Furthermore, the findings that silencing Smad3 further promotes the rise of hTERT promoter activity induced by mutation of the Smad3 binding site argue that Smad3 may also regulate other factor(s) that are involved in regulating hTERT gene activity, such as SIP1,[64] AP-1,[53,76] and/or c-Myc proto-oncogene.[34]

Since c-myc is an effective transcription factor of the hTERT, investigation of the relationship between Smad3 and c-myc is significant. Interestingly, a dual mechanism of molecular cross-talks between the Smad3 and c-myc pathways has emerged from the investigations of the mechanisms of action of Smad3 in regulating hTERT gene expression.[34,66] It appears that Smad3 inhibits the gene transcriptional effect of c-myc on the hTERT gene at both protein and gene levels of c-myc.[34] First, although TGF-β does not induce any discernible change in c-myc binding to the hTERT gene promoter, c-myc appears to be required for the binding of Smad3 to the hTERT gene promoter, as silencing c-myc blocks Smad3 binding to the hTERT promoter. Consistent with the model that c-myc plays an essential role in recruiting Smad3 to get access to the hTERT gene, c-myc does bind Smad3 directly as demonstrated by co-immunoprecipitation. Second, Smad3 appears to repress the c-myc gene in a relatively delayed time course.[34,66] Thus, Smad3 may repress the hTERT gene in three steps: (1) the binding of Smad3 to c-myc recruits Smad3 to the hTERT gene, and this may initiate the inhibition of c-myc in recruiting the transcriptosome for hTERT gene transcription; (2) subsequently, Smad3 physically binds to the CAGA box adjacent to the c-myc binding element E box on the hTERT gene promoter, to further prevent c-myc activity on hTERT gene transcription; and (3) to sustain the repression against the role of c-myc transcription factor in hTERT gene transcription, Smad3 binds the c-myc gene to inhibit c-myc gene expression and in turn inhibit the c-myc-driven hTERT gene transcription.

CONCLUSIONS

TGF-β promotes cell differentiation and senescence by a direct transcriptional inhibitory pathway to the hTERT gene in human and rodent cells. This mechanism requires Smad3 interactions with the hTERT gene promoter as well as with c-myc protein and gene, to repress hTERT gene transcription by both *cis* and *trans* gene transcription-dependent mechanisms. Future studies are required to delineate potential relationships between TGF-β family members, between Smad protein members, and between Smad3 and molecules, such as AP-1 and SIP1, that are also implicated in TGF-β-induced cell senescence.

ACKNOWLEDGMENTS

This work was supported by grants from the National Health and Medical Research Council of Australia, the Australia Research Council, and the Cancer Council of Victoria, Australia.

REFERENCES

1. Moses, H.L. *et al.* 1990. TGF-beta stimulation and inhibition of cell proliferation: new mechanistic insights. Cell **63:** 245–247.
2. Lin, S.Y. *et al.* 2003. Regulation of ovarian function by the TGF-beta superfamily and follistatin. Reproduction **126:** 133–148.
3. Findlay, J.K. *et al.* 2002. Recruitment and development of the follicle; the roles of the transforming growth factor-beta superfamily. Mol. Cell Endocrinol. **191:** 35–43.
4. Russo, L.M. *et al.* 2002. Albuminuria in hypertension is linked to altered lysosomal activity and TGF-beta1 expression. Hypertension **39:** 281–286.
5. Wakefield, L.M. & A.B. Roberts. 2002. TGF-[beta] signaling: positive and negative effects on tumorigenesis. Curr. Opin. Genet. Dev. **12:** 22–29.
6. Fanayan, S. *et al.* 2002. Signaling through the Smad pathway by insulin-like growth factor-binding protein-3 in breast cancer cells. Relationship to transforming growth factor-beta 1 signaling. J. Biol. Chem. **277:** 7255–7261.
7. Shi, Y. & J. Massague. 2003. Mechanisms of TGF-beta signaling from cell membrane to the nucleus. Cell **113:** 685–700.

8. KIM, S.J. *et al.* 2000. Molecular mechanisms of inactivation of TGF-[beta] receptors during carcinogenesis. Cytokine Growth Factor Rev. **11:** 159–168.
9. HAHN, S.A. *et al.* 1996. DPC4, a candidate tumor suppressor gene at human chromosome 18q21.1. Science **271:** 350–353.
10. YANAGISAWA, K. *et al.* 2000. Heterogeneities in the biological and biochemical functions of Smad2 and Smad4 mutants naturally occurring in human lung cancers. Oncogene **19:** 2305–2311.
11. HAN, S.U. *et al.* 2004. Loss of the Smad3 expression increases susceptibility to tumorigenicity in human gastric cancer. Oncogene **23:** 1333–1341.
12. ZAVADIL, J. *et al.* 2004. Integration of TGF-beta/Smad and Jagged1/Notch signalling in epithelial-to-mesenchymal transition. EMBO J. **23:** 1155–1165.
13. ZHU, Y. *et al.* 1998. Smad3 mutant mice develop metastatic colorectal cancer. Cell **94:** 703–714.
14. DATTO, M.B. *et al.* 1999. Targeted disruption of Smad3 reveals an essential role in transforming growth factor beta-mediated signal transduction. Mol. Cell Biol. **19:** 2495–2504.
15. ASHCROFT, G.S. *et al.* 1999. Mice lacking Smad3 show accelerated wound healing and an impaired local inflammatory response. Nat. Cell Biol. **1:** 260–266.
16. YANG, X. *et al.* 1999. Targeted disruption of SMAD3 results in impaired mucosal immunity and diminished T cell responsiveness to TGF-beta. Embo J. **18:** 1280–1291.
17. WOLFRAIM, L.A. *et al.* 2004. Loss of Smad3 in acute T-cell lymphoblastic leukemia. N. Engl. J. Med. **351:** 552–559.
18. SHAY, J.W. & S. BACCHETTI. 1997. A survey of telomerase activity in human cancer. Eur. J. Cancer **33:** 787–791.
19. BLACKBURN, E.H. 2001. Switching and signaling at the telomere. Cell **106:** 661–673.
20. DE LANGE, T. 1998. Telomeres. Nature **392:** 753–754.
21. COLLINS, K. & J.R. MITCHELL. 2002. Telomerase in the human organism. Oncogene **21:** 564–579.
22. HARLEY, C.B. 2002. Telomerase is not an oncogene. Oncogene **21:** 494–502.
23. BLASCO, M.A. 2002. Opinion: telomerase beyond telomeres. Nat. Rev. Cancer **2:** 627–633.
24. BLACKBURN, E.H. 2005. Telomeres and telomerase: their mechanisms of action and the effects of altering their functions. FEBS Lett. **579:** 859–862.
25. MEYERSON, M. *et al.* 1997. hEst2, the putative human telomerase catalytic subunit gene, is up-regulated in tumor cells and during immortalization. Cell **90:** 785–795.
26. NAKAMURA, T.M. *et al.* 1997. Telomerase catalytic subunit homologs from fission yeast and human. Science **277:** 955–959.
27. SHAY, J.W. & W.E. WRIGHT. 2005. Mechanism-based combination telomerase inhibition therapy. Cancer Cell **7:** 1–2.
28. KELLAND, L.R. 2005. Overcoming the immortality of tumour cells by telomere and telomerase based cancer therapeutics-current status and future prospects. Eur. J. Cancer **41:** 971–979.
29. LI, S. *et al.* 2005. Cellular and gene expression responses involved in the rapid growth inhibition of human cancer cells by RNA interference-mediated depletion of telomerase RNA. J. Biol. Chem. **280:** 23709–23717.
30. DIKMEN, Z.G. *et al.* 2005. *In vivo* inhibition of lung cancer by GRN163L: a novel human telomerase inhibitor. Cancer Res. **65:** 7866–7873.

31. GELLERT, G.C. *et al.* 2006. Effects of a novel telomerase inhibitor, GRN163L, in human breast cancer. Breast Cancer Res. Treat. **96:** 73–81.
32. CAO, Y. *et al.* 2002. Telomerase activation causes vascular smooth muscle cell proliferation in genetic hypertension. FASEB J. **16:** 96–98.
33. LI, H. *et al.* 2005. Telomerase down-regulation does not mediate PC12 pheochromocytoma cell differentiation induced by NGF, but requires MAP kinase signalling. J. Neurochem. **95:** 891–901.
34. LI, H. *et al.* 2006. Transforming growth factor beta suppresses human telomerase reverse transcriptase (hTERT) by Smad3 interactions with c-Myc and the hTERT gene. J. Biol. Chem. **281:** 25588–25600.
35. BESTILNY, L.J. *et al.* 1996. Selective inhibition of telomerase activity during terminal differentiation of immortal cell lines. Cancer Res. **56:** 3796–3802.
36. HOLT, S.E. *et al.* 1996. Regulation of telomerase activity in immortal cell lines. Mol. Cell Biol. **16:** 2932–2939.
37. TZUKERMAN, M. *et al.* 2000. Identification of a novel transcription factor binding element involved in the regulation by differentiation of the human telomerase (hTERT) promoter [in process citation]. Mol. Biol. Cell **11:** 4381–4391.
38. PARSCH, D. *et al.* 2004. Telomere length and telomerase activity during expansion and differentiation of human mesenchymal stem cells and chondrocytes. J. Mol. Med. **82:** 49–55.
39. ARMSTRONG, L. *et al.* 2004. A role for nucleoprotein Zap3 in the reduction of telomerase activity during embryonic stem cell differentiation. Mech. Dev. **121:** 1509–1522.
40. NOZAWA, K. *et al.* 2001. Mechanism for the reduction of telomerase expression during muscle cell differentiation. J. Biol. Chem. **276:** 22016–22023.
41. LIU, L. *et al.* 2004. Telomerase inhibition by retinoids precedes cytodifferentiation of leukemia cells and may contribute to terminal differentiation. Mol. Cancer Ther. **3:** 1003–1009.
42. LIU, J.P. 1999. Studies of the molecular mechanisms in the regulation of telomerase activity. FASEB J. **13:** 2091–2104.
43. LIU, J.P. 2001. Molecular mechanisms regulating telomerase activity. *In* Telomerase, Aging and Disease. M.P. Mattson & T. Pandita, Eds.: 33–59. Elsevier Science. Amsterdam, the Netherlands.
44. LIU, J.P. *et al.* 2002. Regulation of telomerase activity. *In* Telomeres and Telomerases: Cancer and Biology. G. Krupp & R. Parwaresch, Eds.: 134–159. Landes Bioscience/Eurekah. Austin, TX.
45. CONG, Y.S. *et al.* 1999. The human telomerase catalytic subunit hTERT: organization of the gene and characterization of the promoter [in process citation]. Hum. Mol. Genet. **8:** 137–142.
46. MAC, S.M. *et al.* 2000. Direct recruitment of N-myc to target gene promoters. Mol. Carcinog. **29:** 76–86.
47. EBERHARDY, S.R. *et al.* 2000. Direct examination of histone acetylation on Myc target genes using chromatin immunoprecipitation. J. Biol. Chem. **275:** 33798–33805.
48. KYO, S. *et al.* 2000. Sp1 cooperates with c-Myc to activate transcription of the human telomerase reverse transcriptase gene (hTERT) [in process citation]. Nucleic Acids Res. **28:** 669–677.
49. ADHIKARY, S. & M. EILERS. 2005. Transcriptional regulation and transformation by Myc proteins. Nat. Rev. Mol. Cell Biol. **6:** 635–645.

50. Wu, K.J. *et al.* 1999. Direct activation of TERT transcription by c-MYC. Nat. Genet. **21:** 220–224.
51. Xu, D. *et al.* 2001. Switch from Myc/Max to Mad1/Max binding and decrease in histone acetylation at the telomerase reverse transcriptase promoter during differentiation of HL60 cells. Proc. Natl. Acad. Sci. USA **98:** 3826–3831.
52. Yago, M. *et al.* 2002. Variant forms of upstream stimulatory factors (USFs) control the promoter activity of hTERT, the human gene encoding the catalytic subunit of telomerase. FEBS Lett. **520:** 40–46.
53. Takakura, M. *et al.* 2005. Function of AP-1 in transcription of the telomerase reverse transcriptase gene (TERT) in human and mouse cells. Mol. Cell. Biol. **25:** 8037–8043.
54. Armstrong, L. *et al.* 2005. Overexpression of telomerase confers growth advantage, stress resistance, and enhanced differentiation of ESCs toward the hematopoietic lineage. Stem Cells **23:** 516–529.
55. Lee, M.K. *et al.* 2005. Ectopic mTERT expression in mouse embryonic stem cells does not affect differentiation but confers resistance to differentiation- and stress-induced p53-dependent apoptosis. J. Cell Sci. **118:** 819–829.
56. Wang, J. *et al.* 2005. Human telomerase reverse transcriptase immortalizes bovine lens epithelial cells and suppresses differentiation through regulation of the ERK signaling pathway. J. Biol. Chem. **280:** 22776–22787.
57. Katakura, Y. *et al.* 1999. Transforming growth factor beta triggers two independent-senescence programs in cancer cells. Biochem. Biophys. Res. Commun. **255:** 110–115.
58. Vijayachandra, K. *et al.* 2003. Smad3 regulates senescence and malignant conversion in a mouse multistage skin carcinogenesis model. Cancer Res. **63:** 3447–3452.
59. Debacq-Chainiaux, F. *et al.* 2005. Repeated exposure of human skin fibroblasts to UVB at subcytotoxic level triggers premature senescence through the TGF-beta1 signaling pathway. J. Cell Sci. **118:** 743–758.
60. Nozaki, Y. *et al.* 2000. Induction of telomerase activity in fibroblasts from bleomycin-injured lungs. Am. J. Respir. Cell Mol. Biol. **23:** 460–465.
61. Yang, H. *et al.* 2001. Autocrine transforming growth factor beta suppresses telomerase activity and transcription of human telomerase reverse transcriptase in human cancer cells. Cell Growth Differ. **12:** 119–127.
62. Cerezo, A. *et al.* 2002. Dual regulation of telomerase activity through c-Myc-dependent inhibition and alternative splicing of hTERT. J. Cell Sci. **115:** 1305–1312.
63. Liu, T. *et al.* 2002. Regulation of telomerase activity in rat lung fibroblasts. Am. J. Resp. Cell Mol. Biol. **26:** 534–540.
64. Lin, S.Y. & S.J. Elledge. 2003. Multiple tumor suppressor pathways negatively regulate telomerase. Cell **113:** 881–889.
65. Li, H. *et al.* 2006. TGF-beta and cancer: is Smad3 a repressor of hTERT gene? Cell Res. **16:** 169–173.
66. Hu, B. *et al.* 2006. Role of Smad3 in the regulation of rat telomerase reverse transcriptase by TGFbeta. Oncogene **25:** 1030–1041.
67. Matsuura, I. *et al.* 2004. Cyclin-dependent kinases regulate the antiproliferative function of Smads. Nature **430:** 226–231.
68. Kamaraju, A.K. & A.B. Roberts. 2005. Role of Rho/ROCK and p38 MAP kinase pathways in transforming growth factor-beta-mediated Smad-dependent growth

inhibition of human breast carcinoma cells *in vivo*. J. Biol. Chem. **280:** 1024–1036.

69. MASSAGUE, J. *et al.* 2005. Smad transcription factors. Genes Dev. **19:** 2783–2810.
70. DANIELPOUR, D. & K. SONG. 2006. Cross-talk between IGF-I and TGF-beta signaling pathways. Cytokine Growth Factor Rev. **17:** 59–74.
71. STAMPFER, M.R. *et al.* 1997. Gradual phenotypic conversion associated with immortalization of cultured human mammary epithelial cells. Mol. Biol. Cell. **8:** 2391–2405.
72. STAMPFER, M.R. *et al.* 2001. Expression of the telomerase catalytic subunit, hTERT, induces resistance to transforming growth factor beta growth inhibition in p16INK4A(-) human mammary epithelial cells. Proc. Natl. Acad. Sci. USA **98:** 4498–4503.
73. ZHANG, H. & S.N. COHEN. 2004. Smurf2 up-regulation activates telomere-dependent senescence. Genes Dev. **18:** 3028–3040.
74. XU, D. *et al.* 2006. Inhibition of telomerase targeting MAP kinase signaling. *In* Methods in Molecular Biology: telomerase Inhibition: strategies and Protocols. L.G. Andrews & T.O. Tollefsbol, Eds.: in press. Humana Press Inc. Totowa; NJ.
75. DWYER, J., LI, H., XU, D. & LIU, J-P. 2007. Transcriptional regulation of telomerase activity: roles of the Ets transcription factor family. Ann. N.Y. Acad. Sci. This volume.
76. LIBERATI, N.T. *et al.* 1999. Smads bind directly to the Jun family of AP-1 transcription factors. Proc. Natl. Acad. Sci. USA **96:** 4844–4849.

The Role of Autophagy in Aging

Its Essential Part in the Anti-Aging Mechanism of Caloric Restriction

ETTORE BERGAMINI, GABRIELLA CAVALLINI, ALESSIO DONATI, AND ZINA GORI

Centro di Ricerca Interdipartimentale di Biologia e Patologia dell'Invecchiamento, Università degli Studi di Pisa, Scuola Medica,. Via Roma 55, 56126 Pisa, Italy

ABSTRACT: Aging denotes a postmaturational deterioration of cells and organisms with the passage of time, an increased vulnerability to challenges and prevalence of age-associated diseases, and a decreased ability to survive. Causes of this deterioration may be found in an enhanced production of reactive oxygen species (ROS) and oxidative damage and incomplete "housekeeping." Caloric restriction is the most robust anti-aging intervention known so far. Similar beneficial effects on median and maximum life span were obtained by feeding animals a 40%-reduced diet or by every-other-day *ad libitum* feeding. In both instances, animals are forced to spend a great part of their time in a state of fasting and activated autophagy. Autophagy is a highly conserved process in eukaryotes, in which the cytoplasm, including excess or aberrant organelles, is sequestered into double-membrane vesicles and delivered to the lysosome/vacuole, for breakdown and eventual recycling of the resulting macromolecules. This process has an essential role in adaptation to fasting and changing environmental conditions, cellular remodeling during development, and accumulation of altered ROS-hypergenerating organelles in older cells. Several pieces of evidence show that autophagy is involved in aging and is an essential part of the anti-aging mechanism of caloric restriction. As an application, intensification of autophagy by the administration of an antilipolytic drug rescued older cells from accumulation of altered mtDNA in less than 6 hours. It is concluded that the pharmacologic intensification of autophagy (PISA treatment) has anti-aging effects and might prove to be a big step toward retardation of aging and prevention of age-associated diseases in humans.

KEYWORDS: aging; autophagy; caloric restriction; antilipolytic drugs

Address for correspondence: Ettore Bergamini, M.D., Ph.D., Centro di Ricerca di Biologia e Patologia dell'Invecchiamento, Dipartimento di Patologia Sperimentale–Scuola Medica, Via Roma 55, 56123 Pisa, Italy. Voice: +39 050 2218584; fax: +39 050 2218582.
ebergami@med.unipi.it

Ann. N.Y. Acad. Sci. 1114: 69–78 (2007).
doi: 10.1196/annals.1396.020

AGING

According to Masoro,[1] most gerontologists would probably accept this definition of aging as being free of errors of commission: deteriorative changes with time during postmaturational life that underlie an increasing vulnerability to challenges, thereby decreasing the ability of an organism to survive. The many theories of aging proposed over the years have been aimed at identifying the nature of the primary aging process(es). The basic concept is that imperfections of the genetic program cause the inability of homeostasis to prevent deterioration of the living system. In many cellular membranes, fluidity decreases with age because of changes in phospholipids and cholesterol content, altered lipid structure, and altered protein and protein–lipid interactions. Many investigators have reported that decrease in mitochondrial number and accumulation of injured mitochondria may occur with age. Accumulation of altered free-radical-hypergenerating mitochondria may increase oxidative stress and accelerate rate of aging beyond a given age, akin to the "error catastrophe" theory of Orgel.[2] It is now very clear that aging and death have multiple causes,[3] including the accumulation of nuclear and mitochondrial genetic damage and altered reactive oxygen species (ROS)-generating organelles, abnormal modifications of cytosolic and membrane proteins, changes in membrane lipid, and cross-linking of collagen and elastin. Lesions may be secondary to damage by ROS[4] and formation of advanced glycation endproducts (AGEs).[5] Basal oxygen and sugar consumption are quite constant in all cells, however, and are dictated by requirement of energy; thus, the rate of aging and longevity in any given species may rather depend on the failure of cell, tissue, and organ maintenance. Mechanisms of cell maintenance operate at the molecular, subcellular, and cellular levels[6]: they include nuclear DNA repair, protein chaperoning, ubiquitinization, degradation and replacement, fatty acid repair and/or replacement in membrane lipids; degradation of irreversibly altered mitochondria, peroxisomes, and cytomembranes by autophagy; and degradation of irreversibly altered cells by apoptosis. Cell and extracellular changes eventually lead to a deterioration with advancing age that occurs in most physiological systems and was quantified some 40–50 years ago by Nathan Shock's study (the Baltimore Longevity Study), although the extent may vary from person to person.

ANTI-AGING CALORIE RESTRICTION (CR)

Restricting the food intake by laboratory organisms, vertebrates, and invertebrates, increases both the mean and maximum life span of these species.[7] The dietary manipulation also retards the rate of increase in age-specific mortality and slows many age-associated disease processes.[7] Dietary restriction–induced life extension is a broadly based biological phenomenon, whose occurrence has been documented in species of the three kingdoms: Animalia, Fungi, and

Protoctista. In the class of Mammalia, dietary restriction has been shown to extend the life of rats, mice, hamsters, and dogs.[8] Rhesus monkeys tolerate an imposed 30% reduction in calories very well without any unhealthy side-effects.[9] Preliminary reports indicate higher survival rates among monkeys in which a 30% CR was initiated at adult (but not at old) age.[10] Robust survival data for monkeys subjected to highly controlled CR will be available in approximately 25 years.[11] The available evidence suggests that it is highly probable that CR will increase the span of good health, since several measures of health, including indices of risk for cardiovascular disease and diabetes, are enhanced in monkeys on CR.[9,12,13] In regard to humans, these indices—such as blood pressure, blood lipids, glucose tolerance, and insulin sensitivity—have also been reported to be improved in humans who practice CR.[14,15] An additional support for the belief that CR should increase health and life span in humans is the report that key metabolic and hormonal indices, such as reduced body temperature and tri-iodothyronine, as well as reduced insulin and enhanced DHEA, which are observed in rodents and monkeys on CR, were predictive of better survival among healthy men involved in the Baltimore Longitudinal Study of Aging.[16] Epidemiologic evidence indicates that CR may have contributed to an extension of average and maximum life span in the Okinawan population.[17] Very interestingly, a 1-day per week fast, akin to that prescribed by religion, may have anti-aging[18] and anti-cancer[19] effects in rodents. Importantly, anti-aging effects by dietary restriction can be obtained by two different approaches: 30–50% daily reduction of food consumption or intermittent feeding (or every-other-day *ad libitum* feeding), and effects of treatments on health and life span are similar, but effects on metabolism may be different.[20] Thus, effects could be compared in order to get a better understanding of the mechanisms which are specifically involved in the anti-aging effects.

THE ANTI-AGING MECHANISM OF CALORIE RESTRICTION

The mechanisms underlying life-extension and other anti-aging effects of CR have been discussed extensively.[7] CR protects rodents and worms from oxidative stress without affecting metabolic rate per unit of lean body mass, and enhances repair of oxidatively damaged molecules. CR has a profound effect on the glucose–insulin system: with 40% CR rats, daily maximal glucose concentration was similar, but the plasma insulin concentration was about 50% that of the *ad libitum*–fed rats; and 24-hour plasma glucose concentration was lower and plasma insulin concentration was markedly lower, because of a 15-hour-long hypoglycemia during the last part of the dark phase and the entire light phase of the day.[21] On the basis of these effects, an alteration of the glucose–insulin system was hypothesized as a mechanistic explanation of the anti-aging and life-prolonging action of CR.[22] Similar results (under less-controlled conditions) had been obtained by our group[20] and explanation was that 40% CR rats are very hungry and consume all their given food in less than 6 hours after feeding (in other words, the feeding pattern of 40% CR rats resembles that of every-other-day *ad libitum* feeding (EOD) rats, which spend 24 hours of their time fed (during the day of feeding) and the next 24 hours in the state of fasting. These observations led two of us to propose a different hypothesis, based on the well-known stimulatory effect of fasting and lower insulin levels on autophagy,[23] a cell-repair mechanism which is essential for keeping cells "clean" (FIG. 1). Additional support for this hypothesis came by discovery that loss-of-function mutations in genes coding for components of the cellular insulin-like signaling pathway markedly increase the longevity of worms, and that the target-of-rapamycin (TOR) pathway, which regulates protein synthesis and autophagic degradation rates, may be involved.[24]

AUTOPHAGY

Macroautophagy, hereafter called autophagy, is a universal, dynamic process which takes place in all eukaryotic cells involving a rearrangement of subcellular membranes to sequester cytoplasm and organelles for delivery to the lysosome or vacuole, where the sequestered cargo is degraded and recycled.[25,26]

The primary roles of autophagy are baseline turnover of intracellular proteins and organelles, production of amino acids in nutrient emergency, and regression of retired tissues. These functions guarantee rejuvenation and adaptation to adverse conditions, and even underlie dynamic processes such as development/metamorphosis.[27]

Autophagy is involved in starvation response in order to degrade cell macromolecules and organelles and produce substrates for ATP production: extensive research on mammalian cells during the last three decades showed that autophagy and lysosomal proteolysis are induced by lower amino acid and insulin levels during fasting, whereas the higher (postprandial) levels of insulin fully suppress autophagy in the physiological range of plasma amino acid concentration.[28–30] Autophagy has been demonstrated to play an important role in the degradation of excess or injured organelles.[31] The process is involved in the degradation of redundant peroxisomes[32] and of damaged mitochondria[33] and endoplasmic reticulum.[34]

In conclusion, autophagy is a mechanism by which cells, in times of nutrient shortage, can degrade protein for metabolic needs and simultaneously eliminate unwanted, damaged, or redundant cell membranes and organelles including mitochondria, endoplasmic reticulum, and peroxisomes; in this perspective autophagy may contrast with the cell damage accumulation during aging.[35]

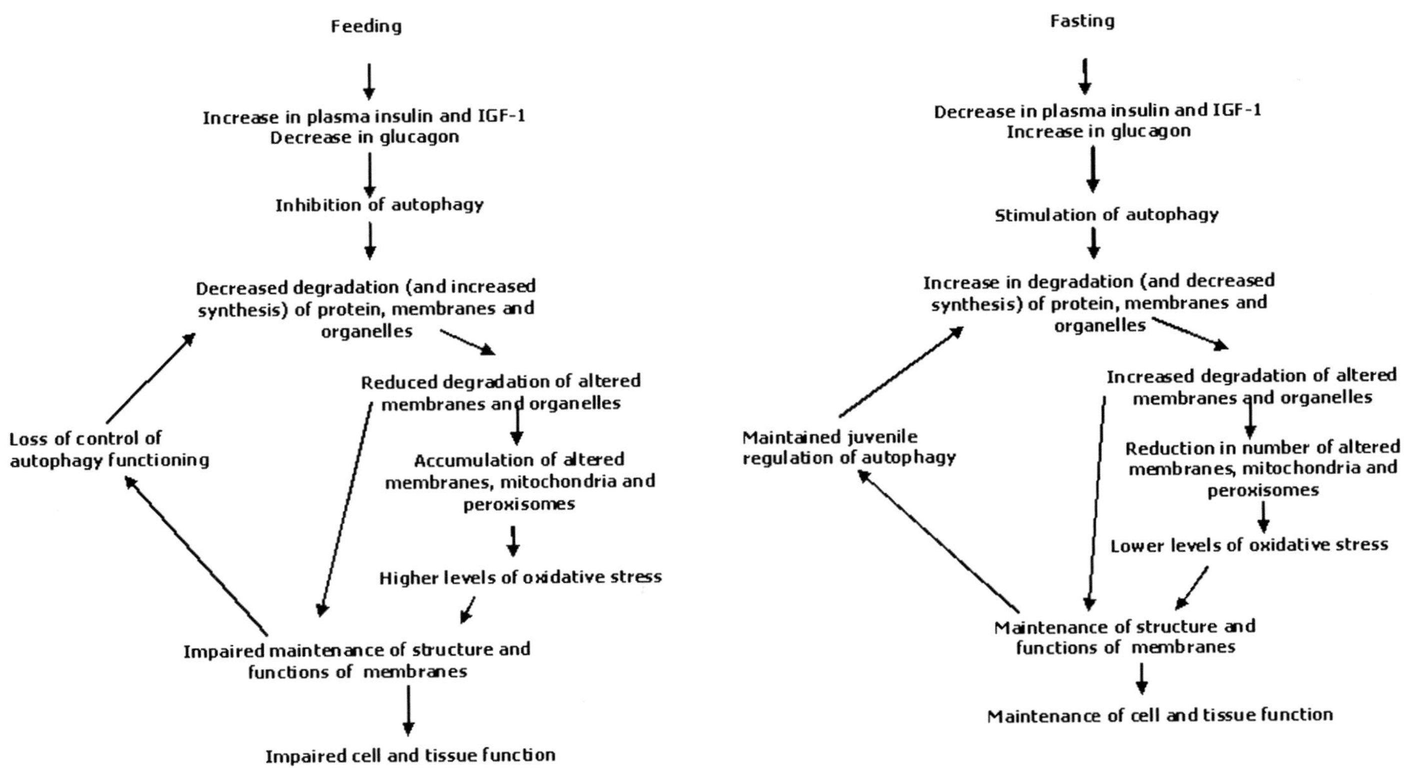

FIGURE 1. Alternation of fast and feeding-time may enhance turnover rate and rejuvenate cells in calorie-restricted rats.

EVIDENCE THAT AUTOPHAGY IS INVOLVED IN AGING AND ANTI-AGING CR

At least six pieces of evidence were found to support the hypothesis that autophagy is involved in aging and is an essential part in the anti-aging mechanism of CR.

1. *The* in vivo *and* in vitro *function of autophagy decline with increasing age.*

Age-related changes in functioning of autophagy were studied *in vivo* by pharmacologic stimulation of the process. Older *ad libitum*–fed rats were less sensitive to the metabolic and endocrine effects of fasting (response was paramount in 1-month-old rats, high but delayed in 2-month-old rats, remarkably smaller in 6-month-old rats, and almost negligible at older age).[36,37] Regulation by amino acids was studied *in vitro* and declined after age 6 months. Sensitivity to hormones decreased in older age: alteration in the response to glucagon was earlier than that to insulin.[38] It was concluded that the function of autophagy may be tuned down with increasing age and lead to the accumulation of altered organelles and membranes in older cells.

2. *The temporal pattern of decline parallels changes in biomarkers of membrane aging and in the transduction of amino acid and hormone signaling.*

The decline in liver autophagy is accompanied by the accumulation of dolichol,[39] a fat-soluble antioxidant that satisfies all proposed primary and secondary applicable criteria and the desirable features required to be qualified as a biomarker of aging of biological membranes.[40]

3. *The age-dependent changes in autophagy are prevented by CR.*

Anti-aging CR prevented these age-associated changes. Two anti-aging dietary regimens were tested, with different effects on metabolism and similar effects on longevity (namely, 40% dietary restriction and EOD, both initiated in rats at the age of 2 months). The protective effects of dietary restriction and EOD are partially different in 24-month-old rats (but benefits on regulation of autophagic proteolysis were always similar). With regard to endocrine regulation, both types of CRs restored sensitivity to glucagon to a juvenile level and had minor effects on insulin.[41]

4. *Prevention of changes in autophagy and biomarkers of membrane aging co-varies with duration and level very like the effects of CR on life span.*[42,43]

5. *A long-lasting inhibition of autophagy accelerates the process of aging.*

Autophagic lysosomal function was reduced by the administration of lysosomotropic agents like chloroquine, which is a general lysosomal enzyme inhibitor.[44] In several species and organ systems the treatment induced the formation of lysosomal aggregates closely resembling ceroid-lipofuscin.[45,46] The life-long administration of chloroquine was shown to reduce life span, to impair function and regulation of autophagy, and to cause an early accumulation of dolichol in liver cells (data not published).

6. *A long-lasting stimulation of autophagy retards the process of aging in rats.*

Treatment with antilipolytic drugs can decrease plasma free fatty acids, glucose, and insulin and stimulate autophagy.[47] Life-long once-a-week administration of ACIPIMOX (an antilipolytic drug licensed for human use) by gastric tube prevented the accumulation of dolichol and the decline in autophagic proteolysis in older liver cells. Speculation was that transiently lower plasma insulin levels might stimulate the anti-aging cell-repair mechanism autophagy, which has longer-lasting effects on cell housekeeping.[48]

AN APPLICATION: THE PISA TREATMENT

It is known that autophagy is the only tier of defense against the accumulation of ROS-hypergenerating mitochondria and peroxisomes, whose accumulation in older cells might accelerate the process of aging. The acute stimulation of macroautophagy, by the injection of an antilipolytic agent, rescued older cells from the age-related accumulation of oxidatively altered mitochondrial DNA (mtDNA) in less than 6 hours.[49] It was concluded that the age-related changes in mtDNA and function are likely the consequence of a failure of macroautophagy in the recognition and disposal of a (small) number of severely injured mitochondria. Thus, the pharmacologic stimulation of autophagy may suppress age-associated changes. The final conclusion is that the PISA (pharmacologic intensification of suppression of aging) treatment might prove to be a big step toward retardation of aging and prevention and therapy of age-associated diseases.

REFERENCES

1. Masoro, E.J. 1995. Aging: current concepts. *In* Handbook of Physiology, Section 11: Aging. E.J. Masoro, Ed.: 1–21. Oxford University Press, New York.
2. Orgel, L.E. 1963. The maintenance of the accuracy of protein synthesis and its relevance to aging. Proc. Natl. Acad. Sci. USA **49:** 512–517.
3. Hollyday, R. 2006. Aging is no longer an unsolved problem in biology. Ann. N. Y. Acad. Sci. **1067:** 1–9.

4. HARMAN, D. 1956. Aging: a theory of free radical and radiation chemistry. J. Gerontol. **11:** 298–302.
5. MONNIER, V.M. 1990. Minireview: non-enzymatic glycosylation, the Maillard reaction and the aging process. J. Gerontol. Biol. Sci. **45:** B105–B111.
6. BERGAMINI, E. 2005. Targets for anti-aging therapy. Expert Opin. Ther. Targets **9:** 77–82.
7. MASORO, E.J. 2002. Caloric restriction: a key to understanding and modulating aging. *In* Research Profile in Aging, Vol. 1. J. Vijg, Ed. Elsevier. New York.
8. MASORO, E.J. 2006. Dietary restriction-induced life-extension: a broadly based biological phenomenon. Biogerontology **7:** 153–155.
9. MATTISON, J.A. *et al.* 2003. Calorie restriction in Rhesus monkeys. Exp. Gerontol. **38:** 35–46.
10. ROTH, G.S., D.K. INGRAM & M.A. LANE. 1999. Calorie restriction in primates: will it work and how will we know? J. Am. Geriatr. Soc. **47:** 896–903.
11. WEINDRUCH, R. 2006. Will dietary restriction work in primates? Biogerontology **7:** 169–171.
12. BODKIN, N.L. *et al.* 2003. Mortality and morbidity in laboratory-maintained Rhesus monkeys and affects of long-term dietary restriction. J. Gerontol. Biol. Sci. **58:** 212–219.
13. ROTH, G.S. *et al.* 2004. Aging in Rhesus monkeys: relevance to human health interventions. Science **305:** 1423–1426.
14. FONTANA, L. *et al.* 2004. Long-term calorie restriction is highly effective in reducing the risk for atherosclerosis in humans. Proc. Natl. Acad. Sci. USA **101:** 6659–6663.
15. MEYER, T.E. *et al.* 2006. Long-term calorie restriction ameliorates the decline in diastolic function in humans. J. Am. Coll. Cardiol. **47:** 398–402.
16. INGRAM, D.K. *et al.* 2006. The potential for dietary restriction to increase longevity in humans: extrapolation from monkey studies. Biogerontology **7:** 143–148.
17. WILLCOX, D.C. *et al.* 2006. Caloric restriction and human longevity: what can we learn from Okinawans? Biogerontology **7:** 173–177.
18. CAVALLINI, G. *et al.* 2002. Low-level dietary restriction retards age-related dolichol accumulation. Aging Clin. Exp. Res. **14:** 152–154.
19. BARRIGAN, D. *et al.* 2002. Adult onset calorie restriction and fasting delay spontaneous tumorigenesis in P53-deficient mice. Carcinogenesis **23:** 817–822.
20. BERGAMINI, E. *et al.* 1990. Different circadian variations of plasma glucose and insulin concentrations in rats submitted to 60% food restriction or intermittent feeding. *In* Protein Metabolism in Aging. H.L. Segal, M. Rothstein & E. Bergamini, Eds.: 295–300. Wiley-Liss. New York.
21. MASORO, E.J. *et al.* 1992. Dietary restriction alters characteristics of glucose fuel use. J. Gerontol. Biol. Sci. **47:** B202–B208.
22. MASORO, E.J. 1996. Possible mechanisms underlying the antiaging actions of calorie restriction. Toxicol. Path. **24:** 738–741.
23. BERGAMINI, E. & Z. GORI. 1995. Towards an understanding of the biological mechanism of dietary restriction: a signal transduction theory of aging. Aging Clin. Exp. Res. **7:** 473–475.
24. BRAECKMAN, B.P., L. DEMETRIUS & J.R. VANFLETEREN. 2006. The dietary restriction effect in *C. elegans* and humans: is the worm a one-millimeter human? Biogerontology **7:** 127–133.

25. Kanazawa, T. *et al.* 2004. Amino acids and insulin control autophagic proteolysis through different signaling pathways in relation to mTOR in isolated rat hepatocytes. J. Biol. Chem. **279:** 8452–8459.
26. Wang, C.W. & D.J. Klionsky. 2003. The molecular mechanism of autophagy. Mol. Med. **9:** 65–76.
27. Mizushima, N. 2005. The pleiotropic role of autophagy: from protein metabolism to bactericide. Cell Death Differ. **12**(Suppl 2): 1535–1541.
28. Mortimore, G.E. & A.R. Poso. 1987. Intracellular protein catabolism and its control during nutrient deprivation and supply. Annu. Rev. Nutr. **7:** 539–564.
29. Mortimore, G.E., A.R. Poso & B.R. Lardeux. 1989. Mechanism and regulation of protein degradation in liver. Diabetes Metab. Rev. **5:** 49–70.
30. Seglen, P.O. & P. Bohley. 1992. Autophagy and other vacuolar protein degradation mechanisms. Experientia **48:** 158–172.
31. Pfeifer, U. 1977. Inhibition by insulin of the physiological autophagic breakdown of cell organelles. Acta Biol. Med. Ger. **36:** 1691–1694.
32. Locci Cubeddu, T. *et al.* 1985. Effects of antilipolytic agents on rat liver peroxisomes and peroxisomal oxidative activities. Biochim. Biophys. Acta **839:** 96–104.
33. Lemasters, J.J. 2005. Selective mitochondrial autophagy, or mitophagy, as a targeted defense against oxidative stress, mitochondrial dysfunction, and aging. Rejuvenation Res. **8:** 3–5.
34. Yorimitsu, T. & D.J. Klionsky. 2007. Endoplasmic reticulum stress: a new pathway to induce autophagy. Autophagy **3:** 160–162.
35. Donati, A. 2006. The involvement of macroautophagy in aging and anti-aging interventions. Mol. Aspects Med. **27:** 455–470.
36. Del Roso, A. *et al.* 1991. Exploring the mechanisms of in vivo regulation of liver protein breakdown in the rat by the use of the new antilipolytic-agent model. Acta Biol. Hung. **42:** 87–99.
37. Del Roso, A. *et al.* 2003. Ageing-related changes in the in vivo function of rat liver macroautophagy and proteolysis. Exp. Gerontol. **38:** 519–527.
38. Donati, A. *et al.* 2001. Age-related changes in the regulation of autophagic proteolysis in rat isolated hepatocytes. J. Gerontol. A Biol. Sci. Med. Sci. **56:** B288–293.
39. Marino, M. *et al.* 1998. Age-dependent accumulation of dolichol in rat liver: is tissue dolichol a biomarker of aging? J. Gerontol. A Biol. Sci. Med. Sci. **53:** B87–93.
40. Parentini, I. *et al.* 2005. Accumulation of dolichol in older tissues satisfies the proposed criteria to be qualified a biomarker of aging. J. Gerontol. A Biol. Sci. Med. **60:** 39–43.
41. Donati, A. *et al.* 2001. Age-related changes in the autophagic proteolysis of rat isolated liver cells: effects of antiaging dietary restrictions. J. Gerontol. A Biol. Sci. Med. Sci. **56:** B375–383.
42. Cavallini, G. *et al.* 2001. The protection of rat liver autophagic proteolysis from the age-related decline co-varies with the duration of anti-ageing food restriction. Exp. Gerontol. **36:** 497–506.
43. Dolfi, C. *et al.* 2003. The age-related accumulation of dolichol in rat liver may correlate with expectation of life. Biogerontology **4:** 113–118.
44. Ivy, G.O. *et al.* 1984. Inhibitors of lysosomal enzymes: accumulation of lipofuscin-like dense bodies in the brain. Science **226:** 985–987.

45. Ivy, G.O. *et al.* 1989. Lipofuscin-like substances accumulate rapidly in brain, retina and internal organs with cysteine protease inhibition. Adv. Exp. Med. Biol. **266:** 31–45.
46. Ivy, G.O. *et al.* 1989. Anomalous accumulation of tau and ubiquitin immunoreactivities in rat brain caused by protease inhibition and by normal aging: a clue to PHF pathogenesis? Brain Res. **498:** 360–365.
47. Bergamini, E. *et al.* 1993. A new method for the investigation of endocrine-regulated autophagy and protein degradation in rat liver. Exp. Mol. Pathol. **59:** 13–26.
48. Donati, A. *et al.* 2004. Anti-aging effects of anti-lipolytic drugs. Exp. Gerontol. **39:** 1061–1067.
49. Donati, A. *et al.* 2006. Stimulation of macroautophagy can rescue older cells from 8-OHdG mtDNA accumulation: a safe and easy way to meet goals in the SENS agenda. Rejuvenation Res. **9:** 408–412.

Age-Related Changes in the Liver Sinusoidal Endothelium

A Mechanism for Dyslipidemia

DAVID G. LE COUTEUR,[a] VICTORIA C. COGGER,[a] ROBERT S. MCCUSKEY,[b,c] RAFAEL DE CABO,[d] BÅRD SMEDSRØD,[c] KAREN K. SORENSEN,[c] ALESSANDRA WARREN,[a] AND ROBIN FRASER[e]

[a]*Centre for Education and Research on Ageing, University of Sydney and Concord RG Hospital, Sydney, NSW, Australia*

[b]*Department of Cell Biology and Anatomy, College of Medicine, University of Arizona, Tucson, Arizona, USA*

[c]*Department of Cell Biology and Histology, Institute of Medical Biology, University of Tromso, Tromso, Norway*

[d]*Laboratory of Experimental Gerontology, National Institute on Aging, Baltimore, Maryland, USA*

[e]*Christchurch School of Medicine, University of Otago, Christchurch, New Zealand*

ABSTRACT: **The liver sinusoidal endothelial cell (LSEC) influences the transfer of substrates between the sinusoidal blood and hepatocytes and has a major role in endocytosis; therefore, changes in the LSEC have significant implications for hepatic function. There are major morphological changes in the LSEC in old age called pseudocapillarization. These changes include increased LSEC thickness and reduced numbers of pores in the LSEC, which are called fenestrations. Pseudocapillarization has been found in old humans, rats, mice, and nonhuman primates. In addition, old age is associated with impaired LSEC endocytosis and increased leukocyte adhesion, which contributes to reduced hepatic perfusion. Given that fenestrations in the endothelium allow passage of some lipoproteins, including chylomicron remnants, age-related reduction in fenestrations impairs hepatic lipoprotein metabolism. In old rats, caloric restriction was associated with complete preservation of LSEC morphology and fenestrations. In conclusion, pseudocapillarization of the LSEC is a newly discovered aging change that, through its effects on lipoproteins, contributes to the association between old age, dyslipidemia, and vascular disease.**

Address for correspondence: Professor David G. Le Couteur, Centre for Education and Research on Ageing, University of Sydney and Concord RG Hospital, Sydney, NSW 2139, Australia. Voice: +612-9767-7212; fax: +612-9767-5419.
dlecouteur@med.usyd.edu.au

Ann. N.Y. Acad. Sci. 1114: 79–87 (2007).
doi: 10.1196/annals.1396.003

KEYWORDS: liver; aging; liver sinusoidal endothelial cell; chylomicron remnant; endocytosis; caloric restriction

INTRODUCTION

Liver Function in Old Age

In the past, the liver has been considered to be relatively unaffected by aging and age-related diseases.[1] Functional changes, particularly related to impaired drug metabolism (which is often reduced by 40–50% in old age), were attributed primarily to age-related reduction in blood flow and liver mass.[2,3] Because such mechanisms cannot fully explain age-related impairment of hepatic drug metabolism,[2] we examined the effects of old age on the structures lying between the blood and the hepatocyte that could impose a barrier to substrate transfer, specifically the liver sinusoidal endothelium and the extravascular space of Disse.

STRUCTURE AND FUNCTION OF THE LIVER SINUSOIDAL ENDOTHELIAL CELL

The blood delivered to the liver via the portal vein from the gut, spleen, and pancreas contains a vast array of nutrients, toxins, and hormones as well as waste macromolecules originating from metabolic processes in other tissues. The portal blood containing this rich mixture of substrates circulates between the plates of hepatocytes through an intricate network of sinusoids. Liver sinusoidal endothelial cells (LSECs) are unique endothelial cells that line the hepatic sinusoids. These cells are very thin, lack any basal lamina or supportive connective tissue, and are perforated with pores called fenestrations. LSECs occupy a critical and strategic position in the hepatic sinusoid because they facilitate the bi-directional transfer of substrates between blood and hepatocytes.[4] The great vascularity of the liver also generates a huge surface area for LSEC interactions with blood cells and particulate substrates for endocytosis.[5,6] Such structural features underpin two important physiological roles: filtering lipoproteins and endocytosis.

Filtering Lipoproteins—The Liver Sieve

Fenestrations are pores approximately 100 nm in diameter that permit the passage of macromolecules, such as lipoproteins, into the extravascular space for subsequent hepatocellular uptake and metabolism. Our research has focused on the role of the LSEC in lipoprotein uptake by the liver—a function that has led to LSECs being termed *the liver sieve*.[7]

We and others investigated the role of fenestrations in the disposition of chylomicrons (diameter, 100–1000 nm) and chylomicron remnants (50–100 nm).[7–9] Isolated hepatocytes cannot differentiate between chylomicrons and chylomicron remnants, yet intact livers selectively take up chylomicron remnants.[10] Further, large chylomicrons were only detected in the sinusoid, whereas smaller chylomicron remnants were also observed in the space of Disse.[11] Small chylomicrons were trapped to a greater extent in the liver than were larger chylomicrons (>100 nm).[11]

Similar results indicating exclusion from the space of Disse on the basis of size have been reported for large and small liposomes (298–337 nm versus 78–157 nm)[12] and colloidal gold particles of different diameters (17 nm versus 79 nm).[13] Likewise we have shown that liposomal doxorubicin (diameter, 100–200 nm) is excluded from the space of Disse, while nonliposomal doxorubicin is freely taken up by hepatocytes and extracted.[14] Thus, macromolecules with diameters greater than that of a typical fenestration do not traverse the endothelium, whereas particles with a smaller diameter enter the space of Disse. This is consistent with a sieving process or, more technically, ultrafiltration.[15] Recently we have shown that conditions associated with reduced fenestrations cause impaired lipoprotein uptake and hypertriglyceridemia: old age,[16] VEGFR knockout mice,[17] and treatment with poloxamer 407.[18]

Endocytosis—Scavenger Activity

The most striking functional characteristic of LSECs is their high endocytic activity. On the basis of this activity, LSECs are often termed *scavenger endothelial cells*.[5,6,19] It is now recognized that LSECs play a critical role in the removal of colloids and soluble macromolecular waste products from the systemic circulation and indeed are the largest endocytotic tissue in the body. The cells are equipped with high-affinity endocytosis receptors and an unusually well-developed clathrin-mediated endocytic machinery and intracellular endo/lysosomal trafficking system that together allow highly efficient uptake and degradation of an array of foreign and physiological waste products, such as major connective tissue macromolecules (collagen alpha-chains, C- and N-terminal propeptides of collagen propeptides, hyaluronan, chondroitin sulfate, nidogen), glycoconjugates with terminal mannose residues (i.e., lysosomal enzymes and tissue plasminogen activator), soluble IgG–immune complexes, modified low-density lipoprotein, advanced glycation end products, and microbial CpG motifs.[6,19–21]

AGING CHANGES IN MORPHOLOGY OF LSECs

Recently we reported that aging is associated with marked changes in the hepatic sinusoidal endothelium and space of Disse in the rat,[22] human,[23] mouse,[24,25] and the nonhuman primate, *Papio hamadryas*.[26] These changes were termed *pseudocapillarization* because the aging sinusoidal endothelium had become more like capillaries seen in other nonfenestrated vascular beds.[22] On electron microscopy we found a 40–80% increase in endothelial thickness and a 60–80% reduction in porosity and fenestrations ("defenestration"). This was associated with basal lamina deposition in 25–40% of old livers. The presence of these changes in at least four species indicates this is a consistent change.

The age-related change in the ultrastructure of the LSECs was associated with an increased expression of von Willebrand's factor (vWF, a capillary marker) in all species and collagen and laminin in humans and rats. Extensive expression of vWF, not seen in LSECs from normal young livers, confirms that significant aging changes are occurring in this endothelium. Changes were even apparent in the standard Masson's trichrome stain (a marker of connective tissue) in humans, consisting of obvious perisinusoidal staining. Major age-related changes in the liver were confined to the perisinusoidal regions.

Old age was associated with some deposition of lipofuscin and multinucleate cells in the hepatic parenchyma; however there was no other indication of liver disease.[22–26] Although lipid-laden stellate cells were observed in old age, there was no morphological evidence of stellate cell activation[24,25] and the expression of smooth muscle actin was not increased in the livers of old baboons.[26] In addition, caveolin-1 expression was reduced in the perisinusoidal regions in old age.[27] These findings indicate that pseudocapillarization is likely to be a primary aging change in the LSECs rather than related to subclinical or undiagnosed liver disease.

FUNCTIONAL IMPLICATIONS OF AGE-RELATED CHANGES IN THE LSEC

Lipoprotein Disposition

After discovering the structural change of pseudocapillarization, we were immediately attracted to the possibility that the loss of fenestrations in old age might impair the hepatic uptake of lipoproteins, thereby contributing to age-related vascular disease.[4,28] This is highly significant because old age is the major risk factor for vascular disease, and vascular disease is the major cause of mortality and morbidity in modern society. Old age is associated with impaired hepatic clearance of chylomicron remnants,[29] which are highly atherogenic lipoproteins.[30] Furthermore, old age is associated with postprandial hypertriglyceridemia,[29,31] which is further evidence of impaired hepatic clearance of remnants.[15]

Recently we confirmed that pseudocapillarization of the LSECs in old rats completely prevents the transfer of lipoproteins with diameters of 56 ± 6 nm across the sinusoidal endothelium.[16] Using the multiple indicator dilution method, we showed that in the livers of young rats, the ratio of the volume of distribution of small chylomicrons to that of sucrose was 1.02 ± 0.14 ($n = 12$) and the recovery ratio to sucrose was 1.05 ± 0.08. However, in livers of old rats the ratio of the volumes of distribution was significantly reduced compared with that of the livers in young rats (0.92 ± 0.07, $n = 11$, $P = 0.02$), with no age-related change in recovery (0.99 ± 0.13). This indicated that in old livers, the access of small chylomicrons into the space of Disse was totally impeded.[32] This result shows that pseudocapillarization is a novel mechanism for dyslipidemia in old age. The therapeutic implications are very evident: Age-related loss of fenestration now represents a novel target for the treatment and prevention of age-related dyslipidemia and hence for cardiovascular disease.

Endocytosis

Recently the effects of old age on LSEC endocytosis were reported by Ito *et al.*[25] Using *in vivo* microscopy, we found that the uptake of a frequently used probe for the scavenger receptor formaldehyde-treated serum albumin (FITC-FSA) by LSECs was diminished, especially in the pericentral sinusoids. The LSEC has a crucial role in the endocytosis of many substrates of relevance to aging and age-related disease (e.g., advanced glycation end products and oxidized low-density lipoproteins).[5] Therefore any reduction in LSEC endocytosis in old age may have important systemic effects for aging.

Blood Flow

There is a marked reduction in hepatic blood flow in old age, which reduces the hepatic clearance of many substrates, particularly those that are highly extracted by the liver.[2,3] Recently we found an increase in ICAM-1 expression in LSECs in old mice associated with a fivefold increase in leukocyte adhesion within the sinusoids. Adherent leukocytes may be partly responsible for a 14% reduction in the numbers of perfused sinusoids and a 35% reduction in sinusoidal blood flow.[25] These results provide a novel mechanism for the effects of old age on hepatic blood flow.

CALORIC RESTRICTION AND THE LSEC

It has been recognized for more than 70 years that caloric restriction promotes longevity. In general, a reduction of caloric intake by about 40% (with appropriate nutritional supplementation and usually commenced after weaning) increases maximum life expectancy by about 40%, and this is associated with a delay in the onset of most age-associated disorders.[33] Understanding the effects of caloric restriction is now a critical component of aging research.[34,35] One of the most important effects of caloric restriction is the reduction in risk factors for, and diseases of, the cardiovascular system, and the beneficial vascular effects of caloric restriction far exceed those of established therapies, such as antihypertensive agents and statins.[33,36] In humans who sustained caloric restriction for more than 6 years, there was improvement in all cardiovascular parameters, including blood pressure and diastolic function,[37] and caloric restriction was highly protective against atherosclerosis.[38] Therefore, we were very keen to establish whether caloric restriction had any effects on pseudocapillarization, which could provide one mechanism for the effects of caloric restriction on cardiovascular disease.

Livers from young (6 months) and old (24 months) calorie-restricted and *ad libitum*–fed F344 rats were examined by means of electron microscopy and immunohistochemical studies. Old-age changes included increased thickness of the liver sinusoidal endothelium and reduced endothelial fenestration porosity. In old calorie-restricted rats, endothelial thickness was significantly less (190 ± 7 nm versus 211 ± 6 nm, $P < 0.05$), and fenestration porosity was significantly greater ($3.9 \pm 0.3\%$ versus $2.4 \pm 0.1\%$, $P < 0.001$) than in old *ad libitum*–fed rats. Furthermore caloric restriction prevented the age-related decrease in caveolin-1 expression and increase in perisinusoidal collagen IV staining. Given the probable role of caveolin-1 in maintaining fenestral integrity,[39] it is fascinating that old age is associated with reduced fenestrations in association with reduced caveolin-1 expression, and that these changes are both reversed by caloric restriction. Whether caloric restriction reverses the effects of pseudocapillarization on lipoprotein metabolism is unknown, although plasma triglycerides are substantially lower in old calorically restricted rats compared with old *ad libitum*–fed rats (0.43 ± 0.04 versus 1.31 ± 0.25 mmol/L, $P < 0.0001$).[40] The results indicate that age-related changes in the LSECs are preventable and thus a plausible therapeutic target.

CONCLUSIONS

There are structural changes in the LSEC including loss of fenestrations that have been called pseudocapillarization. These are associated mechanistically with impaired uptake of lipoproteins and reversed by caloric restriction. Therefore, maintaining the patency of fenestrations in the LSEC is a novel therapeutic strategy for the prevention of age-related dyslipidemia. In addition, old age is associated with impaired endocytosis and increased leukocyte adhesion, which are also likely to have significant systemic implications.

ACKNOWLEDGMENTS

Our thanks are extended to Yoshiyo Ito, Sarah Hilmer, Hamish Jamieson, Karen Sorensen, Dmitri Svistounov, Allan McLean, and Margaret McCuskey for their crucial input into these studies. The research has been funded by the National Health and Medical Research Council of Australia, Ageing and Alzheimer's Research Foundation (a division of the Medical Foundation of the University of Sydney), the National Institutes of Health/National Institute of Aging (Grant R21 AG-02582), and the Norwegian Research Council (Grant 153483/V50).

REFERENCES

1. POPPER, H. 1986. Aging and the liver. Prog. Liver. Dis. **VIII:** 659–683.
2. LE COUTEUR, D.G. & A.J. MCLEAN. 1998. The aging liver: drug clearance and an oxygen diffusion barrier hypothesis. Clin. Pharmacokinet. **34:** 359–373.
3. SCHMUCKER, D.L. 1998. Aging and the liver: an update. J. Gerontol. **53A:** B315–B320.
4. LE COUTEUR, D.G. *et al.* 2005. The hepatic sinusoid in aging and cirrhosis: effects on hepatic substrate disposition and drug clearance. Clin. Pharmacokinet. **44:** 187–200.
5. SETERNES, T., K. SORENSEN & B. SMEDSROD. 2002. Scavenger endothelial cells of vertebrates: a non-peripheral leukocyte system for high capacity elimination of waste macromolecules. Proc. Natl. Acad. Sci. USA **99:** 7594–7597.
6. SMEDSROD, B. *et al.* 1994. Cell biology of liver endothelial and Kupffer cells. Gut **35:** 1509–1516.
7. FRASER, R., B.R. DOBBS & G.W. ROGERS. 1995. Lipoproteins and the liver sieve: the role of fenestrated sinusoidal endothelium in lipoprotein metabolism, atherosclerosis, and cirrhosis. Hepatology **21:** 863–874.
8. YU, K.C. & A.D. COOPER. 2001. Postprandial lipoproteins and atherosclerosis. Frontiers Biosci. **6:** 332–354.
9. WISSE, E. 1970. An electron microscopic study of the fenestrated endothelial lining of rat liver sinusoids. J. Ultrastruct. Res. **31:** 125–150.
10. FLOREN, C.H. 1984. Binding of apolipoprotein E-rich remnant lipoproteins to human liver membranes. Scand. J. Gastroenterol. **19:** 473–479.
11. FRASER, R., A.G. BOSANQUET & W.A. DAY. 1978. Filtration of chylomicrons by the liver may influence cholesterol metabolism and atherosclerosis. Atherosclerosis **29:** 113–123.
12. ROMERO, E.L. *et al.* 1999. On the mechanism of hepatic transendothelial passage of large liposomes. FEBS Lett. **448:** 193–196.
13. HARDONK, M.J., G. HARMS & J. KOUDSTAAL. 1985. Zonal heterogeneity of rat hepatocytes in the in vivo uptake of 17 nm colloidal gold granules. Histochemistry **83:** 473–477.
14. HILMER, S.N. *et al.* 2004. The hepatic pharmacokinetics of doxorubicin and liposomal doxorubicin. Drug Metabol. Dispos. **32:** 794–799.
15. LE COUTEUR, D.G. *et al.* 2006. Aging, atherosclerosis and the liver sieve. *In* New Research in Atherosclerosis. L.V. Clark, Ed.: 19–44. Nova Science. New York.

16. HILMER, S.N. *et al.* 2005. Age-related changes in the hepatic sinusoidal endothelium impede lipoprotein transfer in the rat. Hepatology **42:** 1349–1354.
17. CARPENTER, B. *et al.* 2005. VEGF is crucial for the hepatic vascular development required for lipoprotein uptake. Development **132:** 3293–3303.
18. COGGER, V.C. *et al.* 2006. Surfactants and hyperlipidemia: a link is the liver sieve. Atherosclerosis **189:** 273–281.
19. SMEDSROD, B. *et al.* 1997. Advanced glycation end products are eliminated by scavenger-receptor-mediated endocytosis in hepatic sinusoidal Kupffer and endothelial cells. Biochem. J. **322:** 567–573.
20. HANSEN, B. *et al.* 2005. Stabilin-1 and stabilin-2 are both directed into the early endocytic pathway in hepatic sinusoidal endothelium via interactions with clathrin/AP-2, independent of ligand binding. Exp. Cell Res. **303:** 160–173.
21. MARTIN-ARMAS, M. *et al.* 2006. Toll-like receptor 9 (TLR9) is present in murine liver sinusoidal endothelial cells (LSECs) and mediates the effect of CpG-oligonucleotides. J. Hepatol. **44:** 939–946.
22. LE COUTEUR, D.G. *et al.* 2001. Pseudocapillarization and associated energy limitation in the aged rat liver. Hepatology **33:** 537–543.
23. MCLEAN, A.J. *et al.* 2003. Age-related pseudocapillarization of the human liver. J. Pathol. **200:** 112–117.
24. WARREN, A. *et al.* 2005. Hepatic pseudocapillarization in aged mice. Exp. Gerontol. **40:** 807–812.
25. ITO, Y. *et al.* 2007. Age-related changes in the hepatic microcirculation of mice. Exp. Gerontol. **48:** 789–797.
26. COGGER, V.C. *et al.* 2003. Hepatic sinusoidal pseudocapillarization with aging in the non-human primate. Exp. Gerontol. **38:** 1101–1107.
27. JAMIESON, H. *et al.* 2007. Caloric restriction reduces age-related pseudocapillarization of the hepatic sinusoid. Exp. Gerontol. **42:** 374–378.
28. LE COUTEUR, D.G. *et al.* 2002. Hepatic pseudocapillarisation and atherosclerosis in ageing. Lancet **359:** 1612–1615.
29. KRASINSKI, S.D. *et al.* 1990. Postprandial plasma retinyl ester response is greater in older subjects compared with younger subjects: evidence for delayed plasma clearance of intestinal lipoproteins. J. Clin. Invest. **85:** 883–892.
30. WEINTRAUB, M.S. *et al.* 1996. Clearance of chylomicron remnants in normolipidaemic patients with coronary heart disease: case control study over three years. Br. Med. J. **312:** 935–939.
31. COHN, J.S. *et al.* 1988. Postprandial plasma lipoprotein changes in human subjects of different ages. J. Lipid Res. **29:** 469–479.
32. HUET, P.M. & J.P. VILLENEUVE. 2005. Microcirculation of the aging liver: is getting old like having cirrhosis? Hepatology **42:** 1248–1251.
33. EVERITT, A. *et al.* 2005. Calorie restriction versus drug therapy to delay the onset of aging diseases and extend life. Age **27:** 1–10.
34. SINCLAIR, D.A. 2005. Toward a unified theory of caloric restriction and longevity regulation. Mech. Ageing Dev. **126:** 987–1002.
35. MASORO, E.J. 2005. Overview of caloric restriction and ageing. Mech. Ageing Dev. **126:** 913–922.
36. GERSTENBLITH, G. 2006. Cardiovascular aging: what we can learn from caloric restriction. J. Am. Coll. Cardiol. **47:** 403–404.
37. MEYER, T.E. *et al.* 2006. Long-term caloric restriction ameliorates the decline in diastolic function in humans. J. Am. Coll. Cardiol. **47:** 398–402.

38. FONTANA, L. *et al.* 2004. Long-term calorie restriction is highly effective in reducing the risk for atherosclerosis in humans. Proc. Natl. Acad. Sci. USA **101:** 6659–6663.
39. YOKOMORI, H. *et al.* 2001. Endothelial nitric oxide synthase and caveolin-1 are co-localized in sinusoidal endothelial fenestrae. Liver **21:** 198–206.
40. ZHU, M. *et al.* 2004. Circulating adiponectin levels increase in rats on caloric restriction: the potential for insulin sensitization. Exp. Gerontol. **39:** 1049–1059.

Effects of Old Age on Hepatocyte Oxygenation

RAJKUMAR CHELUVAPPA,[a] SARAH N. HILMER,[a,b]
SUN YOUNG KWUN,[c] VICTORIA C. COGGER,[a]
AND DAVID G. LE COUTEUR[a]

[a]*Centre for Education and Research on Ageing and ANZAC Research Institute, University of Sydney and Concord RG Hospital, Concord, NSW 2139, Australia*

[b]*Departments of Aged Care and Clinical Pharmacology, Royal North Shore Hospital, St Leonards, NSW 2065, Australia*

[c]*Department of Anatomical Pathology, Concord RG Hospital, Concord, NSW 2139, Australia*

ABSTRACT: **Hepatic phase I drug metabolism is diminished in old age. It has been suggested that hepatocyte hypoxia and impaired bioenergetics in old age may contribute to this aging change. Therefore, we sought to determine whether old age was associated with *in vivo* hypoxia in the aged rat liver. Immunohistochemical studies with the nitroimidazole hypoxia marker, pimonidazole, were carried out in livers from young and old rats. Preliminary studies were performed on four young (4-month-old) and six old (2-year-old) F344 rats to directly visualize the distribution and intensity of pimonidazole staining. There were no significant differences in the distribution or in the intensity of pimonidazole immunohistochemical staining between young and aged rat livers. In conclusion, no major changes in hepatocyte oxygenation were seen in the aged rat liver, and the ATP changes are unlikely to be secondary to hepatocyte hypoxia or impaired oxygen diffusion into the liver. It is thus more likely that age-related reduction in liver ATP is attributable to mitochondrial dysfunction.**

KEYWORDS: **hepatocyte; aging; hypoxia; pimonidazole; immunohistochemistry**

INTRODUCTION

It has been well established that phase I hepatic drug metabolism, which involves oxidation, reduction, or hydrolysis, is diminished in old age.[1,2] We

Author for correspondence: Dr. Rajkumar Cheluvappa, ANZAC Research Institute, University of Sydney and Concord RG Hospital, Hospital Road, Concord NSW 2139, Australia. Voice: +612 9767 9104; fax: +612 9767 9101.
rcheluvappa@med.usyd.edu.au

Ann. N.Y. Acad. Sci. 1114: 88–92 (2007).
doi: 10.1196/annals.1396.007

have previously suggested that hepatocyte hypoxia may contribute to this age-associated reduction in phase I drug clearance.[3] Using ^{31}P nuclear magnetic resonance studies on freeze-clamped samples, we reported that aged rat livers have decreased adenosine triphosphate (ATP) and high-energy phosphate metabolite pools, suggestive of hypoxia.[4] Others have reported that livers from aged mice have a significantly lower total baseline ATP as well as marked decreases in isolated mitochondrial oxygen consumption and ATP production.[5] There is evidence of hepatocyte hypoxia in old age beyond changes in ATP levels. Upregulation of various markers of hypoxia has been demonstrated using Western blotting and RT-PCR: hypoxia-inducible factor-1α, heme oxygenase-1, vascular endothelial growth factor, erythropoietin, and inducible nitric oxide synthase.[6] These age-related changes could be caused by decreased liver perfusion,[7] an oxygen-diffusion barrier secondary to age-associated liver pseudocapillarization of the liver sinusoidal endothelial cell,[4] mitochondrial oxidative stress,[8] or mitochondrial dysfunction.[9]

Therefore, we sought to determine whether aging was associated with *in vivo* hypoxia in the aged rat liver by performing immunohistochemical studies with the nitroimidazole hypoxia marker, pimonidazole. 2-Nitroimidazoles including pimonidazole have been reliably used as *in vivo* hypoxia markers in a number of systems.[10] Nitroimidazole-adduct formation, which is increased in regions with low oxygen tension, is determined by immunohistochemical testing. Staining is substantially increased when the intracellular oxygen tension falls below 10 mmHg.[11] In these preliminary studies, pimonidazole immunohistochemical investigation was carried out in four young (4-month-old) and six old (2-year-old) rats in an attempt to directly visualize and compare the distribution and intensity of hypoxic areas in young and old rat livers.

EXPERIMENTAL METHODS

The study had approval from the Sydney Southwest Area Health Service Animal Ethics Committee. Male Fischer F344 rats obtained from the National Institute of Aging were used. Four young (4 months; 300–350 g) and six old (2 years; 300–500 g) rats were studied. The rats were anesthetized with ketamine and xylazine. Pimonidazole at a concentration of 120 mg per kilogram body weight of rat was injected intraperitoneally 2 hs prior to taking the liver sample. Blood from the inferior vena cava was collected from each rat and the liver dissected out and processed for immunohistochemical study.

Immunohistochemical study for pimonidazole adducts was done by means of a direct immunoperoxidase method using a peroxidase-labeled polymer. This involved fixing liver specimens in 4% paraformaldehyde-buffered saline followed by paraffin embedding. Sections measuring 4 micrometers each were deparaffinized in xylene (3×3 min), taken to absolute ethanol (3×2 min), and incubated with 3% H_2O_2 in absolute methanol for 10 min at room temperature

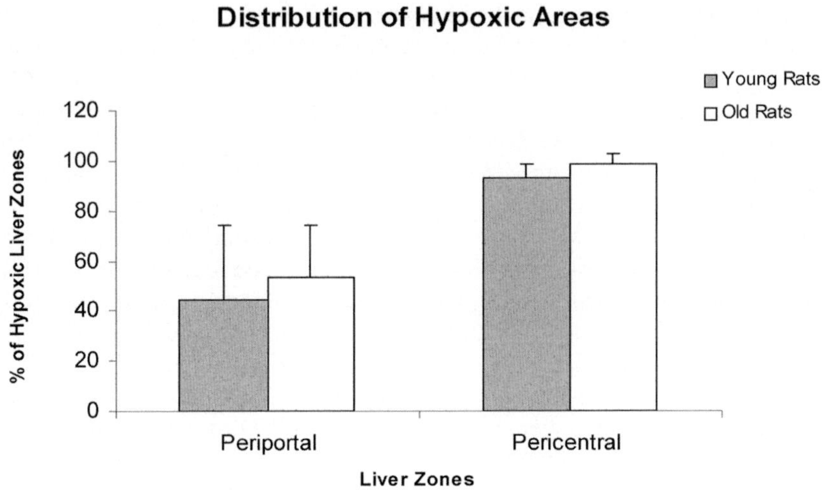

FIGURE 1. Zonal distribution of pimonidazole-adduct immunohistochemical staining in young and old rat livers. The distribution is depicted as the percentage of hypoxic liver zones ± SD.

for blocking endogenous peroxidase. After the sections were hydrated through a decreasing ethanol gradient, the slides were washed in water and heated in a pressure cooker with citrate buffer for 4 mins. After cooling, the slides were treated with goat serum for 5 min before the primary antibody; Hypoxyprobe Mab-1 (HP1-1000; Chemicon, Boronia, Victoria, Australia) was applied in a dilution of 1:200. After overnight incubation and wash, the secondary antibody, goat anti-mouse immunoglobulin polymer conjugated with horseradish peroxidase (TL-060-HL; LabVision, Hallam, Victoria, Australia) was applied for 30 mins. After citrate buffer wash, the sections were treated with diaminobenzidine (DAB) substrate for 5 mins, treated with 1% $CuSO_4$, counterstained with hematoxylin, dehydrated, and mounted. Sixteen random photographs per blinded slide were taken, graded consensually by two or three blinded observers according to staining distribution (periportal, zone-2, pericentral) and intensity of staining (0, +, ++, +++), and semi-quantitatively assessed by chi-square analysis for intensity differentiation.

RESULTS AND DISCUSSION

Pimonidazole staining was most often encountered (FIG. 1) and was most intense (FIG. 2) in the pericentral zones in both young and old livers, which is consistent with the decline in sinusoidal oxygen tension as blood flows from the portal vein via zone-2 to the central vein. No age-related differences in the periportal, zone-2, and pericentral distribution (FIG. 1) or in the intensity (FIG. 2) of

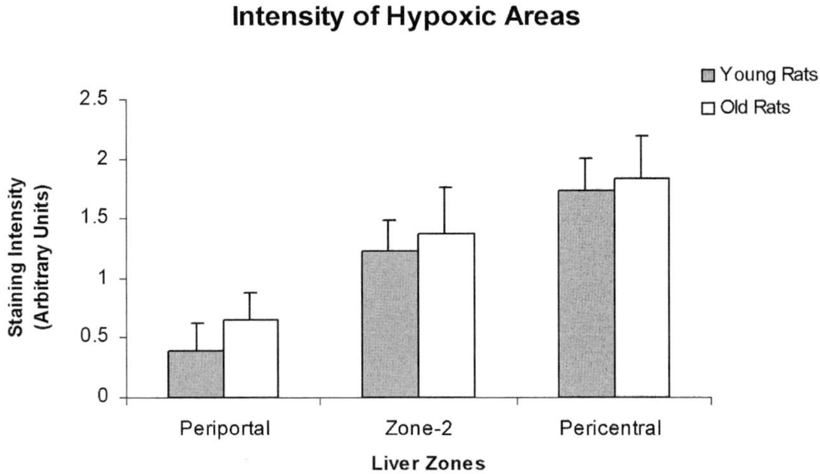

FIGURE 2. Intensity of pimonidazole-adduct immunohistochemical staining in young and old rat livers. The intensity is expressed as arbitrary units from 0 to 3 ± SD.

pimonidazole immunohistochemical staining were detected in the livers. The result indicates that intracellular hypoxia to any major extent is unlikely in the liver in old age and does not support the concept of an oxygen-diffusion barrier generated by age-related changes in the liver sinusoidal endothelial cell.

The question arises as to alternate mechanisms for the age-related decrease in liver ATP levels. This can possibly be ascribed to decreased liver mass,[7] mitochondrial oxidative stress,[8] or mitochondrial dysfunction[9] in aged livers. The increased expression of hypoxia-responsive genes observed in the old rat liver[6] is unlikely to be secondary to hepatocyte hypoxia, and presumably is secondary to other age-related processes such as increased production of free radicals.

In conclusion, hepatocytes are not hypoxic in old age. The age-related decrease in liver ATP is thus probably secondary to impaired mitochondrial function.

REFERENCES

1. KINIRONS, M.T. & M.S. O'MAHONY. 2004. Drug metabolism and ageing. Br. J. Clin. Pharmacol. **57:** 540–544.
2. HERRLINGER, C. & U. KLOTZ. 2001. Drug metabolism and drug interactions in the elderly. Best Pract. Res. Clin. Gastroenterol. **15:** 897–918.
3. LE COUTEUR, D.G. & A.J. MCLEAN. 1998. The aging liver: drug clearance and an oxygen diffusion barrier hypothesis. Clin. Pharmacokinet. **34:** 359–373.
4. LE COUTEUR, D.G., V.C. COGGER, A.M. MARKUS, *et al.* 2001. Pseudocapillarization and associated energy limitation in the aged rat liver. Hepatology **33:** 537–543.

5. SELZNER, M., N. SELZNER, W. JOCHUM, *et al.* 2007. Increased ischemic injury in old mouse liver: an ATP-dependent mechanism. Liver Transpl. **13:** 382–390.
6. KANG, M.J., H.J. KIM, H.K. KIM, *et al.* 2005. The effect of age and calorie restriction on HIF-1-responsive genes in aged liver. Biogerontology **6:** 27–37.
7. SCHMUCKER, D.L. 2001. Liver function and phase I drug metabolism in the elderly: a paradox. Drugs Aging **18**: 837–851.
8. SASTRE, J., F.V. PALLARDO & J. VINA. 2003. The role of mitochondrial oxidative stress in aging. Free Radic. Biol. Med. **35:** 1–8.
9. SASTRE, J., F.V. PALLARDO, R. PLA, *et al.* 1996. Aging of the liver: age-associated mitochondrial damage in intact hepatocytes. Hepatology **24:** 1199–1205.
10. ARTEEL, G.E., R.G. THURMAN & J.A. RALEIGH. 1998. Reductive metabolism of the hypoxia marker pimonidazole is regulated by oxygen tension independent of the pyridine nucleotide redox state. Eur. J. Biochem. **253:** 743–750.
11. GROSS, M.W., U. KARBACH, K. GROEBE, *et al.* 1995. Calibration of misonidazole labeling by simultaneous measurement of oxygen tension and labeling density in multicellular spheroids. Int. J. Cancer **61:** 567–573.

Comparative Analysis of Mitochondrial Genotype and Aging

J. WILLIAM O. BALLARD,[a] SUBHASH D. KATEWA,[a] RICHARD G. MELVIN,[a] AND GRACE CHAN[b]

[a]*School of Biotechnology and Biomolecular Sciences, University of New South Wales, Sydney, NSW, Australia*

[b]*Statistical Consulting Center, Department of Statistics and Actuarial Science, University of Iowa, Iowa City, Iowa*

ABSTRACT: A common feature across all animals, including humans, is that mitochondrial bioenergetics is linked to oxidative stress, but the nature of these relationships with survival is yet to be properly defined. In this study we included 12 *Drosophila simulans* isofemale lines: four of each distinct mtDNA haplogroup (*si*I, -II, and -III). First, we investigated sequence variation in six mtDNA and 13 nuclear encoded genes (nine nuclear-encoded subunits, and the four known isoforms, of complex IV of the electron transport chain). As expected we observed high divergence among the three distinct mitotypes and greatest mtDNA variability in *si*II-harboring flies. In the nuclear encoded genes, no fixed amino acid differences were observed and levels of polymorphism did not differ significantly among flies harboring distinct mtDNA types. Second, 15,456 flies were included in mortality studies. We observed that mtDNA type influenced survival (*si*II $\approx$ *si*III $>$ *si*I), flies harboring *si*II mtDNA had the greatest variation in mortality rates, and in all cases males were longer lived than females. We also assayed maximal rates of hydrogen peroxide (H_2O_2) production from complex III of the electron transport chain in mitochondria isolated from 11-day-old flies. Contrary to our prediction, rates of H_2O_2 production tended to increase with mean survival. This result suggests that higher rates of H_2O_2 production in younger flies may lead to an upregulation of antioxidants, age-dependent increase in the rate of H_2O_2 production differ, and/or flies vary in their mitochondrial uncoupling. Alternatively, the whole organism may not regularly, if ever, experience maximal H_2O_2 production rates.

KEYWORDS: mitochondria; free radicals; aging; longevity; evolution; *Drosophila*

Address for correspondence: Bill Ballard, Ph.D., Ramaciotti Centre for Gene Function Analysis, School of Biotechnology and Biomolecular Sciences, 217B Biology Building, University of New South Wales, Sydney, NSW, Australia 2052. Voice: 61-2-9385-2021; fax: 61-2-9385-1095.
w.ballard@unsw.edu.au

Ann. N.Y. Acad. Sci. 1114: 93–106 (2007).
doi: 10.1196/annals.1396.011

INTRODUCTION

The goal of this study is to identify wild-caught *Drosophila* lines that have significant differences in longevity and rates of hydrogen peroxide (H_2O_2) production from isolated mitochondria. This goal fits into our longer-term objective of determining the role that specific mitochondrial variants play in influencing longevity and fitness. We define a mitochondrial variant as a specific complement of mitochondrial and nuclear-encoded proteins. The conceptual framework for these studies is that mitochondrial variants differentially influence the amounts of reactive oxygen species (ROS) produced and oxidative damage incurred by mitochondria, which in turn is hypothesized to influence mortality rates.[1–4] In *Drosophila* both complexes I and III are known to produce superoxide radicals,[5,6] but the majority is produced by complex III. H_2O_2 produced in mitochondria comes from dismutation of mitochondrial superoxide.

Recent advances make this a particularly exciting time to be studying mitochondria. Gone are the times when mentors actively dissuaded students from studying mitochondria because of the lack of normal recombination in animal mtDNA. It has also been known for several years that most of the 600 or so proteins[7] that reside in the mitochondrion are encoded in the nuclear genome, that the mitochondrion contains a genome, and that each of the products produced by the genome are required for oxidative phosphorylation (OXPHOS). However, the influence of mtDNA variation on ROS production and on aging has scarcely been explored. Mitochondrial variants differentially influence (or are affected by) the amount of oxidative damage incurred by mtDNA, lipids, and cell membranes. This damage affects mitochondrial metabolism and thus susceptibility to degenerative disease and aging. However, with so many gene products being imported from the nucleus, there is great uncertainty as to the specific and relative contributions of the mitochondrial and nuclear components and of mtDNA/nuclear (mitonuclear) interactions.

It is possible that mtDNA is the primary determinant of the efficiency of mitochondrial metabolism. In humans, longevity has been associated with specific mitochondrial DNA types,[8–10] however, the specific mechanism by which this occurs has not been determined. The physical proximity of mtDNA to sites where ROS are produced is likely to make it particularly susceptible to damage as an organism ages. As more copies of the mtDNA sustain mutations, complexes in the electron transport chain become less efficient because of stoichiometric mismatches between the mitochondrial and nuclear-encoded components of each complex. This gradual accumulation of damage to mtDNA is expected to lead to a decline in mitochondrial function.

A second possibility is that nuclear genes are the primary determinants of mitochondrial efficiency, ROS production, and survival. The influence of nuclear genes on mitochondrial metabolism is not well-established, but there are more nuclear-encoded subunits (81 subunits) than mitochondria-encoded subunits (13 subunits) in OXPHOS proteomes, and an increasing number of human mitochondrial diseases are being linked with nuclear mutations.[11]

The third possibility is that mitonuclear interactions are the key to unraveling the link between survival and ROS production because four of the OXPHOS complexes have mitochondria- and nuclear-encoded subunits. De Benedictis *et al.*[12] observed that Italian male centenarians had a significantly higher frequency of the European mitotype J than did sex-matched younger subjects of the same ethnic and geographic origin. This result suggests mitotype-specific effects on the rate and quality of aging. Somewhat surprisingly, however, complete mtDNA sequencing demonstrated that the J mitotype is characterized by a particular suite of six mutations often associated with disease.[13] From these data it has been argued that key mtDNA mutations may induce death or extend lifespan, depending on other mtDNA mutations and on stoichiometric mismatches with nuclear-encoded proteins in each OXPHOS subunit.

Drosophila as a Model to Study Mitochondria and Aging

The utility of *Drosophila melanogaster* as a model organism for studying aging has been firmly established.[14–16] The same 13 mtDNA protein coding genes are found in flies and mammals and the two rRNA gene products can be folded into remarkably similar secondary structures.[17] Further, it is known that key molecular and metabolic pathways in aging are shared by flies and mammals.[18] Not all pathways are shared, but this validates the use of *Drosophila* as a model system.

An advantage of nonvertebrate model systems including *Drosophila* is that wild-caught and introgression strains can be studied where each mtDNA genotype is genetically placed in the nuclear genome of standard strains. In contrast, it is difficult to unravel the influence of the mitochondrial and nuclear genomes in most organisms, including humans. By conducting an initial, detailed study using nonvertebrate model systems, it will be possible to derive specific hypotheses that can be used to interpret existing data and to design and optimize future experiments on a variety of other organisms that are less easy to genetically manipulate. Without initial experiments on nonvertebrate models it would not be possible to design such experiments.

An advantage of *Drosophila* over other nonvertebrate model systems for mitochondrial research, yeasts and nematode worms, is that flies are obligate aerobes.[14] This is important given that aging results at least in part from accumulated oxidative damage.[19,20] Additional aspects of the biology of the fly that are a particular advantage for aging studies include its life history, which is divided into distinct morphologic stages, so that periods of growth and development can be readily distinguished from the sexually mature "aging" adult phase. In many organisms distinguishing mature adults from immature

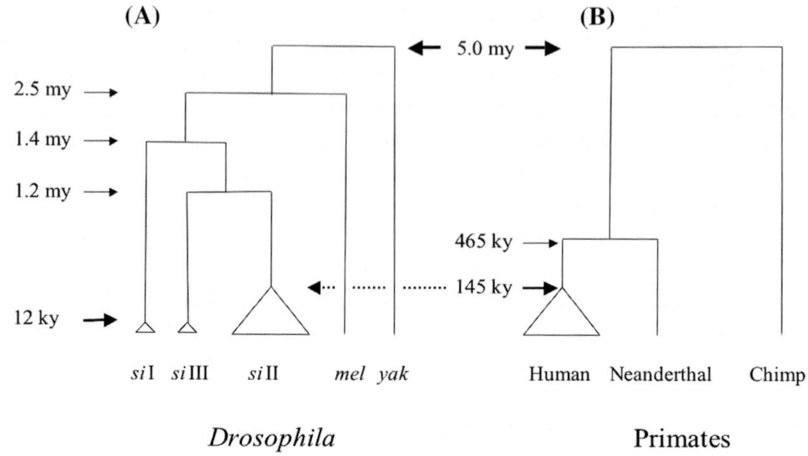

FIGURE 1. Comparison of the mitochondrial DNA divergence times of *Drosophila* and primates. (**A**) Estimates of the divergence times of *Drosophila yakuba* (*yak*), *D. melanogaster* (*mel*), *D. simulans si*I (*si*I), *D. simulans si*II (*si*II), *D. simulans si*III (*si*III) from Ballard[26,40,45] and Moriyama and Powell.[46] Note that the presence of *Wolbachia* is likely to influence the divergence time estimates of *D. simulans ss*II and *ss*III. (**B**) Estimates of the mitochondrial DNA divergence times of chimps, neanderthals and humans from Horai *et al.*[27] and Krings *et al.*[28]

or juvenile stages is not simple. A second feature of note is that the fly consists of almost entirely postmitotic fully differentiated cells. Except for cells in the gonads and a few cells in the gut, no other cells are thought to continue to divide.[21]

A potential disadvantage of employing *D. melanogaster* to unravel the influences of mitochondrial and nuclear-encoded genes on mitochondrial bioenergetics is that its mtDNA exhibits the statistical signature of a recent selective sweep.[22] In addition to reducing variation this has the potential to disrupt co-adapted gene complexes by the process of genetic hitchhiking.[23,24]

As a consequence of the low mtDNA variation in *D. melanogaster*, we propose *D. simulans* as a parallel model for studying the role of mitochondria on aging. *D. simulans* harbors three distinct haplogroups (*si*I, -II, and -III) with high mtDNA diversity.[25,26] We propose that this single model system has the potential to detect both broad-brush differences between mitotypes and subtler differences within mitotypes. The three *D. simulans* mtDNA haplogroups diverged about 1.4 my ago. In contrast, the time to the most recent common ancestor of *si*II mtDNA is about 145 ky years ago, whereas that of *si*I and *si*III mtDNA is 12 ky years ago (FIG. 1A). In comparison, it is hypothesized that chimps diverged from the human and Neanderthal lineage about 5 my ago, whereas modern humans diverged about 465 ky ago[27,28] (FIG. 1B). A disadvantage of the *D. simulans* system is that some of the genetic tools that are available in *D. melanogaster* do not exist in this species.

Wild-Caught Drosophila simulans

For this study, four randomly selected wild-caught *si*I, *si*II and *si*III isofemale lines are included. The mtDNA haplogroup and *Wolbachia* infection status of each line was determined by PCR and sequencing.[29] The flies from Hawaii were 1HW01, 1HW05, 1HW07, and 1HW11. The flies from Kenya were 2KY15, 2KY17, 2KY18, 2KY21 and 3KY10, 3KY12, 3KY14, 3KY20. The 1/2/3 refers to the mtDNA haplogroup; HW designates Hawaii and KY Kenya. The two final numbers designate the specific line number. All lines from Hawaii are infected with the *w*Ha strain of *Wolbachia.* Lines 3KY10 and 3KY20 are infected with the *w*Ma strain. All others were uninfected.

All studies were conducted on flies five and six generations after they were collected from the field. New cultures were started from flies that were less than 14 days old.[30] Fly lines were maintained continuously at constant density, temperature (23°C), humidity (50%RH), diurnal cycle (12 light: 12 dark) and diet (instant food from Carolina Biological Supply Co. [Burlington, NC]). Parents of the experimental flies were released into population cages and allowed to acclimate for 2–3 days. In all cases these adults were less than 2 weeks old. Oviposition resources (petri dish with solidified agar-based medium containing 4% agar and 10% molasses) were provided for 4-h periods to collect cohorts of eggs. Eggs were harvested, washed, and suspended in a small volume of 1× PBS.[31] Fifteen microliters of this suspension, containing about 200 eggs, were added to instant *Drosophila* media. Females of the 1HW11 line laid few eggs, and the larval developmental time extended over 20 days. Combined, these data indicate that the line was sickly and/or had deleterious mutations. This line was excluded from survival and H_2O_2 production assays.

Mitochondrial and Nuclear Variation in Drosophila simulans

We investigated sequence variation in six mtDNA and the 13 nuclear-encoded genes. The mtDNA genes are ND2, COI, COII, COIII, ATPase 6, and ATPase 8. There are 20 nonsynonymous and 127 synonymous fixed differences between *si*I and *si*II; 16 and 108 between *si*I and *si*III; and seven and 108 between *si*II and *si*III. As expected, variability of *si*II mtDNA synonymous nucleotide polymorphism is greater than the variability in either *si*I or *si*III mtDNA (ANOVA $F_{2,15} = 4.83$; $P = 0.02$; FIG. 2A).

The 13 nuclear genes are the nine nuclear-encoded subunits, and four known isoforms, of complex IV of the electron transport chain.[32] We are interested in the variability within and among these genes because complex IV has been hypothesized to be the rate-limiting step of mitochondria bioenergetics.[33] No fixed amino acid differences were observed among flies harboring distinct mtDNA types and no fixed synonymous changes were observed in the

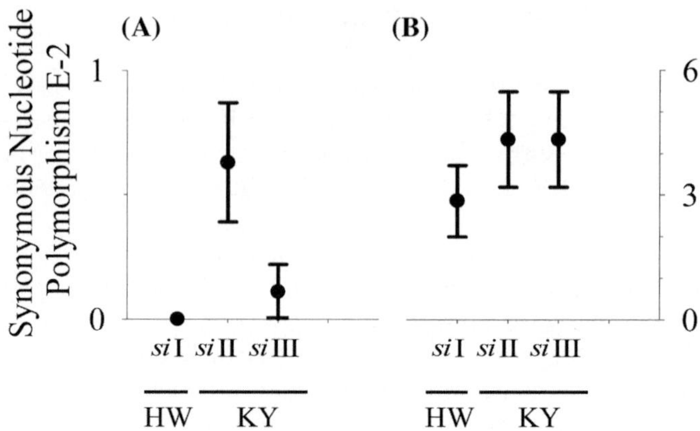

FIGURE 2. DNA sequence variation in 12 *D. simulans* lines (*si*I, -II, and -III) from six mtDNA and 13 nuclear-encoded loci. Flies were collected in Hawaii (HW) and Kenya (KY). (**A**) Mitochondrial ND2, COI, COII, COIII, ATPase 6 and ATPase 8 loci. (**B**) Thirteen nuclear-encoded genes that produce the nine nuclear-encoded subunits of complex IV of the electron transport chain (and the four known isoforms).

nuclear genes of flies harboring *si*II or *si*III mtDNA. Also, there was no significant difference in levels of synonymous nucleotide polymorphism (ANOVA $F_{2,15} = 0.61$; $P = 0.55$; FIG. 2B). However, Hawaiian flies tended to have a unique set of polymorphisms. For example, they have two fixed synonymous changes in COX6C and the line HW01 was found to be heterozygous for a two-amino acid deletion (ΔTrp85, ΔVal86) in the nuclear COX7A locus.

Mitochondrial Genotype and Aging in Drosophila simulans

A total of 15,456 flies were included in four replicate mortality studies of each line and sex. Non-virgin *D. simulans* were collected about 3 days after eclosion (range 2–4 days) and placed into replicate cages in at least two generations such that four replicates were completed for each sex. Deaths were counted, dead flies removed, and fresh Carolina Biological instant food provided at 2- to 3-day intervals. Deaths during the first interval after transfer were considered due to injury while sorting and therefore are not included in calculations. Initial cohort size was calculated as the summed death observations across all ages beginning at the second interval.

The most obvious results are that flies harboring *si*II mtDNA have the greatest variation in survival and in all cases males are longer lived than females (FIG. 3; TABLE 1). ANOVA indicates that mtDNA haplogroup, sex, and line nested within mtDNA influences 50% and 10% survival of *D. simulans* (TABLE 2). The mean survival of *si*II flies is 56.2 ± 0.43 days for males and 50.1 ± 0.39 days for females. The mean survival of *si*III males is

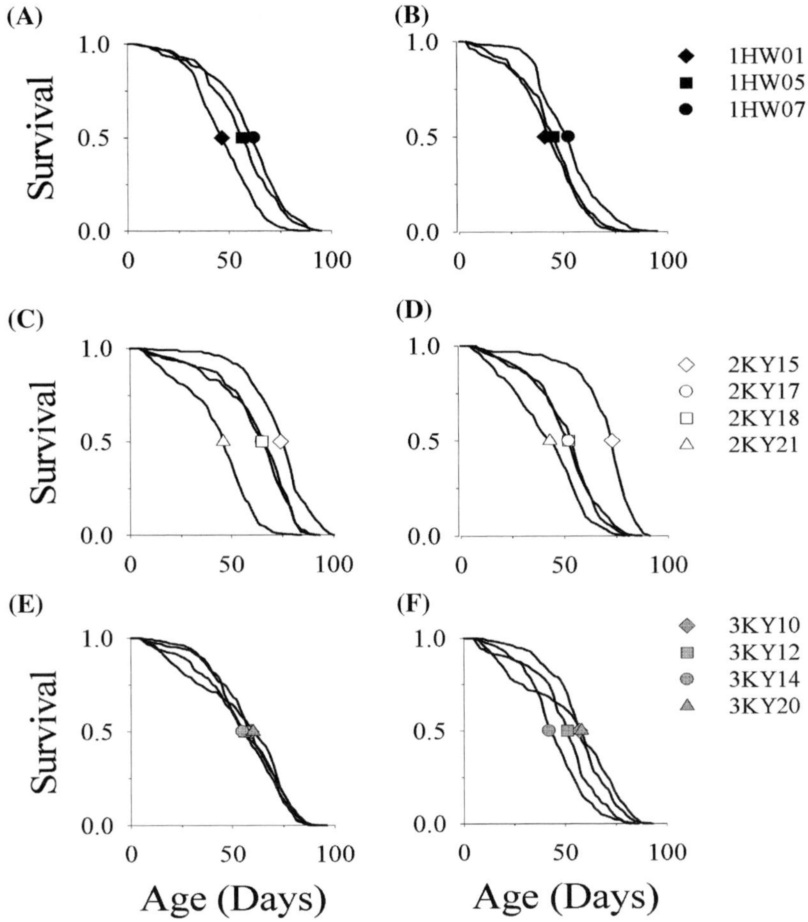

FIGURE 3. Survival of non-virgin male and female *D. simulans* isofemale fly lines from Hawaii (HW) and Kenya (KY). Symbols indicate the point at which 50% of the flies in the study population were still living. **(A)** Male *si*I fly lines. **(B)** Female *si*I fly lines. **(C)** Male *si*II fly lines. **(D)** Female *si*II fly lines. **(E)** Male *si*III fly lines. **(F)** Female *si*III fly lines.

55.9 ± 0.38 days and for females is 51.0 ± 0.33 days. The mean survival of *si*I flies is 54.5 ± 0.36 days for males and 46.8 ± 0.34 days for females. To analyze the mtDNA results further we conducted least-square Tukey post-hoc tests. Tukey's HSD post-hoc test indicates that all mtDNA groups are significantly different at 50% survival ($P < 0.05$). At 10% survival the *si*I group is significantly different from the other two (*si*I *cf si*II, $P < 0.05$; *si*I *cf si*III, $P < 0.05$), but *si*II and *si*III do not differ from each other ($P > 0.1$).

TABLE 1. Life expectancy of *Drosophila simulans* flies from Hawaii and Africa

		Survival (days)			Hydrogen peroxide production (nmol/min/mg protein)		
Line		Mean (d)	SE (d)	Sample size[d]	Mean	SE	Sample size[e]
1HW01[a]	M	47.4	0.55	762	0.62	0.01	4
	F	43.4	0.54	774	0.74	0.08	4
1HW05[a]	M	59.0	0.61	940	0.31	0.03	4
	F	44.3	0.57	871	0.47	0.03	3
1HW07[a]	M	56.2	0.63	763	0.97	0.05	4
	F	53.4	0.57	784	0.95	0.10	4
2KY15[b]	M	73.7	0.96	238	1.36	0.04	4
	F	69.7	0.97	257	1.34	0.14	4
2KY17[b]	M	62.0	0.69	725	1.31	0.04	4
	F	51.3	0.60	762	1.33	0.06	4
2KY18[b]	M	61.9	0.88	469	1.59	0.03	4
	F	50	0.66	582	1.70	0.08	4
2KY21[b]	M	43.6	0.56	886	0.91	0.10	4
	F	41.5	0.67	678	0.88	0.03	4
3KY10[c]	M	58.1	0.72	532	0.81	0.06	4
	F	57.3	0.54	827	0.95	0.02	4
3KY12[c]	M	54.3	0.74	731	0.48	0.04	4
	F	49.5	0.73	583	0.44	0.07	3
3KY14[c]	M	57.1	0.70	690	0.31	0.01	4
	F	44.7	0.50	934	0.34	0.03	4
3KY20[c]	M	54.9	0.85	720	0.64	0.08	4
	F	52.6	0.77	948	0.59	0.04	4

[a]*si*I mtDNA;[b]*si*II mtDNA;[c]*si*III mtDNA.
[d]Sum of four replicates.
[e]Each sample done in duplicate and averaged.

Mitochondrial Genotype, Aging, and Hydrogen Peroxide Production in Drosophila simulans

Non-virgin male and female *D. simulans* were collected about 3 days after eclosion (range 2–4 days) and placed into survival cages until 11 days of age. Food was replaced every 2–3 days. At 11 days of age flies were placed on ice and the heads and abdomens were removed. Mitochondria were isolated from thorax tissue and rates of H_2O_2 production from each line were determined following the method of Katewa and Ballard.[6] All mitochondrial isolation steps were performed on ice or at 4°C. In brief: assays were performed using an Amplex Red H_2O_2/Peroxidase Assay kit (Molecular Probes, Eugene, OR). Standard reactions contained 5–15 μg mitochondrial protein, 10 μM acetylated cytochrome *c*, respiration buffer, complex III–specific respiration substrate *sn*-glycerol 3-phosphate, 30 U/mL superoxide dismutase, and 2 μM rotenone to prevent the back-cycling of electrons to complex I. In

TABLE 2. Nested analyses of variance results showing *F* ratios

	Error *df*	MtDNA $df = 2$	Sex $df = 1$	Line [mtDNA] $df = 8$	Sex × mtDNA $df = 2$
50% survival	74	16.84*	32.26*	15.06*	0.41
10% survival	74	6.16*	33.49*	10.69*	0.78
Hydrogen peroxide	72	309.73*	2.17	38.99*	0.59

$^*P < 0.01$.

all studies, reactions containing known concentrations of H_2O_2 were used to construct a standard curve and run in each assay plate. Controls lacking mitochondrial protein, substrate, or horseradish peroxidase were also included. Specific H_2O_2 production rate was calculated from the initial increase in absorbance at 560 nm measured every 2 min for 1 h using a Molecular Devices, SpectraMax Plus spectrophotometer and SoftMAX Pro software (Molecular Devices Corp., Sunnyvale, CA) at 25°C and is expressed as rate of H_2O_2 production per mg of mitochondrial protein.[6]

Rates of H_2O_2 production were highest in *si*II flies and, in general, increased with mean survival. ANOVA shows that mtDNA type and line nested within mtDNA influences the rate of H_2O_2 production in wild-caught *D. simulans* (TABLES 1 and 2). The *si*II flies had the highest rates of H_2O_2 production (1.29 ± 0.07 and 1.31 ± 0.08 nmol/min/mg protein for males and females, respectively); *si*I flies had intermediate levels (0.63 ± 0.08 for males and 0.74 ± 0.07 for females); and *si*III flies had lowest rates of H_2O_2 production (0.56 ± 0.05 for males and 0.59 ± 0.07 for females). To analyze these results further we regressed H_2O_2 production against the mean survival of each line. Contrary to our expectation, rates of H_2O_2 production increased with mean survival (FIG. 4). To calculate whether the regressions were significantly different from zero we included each H_2O_2 production result, but assumed that the mean survival for each line and sex was estimated without error. When all mitotypes are considered, the slope of the line is significantly different from zero ($T_{84} = 14.95$, $P < 0.001$). The regression is also significantly positive for males ($T_{43} = 11.43$, $P < 0.002$) and females ($T_{40} = 9.32$, $P < 0.004$).

One potential explanation for this result is that higher rates of H_2O_2 production in younger flies lead to an upregulation of antioxidants. Overexpression of Mn-superoxide dismutase has been shown to cause an increase in the lifespan of short-lived *Drosophila*.[34] An alternative or perhaps complementary explanation is that these flies differ in their mitochondrial coupling.[35] Incomplete coupling is mostly due to a natural leak of protons across the mitochondrial inner membrane. In support of this hypothesis Katewa and Ballard[6] observed that 11-day-old males harboring *si*II mtDNA had higher rates of H_2O_2 production by both complexes I and III and higher proton leak from isolated mitochondria than did *si*III males. A second alternative explanation is that the maximal H_2O_2 production rates that we assayed from isolated

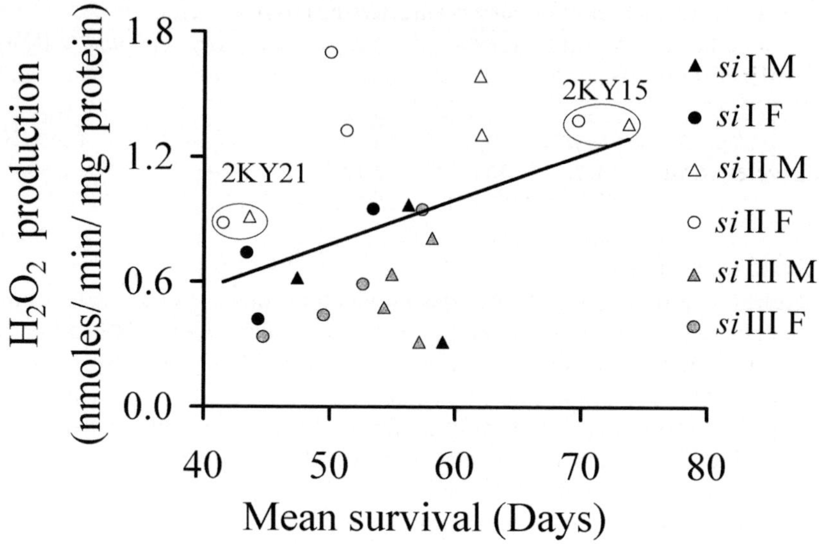

FIGURE 4. Scatterplot showing the relationship between mean survival and rates of hydrogen peroxide (H_2O_2) production of non-virgin male and female *D. simulans* isofemale fly lines harboring *si*I, -II, or -III mitochondrial DNA. Overall, rates of H_2O_2 production increased with mean survival ($y = 0.02x - 0.26$; $R^2 = 0.17$). This increase in H_2O_2 production with survival was true for both males ($y = 0.03x - 0.64$; $R^2 = 0.22$) and for females ($y = 0.03x - 0.40$; $R^2 = 0.22$). The short-lived 2KY21 and long-lived 2KY15 lines from Kenya are shown.

mitochondria are not regularly, if ever, experienced by the whole organism and may be of limited predictive value. Lambert and Merry[36] reported that while isolated mitochondria from liver tissue of calorie-restricted animals display a reduction in the generation of ROS, it was not possible to confirm the same effect in isolated hepatocytes. Use of isolated myocytes or whole thorax with dyes like H_2DCF-DA[37] or DHE[38] could test whether results obtained with isolated mitochondria are representative of H_2O_2 levels in intact cells or tissues.

CONCLUSION

Like humans it has been proposed that *D. simulans* spread out of Africa and now has a cosmopolitan distribution.[39,40] Furthermore, like humans the greatest mtDNA variation occurs in east Africa. A goal of this study was to identify wild-caught *D. simulans* lines with significant differences in longevity and rates of H_2O_2 production so that we can include them in future studies aimed at determining the role that specific mitochondrial variants play in influencing longevity. This study has identified *si*II harboring flies from east

Africa for further study. Flies harboring *si*II mtDNA are globally distributed and diverged at a very similar time to modern humans. Flies with *si*III mtDNA are confined to east Africa, while those harboring *si*I mtDNA are only found on Indo/Pacific islands.

Within the *D. simulans* collected from Kenya, flies harboring *si*II mtDNA had greater variance in mtDNA variation, survival, and rates of H_2O_2 production than those with *si*III mtDNA (FIGS. 1, 3, and 4). Within the lines harboring *si*II mtDNA, 2KY15 and 2KY21 exhibited the largest differences in mean survival and rates of H_2O_2 production (TABLE 1; FIG. 4). It is not known whether the mtDNA, nuclear-encoded gene or mitonuclear interactions cause these differences. The 2KY15 line has a single nonconserved amino acid difference in the ATPase 6 locus, whereas 2KY21 has no unique changes in the six mtDNA genes analyzed to date. No unique amino acid substitutions have been found in 13 nuclear-encoded genes of either line, but nuclear amino acid differences do exist. Line 2KY15 has two substitutions in COX7A that are not present in 2KY21, Ile28→Leu and Thr45→Ala, neither of which we predict will affect enzyme activity. It is notable that the long-lived 2KY15 line has lower cytochrome *c* oxidase activity than any other *si*II line tested to date, while the short-lived 2KY21 line has the highest cytochrome *c* oxidase activity in the haplogroup.[32] Currently, the significance of this result is not known. It is, however, in the opposite direction to the simple prediction that longer survival times would negatively correlate with rate of H_2O_2 production and positively correlate with greater cytochrome *c* oxidase activity from complex IV.

Flies collected from Hawaii had the shortest survival. One possibility is that the *si*I mtDNA type caused the low survival of these flies. James and Ballard[41] found that *si*I-harboring flies had shortest survival in both wild-type and introgression fly lines. The *si*I mtDNA diverged from *si*II and *si*III mtDNA about 1.4 my ago and has accumulated 27 fixed amino acid differences.[25] This large number of fixed differences has the potential to allow us to detect broad-brush differences between mitotypes when genetically introgressed into standard nuclear backgrounds, but will not facilitate the determination of the influence of specific amino acid mutations. An alternative possibility for the low survival of flies from Hawaii is that nuclear-encoded differences in Hawaiian flies caused the low survival. Hawaiian flies tended to have a unique set of polymorphisms, possibly because they are island endemics that have likely been subject to a series of population bottlenecks. In this study we identified a heterozygous two-amino acid deletion in the nuclear-encoded COX7A locus of the HW01 line. Construction of lines homozygous for each allele will enable us to determine the influence of this deletion on cytochrome *c* oxidase activity, rates of H_2O_2 production, and longevity. An alternate possibility for the low survival of *si*I-harboring flies is that the *w*Ha *Wolbachia*-infection increased mortality rates. The symbiont has been shown to influence survival in some but not all *Drosophila*.[42] Future studies will use tetracycline to clear flies of *Wolbachia* infection.[43]

Drosophila is a particularly useful model organism to study healthy aging and longevity. Here we advocate the parallel use of *D. simulans*, where survival in days, very approximately, approaches human survival in years. In *D. melanogoster,* Maklakov and colleagues[44] hypothesized that variation in mtDNA contributes to variation in female mortality and aging.

ACKNOWLEDGMENTS

We thank David Clancy, Mark Tanaka, and Ian Dawes for discussions. Ines Ricafuente helped with sequencing and maintaining the flies. This work was supported by NSF Grant DEB-0444766 and NIH Grant R01 GM067862-01.

REFERENCES

1. HARMAN, D. 1956. Aging: a theory based on free radical radiation chemistry. J. Gerentol. **11:** 298–300.
2. BARJA, G. 2002. Endogenous oxidative stress: relationship to aging, longevity and caloric restriction. Ageing Res. Rev. **1:** 397–411.
3. SOHAL, R.S. 2002. Oxidative stress hypothesis of aging. Free Radic. Biol. Med. **33:** 573–574.
4. SOHAL, R.S. 2002. Mechanisms of aging: an appraisal of the oxidative stress hypothesis. Free Radic. Biol. Med. **33:** 575–586.
5. MIWA, S. *et al.* 2003. Superoxide and hydrogen peroxide production by *Drosophila* mitochondria. Free Radic. Biol. Med. **35:** 938–948.
6. KATEWA, S.D. & J.W.O. BALLARD. 2007. Sympatric *Drosophila simulans* flies with distinct mtDNA show difference in mitochondrial respiration and electron transport. Insect Biochem. Mol. Biol. **37:** 213–222.
7. TAYLOR, M.J. & A. HOERAUF. 1999. *Wolbachia* bacteria of filarial nematodes. Parasitol. Today **15:** 437–442.
8. HARMAN, A.M. 1992. Free radical theory of aging. Mut. Res. **275:** 257–266.
9. SASTRE, J., F.V. PALLARDO & J. VINA. 2000. Mitochondrial oxidative stress plays a key role in aging and apoptosis. IUBMB Life **49:** 427–435.
10. SASTRE, J. *et al.* 1996. Aging of the liver: age-associated mitochondrial damage in intact hepatocytes. Hepatology **24:** 1199–1205.
11. FINSTERER, J. 2004. Mitochondriopathies. Eur. J. Neurol. **11:** 163–186.
12. DE BENEDICTIS, G. *et al.* 2000. Inherited variability of the mitochondrial genome and successful aging in humans. Ann. N. Y. Acad. Sci. **908:** 208–218.
13. ROSE, G. *et al.* 2001. Paradoxes in longevity: sequence analysis of mtDNA haplogroup J in centenarians. Eur. J. Hum. Genet. **9:** 701–707.
14. PARTRIDGE, L., W. PIPER & W. MAIR. 2005. Dietary restriction in *Drosophila*. Mech. Ageing Devel. **126:** 938–950.
15. TATAR, M. 2007. Diet restriction in *Drosophila melanogaster*: design and analysis. Interdiscip. Top. Gerontol. **35:** 115–136.
16. JAFARI, M. *et al.* 2006. The pharmacology of ageing in *Drosophila*. Curr. Drug Targets **7:** 1479–1483.

17. CLARY, D.O. & D.R. WOLSTENHOLME. 1985. The mitochondrial DNA molecular of *Drosophila yakuba*: nucleotide sequence, gene organization, and genetic code. J. Mol. Evol. **22:** 252–271.
18. PARTRIDGE, L. & D. GEMS. 2006. Beyond the evolutionary theory of ageing, from functional genomics to evo-gero. Trends Ecol. Evol. **21:** 334–340.
19. MERRY, B.J. 2000. Calorie restriction and age-related oxidative stres. Ann. N. Y. Acad. Sci. **908:** 180–198.
20. SINCLAIR, D.A. 2005. Toward a unified theory of caloric restriction and longevity regulation. Mech. Ageing Devel. **126:** 987–1002.
21. ITO, K. & Y. HOTTA. 1992. Proliferation pattern of postembryonic neuroblasts in the brain of *Drosophila melanogaster*. Dev. Biol. **149:** 134–148.
22. BALLARD, J.W.O. & M. KREITMAN. 1994. Unraveling selection in the mitochondrial genome of *Drosophila*. Genetics **138:** 757–772.
23. MAYNARD SMITH, J. & J. HAIGH. 1974. The hitchhiking effect of a favourable gene. Genet. Res. **23:** 23–35.
24. KAPLAN, N.L., R.R. HUDSON & C.H. LANGLEY. 1989. The "hitchhiking effect" revisited. Genetics **123:** 887–899.
25. BALLARD, J.W.O. 2000. Comparative genomics of mitochondrial DNA in *Drosophila simulans*. J. Mol. Evol. **51:** 64–75.
26. BALLARD, J.W.O. 2000. Comparative genomics of mitochondrial DNA in members of the *Drosophila melanogaster* subgroup. J. Mol. Evol. **51:** 48–63.
27. HORAI, S. *et al.* 1995. Recent African origin of modern humans revealed by complete sequences of hominoid mitochondrial DNAs. Proc. Natl. Acad. Sci. USA **92:** 532–536.
28. KRINGS, M. *et al.* 1999. DNA sequence of the mitochondrial hypervariable region II from the neandertal type specimen. Proc. Natl. Acad. Sci. USA **96:** 5581–5585.
29. JAMES, A.C. & J.W.O. BALLARD. 2000. Expression of cytoplasmic incompatibility in *Drosophila simulans* and its impact on infection frequencies and distribution of *Wolbachia pipientis*. Evolution **54:** 1661–1672.
30. HERCUS, M.J. & A.A. HOFFMANN. 2000. Maternal and grandmaternal age influence offspring fitness in *Drosophila*. Proc. R. Soc. Lond. Biol. Sci. **267:** 2105–2110.
31. CLANCY, D.J. & W.J. KENNINGTON. 2001. A simple method to achieve consistent larval density in culture bottles. Dros. Inf. Serv. **84:** 168–169.
32. BALLARD, J.W.O. *et al.* 2007. Mitochondrial DNA variation is associated with measurable differences in life history traits and mitochondrial metabolism in *Drosophila simulans*. Evolution **61:** 1735–1747.
33. VILLANI, G. *et al.* 1998. Low reserve of cytochrome c oxidase capacity in vivo in the respiratory chain of a variety of human cell types. J. Biol. Chem. **273:** 31829–31836.
34. SUN, J. *et al.* 2002. Induced overexpression of mitochondrial Mn-superoxide dismutase extends the life span of adult *Drosophila melanogaster*. Genetics **161:** 661–672.
35. BRAND, M.D. 2000. Uncouplng to survive? The role of mitochondrial inefficiency in ageing. Exp. Gerontol. **35:** 811–820.
36. LAMBERT, A.J. & B.J. MERRY. 2005. Lack of effect of caloric restriction on bioenergetics and reactive oxygen species production in intact rat hepatocytes. J. Gerontol.: Biol. Sci. **60:** 175–180.
37. PARIHAR, M.S. & G.J. BREWER. 2007. Simultaneous age-related depolarization of mitochondrial membrane potential and increased mitochondrial reactive oxygen

species production correlate with age-related glutamate excitotoxicity in rat hippocampal neurons. J. Neurosci. Res. **85:** 1018–1032.
38. KEVIN, L.G. *et al.* 2003. Ischemic preconditioning alters real-time measure of O2 radicals in intact hearts with ischemia and reperfusion. Am. J. Physiol. Heart Circ. Physiol. **284:** H566–H574.
39. TEMPLETON, A. 2002. Out of Africa again and again. Nature **416:** 45–51.
40. BALLARD, J.W.O. 2004. Sequential evolution of a symbiont inferred from the host: Wolbachia and *Drosophila simulans*. Mol. Biol. Evol. **21:** 428–442.
41. JAMES, A.C. & J.W.O. BALLARD. 2003. Mitochondrial genotype affects fitness in *Drosophila simulans*. Genetics **164:** 173–186.
42. FRY, A.J., M.R. PALMER & D.M. RAND. 2004. Variable fitness effects of *Wolbachia* infection in *Drosophila melanogaster*. Heredity **93:** 379–389.
43. BALLARD, J.W.O. & R.G. MELVIN. 2007. Tetracycline treatment influences mitochondrial metabolism and mtDNA density two generations after treatment in *Drosophila*. Insect Mol. Biol. In press.
44. MAKLAKOV, A.A. *et al.* 2006. Within-population variation in cytoplasmic genes affects female life span and aging in *Drosophila melanogaster*. Evolution **60:** 2081–2086.
45. BALLARD, J.W.O. 2000. When one is not enough: introgression of mitochondrial DNA in *Drosophila*. Mol. Biol. Evol. **17:** 1126–1130.
46. MORIYAMA, E.N. & J.R. POWELL. 1997. Synonymous substitution rates in *Drosophila*: mitochondrial versus nuclear genes. J. Mol. Evol. **45:** 378–391.

Fit and Well at Eighty

Defying the Stereotypes of Age and Illness

JULIE E. BYLES*

Research Centre for Gender, Health and Aging (RCGHA), The University of Newcastle, Newcastle, NSW, Australia

ABSTRACT: **While aging is associated with physical decline and increased risk of illness, older age is not inevitably a time of ill-being. Data from the Australian Longitudinal Study on Women's Health challenge negative stereotypes of aging and illness. While an accelerating decline in average physical health was observed over the first 6 years of the study, an important and large proportion of the women experienced minimal change in their physical health during this period. Also, while chronic disease was a strong risk factor for declining health, many women aged well in spite of longstanding medical conditions. This paper presents trends in health and illness as women age and explores some of the many physical, social, and healthcare factors that mark out those women who remain "fit and well."**

KEYWORDS: **aging; quality of life; longitudinal studies**

AGING, ILLNESS, AND QUALITY OF LIFE

As people age they have increased risk of disease and disability. Busse[1] described aging as an accumulation of the impact of time, environmental factors, and disease. These factors are not independent. Changes of aging can also be considered in terms of physiology, disease, and social changes. Physiological changes of aging affect all body systems, and can limit reserve, reduce ability to maintain homeostasis, and increase potential for illness.[2] Consequently, older people have a higher incidence and prevalence of many conditions. These include conditions that are rapidly fatal (such as lung cancer) and which contribute greatly to loss of years of life, and those that are chronic and associated with loss of quality of life.

Address for correspondence: Professor Julie Byles, B.Med. Ph.D., AAG Fellow, Director, Research Centre for Gender, Health and Aging (RCGHA), The University of Newcastle, Room 266, DMB, Royal Newcastle Hospital, Newcastle, NSW 2300, Australia. Voice: +61 2 4913 8643; fax: +61 2 4913 8323.
Julie.Byles@newcastle.edu.au

*Professor Byles is a member of the Hunter Medical Research Institute.

Ann. N.Y. Acad. Sci. 1114: 107–120 (2007). © 2007 New York Academy of Sciences.
doi: 10.1196/annals.1396.027

Ischemic heart disease and cerebrovascular disease are leading causes of disability and death among people aged 60 years and over. Other major causes of disability and death in this age group include chronic obstructive lung disease, dementias, vision disorders, and cancer. Older people are also particularly vulnerable to respiratory infections, such as pneumonia. Hearing loss is a major cause of disability that is highly prevalent and has significant impact on quality of life. Likewise, musculoskeletal conditions are common among older people and contribute to disability.[3] These common conditions are associated with a range of functional limitations, such as cognitive impairment, restrictions on mobility, and urinary incontinence.

In addition, older adults have many active health problems that interact with each other and complicate approaches to treatments. In one study of men and women aged 70 years and over, the median number of conditions per person was 7.[4] This multi-morbidity has been shown to be associated with health outcomes, such as quality of life, activities of daily living, health service utilization, and mortality.

Social circumstances add to the difficulties that people face as they age and social inequity is one of the major determinants of ill health at all ages.[5] Poor circumstances have their first impact during pregnancy, when deficiencies in nutrition, maternal stress, smoking and misuse of drugs and alcohol, insufficient exercise, and inadequate prenatal care can inhibit optimal fetal development. The impact of disadvantage continues throughout childhood and adult life and is manifest in terms of psychosocial stress, poor education, poor nutrition, social exclusion, underemployment and poor job security, and work-related stress and injury. Smoking and alcohol and drug use are also higher among people in disadvantaged circumstances. Disadvantaged people also tend to live in poorer neighborhoods with high rates of violence and crime, in overcrowded circumstances, and potentially in areas with high levels of environmental toxins.

Social support is another social factor that affects well-being at older age. A number of studies have demonstrated association between social support and improved survival, reduced disability and health care use, and improved quality of life.[6–9] It is thought that these health effects are mediated through the influence of social support on health behavior, improved access to care, and possibly through effects in physiological pathways.

MEASURING HEALTH IN OLDER AGE

Early research into aging has focused on survival and longevity. More recently great attention has been given to the quantification of quality of life as a measure of general health status. Quality of life in older age is affected not only by disease, but also by symptoms that may not be associated with specific pathology, and by social circumstances. Common symptoms include joint stiffness, constipation, incontinence, and memory complaints. However, despite the presence of symptoms, medical conditions, and disability, older people often report their health as good, very good or excellent, and record high quality-of-life scores. Likewise, while disease is a common but not inevitable consequence of longevity, disease does not always equal "poor health."

LONGITUDINAL STUDIES OF HEALTH AND AGING

In the last century serious old age has become an increasingly expected and predictable part of life and the desire to understand and influence the process and effects of aging has escalated exponentially. The National Institutes of Health database of longitudinal studies includes 54 studies that they consider represent "research initiatives for identifying the physiological and other types of factors across the lifespan, affecting onset and progression of disease with advancing age, as well as elucidation of protective factors contributing to exceptionally healthy aging."[10] One reason for conducting longitudinal studies is to understand trajectories of aging—with and without disability and disease.[11] The aim is to be able to identify the risk of disability among older people, and to understand the impact and natural history of disease in later life.

THE AUSTRALIAN LONGITUDINAL STUDY ON WOMEN'S HEALTH: EVIDENCE TO DEFY THE STEREOTYPES OF AGING

The Australian Longitudinal Study of Women's Health (ALSWH) is a population-based study of changes in the health of a national sample of women in three age cohorts (aged 18–23, 45–50, and 70–75 years in 1996). The study was designed to investigate multiple factors affecting health and well being of women over a 20-year period. Participants in the study were randomly selected from the national Medicare database, which contains name and address details of all Australian citizens and permanent residents. This database is widely regarded as the most up-to-date and complete list of Australians in existence. Further details of the study have been published elsewhere.[12]

The women in the older cohort were aged 70–75 years at the time of the first postal survey in 1996 and have since been invited to complete three follow-up surveys: Survey 2 in 1999 (S2), Survey 3 in 2002 (S3), and Survey 4 in 2005 (when the women were aged 79–84 years). Each survey included many questions about the women's health and lifestyle. Withdrawals and reasons for withdrawal are recorded, and deaths are checked through the National Death Index.

The 1996 baseline data provided by the older cohort provided a *"picture of aging that challenges negative stereotypes of frailty and declining well being"* (p. 56).[13] At the time of that survey the women were aged 70–75 and these women are now into their 80s. They have provided longitudinal data on a wide range of physical, psychological, social, and health service usage variables which allow a comprehensive view of how aging has affected this large cohort of women. This paper presents the magnitude and variation in change in physical function for the women who continued to participate in the first 6-years of the study (S1–S3).

MAIN MEASURES OF HEALTH AND FUNCTION

The Medical Outcomes Study 36-item Short Form Health Survey (SF-36)[14,15] is a generic profile measure which examines self-reported health-related quality of life. The SF-36 is a widely used and well-validated health profile that has been extensively reviewed for use with older populations. A recent structured review of generic self-assessed instruments for older people identified the SF-36 as one of three instruments with extensive evidence of internal consistency, test–retest reliability, construct validity, concurrent validity, and responsiveness.[16] The SF-36 is recommended where a detailed and broad-ranging assessment of health is required, particularly in community-dwelling older people. The SF-36 produces eight subscales and two summary scores (Physical Components Summary score [PCS] and Mental Health Components Score). A low score on the PCS reflects limitations in physical, social, and role activities, severe bodily pain, frequent tiredness and "poor" self-rated health.

Other measures of health included in the surveys are:

- Self-reported diagnoses (arthritis, diabetes, heart disease, hypertension, stroke, thrombosis, low iron level, asthma, bronchitis/emphysema, osteoporosis, cancer, depression, anxiety, Alzheimer's disease/dementia);
- Falls and injuries;
- Days spent in hospital in past 12 months;
- Satisfaction with physical ability (measured using an item taken from the SF-36 Physical Functioning measure[17];
- Self-reported symptoms (breathing difficulty, chest pain, tiredness, stiff or painful joints, back pain, foot problems, dysuria, nocturia, urinary urgency, incontinence, constipation, poor memory, clumsiness, dizziness);
- Vision (difficulty reading a newspaper, even with glasses);
- Hearing (difficulty hearing a conversation);
- Items from the Centre for Epidemiological Studies Depression scale[18];
- Physical activity;
- Body mass index (BMI);
- Nutrition risk[19];
- Stress about health and living arrangements and relationships;
- Items from the abbreviated Duke's Social Support Index[20];

- Ability to use public transport;
- Availability of instrumental support[21];
- Attitudes to life including optimism and hardiness.[22]

Full details of all measures are available at www.alswh.org.au.

SIX-YEAR CHANGE IN HEALTH AND FUNCTION

Survey 1 (S1) included 12,432 women, who were aged 70–75 years at that time. There were 10,434 women who remained in the study at S2, and 8647 women at S3 (69.6% of the original cohort). Between S1 and S3, 1097 women died (9%) and a further 363 withdrew because they had become too frail to complete the survey; a further 1487 (9.5%) withdrew for other reasons, and 838 could not be contacted or otherwise did not complete S3. A total of 8397 women completed S1, S2, and S3.

The most common symptoms reported by women at all surveys were stiff and painful joints (reported by 67% of women at S1), back pain (55%), difficulty sleeping (50%), eyesight problems (49%), poor memory (41%), tiredness (41%), indigestion (39%), allergies (35%), headaches (32%), and constipation (28%). There were few differences between reported prevalence of symptoms at S1 and S3. There were small percentage increases in the prevalence of difficulties with hearing (increasing from 11% at S1 to 13% at S3) and vision (14% at S1, 17% at S3).[23] Prevalence of conditions increased with age. For instance, the prevalence of heart disease was 17% at S1 and increased to 24% by S3. Likewise, 9% of the women reported having diabetes at S1 and this proportion increased to 12% at S3. Arthritis was reported by 41% of the women at S2 and 60% at S3.[24]

The period prevalence of falls did not change, with around 18% falling within the past year at each survey. At S1, 41% of the women were taking three or more medications. This percentage increased to 51% at S2, and 60% at S3. Few older women smoked (4.7% at S2) or drank more than two standard alcoholic drinks per day[25] (4.1% at S2). At S2 around 30% of women were taking no exercise compared to about 40% at S3; the percentage of women undertaking moderate exercise every day of the week was 40% at S2 and 34% at S3.[23]

The percentage regularly needing help with daily tasks increased from 5% at S1 to 13% at S3. The percentage *providing* regular care or assistance to another person increased from 19% at S1 to 24% at S3.[23] SF-36 scores for women who had completed S1, S2, and S3 are presented in TABLE 1. Scores declined for most subscales, and declined more steeply between S2 and S3 than between S1 and S2. For women who answered all three surveys, mean mental health summary scores were high (51.3, 51.8 and 51.4 at S1, S2, and S3, respectively). However, mean physical health summary scores declined from 51.1 at S1 to

TABLE 1. Physical and mental health-related subscales of the SF-36 for the same 8397 women at three different ages[a]

	Mean (95% Confidence Interval)							
Age at each survey	Physical functioning	Physical role limitations	Bodily pain	General health	Vitality	Social functioning	Emotional role limitations	Mental health
70–75 years	66.9 (66.4; 67.5)	62.4 (61.5; 63.3)	67.4 (66.8; 67.9)	68.6 (68.2; 69.0)	62.5 (62.0; 62.9)	84.5 (84.0; 85.0)	80.0 (79.2; 80.7)	78.8 (78.4; 79.1)
73–78 years	64.4 (63.9; 65.0)	58.0 (57.1; 58.9)	66.1 (65.5; 66.7)	68.1 (67.6; 68.5)	59.8 (59.4; 60.3)	84.3 (83.8; 84.8)	80.4 (79.7; 81.1)	79.6 (79.2; 79.9)
76–81 years	58.4 (57.8; 59.1)	50.2 (49.3; 51.1)	62.5 (62.0; 63.1)	65.9 (65.5; 66.4)	56.7 (56.2; 57.1)	78.7 (78.1; 79.3)	76.6 (75.8; 77.4)	78.7 (78.3; 79.0)

[a]Weighted to allow for oversampling of women living in rural and remote areas. Numbers vary on account of missing data (0.3%–7.9%). A higher score indicates better functioning.

Source: Byles J.,J. Powers, C. Chojenta & P. Warner Smith. 2006. Older Australian women: aging in urban, rural, and remote environments. Australas. J. Aging **25:** 151–157.

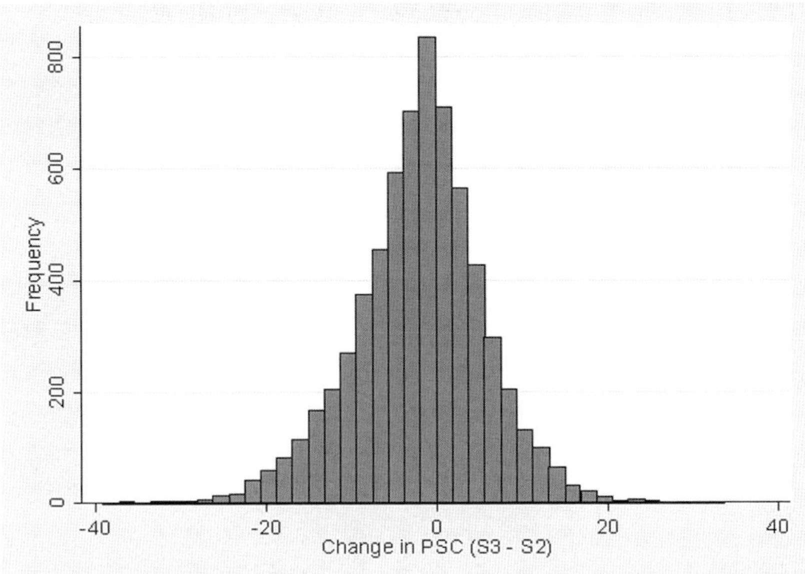

FIGURE 1. Change in physical functional health (PCS) score from Survey 2 to Survey 3.

50.0 at S2 and 48.0 at S3.[23] Variation in the change in PCS between S2 and S3 is shown in FIGURE 1. Further analysis of the variation in change in SF-36 subscale scores reveals three main effects. Depending on the subscale, around 20% of the women had low scores at all three surveys, around 70% had high scores at all three surveys, and 10% had high scores at S1 and low scores at S3 (unpublished data).

FACTORS ASSOCIATED WITH SURVIVAL

The probability that a woman died between S1 and S2 was strongly associated with living in a rural area, having poor self-rated health, needing medications for chronic illness, and smoking. Being married was protective.[26] The number of chronic conditions reported by the women was also associated with survival (unpublished data). Women who never or rarely drank alcohol were also more likely to die than women who have one or two drinks each day.[27] Women with higher socioeconomic status (SES) had lower death rates than women in the low-SES group (relative risk: 0.79, 95% confidence interval 0.64–0.98).[28] Another 22% of the original cohort did not die but either withdrew or otherwise failed to provide data for S3. Analysis of reasons for this attrition indicate that compared to women who stayed in the study, these women had poorer health at S1, were more likely to be taking medication for

TABLE 2. Health characteristics at Survey 2 for women who maintained high PCS scores, women with high scores at S2 and low scores at S3 (declining scores), and women with low score at S2

	Maintained high PCS score at S2-S3	Declining PCS scores at S2-S3	Low PCS scores at S2
Age: mean (SD)	75.4 (1.5)	75.5 (1.5)	75.6 (1.5)
No. of diagnoses (in past 12 months)	1.3 (1.2)	2.0 (1.4)	2.5 (1.6)
No. of symptoms (in past 12 months)	8.7 (7.4)	14.0 (8.6)	17.7 (10.0)
Body mass index mean (SD)	24.8 (4.1)	25.8 (5.1)	26.7 (5.1)
Prescription medications (% taking five or more in past 4 weeks)	12%	29%	46%
No. of GP visits (% >12 times in past 12 months)	7%	19%	30%
Specialist visits (% in past 12 months)	41%	56%	67%
Hospital admission (in past 12 months) (% yes)	21%	31%	45%
Been injured from a fall (% yes)	10%	13%	19%

All comparisons shown in table are statistically significant, $P < 0.001$.

chronic illness, and more likely to be current smokers. They were also more likely to be born overseas, living in a rural area, have difficulty managing on their income, and have poor social support. They were less likely to be married.[26]

FACTORS ASSOCIATED WITH MAINTAINING HIGH LEVELS OF PHYSICAL HEALTH–RELATED QUALITY OF LIFE

We classified women who survived and completed S1, S2, and S3 according to whether their S2 PCS score was above the 1st quintile (PCS < 40.0) or below (1665 women, 24%). Women with high scores at S2 were further classified as maintaining high scores (4803 women (68%), and declining scores between S2 and S3 (557 women, 8%). A large number of variables were significantly associated with whether women who survived and stayed in the study had high PCS scores at S2 and whether they maintained these high scores at S3. Main health factors are shown in TABLE 2. Those women who maintained high PCS scores had fewer conditions and fewer symptoms at S2. Significant differences between groups were also evident for falls, BMI, satisfaction with

TABLE 3. Probability of having and maintaining high PCS scores for women with and without specific conditions at Survey 2

	Condition prevalent at S2	
	No	Yes
Arthritis	80%	54%
Heart disease	72%	45%
Hypertension	72%	62%
Diabetes	70%	51%
Stroke	69%	41%
Low iron level	70%	41%
Asthma	70%	50%
Bronchitis/emphysema	70%	47%
Osteoporosis	71%	51%
Depression	70%	51%
Anxiety/nervousness	69%	54%

physical ability, ability to use public transport, vision and hearing, stress about health, living arrangements and relationships, some items from the abbreviated Duke Social Support Index, all items on instrumental support, optimism and hardiness, nutrition risk, and all items from CES-D.[29]

The most common conditions reported by the women and the probability that women with these conditions will have high PCS scores at S2 and maintain these at S3 are shown in TABLE 3. Women surviving with these conditions are less likely to have and maintain high PCS scores than those without these conditions, although many maintained higher levels of PCS over the 3 years. Other analyses compared PCS scores for women according to whether they reported a chronic condition at S1 (existing case), reported the condition at S2 or S3, or never reported the condition. Trends in PCS for women defined according to their reporting of arthritis and heart disease are shown in FIGURE 2. For both conditions, women who never report the condition have highest PCS scores at all time points. Women classified as existing cases had lowest PCS scores at S1, but women classified as new cases had similarly low PCS scores at S3.

HEALTH PROMOTION FOR OLDER AGE

An understanding of factors associated with successful aging leads to development of opportunities to promote successful aging. Longitudinal studies point to the importance of avoiding risk factors, such as smoking, getting good nutrition, physical activity, social interaction, being productive and engaged, and maintaining a positive outlook. Strategies to promote optimal aging address the need to operate across the life-course to increase redundancy and

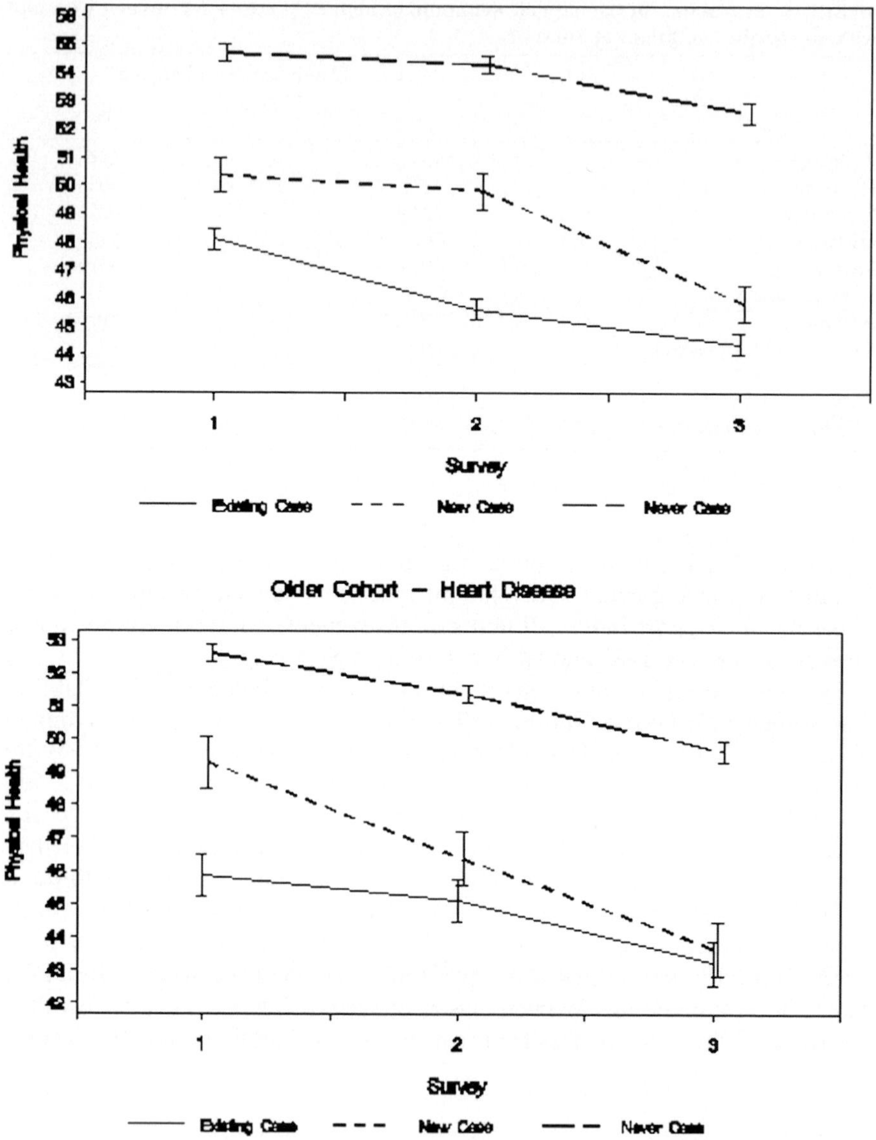

FIGURE 2. Trends in PCS score for Survey 1, Survey 2, and Survey 3 by arthritis and heart disease status.

durability, avoid damage, and improve maintenance and repair at cellular and structural levels, and to enable the individual to adapt and compensate for losses when they occur.

The avoidance of chronic disease appears to be an important factor in ensuring survival and good health in older age. Many risk factors have been shown to have clear and strong association with chronic disease. For example, smoking is a risk factor for cardiovascular disease, stroke, respiratory disease, and many kinds of cancer, and smoking prevention is important in reducing the incidence of these conditions.[30] Physical activity is strongly associated with lower rates of cardiovascular disease, depression, and osteoporosis in many observational studies. Obesity and overweight are associated with increased risk of stroke, coronary artery disease, diabetes, arthritis, and some forms of cancer.[31] Observational studies suggest a clear association between nutrition and health at older ages.[32]

An analysis of major risk factors in ALSWH showed strong associations between lower levels of education and greater risk of chronic disease in later life. Overweight and obesity were strongly associated with prevalence and incidence of hypertension, heart disease, diabetes, asthma, and with the prevalence of arthritis, and smoking was significantly associated with heart disease, diabetes, asthma, and arthritis. A low level of physical activity was associated with hypertension, heart disease, diabetes, osteoporosis, arthritis, and asthma.[24]

These factors are the focus of health-promotion campaigns for young people and older adults, with the implication that they are within the control of the individual. However, other factors that are outside of the individual's control are also important and these include genetic factors and socioeconomic background. Among older women in the ALSWH, lower education was significantly associated with higher prevalence of hypertension, and with prevalence and incidence of diabetes.[24] Socioeconomic factors may also influence people's engagement in healthy and unhealthy behaviors throughout life. There is a complex interplay between factors that might be within an individual's power to change, factors that require change at a systems level, and factors that are difficult to change. Thus, while it is useful to highlight behaviors of individuals as risk factors for ill health at later ages, addressing these problems requires that the social circumstances that underpin these behaviors are also accounted for.

Prevention of some conditions would result in a relative increase in the ratio of healthy life expectancy to life expectancy. This scenario is predicted for elimination of endocrine, nutritional, and metabolic disorders, respiratory and digestive disorders in men, neoplasms in women, and musculoskeletal disorders and mental conditions.[33] The population effects of preventing a condition such as arthritis would be considerable, since so many individuals are affected. As well as the direct effects of pain and stiffness, arthritis has indirect effects by limiting physical activity and thereby increasing risk of other conditions. Boult *et al.*[34] modeled the potential impact of a 0.5% per-annum decrease in prevalence of six chronic conditions, against increasing numbers of older persons. In this model the greatest reductions in the number of people with functional limitations came from decreased prevalence of arthritis, with smaller reductions from decreases in prevalence of mental confusion, diabetes, and stroke.

However, even if half of the incidence of chronic disease could be prevented, population aging means that these conditions will become increasingly prevalent. On current evidence, even with the most effective prevention campaigns people still develop chronic conditions in older age. However, much can be done to reduce the impact of these conditions on quality of life. Many older people who live with chronic disease manage well, and consider themselves to be healthy long after diagnosis. The role of research into healthy aging is not only to find ways to prevent disease, but also to find ways to ensure the best possible health for people aging with chronic disease.

FUTURE RESEARCH

The ALSWH continues to collect and analyze data on the three cohorts of women. Plans are currently in place to undertake Survey 5 of the older women in 2008, when the women will be aged 82–87 years. The study is planned to continue for at least another 10 years until the women are 90–95 years old. As well as being able to describe the changing health of these women, the study will also provide data to determine whether successive cohorts of women are likely to be healthier (or not) as they reach older ages. The need for longitudinal studies on aging increases as the population ages. The environment in which we grow and age is constantly changing and we cannot only examine "old" data to tell us about our own futures.

ACKNOWLEDGMENTS

The Australian Longitudinal Study on Women's Health was developed by groups of interdisciplinary researchers at the Universities of Newcastle and Queensland, and is funded by the Australian Department of Health and Aging. We thank all participants for their valuable contribution to this project.

REFERENCES

1. BUSSE, E.W. 1969. Theories of aging. *In* Behaviour and Adaptation in Later Life. E.W. Busse & E. Pfeiffer, Eds.: 11–32. Little Brown. Boston, MA.
2. EBRAHIM, S. 2002. Health of elderly people. *In* Oxford Textbook of Public Health, 4th ed. R. Detels, J. McEwan, R. Beaglehole, & H. Tanaka, Eds.: Vol. 3, Ch. 11.8.: 1712. Oxford University Press. Oxford, UK.
3. WORLD HEALTH ORGANISATION. The World Health Report 2003: Shaping the Future. Geveva.
4. BYLES, J.E. *et al.* 2005. Single index of multimorbidity did not predict multiple outcomes. J. Clin. Epidemiol. **58:** 997–1005.
5. MARMOT, M. 2005. Social determinants of health inequalities. Lancet **365:** 1099–1104.

6. BERKMAN, L.F & S.L. SYME. 1979. Social networks, host resistance and mortality: a nine year follow-up study of Alameda County residents. Am. J. Epidemiol. **109:** 186–204.
7. BOWLING, A. & P.D. BROWNE. 1991. Social networks, health, and emotional well-being among the oldest old in London. J. Gerontol. **46:** S20–32.
8. STEINBACH, U. 1992. Social networks, institutionalization, and mortality among elderly people in the United States. J. Gerontol. **47:** 5183–5190.
9. MENDES DE LEON, C.P. *et al.* 1999. Social networks and disability transitions across eight intervals of yearly data in the New Haven EPESE. J. Gerontol. **54:** S62–S72.
10. NATIONAL INSTITUTES OF HEALTH. National Institute of Aging Database of Longitudinal Studies. http://www.nia.nih.gov/ResearchInformation/ScientificResources/LongitudinalStudies.htm. Last accessed April 1, 2007.
11. FOZARD, J.L., J. METTER & L.J. BRANT. 1990. Next steps in describing aging and disease in longitudinal studies. J. Gerontol. **45:** 116–127.
12. LEE, C. *et al.* 2005. Cohort profile: the Australian Longitudinal Study on Women's Health. Int. J. Epidemiol. **34:** 987–991.
13. BYLES, J.E. 1999. Over the hill and picking up speed: A profile of older women of the Australian Longitudinal Study on Women's Health. Australas. J. Ageing **18**(3 Suppl): 55–62.
14. WARE, J.E. & C.D. SHERBOURNE. 1992. The MOS 36-item Short Form Health Survey (SF-36): I. Conceptual framework and item selection. Med. Care **30:** 473–483.
15. WARE, J.E., M. KOSINSKI & S.K. KELLER. 1994. SF-36® Physical and Mental Health Summary Scales: A User's Manual. The Health Institute. Boston, MA.
16. HAYWOOD, K.L., A.M. GARRAT & R. FITZPATRICK. 2005. Quality of life in older people: a structured review of generic self-assessed health instruments. Qual. Life Res. **14:** 1651–1668.
17. STEWART, A.L. & J.E. WARE. 1992. Measuring Functioning and Well Being: the Medical Outcomes Study Approach: 375–376. Duke University Press. Durham, NC.
18. SCHEIN, R.L. & H.G. KOENIG. 1997. The Center for Epidemiological Studies Depression (CES-D) Scale: assessment of depression in the medically ill elderly. Int. J. Geriatr. Psych. **12**: 436–446.
19. LIPSKI, P.S. 1996. Australian nutrition screening initiative. Aust. J. Ageing **15:** 14–16.
20. GOODGER, B. *et al.* 1999. Assessment of a short scale to measure social support among older people. Aust. N. Z. J. Public Health **23:** 260–265.
21. SUURMEIJER, T.P.B.M. *et al.* 1995. The measurement of social support in the European research on incapacitating diseases and social support: the development of the social support questionnaire for transactions (SSQT). Soc. Sci. Med. **40:** 1221–1229.
22. POLLOCK, S.E. & M.E. DUFFY. 1990. The health-related hardiness scale: development and psychometric analysis. Nurs. Res. **39:** 218–222.
23. BYLES, J. *et al.* 2006. Older Australian women: aging in urban, rural and remote environments. Australas. J. Ageing **25:** 151–157.
24. LUCKE, J. *et al.* 2007. Trends in women's risk factors and chronic conditions: findings from the Australian Longitudinal Study on Women's Health. Future Medicine—Women's Health. **3**(4): 423–432.
25. NATIONAL HEALTH AND MEDICAL RESEARCH COUNCIL. 2001. Australian Alcohol Guidelines: Health Risks and Benefits. NHMRC. Canberra, ACT, Australia.

26. YOUNG, A.F., J.R. POWERS & S.L. BELL. 2006. Attrition in longitudinal studies: who do you lose? Aust. N. Z. J. Public Health **304:** 353–361.
27. BYLES, J.E. *et al.* 2006. A drink to healthy aging: outcomes associated with recommended, harmful and hazardous levels of alcohol use for older women. J. Am. Geriatric. Soc. **54:** 1341–1347.
28. MISHRA, G.D. 2004. Do socio-economic gradients in women's health widen over time and with age? Soc. Sci. Med. **58:** 1585–1595.
29. SIBBRITT, D.W., BYLES, J.E. & REGAN C. 2007. Factors associated with decline in physical function in a cohort of older women. Age Ageing **36:** 382–388.
30. DOLL, R. *et al.* 1994. Mortality in relation to smoking: 40 years' observations on male British doctors. Br. Med. J. **309:** 901–911.
31. WORLD HEALTH ORGANIZATION. 2000. Obesity: preventing and Managing the Global Epidemic. Report of a WHO Consultation. World Health Organization Technical Report Series. Report No.: 894 i–xii, 1–253.
32. TUCKER, K.L. *et al.* 2005. Baltimore Longitudinal Study of Aging. The combination of high fruit and vegetable and low saturated fat intakes is more protective against mortality in aging men than is either alone: the Baltimore Longitudinal Study of Aging. J. Nutr. **135:** 556–561.
33. MATHERS, C. 1999. Health Differentials Among Older Australians. Health Monitoring Series No. 2. Australian Government Publishing Service. Canberra, ACT, Australia.
34. BOULT, C. *et al.* 1996. Decreasing disability in the 21st century: the future effects of controlling six fatal and nonfatal conditions. Am. J. Public Health **86:** 1388–1393.

Delaying Dementia and Nursing Home Placement

The Dubbo Study of Elderly Australians over a 14-Year Follow-Up

JOHN McCALLUM,[a] LEON A. SIMONS,[b] JUDITH SIMONS,[b] AND YECHIEL FRIEDLANDER[c]

[a]*Senior Deputy Vice Chancellor, Victoria University, Melbourne, Australia*

[b]*University of New South Wales, Lipid Research Department, St. Vincent's Hospital, Sydney, Australia*

[c]*Unit of Epidemiology, Hebrew University–Hadassah School of Public Health, Jerusalem, Israel*

ABSTRACT: **In order to capture the "longevity dividend," modifiable risk factors for a diagnosis of dementia and nursing home placement were examined in a longitudinal study of an elderly cohort living in Dubbo, New South Wales, Australia. One thousand two-hundred thirty-three men and 1572 women 60 years and older living in the community were examined in 1988 and followed to 2002 for diagnosis of dementia and nursing home placement. There were 244 (8.7%) nursing home placements and 44% of these placements were primarily due to dementia, but dementia was a secondary diagnosis in another 20% of cases. In a proportional hazards model for dementia, any intake of alcohol predicted a 34% lower risk, and daily gardening a 36% lower risk. Daily walking predicted a 38% lower risk of dementia in men, but there was no significant prediction in women. The lowest tertile of peak expiratory flow predicted an 84% higher risk of dementia, the upper tertile of depression score predicted a 50% higher risk. The Cox proportional hazards model for nursing home placement, showed placement increased significantly with age, urinary incontinence, impaired peak expiratory flow, physical disability, and depression. The hazard of placement was significantly reduced by alcohol intake and female gender. Socioeconomic factors were not significant. Similar risk factors for dementia and nursing home placement indicate that the continuation of moderate alcohol intake, the maintenance of physical activity, especially daily gardening, and improvement of respiratory function, and the treatment of depression are recommended targets for interventions to delay or prevent major negative late-life experiences.**

Address for correspondence: Professor John McCallum, Senior Deputy Vice Chancellor, Victoria University, PO Box 14428, Melbourne, Victoria 8001, Australia. Voice: 61 3 9919 4255; fax: 61 3 9919 4006.

john.mccallum@vu.edu.au

Ann. N.Y. Acad. Sci. 1114: 121–129 (2007). © 2007 New York Academy of Sciences.
doi: 10.1196/annals.1396.049

Keywords: elderly; cohort study; dementia; nursing home placement

INTRODUCTION

The belief that the effects of aging are immutable, programmed by evolution, is known to be wrong. Olshansky and colleagues have proposed an aggressive pursuit of the "longevity dividend," particularly around genetic and dietary interventions.[1] Dementia and nursing home admission are significant threats to the quality of older life and primary targets for prevention or delay. This report combines two investigations of predictors of nursing home placement and dementia in a longitudinal study of a cohort of Australians 60 years and older, ongoing since 1988.[2]

METHOD

The study population comprised all non-institutionalized residents of the Dubbo Local Government Area in New South Wales who were born before 1930 (i.e., those 60+ years old). The participation rate was 73% of those eligible, based on a reconciliation of family doctor and electoral records. A group of 2805 citizens (1233 men and 1572 women) were first interviewed in 1988 and have been followed to the present with continuous recording of death, hospitalization, and residential-care data. Methods and measures have been described in detail elsewhere.[2–4] The study population was broadly representative of the Australian population born before 1930 by gender, age, employment, socioeconomic status, housing tenure, tobacco usage, blood pressure, and other variables.[3,4]

Dubbo is a major semi-urban center (population 34,000, about 400 km from Sydney) and its health and aged-care services are well developed. Nursing home admissions in Dubbo were routinely collected, while placement outside Dubbo was tracked by two-yearly questionnaires. Placements were followed from late 1988 onward. Medical records were coded according to the *International Classification of Diseases*, 9th edition, Clinical Modification (ICD-9-CM) [World Health Organization and U.S. National Center for Health Statistics] and 10th edition, Australian Modification (ICD-10-AM). The coding was accepted at face value without verification of records. Hospitalization, nursing home admission, and death records were monitored continuously, with postal surveys conducted every 2 years to confirm vital status. The first recorded presentation of dementia (September 1988–September 2004) was included in the analysis. Cases were diagnosed by a geriatric assessment team with access to CT scan, but medical record coding did not reliably distinguish between Alzheimer's disease and vascular dementia, so dementia is analyzed as a single category.

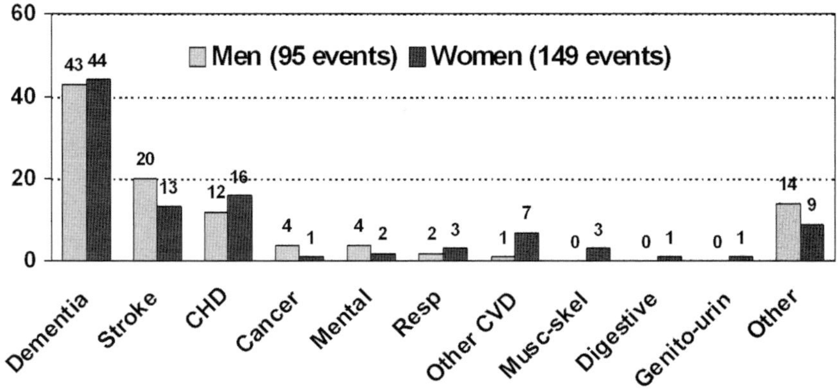

FIGURE 1. Nursing home admissions in Dubbo 1998–2002 by diagnostic group (%).

A large number of biomedical and behavioral variables were entered in a single block using the Cox proportional hazards regression models, and a final model then re-calculated as the most parsimonious version, retaining only variables of statistical significance as well as potential confounders (e.g., marital status, home ownership, self-rated health, education level, the presence of diabetes or prior cardiovascular disease). Dementia was identified on nursing home placement and therefore not used in the nursing home model. Statistical analysis was performed using SPSS Version 11. The study has received ongoing approvals from institutional ethics committees at St. Vincent's Hospital, Sydney, the University of New South Wales, and the University of Western Sydney. All participants gave informed written consent.

RESULTS

Two-hundred forty-four (8.7%) of the 2805 participants in the study required nursing home placement over the 14 years of follow-up; these comprised 95 men (7.7%) and 149 women (9.5%). The principal diagnosis leading to nursing home placement over 14 years was dementia, accounting for approximately 44% of placements and another 20% with a secondary diagnosis (FIG. 1). We have not identified less severe cases managed in the community, and thus we

TABLE 1. Baseline predictors of nursing home placement over 14 years in Cox proportional hazards regression

Variable	Hazard ratio	95% Confidence interval
Age	1.13	1.10–1.16
Female gender	0.55	0.38–0.81
Incontinence	1.66	1.13–2.44
Peak expiratory flow		
Tertile 1	3.19	2.01–5.06
Tertile 2	2.37	1.49–3.78
Blood pressure		
≥160–199/95–99	1.40	0.97–2.02
≥200/100	1.45	0.87–2.43
Activities of daily living		
Severely impaired	1.59	1.07–2.35
Alcohol		
Any	0.55	0.41–0.74
1–7 drinks/week	0.52	0.37–0.72
8–14 drinks/week	0.62	0.40–0.97
15–28 drinks/week	0.66	0.35–1.23
Depression score		
Tertile 2	1.71	1.14–2.58
Tertile 3	1.85	1.22–2.82

NOTE: For categorical variables the opposite or category served as the reference group. Other reference groups were as follows: peak expiratory flow, tertile 3; blood pressure <140/90; activities of daily living, no impairment; alcohol, zero intake; depression score, tertile 1. Hazard ratio for continuous variables refers to each standard deviation of change. There were separate models for any alcohol versus zero and quantity of alcohol versus zero.

have likely underestimated the true incidence of dementia. Recent research from the Rotterdam Study has highlighted how difficult it is to assess the true incidence of dementia, either during short- or longer-term follow-up.[5] Virtually none of the 2805 members of the Dubbo cohort showed evidence of significant cognitive impairment at base line (>3 errors on Short Portable Mental Status Questionnaire). During 16 years of follow-up (relating to 32,284 person–years) there were 115 initial presentations of dementia in men (9.3/100) and 170 presentations in women (10.8/100).

The final proportional hazards models use the baseline measures rather than factors immediately prior to placement (or diagnosis of dementia identified at placement). The hazard of nursing home placement[6] was significantly increased by the following at base line: age, urinary incontinence, impaired peak expiratory flow (PEF), physical disability, and depression (TABLE 1). The hazard of placement was significantly reduced by the presence at base line of a moderate alcohol intake and female gender. The hazard of diagnosis of dementia[7] was increased by age, having poor respiration, and depression. It was significantly reduced by gardening daily, moderate alcohol intake, and walking daily (for men only) (TABLE 2).

TABLE 2. Proportional hazards model for dementia

	Hazard ratio (285/2805, 10.2%)	95% Confidence interval
Age	1.13	1.10–1.15
Alcohol intake (any vs. nil)	0.66	0.51–0.86
Gardening (daily vs. rarely)	0.64	0.50–0.83
Walking (daily vs. rarely)	1.00	0.78–1.28
Peak expiratory flow		
(tertile I vs. III)	1.84	1.30–2.61
(tertile II vs. III)	1.48	1.05–2.10
Depression score		
(tertile II vs. I)	1.36	0.97–1.91
(tertile III vs. I)	1.50	1.07–2.10

NOTE: The number of cases and number at risk are shown. This is the final model, which also controlled for marital status, education, prior history of stroke, and activities of daily living.

DISCUSSION

The focus of gerontology has shifted from disease management and palliative care to "a new longevity science," which concentrates on health quality improvements in later life and is dependent on continuing work in epidemiology among other disciplines.[1] The Dubbo study results derive from a well-defined, longitudinal cohort, initially living independently in the community, free of cognitive impairment.[3,4] However, the study has no information on relevant lifestyle behavior in our subjects over the longer term, prior to baseline assessment. Hence some of the associations found may have arisen through confounding by factors measured or unmeasured. We also need to discount reverse causation effects; in particular, we may be looking at consequences rather than precursors of dementia. This is unlikely, though, as only 52 of the 285 dementia cases presented during the first 5 years of follow-up. Given the paucity of truly effective medical treatments, medical treatment during the study is unlikely to have had a major effect methodologically on incidence and, as a consequence, lifestyle interventions are of major interest.

The protective effect of moderate alcohol intake on dementia adds to accumulating evidence from previous studies.[8–11] Physical activity also has regularly been found to be protective against onset of dementia,[8,12,13] although we have not obtained a clear result with regard to daily walking. By contrast, the result with daily gardening is very strong, but represents a complex of different effects like physical activity,[14] healthy diet from the produce, leisure or mental activity,[15] and even purpose in life. This lack of a simple, single-agent effect does not reduce interest in it as an activity to be maintained in later life and promoted earlier. It also accords with descriptive evidence on gardening activities of centenarians in locations like Okinawa reported elsewhere in this volume and fits with emerging evidence on interventions in therapeutic recreation.[16] In regard to impaired PEF, we note reports showing improved cognitive performances in a physical activity program dedicated to patients with chronic airways disease.[17]

We have previously reported that zero alcohol intake and/or impaired PEF are key risk factors for cardiovascular disease[18,19] and reduced life expectancy.[20] It is possible that these factors increase the risk of dementia through increased cerebrovascular disease. However, given the lack of any association with hypertension, cigarette smoking, diabetes, or lipid abnormalities, this seems to be an unlikely pathway. Other studies have reported associations between dementia and hypertension[21] and diabetes,[22] or have shown reduced incidence of dementia (all types) after treatment of hypertension.[23]

Care burdens, predominantly those resulting from diseases like dementia, are the dominant factors that lead to nursing home placement. As consistently reported in the literature,[24–27] we have found that advancing age is a predictor of placement and is independent of measured health and care needs. Being female in Dubbo reduces placement risk by 45%, which reflects equivocal effects of gender in the international literature, possibly because of different admission practices. American women have a reduced risk of nursing home admission after controlling for the effects of age, poverty, functional status, and social support.[27] Women in Hong Kong had an increased risk of nursing home admission,[28] but in other studies female gender was not significant when other factors were controlled.[25,27,29,30] As well as lifestyle interventions, admission practices related to different cultures in different jurisdictions also deserve more thorough investigation as potential barriers to higher-quality later life.

Considering nursing home placement, impaired physical activity (i.e., severely impaired activities of daily living) conveyed a 59% higher risk in previous Dubbo work.[31] The benefit of alcohol consumption, a 45% reduction in risk at any intake is consistent with previous Australian findings,[29] and highlights the relationship with cardiovascular and all-cause mortality.[32] There is also support for an argument that it is moderate alcohol intake that protects against dementia and therefore nursing home placement.[5,33] Urinary incontinence at base line conveys a 66% higher risk of nursing home placement, and much can be done to prevent this problem,[34] provided we overcome the prejudice and fear associated with reporting and treating the condition. Impaired respiratory function, independent of diagnosed respiratory disease, conveyed in excess of 200% higher risk of placement and other major events in the study.[35] This is a known biomarker of functional age with independent effects across a range of later life outcomes.[36]

Our findings suggest potential public health actions for capturing the longevity dividend in our aging society. Factors identified in the two analyses potentially amenable to intervention, namely moderate intake of alcohol and/or daily gardening activity, seem to offer substantial protection against the onset of dementia and subsequent nursing home placement. Impaired respiratory function is also a strong and consistent predictor of dementia and nursing home placement, but as a potential "biomarker" of aging is more than a simple measure of fitness. A reduced risk of dementia attached to daily walking and an increased risk attached to depression showed inconsistency between the sexes, but this may be related to statistical chance. In general, all the above factors predicted the onset of dementia in a graded relationship and this is suggestive of a causal relationship. With the exception of exercise factors, which are replaced by Activities of Daily Living deficits in the nursing home model, and being female, these factors are also significant for nursing home placement. We have reason to be optimistic about accumulating evidence from descriptive epidemiology for improving the quality of longevity in our aging societies.

ACKNOWLEDGMENTS

This paper draws on material published in two papers, namely, those of J. McCallum *et al.*[6] and L.A. Simons *et al.*[7]

REFERENCES

1. Olshansky, S.J., D. Perry, R.A. Miller & R.N. Butler. 2006. In pursuit of the longevity dividend. Scientist (March): 28–36.
2. The Dubbo Study of the Elderly. http//www.dubbostudy.org.
3. Simons, L.A., J. McCallum, J. Simons, *et al.* 1990. The Dubbo Study: an Australian prospective community study of the health of the elderly. Aust. N. Z. J. Med. **20:** 783–789.
4. Simons, L.A., J. McCallum, Y. Friedlander, *et al.* 1991. Dubbo study of the elderly: sociological and cardiovascular risk factors at entry. Aust. N. Z. J. Med. **21:** 701–709.
5. Ruitenberg, A., J.C. van Sweiten, J.C.M. Witterman, *et al.* 2002. Alcohol consumption and risk of dementia: the Rotterdam study. Lancet **359:** 281–286.
6. McCallum, J., L.A. Simons, J. Simons & Y. Friedlander. 2005. Patterns and predictors of nursing home placement over 14 years: Dubbo Study of elderly Australians. Aust. J. Ageing **24:** 168–172.
7. Simons, L.A., J. Simons, J. McCallum & Y. Friedlander. 2006. Lifestyle factors and risk of dementia: Dubbo Study of the elderly. Med. J. Aust. **184:** 68–70.
8. Flicker, L., O.P. Almeida, J. Acres, *et al.* 2005. Predictors of impaired cognitive function in men over the age of 80 years: results from the Health in Men Study. Age Ageing **34:** 77–80.
9. Letebbeur, L. 2004. Risk of dementia and alcohol and wine consumption: a review of recent results. Biol. Res. **37:** 189–193.
10. Antilla, T., E.-L. Helkala, M. Viitanen, *et al.* 2004. Alcohol drinking in middle age and subsequent risk of mild cognitive impairment and dementia in old age: a prospective population based study. BMJ **329:** 539.
11. Espeland, M.A., L. Gu, K.H. Masaki, *et al.* 2005. Association between reported alcohol intake and cognition: results from the Women's Health Initiative Memory Study. Am. J. Epidemiol. **161:** 228–238.
12. Abbott, R.D., L.R. White, G.W. Ross, *et al.* 2004. Walking and dementia in physically capable elderly men. JAMA **292:** 1447–1453.

13. WEUVE, J., J.H. KANG, J.E. MANSON, *et al.* 2004. Physical activity, including walking, and cognitive function in older women. JAMA **292:** 1454–1461.
14. LAUTENSCHLAGER, N.T., O.P. ALMEIDA, L. FLICKER & A. JANCA. 2004. Can physical activity improve the mental health of older adults? Ann. Gen. Hosp. Psych. **3:** 12.
15. VERGHESE, J., R.B. LIPTON, M.J. KATZ, *et al.* 2003. Leisure activities and the risk of dementia in the elderly. N. Engl. J. Med. **348:** 2508–2516.
16. LEVINE-MIDORI, L. 2007. Therapeutic Thematic Arts Programming for Older Adults. Health Professions Press. Baltimore, MD.
17. EMERY, C.F., R.L. SCHEIN, E.R. HAUCK & N.R. MACINTYRE. 1998. Psychological and cognitive outcomes of a randomized trial of exercise among patients with chronic obstructive pulmonary disease. Health Psychol. **17:** 232–240.
18. SIMONS, L.A., J. MCCALLUM, J. SIMONS & Y. FRIEDLANDER. 1997. Relationship of peak expiratory flow rate with mortality and ischaemic heart disease in elderly Australians. Med. J. Aust. **166:** 526–529.
19. SIMONS, L.A., J. MCCALLUM, Y. FRIEDLANDER, *et al.* 2000. Moderate alcohol intake is associated with survival in the elderly: the Dubbo Study. Med. J. Aust. **172:** 121–124.
20. SIMONS, L.A., J. SIMONS, J. MCCALLUM & Y. FRIEDLANDER. 2005. Impact of smoking, diabetes and hypertension on survival time in the elderly: the Dubbo Study. Med. J. Aust. **182:** 219–222.
21. TZOURIO, C., C. DUFOUIL, P. DUCIMETIERE & A. ALPEROVITCH. 1999. Cognitive decline in individuals with high blood pressure: a longitudinal study in the elderly: EVA Study Group. Neurology **53:** 1948–1952.
22. STEWART, R., M. RICHARDS, C. BRAYUTE & A. MANN. 2001. Vascular risk and cognitive impairment in an older, British, African-Caribbean population. J. Am. Geriatr. Soc. **49:** 263–269.
23. FORETTE, F., M.-L. SEUX, J.A. STAESSEN, *et al.* 1998. Prevention of dementia in randomised double-blind placebo-controlled Systolic Hypertension in Europe (Syst-Eur) trial. Lancet **352:** 1347–1351.
24. KAPLAN, R.M., D.L. WINGARD, J.B. MCPHILLIPS, *et al.* 1992. Cigarette smoking, mortality, institutional and community-based care utilization in an adult community. J. Commun. Health **17:** 53–60.
25. HANLEY, R.J., L. ALECWITH, J. WIENER & D.L. KENNELL. 1990. Predicting elderly nursing home admissions: results from the 1982–1984 National Long-Term Care Survey. Res. Aging **12:** 1999–2028.
26. KERSTING, R. 2001. Impact of social support, diversity and poverty on nursing home utilization in a nationally representative sample of older Americans. Soc. Work Health Care **33:** 67–87.
27. WOLINSKY, F. & R. JOHNSON. 1991. The use of health services by older adults. J. Gerontol. Soc. Sci. **46:** S345–357.
28. WOO, J., S.C. HO, A.L. YU & J. LAU. 2000. An estimate of long-term care needs and identification of risk factors for institutionalisation among Hong Kong Chinese aged 70 years and over. J. Gerontol. **55:** M64–M69.
29. WANG, J.J., P. MITCHELL, W. SMITH, *et al.* 2001. Incidence of nursing home placement in a defined community. Med. J. Aust. **174:** 271–275.
30. HANCOCK, R., A. ARTHUR, C. JAGGER & R. MATTHEWS. 2002. The effect of older people's economic resources on care home entry under the United Kingdom's long-term care financing system. J. Gerontol. **5:** S285–293.

31. Simons, L.A., J. McCallum, Y. Friedlander & J. Simons. 2000. Healthy ageing is associated with reduced and delayed disability. Age Ageing **29:** 143–148.
32. Simons, L.A., J. McCallum, Y. Friedlander, *et al.* 2000. Moderate alcohol intake is associated with survival in the elderly: the Dubbo Study. Med. J. Aust. **173:** 121–124.
33. Mukamal, K.J., L.H. Kuller, A.L. Fitzpatrick, *et al.* 2003. Prospective study of alcohol consumption and risk of dementia in older adults. JAMA **289:** 1405–1413.
34. McCallum, J., L. Millar, T. Dong & J. Parsons. 2006. National Continence Management Strategy Independent Advisor Final Report. Department of Health and Ageing. Canberra. January 2006.
35. Simons, L.A., J. McCallum, J. Simons & Y. Friedlander. 1997. Relationship of peak expiratory flow with mortality and ischaemic heart disease in elderly Australians. Med. J. Aust. **166:** 526–529.
36. Anstey, K.J., S.R. Lord & G.A. Smith. 1996. Measuring human functional age: a review of empirical findings. Exp. Aging Res. **22:** 245–266.

Patients with Dementia and Their Carers

CHARLES BRIDGES-WEBB, BELINDA GILES, CATHERINE SPEECHLY, YVONNE ZURYNSKI, AND NAVAZ HIRAMANEK

Royal Australian College of General Practitioners, NSW Projects, Research and Development Unit, North Ryde, New South Wales, Australia

ABSTRACT: Clinical guidelines for general practitioners' care of patients with dementia still living in the community (not in hostels or nursing homes) were used in an educational program for general practitioners. At the same time information was obtained about the current health status, needs, and management of such patients and their carers. Eight general practitioners audited a total of 25 patients. Carer questionnaires were returned from 30 carers recruited by the general practitioners and 77 recruited by Alzheimer's Australia NSW. Many of the audit results are as expected, but some raise important issues. The diagnosis was not fully established for 20% of patients and the MMSE score not known by the general practitioner for 56%. Home safety had not been assessed for 44%. Legal matters such as driving or guardianship had not been considered for more than half. Most carers felt supported by their general practitioner in their role as carer, but services that would have been useful were not available to 23%, and most felt that being a carer had affected their own health. The small number of participants means that the audit results come from too few general practitioners to be reliably representative, but they do provide an indication of likely general results, and indicate priority areas for improved care.

KEYWORDS: Alzheimer's disease; dementia; carer; general practice

INTRODUCTION

When developing clinical guidelines for general practitioners' care of patients with dementia still living in the community (not in hostels or nursing homes),[1] we found that there was little information available about the current health status, needs, and management of such patients and their carers. We initiated a project to develop and use an education program using these guidelines for general practitioners, and at the same time obtain more information about the status and needs of patients and carers by including an audit and carer questionnaire. The project was given approval by the Royal Australian

Address for correspondence: Professor C. Bridges-Webb, RACGP NSW Projects, Research and Development Unit, P.O. Box 145, North Ryde, NSW 1670, Australia. Voice: 61 2 9886 4749; fax: 61 2 9888 3154.
nsw.research@racgp.org.au

Ann. N.Y. Acad. Sci. 1114: 130–136 (2007).
doi: 10.1196/annals.1396.012

College of General Practitioners Research and Evaluation Ethics Committee in July 2003.

BACKGROUND

In 2004, 6.5% of persons over 65 were estimated to have dementia and the prevalence of dementia is rising as a result of our aging population.[2] This increase in the number of patients with dementia will have a considerable impact on general practice.

There is evidence to show that carer wellbeing is important for the wellbeing of the patient. The increasing burden of care may lead to depression and anxiety and has been associated with more frequent physical illness in the carer.[3] Higher levels of depression in carers are associated with earlier institutionalization of patients with dementia.[4]

Bettering the Evaluation and Care of Health (BEACH) data[5] shows that during 100 consecutive encounters only 21% of general practitioners reported having any encounter with a patient where dementia was treated, and dementia was treated at only 43 per 1000 encounters. It is likely that general practitioners see more patients with dementia than is suggested by the BEACH data collection, since BEACH collects information on encounters where a condition is treated, and at many encounters dementia might have been present but not treated.[5]

The general practitioner's role in dementia care includes recognizing early dementia, assessment to confirm the diagnosis, management, health promotion, and prevention for patient, family, and carer, and follow-up. Little information exists about the extent to which general practitioners currently fulfill this role. Our project aimed to collect information about these issues from general practitioners and carers of patients.

METHODS

The original proposal was to further develop and put to trial an educational module which included a clinical audit of five patients and an educational workshop for each of the randomly recruited general practitioners. The general practitioners would also distribute a questionnaire to carers to be returned anonymously to the project team by prepaid post.

Patients were those with a diagnosis of dementia still living in the community (not in a nursing home or hostel). Carers were the primary carer of the patient with dementia, known to the general practitioner or to the general practitioner able to make contact, and giving consent to participate.

The clinical audit on the current status and level of support provided to people with dementia and their carers was developed in consultation with several practicing general practitioners. The standards were the RACGP

TABLE 1. Dementia in 25 general practitioner patients with dementia living at home

Diagnosis	%	Severity	%	MMSE[a]	%
Alzheimer's disease	40	Mild	32	12–20	21
Vascular disease	28	Moderate	52	21–24	17
Parkinson's disease	8	Severe	12	25+	4
Other	4	Uncertain	4	Not known	58
Dementia unspecified	20				

[a]MMSE = score on Mini-Mental State Examination done in last 12 months.

Dementia Guidelines.[1] General practitioners were asked to report the following data on five consecutive eligible patients: age and sex, diagnosis and severity of dementia, co-morbidities (including mental health), level of ability/disability (Activities and Instrumental Activities of Daily Living),[6] home safety, level of communication between the general practitioner and carer, and extent of assessment of carer health and support.

The carer questionnaire focused on level of burden and was a modification of the Zarit Carer Burden questionnaire, which gives information on the patients' mental health, social support, resources, behavioral management, and coping.[7] It also included items on carer's health and quality of life, and use of and level of satisfaction with support services.

It did not prove possible to recruit the 100 general practitioners planned, despite obtaining quality assurance and continuing education points from the RACGP, so a reduced number of general practitioners were recruited in collaboration with educational programs in two Divisions of General Practice. Since not enough carers were enlisted by the general practitioners, further questionnaires were also sent out to carers registered with Alzheimer's Australia NSW.

RESULTS

From 60 general practitioners expressing interest in a workshop, only 24 attended one, and only eight general practitioners undertook an audit before the workshop, four completing the full five patients, and four doing one or two, for a total of 25 patients. Thirty carers submitted by free post questionnaires given to them by participating general practitioners. A further 108 carers approached through Alzheimer's Australia NSW submitted another 77 questionnaires, giving a total of 107 carer questionnaires.

Patients

The patients audited were aged from 72 to 91 years, and 56% were male. The kind of dementia was established in only 80%, with 40% having Alzheimer's

TABLE 2. Co-morbidity in 25 patients with dementia living at home and seen by a general practitioner

Condition		%	Number of other conditions	%
Hypertension		44	0	8
Depression		38	1 or more	92
Mild	34%		2 or more	72
Moderate	4%		3 or more	52
Heart disease		28	4 or more	24
Ischemic	12%		5 or more	5
Other	16%			
Malignancy		24		
Cerebrovascular disease		20		
Diabetes		12		

disease and 28% vascular dementia (TABLE 1). Reversible causes had been excluded for 86%. The dementia was mild for 33%, moderate for 54%, and severe for 12%, and had been present for 3+ years for 31%. The Mini-Mental State Examination (MMSE) score was not known to the general practitioner for 56%. There was associated depression for 38% of patients and other psychiatric problems for 20%. Only 24% of patients were receiving medication for dementia.

Co-morbidity was present for 92% of patients, with 52% having three or more other conditions, the most common of which were hypertension (44%), depression (38%), and heart disease (28%) (TABLE 2).

All medications had been reviewed for 88% of patients within the past 6 months. General health assessments, including vision, hearing, nutritional status, continence of bladder and bowels, and sleep patterns, had each been done within a year for about 70% of patients, but home safety had been assessed for only 56%.

The most frequent problems with behavior were poor memory (44%) and poor thinking and planning (36%), and for fewer than half of these patients was a management strategy in place for the problem.

While 36% of patients could use the telephone normally, few could take medicines, prepare food, or shop without help. Most patients needed at least some assistance with activities of daily living, 52% with bathing and dressing, 40% with walking, 36% with toileting, and 28% with feeding. For less than half had legal matters, such as driving, enduring power of attorney or guardianship, or a will been discussed with patient or family, though for all patients the general practitioner had been in touch with the carer.

Carers

Most carers were female (80%) and married (82%); 68% were spouses, and 26% daughters/sons. Most (62%) had been caring for 2–5 years, 23% for

TABLE 3. Community services used by carers for patients with dementia

Service	AA $n = 77$ %	GP $n = 30$ %	$P < 0.05$
Alzheimer's Association	60	0	*a*
ACAT	52	48	
Carer support group	35	7	*a*
Respite care	32	23	
Dementia support group	31	3	*a*
Daycare	27	17	
Geriatrician	27	13	
Home help	13	17	
Community nurse	8	23	*a*
Meals on wheels	1	10	

[a]Significant difference between carers recruited by Alzheimer's Association (AA) and those recruited by general practitioners (GP).

more than 5 years and 15% for under 2 years; and 70% had the same general practitioner as the patient.

The most commonly used community services were an aged-care assessment team (ACAT, 48%), Alzheimer's Australia (44%) (more often carers were recruited from their list), respite care (30%), daycare (24%) and a carer support group (24%) (TABLE 3). Services that would have been useful were not available to 23%, most commonly day care, home help, transport, and respite care. Most carers felt supported by their general practitioner in their carer role (86%).

Most carers said that being a carer had affected their health (61%), and nearly all said they had other medical problems, including arthritis (54%), high blood pressure (47%), anxiety (44%), and depression (36%).

The most common feelings about their carer role were that the person with dementia was dependent on them (85%) and expected their care (60%) (TABLE 4). They were afraid of the future (59%), without enough time for themselves and their own social life (45%), frequently or always burdened by caring (39%) and stressed (35%), and felt they had lost control of their lives (34%).

DISCUSSION

The project achieved its main aim of developing an educational module on dementia for general practitioners. A major hurdle for the research study was obtaining enough general practitioner participation in the audits and the workshops. This was thought to be greatly influenced by its being done late in the last year of the quality assurance and continuing education triennium, when most general practitioners had achieved the necessary points and incentive to participate was low. This particularly applied to the audits, which take more time.

TABLE 4. Feelings of carers about their caring role

	%
Patient is dependent upon them	85
Carer is the only one the patient can depend on	60
Carer is afraid of the future for patient	59
Not enough time for self, and social life suffers	45
Frequently or nearly always burdened	39
Stressed between care and other responsibilities	35
Lost control of life	34
Strained when with patient	32
Lack privacy	26
Uncertain about what to do	24

Many of the audit results are as expected, but some raise important issues. The diagnosis was not fully established for 20% of patients and the MMSE score not known by the general practitioner for 56%. Home safety had not been assessed for 44%, and a management strategy was in place for fewer than half the patients with behavior problems. Legal matters, such as driving or guardianship, had not been considered for more than half. For all patients the general practitioner was in contact with the carer.

It was reassuring that the carer questionnaire results were not greatly different for those recruited by general practitioners and those recruited through Alzheimer's Australia, and that most carers felt supported by their general practitioner in their role as carer. It is of concern that services that would have been useful were not available to 23%, and that most carers felt that being a carer had affected their own health. This emphasizes the need for careful monitoring the health and coping ability of carers, and provision of support as needed.

The small numbers of participants means that the audit results come from too few general practitioners to be reliably representative, but they do provide an indication of likely general results. In the case of carer questionnaires, the same applies to those given out by general practitioners, but in this case we were able to distribute more through Alzheimer's Australia, providing more information, though its representativeness is also less than ideal. However, the results relating to patients' age and sex and diagnosis are not greatly different from those obtained by a recent national but as yet unpublished survey of medication for dementia in 1226 patients by Alzheimer's Australia and the Rehabilitation Studies Unit, University of Sydney. A greater proportion of national patients had Alzheimer's disease (82%) and fewer had vascular dementia, and the carers' use of services was in general greater, particularly for home help, carer support group, and daycare.

CONCLUSIONS

The implications of these results about patients with dementia still living in the community and their carers are addressed in the educational module for general practitioners and were taken into consideration in the preparation of a pamphlet for carers detailing the kind of support they might seek from general practitioners. The pamphlet is now available on the RACGP website (www.racgp.org.au/guidelines/dementia) and Alzheimer's Australia NSW website (www.alzheimers.org.au/content.cfm?infopageid=2209), but printing and distribution is still under consideration.

ACKNOWLEDGMENTS

This project was made possible with a grant from the Bernard Judd Foundation, an initiative of the Hammond Care Group. The Illawarra and Central Sydney Divisions of General Practice assisted with recruitment of general practitioners and Alzheimer's Australia NSW assisted in recruiting carers. We thank those general practitioners who participated in the development of the project and in its implementation.

A brief article on the project, including TABLE 1, has already been published in part in *Australian Family Physician* (**35:** 924–925 [November 2006]); they have consented to this further more-detailed publication.

REFERENCES

1. BRIDGES-WEBB, C. *et al.* 2002. Guidelines for the Care of Patients with Dementia in General Practice, www.racgp.org.au/guidelines/dementia.
2. AUSTRALIAN INSTITUTE OF HEALTH AND WELFARE. 2004. The impact of dementia on the health and aged care systems. AIHW Cat. No. AGE 37. AIHW. Canberra, ACT.
3. BROWN, L.J. *et al.* 1990. Caregiver burden should be evaluated during geriatric assessment. J. Am. Geriatr. Soc. **38:** 455–460.
4. PATTERSON, C. *et al.* 2001. The recognition, assessment and management of dementing disorders: conclusions from the Canadian Consensus Conference on dementia. Can. J. Neurol. Sci. **28**(Suppl 1): s3–s16.
5. BRIDGES-WEBB, C. 2002. Dementia care in general practice: what can the BEACH survey tell us? Aust. Fam. Phys. **31:** 381–383.
6. SKURLA, E. *et al.* 1998. Direct assessment of activities of daily living in Alzheimer's disease. J. Am. Geriatr. Soc. **36:** 97–103.
7. PARKS, M.S. & K.D. NOVIELLI. 2000. A practical guide to caring for caregivers. Am. Fam. Phys. **62:** 2613–2622.

Healthy Aging

Can Clinical Trials Deliver?

ERICA BELL

University Department of Rural Health, University of Tasmania, Hobart, Tasmania

ABSTRACT: The potential of a transdisciplinary research method—Quali-Quantitative Analysis (QQA)—for delivering better evidence for healthy aging is explored. Illustrative reviews from the literature on healthy aging are summarized, as are papers on methodology by Charles Ragin and colleagues. The evidence needs for healthy aging are explored, particularly in relation to the ability of randomized controlled trials (RCTs) to deliver rich, diversity-oriented evidence. The techniques of QQA, which involves treating cases as configurations, are described. The QQA method has potential value for healthy aging research in that: (*a*) it could provide evidence in combinatorial forms useful to holistic practice; (*b*) it could enrich the evidence for practice by providing more rigorous information from small-*N* groups, within RCTs and where RCTs cannot be used; and (*c*) it could add to the ability of RCTs to generate theoretical advances.

KEYWORDS: healthy aging; transdisciplinary research methods; randomized controlled trials; evidence for practice

AIM

The gold standard for evaluating interventions in aging research is large randomized controlled trials (RCTs) using quantitative research techniques. Yet this approach has serious limitations for studying small-*N* populations and considering cases as holistic configurations. This paper describes the potential of a new research method—Quali-Quantitative Analysis (QQA)—for delivering that evidence.

METHOD

A total of 1405 papers and abstracts were hand-searched. Papers on interventions for older clients were selected from searches of systematic literature

Address for correspondence: Erica Bell, University Department of Rural Health, University of Tasmania, Private Bag 103 Hobart, Tasmania 7000 Australia. Voice: +61 3 6226 6345; fax: +61 3 6231 6601.
Erica.Bell@utas.edu.au

Ann. N.Y. Acad. Sci. 1114: 137–143 (2007). © 2007 New York Academy of Sciences.
doi: 10.1196/annals.1396.006

reviews on PubMed, using the terms "aging interventions," "geriatrics," and "chronic disease" for the period from 2000 to 2006. The Cochrane Database of Systematic Reviews was also searched for the same period using chronic-disease-specific terms. The criterion for inclusion was whether papers were illustrative of methodological needs for future healthy aging research.

The selection of references to Ragin's work was made from his website at http://www.u.arizona.edu/~cragin/cragin/ and the Compass Small-N website located at http://www.compasss.org.

RESULTS

Evidence Needs of Healthy Aging

The classical experimental model dominates the evaluation of interventions for older clients, and the whole notion of evidence-based health practice. Yet much of the evidence required in the future is for holistic information about bio-psycho-social elements or client-service-community elements, that is, for research capturing configurational, case-based evidence for healthy aging.

For example, chronic disease management is moving to a "model of care" approach that goes beyond traditional clinical approaches to incorporate multidisciplinary, multi-service, interprofessional, "whole-of-patient" and similar approaches. To go beyond such rhetoric, new kinds of evidence are needed,[1] including evidence that better theorizes and captures complex causalities and outcomes for co-morbidities and multi-morbidities.[2,3]

Dementia research faces challenges to do with capturing the interaction of complex pathogenic mechanisms of underlying disease, and the effectiveness of a range of innovative treatment strategies, such as immunotherapies.[4] It also faces a future in which better evidence will be needed for psychosocial interventions such as "reminiscence therapy" and how these treatments work to affect complex whole-of-client configurations of mood, cognition, behavior, and well-being.[5] The mental health literature generally includes calls for better research into how psychotherapies work for medically ill clients: research that recognizes the diversity of conditions and complexity of outcomes relevant to understanding mental health interventions.[6]

Cancer research faces related challenges to do with multidisciplinary and multidimensional evidence needed to decide the most optimal pattern of care of clients with different health statuses.[7] Increasingly, case management in cancer intervention focuses upon whole-of-patient care in the context of services and communities.[8]

In the field of vascular research, there are now calls for "bio-psychosocial models of care"[9,10] and evidence that captures the complex causalities of morbidity and mortality in cardiovascular disease, such as negative emotional states, psychosocial stressors, and social connections important to integrative medicine approaches.[11]

Nutrition and its role in longevity and quality of life presents particular challenges of identifying underlying mechanisms of, for example, the elements of the Mediterranean diet and its relationship to specific conditions, such as age-related cognitive impairment, cardiovascular disease, and diabetes—essentially the challenge of complex corroboration.[12] What is needed is robust research into not just one ideal configuration, but also multiple optimal micronutrient combinations, doses, the timing of interventions, and so on.[13] The nutritional management of disease, or clinical nutrition, is an area where multiple optimal configurations of information about dietary elements, and whole-of-client information will be increasingly required in the future.[14]

Across the different areas of "geriatric" and "chronic disease" research there are calls for rigorous diversity-oriented research approaches that move beyond narrow clinical models of health care and reductive research methods. Why then do so many of the studies just cited also call for more and better RCTs? Is the solution really more of the same?

Clinical trials do not produce diversity-oriented evidence largely because they are characterized by statistical methods that require a high ratio of cases to variables. Statistical methods used in clinical trials too often produce little or very poor information about subgroups.[15,16] This has much to do with the fact that the more we divide our data into subgroups, the increase in variance around the mean estimate means there is an increased statistical likelihood of a false finding (or a positive result in subgroups that is, in fact, due to chance).[16]

The main two strategies for overcoming the limitations of RCTs have been to extend their methods or to reject RCTs in favor of observational methods.[17] However, recent transdisciplinary advances in research methods suggest how health researchers can find a middle path between generality and complexity, and between internal (within-trial) and external (real bio-psycho-social) validity. They also suggest why the methods of the "variable-oriented researcher"[18] may not be able, by themselves, to deliver the rich evidence needed for healthy aging.

Quali-Quantitative Analysis

A global research movement began with Charles Ragin's seminal 1987 work *The Comparative Method*,[19] now represented in multiple disciplines by more than 250 applications. The term "Quali-Quantitive Analysis" is preferred in this paper over Ragin's term "Qualitative Comparative Analysis" (QCA) because it emphasizes the transdisciplinary use of both qualitative and quantitative techniques.

In a sentence, QQA represents a set of techniques for making holistic comparisons of cases as configurations to explore causality.[20] The core methods are given in a body of writings by Ragin and colleagues,[18,19,21–25] though *Fuzzy-Set Social Science* is the clearest and fullest elaboration of this diversity-oriented method.[26] Ragin argues that viewing cases as configurations keeps the researcher closer to the evidence, and points to the "veiling" of cases created by conventional quantitative correlational analysis arising from the reductive focus on the net effects of independent variables and explanations of variations in dependent variables.[27,28] Ragin's techniques attempt to make sense of cases using techniques for structuring case-based enquiry.[28,29]

There are three basic steps in a QQA approach:

- selection of cases and their tabulation in "property spaces,"
- undertaking tests of "necessity" and "sufficiency" of possible causal conditions, and
- evaluation of the results of these tests.[23]

Numerical representations of case attributes are used to cross-tabulate the presence or absence (crisp sets), or degrees of membership (fuzzy sets) of the attributes of cases (causal conditions and outcomes). Larger sets and fuzzy sets can be analyzed using software[30] supported by related explanatory papers.[25–27,31]

Having so tabulated each case, the researcher using the version of QCA methods limited to categorical data simplifies the property spaces or "truth table" through a process of "logical minimization" that reduces information about case attributes to shorthand expressions of combinations. The detail of case attributes is reduced to "logical equations" for the outcome/s being examined.[29,32,33]

The necessity and sufficiency of causal conditions are then considered: a cause is necessary if all instances of the outcome occur with that cause; a cause is sufficient if all instances of the cause occur with the client outcome.[23,34,35]

Later elaborations of the approach discuss how Bayesian-influenced "probabilistic criteria" may be used for fuzzy sets using numbers on a continuous scale for linguistic qualifiers (such as "0.8" for "almost always sufficient" and "0.5" for "sufficient more often than not").[20,23,35,36] However, much of Ragin's techniques are also about qualitative rigor and the use of iterative, interrogative techniques conducive to theorizing.[23]

The approach offers a way of more closely exploring cases in small-*N* populations, including subgroups from clinical trials: single or very few unusual cases or subgroups of positive or negative cases. In this way, and in many refinements, Ragin's techniques facilitate systematic consideration of cases as holistic, configurational entities.

CONCLUSIONS

How could QQA help deliver better evidence for healthy aging?

First, QQA could provide evidence in a form that is usable for holistic practice. Traditional quantitative approaches used in RCTs, emphasizing linearity, additive effects, and unifinality (one effective configuration), fail to capture the nonlinearity, synergy, and equifinality (multiple effective configurations)[37] important to holistic practice for healthy aging. Ragin has argued that much social theory relies on set-theoretic relations that are poorly explored by the correlational "indicator" measurement approaches of traditional quantitative techniques.[27] Similarly, evidence for healthy aging practice that is responsive to bio-psycho-social or client-service-community information requires verbal configurational evidence for nuanced holistic judgments.

Second, QQA could enrich the evidence base for practice by providing more rigorous information from small-*N* groups, within RCTs, and where RCTs cannot be used. In classical experimental methodology, small-group studies are regarded as "underpowered," with severely restricted relevance to evidence-based health care.[38] Much information from small groups that could advance knowledge about healthy aging is therefore lost.

Third, QQA could add to the ability of clinical trials to generate theoretical advances. Traditional quantitative methods tend to focus on testing existing theories or hypotheses and often do not push the theoretical envelope. Recent developments in RCT methodologies aimed at better understanding treatment moderators (for whom and under what conditions treatment works) and mediators (possible mechanisms by which a treatment produces its effects) suggest pressures to make RCTs theoretically richer.[39] Ragin has repeatedly demonstrated how his approach can be used to add value to traditional statistical methods[19,20,40–42]—its potential as a device for theoretical enrichment in RCTs is untapped.

While QQA is a developing methodology, it could reinvigorate the evidence base for healthy aging through a focus on different effective client-service-community configurations or bio-psycho-social combinations in small-*N* studies and analyses of clinical trial subgroups. As such, it has potential for the holistic, diversity-oriented evidence needs of healthy aging.

REFERENCES

1. Davidson, P., E. Halcomb, L. Hickman, *et al.* 2006. Beyond the rhetoric: what do we mean by a 'model of care'? Aust. J. Adv. Nurs. **23:** 47–55.
2. Fortin, M., L. Lapointe, C. Hudon, *et al.* 2004. Multimorbidity and quality of life in primary care: a systematic review. Health Qual. Life Outcomes **2:** 51.
3. Fortin, M., L. Lapointe, C. Hudon, *et al.* 2005. Multimorbidity is common to family practice: is it commonly researched? Can. Fam. Physician **51:** 244–245.
4. Feldman, H.H., S. Gauthier, H. Chertkow, *et al.* 2006. Progress in clinical neurosciences: Canadian guidelines for the development of antidementia therapies: a conceptual summary. Can. J. Neurol. Sci. **33:** 6–26.
5. Woods, B., A. Spector, C. Jones, *et al.* 2005. Reminiscence therapy for dementia. Cochrane Database Syst. Rev. CD001120.
6. Vamos, M. 2006. Psychotherapy in the medically ill: a commentary. Aust. N. Z. J. Psychiatry **40:** 295–309.

7. TERRET, C. 2004. Management and geriatric assessment of cancer in the elderly. Expert Rev. Anticancer. Ther. **4:** 469–475.
8. BOURBONNIERE, M. & S. H. KAGAN. 2004. Nursing intervention and older adults who have cancer: specific science and evidence based practice. Nurs. Clin. North Am. **39:** 529–543.
9. PITERMAN, L., H. ZIMMET, H. KRUM, *et al.* 2005. Chronic heart failure—optimising care in general practice. Aust. Fam. Physician **34:** 547–553.
10. MCDONAGH, T.A. 2005. Lessons from the management of chronic heart failure. Heart **91**(Suppl 2): ii24–27; discussion ii31, ii43–48.
11. EVERSON-ROSE, S.A. & T.T. LEWIS. 2005. Psychosocial factors and cardiovascular diseases. Annu. Rev. Public Health **26:** 469–500.
12. SERRA-MAJEM, L., B. ROMAN & R. ESTRUCH. 2006. Scientific evidence of interventions using the Mediterranean diet: a systematic review. Nutr. Rev. **64**(2 Pt 2): S27–47.
13. BERGER, M. 2005. Can oxidative damage be treated nutritionally? Clin. Nutr. **24:** 172–183.
14. MADDEN, A.M. 2003. Changing perspectives in the nutritional management of disease. Proc. Nutr. Soc. **62:** 765–772.
15. ASSMAN, S., S. POCOCK, L. ENOS, *et al.* 2000. Sugroup analysis and other (mis)uses of baseline data in clinical trials. Lancet **355**(March 25): 1064–1069.
16. SLEIGHT, P. 2000. Debate: Subgroup analyses in clinical trials: fun to look at—but don't believe them! Curr. Controlled Trials Cardiovasc. Med. **1:** 25–27.
17. HOTOPF, M. 2002. The pragmatic randomised controlled trial. Advan. Psychiatric Treat. **8:** 326–333.
18. RAGIN, C. 1997. Turning the tables: how case-oriented research challenges variable-oriented research. Compar. Soc. Res. **16:** 27–42.
19. RAGIN, C. 1987. The Comparative Method. University of California Press. Los Angeles, CA.
20. RAGIN, C.C. 1998. The logic of qualitative comparative analysis. Int. Rev. Soc. Hist. **43**(Suppl): 105–124.
21. RAGIN, C.C. & H.S. BECKER. 1992. What is a Case? Exploring the Foundations of Social Inquiry. Cambridge University Press. Cambridge, UK.
22. RAGIN, C.C. 1994. Constructing Social Research: The Unity and Diversity of Method. Pine Forge Press. Thousand Oaks, CA.
23. RAGIN, C. 2000. Fuzzy-Set Social Science. The University of Chicago Press. Chicago, IL.
24. RAGIN, C. 2007. Recent advances in fuzzy-set methods and their application to policy questions. http://smalln.spri.ucl.ac.be/WPFull.htm. [Accessed 27 March 2007].
25. RAGIN, C. & J. SONNETT. 2007. Between complexity and parsimony: limited diversity, counterfactual cases, and comparative analysis. http://www.u.arizona.edu/~cragin/fsQCA/software.shtml. [Accessed 27 March 2007].
26. RAGIN, C. 2007. Qualitative comparative analysis using fuzzy sets (fsQCA). *In* Configurational Comparative Analysis. B. Rihoux & C. Ragin, Eds. Sage Publications. Thousand Oaks, CA. In press. Available at http://www.u.arizona.edu/~cragin/fsQCA/software.shtml.
27. RAGIN, C. 2006. Set relations in social research: evaluating their consistency and coverage. Political Analysis **14:** 291–310.
28. RAGIN, C. 2006. The limitations of net effects in thinking. *In* Innovative Comparative Methods for Policy Analysis. H. Grimm & B. Rihoux, Eds.: 13–41. Springer. New York.

29. RAGIN, C.C. 1999. Using qualitative comparative analysis to study causal complexity. Health Services Res. 34(5 Part II): 1225–1239.
30. RAGIN, C., K. DRASS & S. DAVEY. 2007. Fuzzy-Set/Qualitative Comparative Analysis 2.0. http://www.u.arizona.edu/~cragin/fsQCA/software.shtml. [Accessed 27 March 2007].
31. RAGIN, C. 2007. Calibration versus measurement. *In* Oxford Handbook of Political Methodology. J.M Box-Steffensmeier, H.E. Brady, and D. Collier, Eds. Oxford University Press. Oxford, UK. In press. Available at http://www.u.arizona.edu/~cragin/fsQCA/software.shtml.
32. RAGIN, C. 1995. Using qualitative comparative analysis to study configurations. *In* Computer-Aided Qualitative Data Analysis: Theory, Methods and Practice. U. Kelle, Ed.: 177–189. SAGE Publications. London.
33. RAGIN, C.C., D. BERG-SCHLOSSER & G. DE MEUR. 1996. Political methodology: qualitative methods. *In* A New Handbook of Political Science. R.E. Goodin & H.D. Klingemann, Eds.: 749–768. Oxford University Press. Oxford, UK.
34. RAGIN, C.C. 1999. The distinctiveness of case-oriented research. Health Services Res. **34**(5 Part II): 1137–1151.
35. RAGIN, C.C. & H.M. GIESEL. 2002. User's Guide to Fuzzy-Set/Qualitative Comparative Analysis 1.1. Department of Sociology, University of Arizona. Tucson, Arizona. http://www.u.arizona.edu/c̃ragin/software.htm.
36. RAGIN, C.C. 2002. Fuzzy-set analysis of necessary conditions. *In* Necessary Conditions: Theory, Methodology and Applications. G. Goertz & H. Starr, Eds.: 179–196. Rowman and Littleford. Lanham, MD.
37. FISS, P.C. 2005. A Set-Theoretic Approach to Organisational Configurations. Queen's University. Kingston, Ontario, Canada. http://smalln.spri.ucl.ac.be/WPFull.htm.
38. HALPERN, S., J. KARLAWISH & J. BERLIN. 2002. The continuing unethical conduct of underpowered clinical trials. JAMA **288:** 358–362.
39. KRAEMER, H., T. WILSON, C. FAIRBURN, *et al.* 2002. Mediators and moderators of treatment effects in randomized clinical trials. Arch. Gen. Psychiatry **59**(Oct): 877–883.
40. RAGIN, C.C. 1994. Introduction to qualitative comparative analysis. *In* The Comparative Political Economy of the Welfare State. T. Janoski & A.M. Hicks, Eds.: 299–319. Cambridge University Press. Cambridge, UK.
41. RAGIN, C.C. 2001. Case-oriented research. *In* International Encyclopedia of the Social and Behavioural Sciences, Vol 3. N.J. Smelser & P.B. Baltes, Eds.: 1519–1525. Elsevier. Oxford.
42. GRIFFIN, L. & C.C. RAGIN. 1994. Some observations on formal methods of qualitative analysis. Sociol. Meth. Res. **23:** 4–21.

Debating the Influence of Self-Reports by People Living with Chronic Disease on Healthy Aging and Longevity

KAY PRICE

University of South Australia, Adelaide, South Australia, Australia

ABSTRACT: For people living with chronic disease, decision-making in relation to their health and lifestyle choices is an integral aspect to managing the complexity that living with chronic disease brings into their lives. Research findings reveal that people living with chronic disease can self-report what may seem to be appropriate explanations for their health status and seem to be in control of their health in their efforts to master the constraints imposed by their life situation. Precisely because ignoring disease-related changes may actually contribute to disease progression, it is imperative to know whether self-reports are made linking biomedical understandings of their disease with the person's social situation and life events. To understand how to influence choices made by people living with chronic disease to promote healthy aging and longevity, it is imperative for health and social care professionals to learn how to question people's self-reports, given that self-reports are often used as evidence in practice and research to inform clinical solutions.

KEYWORDS: self-reports; chronic disease; decision-making; healthy aging; longevity

INTRODUCTION

In health and social care practice and research, seeking self-reports or oral accounts from people living with chronic disease gives primacy to the role of the individual to gain insight into preventing and managing chronic disease. Symptoms and decisions about ongoing management options are usually assessed by self-reports.[1] The paradox of the self-report is that it may be the principal method to obtain a report/history or account of any event for an individual person, but it can also be a vehicle through which people living with

Address for correspondence: Dr. Kay Price, School of Nursing & Midwifery, University of South Australia, City East Campus, North Terrace, Adelaide, South Australia 5000, Australia. Voice: +61 8 83202721; fax: +61 8 83022168.
Kay.price@unisa.edu.au

**Ann. N.Y. Acad. Sci. 1114: 144–153 (2007). © 2007 New York Academy of Sciences.
doi: 10.1196/annals.1396.039**

chronic conditions may hide their struggles from themselves in an attempt to portray and preserve a valued sense of self and identity.[2] The intent of this theoretical paper is to present an argument why health and social care professionals need to critically explore *with* people living with chronic disease, in an ongoing reflective and collaborative approach, their self-reported evidence if these social and healthcare professionals are to ensure healthy aging and longevity for people living with chronic disease.

WAYS OF THINKING TO PROMOTE HEALTHY AGING AND LONGEVITY

The State of Aging and Health in America 2007 report[3] acknowledges that improved medical care, access to this care, and efforts at prevention have contributed to dramatic increases in life expectancy in the United States of America, while at the same time about 80% of older adults have at least one chronic condition and 50% have at least two. Basing its conclusions on self-reporting, the Australian Institute of Health and Welfare (AIHW) estimated that between 2004 and 2005, 77% of the Australian population had one or more long-term health conditions (which had lasted or was expected to last for at least 6 months).[4] According to the AIHW, health is a growing sector of the economy and is growing faster than the economy. *The State of Aging and Health in America 2007* report[3] states that almost 95% of healthcare expenditures in the United States is for chronic diseases.

In view of the increasing expenditure associated with longevity and living with chronic disease, preventing chronic disease and lowering healthcare costs has been identified as essential. For health and social care professionals, the goal of healthy aging and longevity (not dying) translates into working with people to ensure they do not suffer from a chronic disease. To be in good health has been turned into a permanent war against disease.[5]

The Australian government has introduced a National Chronic Disease Strategy[6] and an Australian Better Health Initiative.[7] These initiatives are about "strengthening the focus of the health system on prevention and health promotion, and management of chronic disease," given the view that "many Australians suffer from chronic illness, which is a growing problem facing the health system as the population ages" (p. 1).[7] In the United Kingdom, the white paper "Choosing Health: Making Healthy Decisions Easier"[8] places an emphasis on informed choice. The national chronic disease prevention agenda in the United States has been premised on the view that "most Americans have not changed their lifestyles sufficiently to reduce the risk of death or illness" and that Americans "cannot afford to ignore the power of prevention" (pp. 1, 6).[9] *The State of Aging and Health in America 2007* report[3] focuses on ways to reduce behaviors "that contribute to premature death and disability" (Section 2, Executive Summary, p. III). Government policies and strategies increasingly focus attention on people "making good choices, only good choices and even better choices" (p. 4).[10] Consequently people living with chronic conditions are expected to take individual responsibility for the choices they make to being actively involved in their own health care and to self-manage in an effective and productive way.

Enacting individual responsibility can also mean people wanting the freedom to make their own lifestyle choices and at the same time have the "freedom from bearing the consequences of wrong choices" (p. 89),[5] Sherman, Nelson, and Steele[11] have previously argued that health campaigns, such as those implemented by Western governments, are unlikely to produce anticipated change in behaviors. They say that

> . . . research in social psychology reveals a potential obstacle for these health campaigns. Individuals often are motivated to interpret information in a self-serving manner that may prevent them from accepting their risk for disease. . . . If recipients of a health message fail to accept the information, then they will be unlikely to change their risky behaviors. This problem is compounded because health messages are often the least persuasive among the individuals for whom the issue is of high personal relevance (p. 1046).[11]

Additionally, as Midgley[12] says, people are inherently conservative (this includes health and social care professionals as much as people living with chronic disease) and they tend only to move into new ways of thinking/doing when there is a good reason (with an associated emotional commitment) to do so.

Therefore, on the one hand we have an aging society that has access to diverse health services, many likely to be suffering from at least one chronic condition, among which the fastest growing age groups are those 75 years and over. But on the other hand, these very conditions are creating such a burden that the associated healthcare costs are being challenged. At the same time, given that the majority of older people *do* live with a chronic condition yet are expected to age in a way that does not increase health expenditure, another important question focuses on what evidence will inform decisions about how best to assist healthy aging and longevity for people living with chronic conditions.

LIVING WITH CHRONIC DISEASE

As Paterson[13] has identified, for people living with chronic disease, precisely because the disease always requires attention, ignoring disease-related changes in order to sustain a wellness or healthy aging approach to living may actually contribute to disease progression.

In urging healthcare professionals to be aware of their mindsets when interacting with people who live with chronic illness, Telford, Kralik, and Koch say:

> If we understand the fluctuating response to long-term illness from the perspective of striving to fit with social norms and maintaining a sense of power and control, it becomes possible to understand behaviors as a person's attempt to preserve a valued sense of self and identity. The fluctuating illness experience is perceived as the individual navigating their way through an often unsupportive and sometimes hostile social context. We contend that labels of acceptance and denial are not helpful and may obstruct this process. . . . Rather than interpreting behavior as denial and promoting acceptance as a desirable end-point, we suggest actively listening to the stories people tell. In doing so, they will have opportunities to share inside views of illness experiences and be truly helpful to people (pp. 462–463).[2]

The way in which persons share their "inside views of illness experiences" is through self-reports. At the same time, the way that any person living with chronic disease can understand his or her own illness experience, and in turn self-report this experience, will be influenced by his or her capacity to have some biomedical understanding of their disease and to link this with their health and social situation and life events. It cannot be assumed that a person living with chronic disease has the necessary knowledge and critical thinking skills to appropriately inform his or her decision-making in self-care. While listening to what people self-report is important, so too is asking probing, reflective questions to explore how people living with chronic conditions have come to say what they self-report. To be able to ask these probing questions, health and social care professionals must establish a collaborative and reflective relationship with people living with chronic disease and have explored their own ways of thinking about living with chronic disease that informs how they listen and reflect on what is self-reported.

Campbell and McGauley[14] state that some doctors may have to redefine their expectations and reformulate their own attitudes when interacting with people living with chronic conditions. Citing Main,[15] who suggested "the best kind of patient is one who, from great suffering and danger of life or sanity, responds quickly to a treatment that interests his doctor and thereafter remains completely well" (p. 668),[14] Campbell and McGauley say they can see how treating chronic illness can threaten the sense of satisfaction that comes from treating people who more closely resemble Main's description. Consequently, for some people living with chronic conditions, their very experience of living with chronic illness may mean they are unable to function in the role that their doctor thinks best.

Through research with colleagues,[16] we know that people who live with chronic conditions often feel they are different, confronted with diverse and difficult choices, and who struggle making choices given the wide and diverse range of services and information that is available. Health and social care professionals must themselves understand that living with chronic disease involves the person's making sense of their illness, accepting that that they will always have to live with their chronic disease and its ensuing complications:

their chronic disease is not curable. Consequently, where health professionals are inattentive to, or do not explore, the person's views and understandings about their disease and health status and its implications on their social situation, they risk not hearing important clues on how to assist healthy aging and longevity.

Because people living with chronic disease seek to preserve a sense of self and maintain a sense of control over their lives, it is necessary for the health and social care professionals to help them to make sense of their illness and to discern whether this sense-making is appropriately informed by a biomedical understanding of their specific disease and of their social situation. Unless health and social care professionals actively explore with the person just how they make sense of their life, then the potential exists for them to make ill-informed judgments about their clients' health status.

EXPLORING THE THINKING THAT INFORMS SELF-REPORTS

To enable healthy aging and longevity for people living with chronic disease, the ill person needs to know how their decision making must be fully informed by a biomedical understanding of their disease and by their self-surveillance of their health status. Being fixed on their body and their personal response to their chronic disease needs to be understood as an important functional response[13] to living well with chronic disease.

To preserve a sense of self and to exert control over their daily activities, persons living with chronic disease may attempt to show that they are capable of knowing their health status and make sense of "who they are" by what they self-report. While people living with chronic disease may therefore appear to have appropriate explanations for their health status informed by "inside views of their illness experiences" and seem to take responsibility for their health "to preserve a valued sense of self and identity," they may also be deceiving themselves or masking their conditions in such a way that is detrimental to achieving their desired health outcomes.

A mechanism that people living with chronic disease may use is to not let their body (the disease) control their lives.[13] Like biomedicine, the body is objectified as a means of mediating the effects of the disease.[13] Denial of what they may be experiencing at any given moment is used in an attempt to maintain self control rather than control of their disease. This is not the same as denial of their disease.

Russell[17] says denial is a form of self-deception that protects the individual from threats to the self and involves exaggerated perceptions of control and self-efficacy. Kreitler[18] says that denial has a positive effect when applied to the first phase of coping, say after diagnosis with cancer at the terminal stage, because it reduces anxiety, but the negative effects are that it may interfere with getting treatment and disrupt processes to assimilate the stressful event. Levine *et al.,* reporting on findings from a longitudinal study of the relationship between denial of illness and the course of recovery in patients with coronary heart disease, say:

> . . . it is possible to view denial as a dynamic coping "process" that may vary in the precise nature and intensity of its expression over time, and as a function of a variety of contextual factors (p. 116).[19]

How people living with chronic conditions make decisions, how they decide what and what not to inform a health professional about, what ways of thinking affect their decision-making and the choices they make about what they know about their health and health status are all important to explore in the context that is their life—living with chronic disease. This then focuses attention on the cognitive and social skills that people need to gain access to, to understand, and to use information in ways that promote and maintain health.

Health and social care professionals and people living with chronic disease need to explore influences on decision-making, knowing that any decision made about activities to perform or not perform must be made from the perspective that the person who has a chronic disease cannot achieve an end point of *not* living with the chronic disease. That is, a curative approach or an attitude of returning to "normal" is not going to assist people living with chronic disease; their life is always going to be living with chronic disease. This does not discount the fact that people *can* live well with chronic disease. Rather it is that the way of thinking that informs decision-making may need to different.

THE IMPERATIVE OF CRITICAL THINKING

The critical thinking skills needed to know how to live life with chronic disease (or prevent chronic disease) are not skills that every person has or has the capacity to develop. Yet these critical thinking skills would seem to be a necessity, given expectations placed on people themselves and the reliance by health and social care professionals in believing people living with chronic disease are capable of self-care and self-management. People living with chronic conditions may attempt to master their life living with a chronic condition and demonstrate through self-reports or oral accounts how they are capable of taking in hand a self-analytical stance about their health status. But it cannot be uncritically assumed that these people, and health and social care professionals, have the necessary critical thinking skills to be self-analytical in a way that promotes healthy aging; they all may be denying to themselves the implications their actions may be having as they attempt to preserve their sense of self and identity.

To be helpful to people living with chronic disease, health and social care professionals must not uncritically accept these persons' self-reports or interpret their behaviors as either denial or acceptance of living with their disease. Working with people so that they can preserve a sense of self and identity does not discount that they need help to understand the implications of their actions. Consequently, the extent to which self-reported evidence uncritically informs medical and lifestyle research and clinical care is important to debate. Adapting Bauman:

> Under the new circumstances [living with chronic disease (author insert)], the odds are that most of human life and most of human lives will be spent agonizing about the choice of goals, rather than finding the means to the ends which do not call for reflection (p. 61).[5]

Establishing an ongoing reflective and collaborative relationship with people living with chronic disease is of critical importance for health and social care professionals. Health and social care is a continual interplay of different moral, ethical, legal, professional, economic, policy, health and social questions and tensions: such is the messiness of life, and consumers and professionals will always need to understand that they are personally, as well as professionally, morally responsible for their actions. But knowing this does not change knowing that human beings use knowledge to fit their individual purposes. As Flax writes:

> We adopt the knowledge that fits our uses. Humans are very good at creating rational reasons for rejecting knowledge that does not fit our purposes or would make us doubt them (p. 31).[20]

If health and social care professionals promise a "better life" through conformity to treatment or management plans that become a series of tasks to be performed, conditions for doubt and noncompliance are made possible. As well, asking people to be obedient and to conform to set tasks has the potential to set up the conditions for nonconfidence and a lack of self-awareness. These are the very conditions that can make people not take responsibility for their own lives and set them in a personal path of resistance and/or denial to measures purported by health and social care professionals to lead to healthy aging and longevity.

Having the cognitive and social skills to gain access to, understand, and use information to inform choices that are made in ways that promote and maintain good health living with chronic conditions is obviously no easy task. It is known that not all people have the personal characteristics or capacity for reflection or critical thinking enabling them "to use thought to understand thought, which is another way of saying that they have not learned to use thought to gain knowledge" (p. 330).[21] Critical thinking in this sense means the disciplined art of ensuring that you use "the best thinking you are capable of in any set of circumstances" (p. 7).[21] What can be assumed is that people make lifestyle decisions without having the skills to link biomedical understanding of their body and/or the pathology of their chronic conditions with their specific social understandings of their lives. At the same time, it needs questioning whether or not health and social care researchers and practitioners interpret self-reported evidence understanding or linking social understandings of people's lives, especially those living with chronic conditions. Health and social care services continue to foster resistance to management plans where they discount the influence of knowing that people living with chronic disease make their own decisions and at all times will seek to preserve their sense of self and self identity.

CONCLUDING COMMENTS

The human and social consequences of people's actions will always be under scrutiny. Being in good health has become the expectation and requirement of persons living in Western societies. Indeed, in contemporary Western societies, the daily management of lifestyle by health and social care professionals is no longer restricted to those people with a recognized disease, but must be followed by those deemed healthy. A shift away from disease management to disease prevention and health promotion is now argued from the perspective of ensuring people's freedom from ill health and a mechanism for healthy aging and longevity.

To believe that all people have the necessary skills to make healthy choices both negates the uncertainty and unpredictability of health and ill health and the knowledge that education and income levels also influences their capacity to do so.[22] Precisely because of the impact living with chronic conditions has on the lives of people, they attempt to overcome the constraints implicit in their life situation by showing that they are capable of taking in hand their own objectification and adopting toward themselves a self-analytical stance about their health status.

This theoretical paper raises concern about self-reports made by people living with chronic disease to health and social care professionals, not to discount these self-reports, but to challenge the uncritical acceptance of these self-reports as though these reports can only be right or wrong. To understand how to influence choices people make so as to promote healthy aging and longevity, we need to recognize that people living with chronic disease have the capacity for denial as a means to preserve their sense of self and to have a sense of control in their daily lives. At the same time, these people may not able to link biomedical understanding of their disease with their self-surveillance with regard to their health status, social situation and life events, and thus they may unknowingly ignore disease-related changes. Self-reports provide opportunities to explore in collaborative reflective dialogue how best to assist healthy aging and longevity for people living with chronic disease.

REFERENCES

1. DE RIDDER, D., M. FOURNIER & J. BENSING. 2004. Does optimism affect symptom report in chronic disease? What are the consequences for self-care behaviour and physical functioning? J. Psychosom. Res. **56:** 341–350.
2. TELFORD, K., D. KRALIK, & T. KOCH. 2006. Acceptance and denial: implications for people adapting to chronic illness: literature review. J. Advan. Nurs. **55:** 457–464.
3. CENTERS FOR DISEASE CONTROL AND PREVENTION AND THE MERCK COMPANY FOUNDATION. 2007. The State of Aging and Health in America 2007. The Merck Company Foundation. Whitehouse Station, NJ.
4. AUSTRALIAN INSTITUTE OF HEALTH AND WELFARE. 2006. Australia's Health 2006. Cat. No. AUS 73. AIHW. Canberra.
5. BAUMAN, Z. 2000. Liquid Modernity. Polity Press. Cambridge, UK.
6. NATIONAL HEALTH PRIORITY ACTION COUNCIL. 2005. National Chronic Disease Strategy. Australian Government Department of Health and Ageing. Canberra.
7. DEPARTMENT OF HEALTH AND AGEING. 2006. Australian Better Health Initiative. Available at http://www.health.gov.au/internet/wcms/publishing.nsf/Content/feb2006coag03.htm.
8. DEPARTMENT OF HEALTH FOR ENGLAND. 2004. Choosing Health: making Healthy Choices Easier. Department of Health. London.
9. CENTERS FOR DISEASE CONTROL AND PREVENTION. 2003. The Power of Prevention. Reducing the Health and Economic Burden of Chronic Disease. U.S. Department of Health and Human Services. Atlanta. Available at: http://www.cdc.gov/nccdphp/publications/PowerOfPrevention.
10. BAUMAN, Z. 1992. Survival as a social construct. Theory, Culture and Society **9:** 1–36.
11. SHERMAN, D.A.K., L.D. NELSON, & C.M. STEELE. 2000. Do messages about health risks threaten the self? Increasing the acceptance of threatening health messages via self-affirmation. Pers. Soc. Psychol. Bull. **26:** 1046–1058.
12. MIDGLEY, G. 2000. Systemic Intervention: philosophy, Methodology and Practice. Kluwer/Plenum Publishers. New York.
13. PATERSON, B.L. 2001. The shifting perspectives model of chronic illness. J. Nurs. Schol. **33:** 21–26.
14. CAMPBELL, C. & G. MCGAULEY. 2005. Doctor-patient relationships in chronic illness: insights from forensic psychiatry. BMJ **330:** 667–670.
15. MAIN, T. 1957. The ailment. Br. J. Med. Psychol. **30:** 129–145.
16. AUSTRALIAN RESEARCH COUNCIL DISCOVERY PROJECTS DP0346092. Funded 2003 and awarded to University of South Australia; DP0452805 funded 2004 and awarded to University of South Australia. Available at: http://www.arc.gov.au/ncgp/dp/dp_outcomes.htm.
17. RUSSELL, G.C. 1993. The role of denial in clinical practice. J. Advan. Nurs. **18:** 938–940.
18. KREITLER, S. 1999. Denial in cancer patients. Cancer Invest **17:** 514–534.
19. LEVINE, J., S. WARRENBURG, R. KERNS, *et al.* 1987. The role of denial in recovery from coronary heart disease. Psychosom. Med. **49:** 109–117.
20. FLAX, J. 1993. Disputed Subjects: essays on Psychoanalysis, Politics and Philosophy. Routledge. New York.

21. PAUL, R.W. & L. ELDER. 2002. Critical Thinking: tools for Taking Charge of Your Professional and Personal Life. Pearson Education (publishing as Financial Times Prentice Hall). Upper Saddle River, NJ.
22. PRICE, K. 2006. Health promotion and some implications of consumer choice. J. Nurs. Manage. **14:** 494–501.

Health and Aged Care Enabled by Information Technology

JEFFREY SOAR[a] AND YOUNGJOON SEO[b]

[a]*Collaboration for Ageing and Aged-care Informatics Research (CAAIR), University of Southern Queensland, Toowoomba, Australia*

[b]*Department of Health Administration, Yonsei University, Wonju, Korea*

ABSTRACT: One of the challenges facing health and welfare policymakers as well as researchers in most developed countries is the increasing demand for aging services and aged care. Low birth rates and rapid increases in the percentages of elderly people make aging and aged care one of the top-priority issues among the national agenda of many countries. The responses of governments have included initiatives to extend productive working lives and promote self-funded retirement; to promote healthy, active aging; and to encourage more care to be delivered in home and community settings. Technology will be a major enabler of these strategies. People requiring health services are increasingly being offered more care in their own homes and community settings as an alternative to hospital admission and to delay or avoid moving into institutional care. Research is providing intelligent technology to enable care in the home as well as to monitor safety, security, and quality. Innovation will provide greater independence and better access to care in their own homes for the elderly, sufferers of chronic illness, and persons with disability and reduce the incidence of hospital admissions and the length of stay when admissions do occur. Technologies will support families and professional caregivers and are expected to reduce costs. This paper reports on developments in technology to support care for the aged in home and community settings.

KEYWORDS: health care; aged-care; information technology; smart home; telemedicine

BACKGROUND

The age structure of the Australia population has been gradually changing from a pyramid shape in the past, with large numbers in younger age groups and few people reaching old age, to more of a column or coffin-shape, with few people dying before reaching old age.[1] Life expectancy at birth in Australia in

Address for correspondence: Associate Professor Jeffrey Soar, CAAIR, University of Southern Queensland, Toowoomba, QLD, Australia 4350. Voice: +61400 746657; fax: +617 4635 2512. soar@usq.edu.au

Ann. N.Y. Acad. Sci. 1114: 154–161 (2007). © 2007 New York Academy of Sciences. doi: 10.1196/annals.1396.040

2003 was 78 years for males and 83 for females, with a continuing reduction in mortality rates across younger age groups. As the population continues to age, the incidence of dramatic disability restriction and hospital utilization spirals upward.[2] In addition the rapid increase of percentages of consumers aged 65 and over will result in the substantial gap between the availability of care-providers and the demand for care of the aged in the future.[3] There is a worldwide shortage in some areas of professional carers, such as nurses, and there are no longer as many informal carers for elderly parents or others requiring support as there had been in the past. This means we need alternative approaches to providing appropriate health care for the elderly at present as well as in the future.

In response to the challenges posed by aging populations, governments of many developed countries have taken such diverse measures as deferring retirement, promoting more self-funded retirement, increasing the provision of home care, implementation of public insurance schemes for aged care, providing research grants to academics and technologists working in the field of aged care, and providing more support for family carers.

Among these responses is the recognition of the need to explore the application of information technology to health and aged care. Information technology can be expected to better enable older people in independent living and remaining in their own homes for longer and also will aid care delivery in an environment of a shortage of carers. This paper will address what changes might be feasible as a consequence of technology for health and aged care derived from experiences of the application of information technology to the healthcare field. It will discuss current research and challenges, and provide a picture of the future of health and aged care enabled by information technology.

TECHNOLOGY

Technology is expected to enhance the safety and independence of frail older people, enable access to quality care services, and to extend their ability to remain in their own homes.[4] Intelligent systems can monitor patients' vital signs, activity levels, and their safety and security. The technology can monitor indicators of their health status, provide alerts to events such as falls, and give early warnings of potential problems.

Monitors include personal alarms, motion and smoke sensors, and fall detectors. These can be supported by intelligent software that can alert carers or call centers as required. Some monitors keep track of household equipment and how it is being used. This could indicate that a person is no longer cooking for him- or herself or could turn off appliances that have been left on. Sensors are available to turn lights on. They can track and alert for wandering by patients with dementia, and can monitor arrivals and departures of people including the patient or authorized visiting carers. Bed pressure sensors can detect when patients leave their bed during the night, and will trigger an alarm if they do not return within a reasonable time. Monitors can record visits to the bathroom and how long the person might spend there as well as any change in physical activity, such as using the stairs. The technology can alert patients to their own levels of physical activity; and it could set goals, prompting patients to undertake more activity to maintain muscle, bone, and mental fitness.

Older people living alone can be at risk of inadequate nutrition. Sensors on kitchen cupboards, refrigerators and other electric appliances can indicate when the patient is eating and drinking and pass this information to carers. Systems can record patterns of typical behavior over time, note changes, and alert carers when appropriate. Monitoring devices can be more accurate guides to health risks than are patients' own awareness of their symptoms. Devices can monitor important indicators for disease management.

Existing home security monitors in Australia use the simple landline telephones and some additional functionality can still be provided using that service. Broadband, however, will be required for always-on continuous monitoring, especially for image capture and transmission. RFID, Bluetooth, and Wi-Fi technology will allow connectivity of devices around the home and in the community without cabling. Passive RFID tags can absorb enough power from the sensing device to perform simple functions. Devices that are powered electrically can use Ethernet Over Power.

RESEARCH

While the number and sophistication of available technologies continue to increase, there is a need for research into adoption issues, return on investment, realization of benefits, integration, and interoperability. There are gaps in the range of technologies and particularly in intelligent software and interfacing.

As health information technology has developed rapidly in the past two or three decades, it is expected to have a positive impact on the access, quality, and cost of health and medical care.[5] For example, it would be easier for people living in rural areas, nursing homes, and mental institutions to receive care, including consultations, through telemedicine. Health information technology is also expected to help facilitate high-speed communication and transfer of patient records between distant medical centers.[6]

Information technology is expected to have diverse positive outcomes on the quality of health care, such as the reduction of time-to-treatment and improved awareness of clinicians, which means clinicians are better informed and updated with recent advances by telemedicine.[7] It is also expected to have positive impacts on the quality of health care by improving working conditions for clinicians particularly, in rural and remote areas.[7]

There are analyses of the effectiveness and cost savings of information technology and reports on particular aspects of telemedicine, such as teleradiology and telepathology, but not for more comprehensive networks.[7] Of particular interest are the impacts on care and the financial benefits of monitoring and management of patients in remote locations.[7]

According to recent literature which reviewed systematically the impact of health information technology on quality, efficiency, and costs of health care, health information technology has been shown to improve quality by increasing adherence to guidelines, enhancing disease surveillance, and by decreasing errors.[5] In primary and secondary care, the impact of information technology on quality has been through facilitating care guidelines or protocols. Information technology enables management and analysis of massive healthcare databases that would not be feasible with paper-based systems.[5]

There are some studies on the effect of information technology applied to health care in Korea. A result of a study conducted in Korea revealed that hospitals that implemented more information technology-related strategies (for example, computerized physician order entry, picture archiving and communication systems [PACS], Enterprise Resource Planning [ERP], billing and medical data transportation by Electronic Data Interchange [EDI] and other similar systems) showed a higher financial performance than those that did not.[8] Another study examined the effects of a Web-based health information service system for the elderly and revealed that users rated it very useful, relevant, convenient, and efficient for preventing disease and promoting health.[9] A study revealed that most users were comfortable and satisfied with the telemedicine in primary care.[10]

SMART HOMES

Efforts to apply information technology for aging and aged care are recently being made to workplaces, homes, vehicles, wearable or pervasive goods, and personal care assistants. There is worldwide interest in so-called "smart homes" for both enhancing wellness as well as supporting people with disabilities and/or healthcare needs. Smart homes involve embedding assistive technologies with communication links to carers and clinicians to support the frail elderly, particularly in staying in their own homes rather than in institutions.[11] A smart home is usually embedded with the following technologies: sensor networks, home networks, activity tracking, ambient displays, context-aware memory aids, and location tracking and alert systems.

An example of a smart home environment includes the following:

1. Sensor networks build environments equipped with sensors for movement, location, etc. In home-care settings these may use in-building location services. In aged care, it can refer to:

a. bed-sensors that monitor when a person is in bed and how many times he gets up in the night, and it will send an alert if he does not return to bed;
b. monitors for falls; and
c. movement detectors that build the normal patterns of movement and can provide an alert if someone deviates from her usual pattern, or these can provide incentives for people to maintain physical activity.

2. Home network is sensor networks in the home.
3. Activity tracking, that is, tracking the activity of a patient will provide information on levels of physical activity, including changes in pace in, for example, navigating stairs; how many times and how long someone needs to visit the toilet at night; whether he or she is using devices, such as the stove and refrigerator, which might indicate whether he or she is eating and whether he or she was deviating from daily patterns of activity.
4. Ambient displays include lights or other indicators that provide background information. For example in aged care, the Intel project has a lamp that goes on to show the elderly person when her daughter's family are in their own home; the elderly person then she knows they are there and can call them by telephone if necessary.
5. Context-aware memory aids are applications that watch a user's context and proactively suggest information that may be of use, such as prompts or other reminders specific to the person's activity. For example, if someone switches on the coffee machine, a screen might display the steps involved in making a cup of coffee—it is common things like this that people with dementia forget how to do—or reminders to take medication. Sensors that become aware of what a person is doing can reminds the person of where they were up to and what the next step is.

Additional technology might include location and alert systems for tracking wandering patients and providing an alert to someone responsible if they were considered to be at risk.

Telehealth is the use of telecommunications to deliver care services across distances and can enhance communications between patients, their families, and clinicians. Videophones have been used in tele-hospice programs for the terminally ill,[12] home telecare studies,[13] telerehabilitation support,[14] and disease-specific interventions.[15] Examples such as these indicate the feasibility of the technology and other studies have indicated cost-effectiveness.[16,17]

Videophone technology has been found to assist communication and the participation of families in care planning when they live some distance from a family member in a nursing home. Videophones improved communication between nursing home residents and their family members, reduced isolation, and increased family support.

Nevertheless, the market for smart-home services for senior housing is still barely existent in many developed countries including Australia and Korea.

This phenomenon may have resulted from the fact that providers of smart homes have been in practice very focused on the technical possibilities, rather than solving the everyday problems of end-users. Important reasons for this are insufficient capital, little knowledge about how to involve the user, lacking ambition to work in a multidisciplinary way, and unprofitable market conditions. Thus the challenges we are facing in aging and aged care is the development of user-oriented products, education and training of users for new products and services, internal business organization so that the provider of the product draws on the resources of a multidisciplinary team, and the profitability of market, not the technology.

Aging and aged care as we know it will be transformed as technology enables extending active and productive lives and facilitates more care to be delivered according to consumer preferences for place and time. These changes will reflect the impacts of technologies that have transformed other industries, enabled new products and services, and provided delivery at the convenience of the consumer.

Just as assistive technology aids airline pilots, it will compensate for decline in physical or cognitive fitness, allowing professional lives to be extended and to extend the reach of clinical care into homes. It will enable seniors to choose how long and how much they will work. Health and aged care will be observable and safe when and where it is wanted. The systems will provide a virtual personal care assistant; it will know how to find your records, it will assist clinicians in linking to research, and it will help the elderly person to stay independent and connected. It will also support families as part of the care team for their frail elderly parents, even when they live large distances apart.[18] From these perspectives, it is not unreasonable to anticipate that information technology will become the core discipline of health and aged care.

SUMMARY

There will be many opportunities for new and imaginative research from aging of societies. Some urgent research requirements include developing standards of elder care, identifying the issues, proof-of-concept projects, real-world testing, proving the business case,[19] promoting technology for quality care, consumer-oriented care, and electronic health records. While some of the devices are emerging we do not yet know where or how to deploy technology effectively to deliver both the care and the financial benefits. Research is needed to guide changes that will be required in policy, strategy, funding, work practices, integration of technology into care, future roles for the professions, and new models of care. A multi-discipline consortium is one approach that is already being pursued to provide a platform for enhancing the scale of related research activity, stimulate and commercialize innovation, and support collaboration between researchers to increase their impact.

The Collaboration for Aging and Aged Informatics Research at the University of Southern Queensland in Australia is an example of a multi-discipline consortium: that is, it brings together researchers from across the Faculties of Business, Science and Engineering; as well as partners from universities across Australia and from Japan and Korea, from state and federal governments, and from aged care providers and ICT companies. Current research projects include a hospital-avoidance pilot and the establishment of a smart home as a research facility.

We are on the cusp of a major social change in healthcare delivery across the world. Technology has enabled transformation in many industries. There is interest in technology to enable societies to manage the pressures of aging populations. Consequently it is not unreasonable that greater deployment of technology will result in a transformation of health care from being provider-driven to consumer focused, from delivery at the provider's convenience to the time and place of convenience of the consumer, from a focus on the acute hospital care more toward community care, prevention, and maintenance of wellness. Technology can be expected to enable a vision of care that is more evidence-based and safe.

REFERENCES

1. PRODUCTIVITY COMMISSION. 2005. Economic Implications of an Ageing Australia, Research Report. Canberra.
2. AUSTRALIAN INSTITUTE OF HEALTH AND WELFARE. 2006. Australian Hospital Statistics 2004–05. http://www.aihw.gov.au/hospitals/ataglance/2004/index.cfm (accessed July 16, 2007).
3. National Centre for Social and Economic Modelling. 2006. Who cares? The cost of caring in Australia 2002 to 2005. AMP. NATSEM Income and Wealth Report Issue, May 13, p. 1.
4. CHEEK, P., L. NIKPOUR & H. NOWLIN. 2005. Aging well with smart technology. Nurs. Admin. Q. **29:** 329–338.
5. CHAUDHRY, B., J. WANG, S. WU, *et al.* 2006. Systematic review: impact of health information technology on quality, efficiency, and costs of medical care. Ann. Intern. Med. **144:** 742–752.
6. PARSONS, D. 1994. The impact of information technology on health care: a practitioner's perspective. Telematics Informatics **11:** 127–135.
7. COURTNEY, K., G. DEMIRIS & G. ALEXANDER. 2005. Information technology: changing nursing processes at the point-of-care. Nurs. Admin. Q. **29:** 315–322.
8. SEO, Y. 2001. Strategic orientation and behaviors of Korean hospitals. Korean Journal of Hospital Management **6:** 173–201.
9. PARK, H.A., H. KIM, M. SONG, *et al.* 2002. Development of web-based health information service system for the elderly. Korean Journal of Hospital Management **8:** 37–45.
10. LIU, T.W., H. KO & S. OH. 2003. Tele-primary care and patient satisfaction in Korea. Korean Journal of Hospital Management **9:** 17–24.
11. ESSEN, A. & M. CONRICK. 2007. Visions and realities: developing 'smart' homes for seniors in Sweden. eJ. Health Informatics **2(1):** e2.

12. PARKER, D., G. DEMIRIS, H. DAY, *et al.* 2005. Tele-hospice support for elderly caregivers of hospice patients: two case studies. J. Palliat. Med. **9:** 264–267.
13. JOHNSON, B., L. WHEELER, J. DUESER & K. SAUSA. 2000. Outcomes of the Kaiser Permanente tele-home health research project. Arch. Family Med. **9:** 44–45.
14. HAUBER, R.P. & M.L. JONES. 2002. Telerehabilitation support for families at home caring for individuals in prolonged states of reduced consciousness. J. Head Trauma Rehab. **17:** 535–451.
15. DEMAIO, J., L. SCHWARTZ, P. COOLEY & A. TICE. 2001. The application of telemedicine technology to a directly observed therapy program for tuberculosis: a pilot project. Clin. Infect. Dis. **33:** 2082–2084.
16. DOOLITTLE, G. 2000. A cost measurement study for a home-based telehospice service. J. Telemedicine Telecare **6**(Suppl 1): s193–s195.
17. NAKAMURA, K., T. TAKANO & C. CAKAO. 1999. The effectiveness of videophones in home health care for the elderly [see comment]. Med. Care **37:** 117–125.
18. MICKUS, M.A. & C.C. LUZ. 2002. Televisits: sustaining long distance family relationships among institutionalized elders through technology. Aging Ment. Health **6:** 387–396.
19. VIMARLUND, V. & N.G. OLVE. 2005. Economic analysis for ICT in elderly health care: questions and challenges. Health Informatics J. **11:** 309–313.

Healthy Aging as an Intervention to Minimize Injury from Falls among Older People

NANCYE MAY PEEL,[a] HELEN PATRICIA BARTLETT,[b] AND RODERICK JOHN McCLURE[b]

[a]*Australasian Centre on Ageing, The University of Queensland, St. Lucia, Queensland 4072, Australia*

[b]*School of Medicine, Griffith University, Meadowbrook, Queensland 4131, Australia*

ABSTRACT: With global trends toward population aging, many countries are adopting healthy aging policies to minimize disability and increase quality in the extended years of life. Falls in older people are a major contributor to functional decline generally associated with aging. Based on a study quantifying the relationship between healthy aging factors and risk of fall-related hip fracture in community-dwelling older people, this paper discusses evidence for the promotion of healthy aging as a population-based intervention for prevention of injuries from falls. To examine the protective effect of healthy aging on the risk of fall-related hip fractures, a case–control study was conducted with 387 participants. Persons aged 65 and over hospitalized with a fall-related hip fracture were matched with community-based controls recruited via electoral roll sampling. A questionnaire designed to assess lifestyle risk factors, identified as determinants of healthy aging, was administered during face-to-face interviews. After adjustment for health status and demographic factors, a number of lifestyle factors were seen to have a significant independent protective effect on the risk of hip fracture. These included never smoking, moderate alcohol consumption, being active, maintaining normal weight, and being proactive in preventive health care. Psychosocial factors included having supportive environments and personal resources to cope with stress. This study identified a range of modifiable lifestyle factors associated with fall-related hip fracture, suggesting that the "healthy aging" paradigm offers a comprehensive approach to falls injury prevention, and thus supports the adoption of healthy aging policies to extend years of quality life among older persons.

KEYWORDS: healthy aging; hip fracture; case–control study; behavioral and psychosocial factors; aged population; falls injury prevention

Address for correspondence: Dr. Nancye May Peel, Postdoctoral Research Fellow, Australasian Centre on Ageing, The University of Queensland, St. Lucia, Queensland 4072, Australia. Voice: + 61 7 33467577; fax: +61 7 33657503.

n.peel@uq.edu.au

Ann. N.Y. Acad. Sci. 1114: 162–169 (2007).
doi: 10.1196/annals.1396.026

INTRODUCTION

With global trends in population aging, many countries are developing and implementing healthy aging policies to minimize disability and increase quality in the extended years of life.[1] While a number of terms have been used interchangeably with "healthy aging," such as "successful," "active," or "positive aging,"[1] the notion of healthy aging has been characterized as a multidimensional concept, encompassing the avoidance of disease and disability, the maintenance of high physical and cognitive function, and sustained engagement in social and productive activities.[2] The World Health Organization emphasizes strategies that aim to promote well-being and to minimize disability by reducing risk and increasing factors that protect health throughout the life-course, using population health measures.[3]

A priority area for promotion of health in aging is the prevention of falls and fall-related injuries, which cause substantial disability and have a great impact on the functional independence and quality of life of older people.[4] Hip fractures are one of the most serious of fall-related injuries in terms of morbidity and mortality. The incidence of hip fractures in both the developed and developing world is increasing with aging of the population,[5] and these injuries present a serious challenge to healthcare systems in future years.[6]

Research suggests that falls prevention is an effective intervention in minimizing the functional decline that is generally associated with aging.[7] Currently, the majority of falls prevention initiatives are based on screening patients at high risk of falling and implementing targeted multi-factorial interventions aimed at lessening both intrinsic and extrinsic risk factors present at the time of the fall.[8,9] In contrast to the high-risk strategy, population strategies aim to change the distribution of risk in the whole population.[10] However, such approaches to falls injury prevention are rare,[11] due in part to lack of evidence of population-wide risk factors which can be addressed by community-based strategies.

A framework that has the potential to address population-based risk factors is a comprehensive hierarchical health determinants model, which allows for the examination of more distal factors that influence health outcomes over the life-course and contribute to falls risk. Such a model also has the potential to tailor interventions appropriate to life stages using healthy aging strategies. The conceptual framework for this study explored as an *a priori* aim the protective effects of healthy aging on the risk of fall-related hip fracture injury in older community-dwelling people. A broad-based hierarchical analytic model was developed, organizing health determinants distally from the behavioral and psychosocial determinants (factors of interest) to the more proximal biological and medical factors (confounders) which resulted in the adverse health outcome (fall-related hip fracture).

Using data from the study, this paper discusses the application of the healthy aging paradigm to the prevention of fall-related injury.

METHODS

The study design and results have been published previously.[12,13] In brief, a case–control study with 387 participants was conducted in Brisbane, Queensland, on the eastern coast of Australia. Eligible subjects were aged 65 years or over and admitted to one of six metropolitan hospitals for treatment of a fall-related hip fracture. For each case, at least two controls were randomly selected from the electoral roll, after matching for sex, age (within a 5-year age range), and postcode. Residents of high-care institutions were ineligible.

On the basis of literature review of the measurement and determinants of healthy aging,[1,14] lifestyle factors that might plausibly be related to hip fracture were included in a questionnaire administered during face-to-face interviews. Behavioral factors included smoking status and history, alcohol consumption patterns, body weight and perceptions of weight, diet, physical activity and participation in sport, and preventive healthcare actions. Psychosocial factors included living arrangements, community integration, social support, and social resources. Psychological well-being was measured by questions on self-rated health, emotional health, stress and coping mechanisms, and life satisfaction. Life engagement measures included productive and social activities. Where applicable, data were collected over life stages, defined as older age (65 and over), middle age (40–64), and young adulthood (15–39).

In addition to the behavioral and psychosocial factors of interest, health status measures included functional ability, co-morbidities, sensorimotor impairments, and history of falls. A comprehension test assessing orientation to person, place, and time was a measure of cognitive ability. Demographic data, such as age, sex, education, financial status, and country of birth, were also recorded. Details of the measures for these self-reported variables have previously been described.[12,13]

Results were analyzed using conditional logistic regression models to produce odds ratios (ORs) with 95% confidence intervals (CIs) to examine differences between cases and controls on distribution of healthy aging factors. Using model building guidelines,[15] multivariate models were developed for behavioral and psychosocial factors separately by adding, in step-wise progression, factors significant in univariate analysis at the probability (P) value of less than 0.05. Potential confounders to add to the models were chosen from health status factors shown to be significant predictors of outcome in univariate analysis and on the basis, from understanding of the literature, that they could plausibly be associated with the factors of interest. The purpose of the study was to develop an investigative model rather than a predictive model of all risk factors, and the study factors were based on potentially modifiable behavioral and psychosocial determinants of healthy aging.

RESULTS

After adjustment for health status, behavioral factors that had a significant independent protective effect on the risk of hip fracture included never smoking [OR: 0.33 (0.12–0.88)], moderate alcohol consumption in mid- and older age [OR: 0.49 (0.25–0.95)], not losing weight between mid- and older age [OR: 0.36 (0.20–0.65)], playing sports in older age [OR: 0.49 (0.29–0.83)], and practicing a greater number of preventive medical care [OR: 0.54 (0.32–0.94)] and self-health behaviors [OR: 0.56 (0.33–0.94)].[12]

Psychosocial factors that had a significant independent protective effect on the risk of hip fracture included currently being married [OR: 0.44 (0.22–0.88)], living in present residence for 5 years or more [OR: 0.43 (0.22–0.84)], having private health insurance [OR: 0.49 (0.27–0.90)], using proactive coping strategies in response to stress [OR: 0.52 (0.29–0.92)], having a higher level of life satisfaction [OR: 0.47 (0.27–0.81)], and engagement in social activities in older age [OR: 0.30 (0.17–0.54)].[13]

DISCUSSION

This study provides evidence that the behavioral and psychosocial determinants of healthy aging are protective of the special case adverse outcome of fall-related hip fracture injury in older people. Previously, there have been few studies which examined as an *a priori* aim, the association between healthy aging determinants and hip fracture in older people, particularly of protective factors over the life-course. The case–control study design has its limitations; in particular, causal relationships are difficult to establish. The results would need to be corroborated in prospective studies to draw causal inferences.

Much of the corroborative evidence for the association of behavioral factors with hip fracture risk comes from epidemiologic studies of osteoporosis, which have shown an effect of lifestyle factors on skeletal health, mediated mainly through lowered bone mineral density.[16–19] While the evidence supporting the association of behavioral factors with hip fracture risk is strong and biologically plausible, the mechanisms by which psychosocial factors protect against hip fracture are less well documented.

The possible explanation is that psychosocial support has beneficial effects on health behaviors, protecting against, for example, exposure to poor diet, medication misuse, physical inactivity, and social isolation, as well as increased frailty,[20] all of which are associated with increased risk of hip fracture.[20,21] Having higher levels of social resources could conceivably influence access to a range of opportunities for support in times of need, which in turn could lead to better health outcomes and protection against falls.[22] Psychological well-being indicators such as absence of psychological distress, high self-efficacy, proactive adaptability, and satisfaction with life provide a buffer against adverse health outcomes.[23] Psychological distress, especially depression, is a risk factor for fall-related fractures through either direct effect on bone mass or use of medication.[24] Low perceived self-efficacy is a falls risk because of associated fear of falling and consequent activity restriction.[25] Engagement in social and productive activities has been associated with positive psychological well-being,[26,27] as well as protecting against cognitive decline[28] and physical functioning limitations.[29] Physical functioning and cognitive status are significantly associated with hip fracture risk.[30]

IMPLICATIONS FOR POLICY AND PRACTICE

Healthy aging and falls injury prevention have been identified as public health priorities in response to the challenges of population aging. Strategies for addressing these priorities are more likely to be assisted by a public health model targeting the population as a whole, rather than a medical model targeting the high-risk individual. The medical model seeks to diagnose (or screen for) the risk condition and then to treat those found to have the adverse condition. The public health model, in contrast, seeks to prevent adverse health in all who may be susceptible.[31]

While treating risk factors can substantially reduce an individual's risk of hip fracture, it is difficult to reduce substantially the rate of hip fractures in a population by programs aimed at changing risk factors at the individual level.[32] The whole-of-population approach can lead to greater population health gains than those achieved by strategies which target only high-risk individuals.[10] Since almost everyone in the population is at risk for sustaining a fracture, and the pathophysiological processes have their genesis in early ages,[33] population-based interventions across the life-course may offer the greatest potential for prevention.

In devising health promotion programs for older adults, most attention has been paid to the role of the behavioral risk/protective factors of healthy aging.[34] Population-based interventions aimed at smoking cessation, preventing alcohol, drug, and medication misuse, promoting physical activity, preventing obesity, and discouraging excessive exposure to sunlight have been designed to raise public awareness of health risks associated with behavioral risk factors. Such programs mostly target the young and middle-aged populations, for these precursors to morbidity have their genesis in early ages and are cumulative in their impact.[35,36] However, while people's health and ability to function in older age depend on their exposure to various influences throughout their life, it is recognized that it is never too late to start, and that older people also benefit from health promotion activities.

The evidence has also shown that personal, social, and environmental factors play a key role in promoting healthy aging and preventing fall-related hip fractures, so that the importance of interventions to enhance the social, work, and built environments for older people are increasingly being recognized. Opportunities for healthy and productive aging can be increased substantially by facilitating participation in both the paid and volunteer workforce. Community development programs contribute to societal engagement,[37] build social capital,[38,39] and reduce social isolation.[40] Creating "age-friendly" built environments, and access to assistive technology provides opportunities for independence, "aging in place," and healthy lifestyles through increased physical and social activity. Valuing the significant contributions made by older people to society and emphasizing the positive aspects of aging through intergenerational programs can assist in breaking down age stereotypes and ageist attitudes.[41]

The major challenge is to discover the best ways to apply this evidence about behavioral change and supportive environments to encourage people to adopt healthy actions. It is also important to identify the key points in the life span at which interventions have the greatest benefits. Further research will be needed to evaluate interventions, to understand facilitators and barriers to their adoption, and to determinate the best means for dissemination of good practice.

CONCLUSION

It is opportune now to take a more unified approach to the prevention of geriatric syndromes (including falls),[42] using common pathways to preserving health and promoting disability-free life expectancy. The research presented in this paper suggests that by optimizing opportunities for improving and preserving physical, social, and mental wellness, independence and quality of life—that is, using the goals of "healthy aging"—reduction in the public health burden of hip fractures as a result of falls may be achieved.

REFERENCES

1. PEEL, N.M., H.P. BARTLETT & R.J. MCCLURE. 2004. Healthy ageing: how is it defined and measured? Australas J. Ageing **23:** 115–119.
2. ROWE, J.W. & R.L. KAHN. 1997. Successful aging. Gerontologist **37:** 433–440.
3. WORLD HEALTH ORGANIZATION. 2002. Active Ageing: a Policy Framework. World Health Organization. Geneva.
4. KANNUS, P. *et al.* 2005. Prevention of falls and consequent injuries in elderly people. Lancet **366:** 1885–1893.
5. CUMMINGS, S.R. & L.J. MELTON. 2002. Epidemiology and outcomes of osteoporotic fractures. Lancet **359:** 1761–1767.
6. MOLLER, J. 2005. Current costing models: are they suitable for allocating health resources? The example of fall injury prevention in Australia. Accid. Anal. Prev. **37:** 25–33.

7. FREEDMAN, V.A. *et al.* 2006. Promoting declines in the prevalence of late-life disability: comparisons of three potentially high-impact interventions. Milbank Q. **84:** 493–520.
8. AMERICAN GERIATRICS SOCIETY, BRITISH GERIATRICS SOCIETY & AMERICAN ACADEMY OF ORTHOPAEDIC SURGEONS PANEL ON FALLS PREVENTION. 2001. Guideline for the prevention of falls in older persons. J. Am. Geriatr. Soc. **49:** 664–672.
9. GILLESPIE, L.D. *et al.* 2003. Interventions for preventing falls in elderly people. Cochrane Database Syst. Rev. DOI: 10.1002/14651858.CD000340.
10. ROSE, G. 1992. The Strategy of Preventive Medicine. Oxford University Press. Oxford.
11. MCCLURE, R. *et al.* 2005. Population-based interventions for the prevention of fall-related injuries in older people. Cochrane Database Syst. Rev. DOI: 10.1002/14651858.CD004441.pub2.
12. PEEL, N.M., R.J. MCCLURE & J.K. HENDRIKZ. 2006. Health-protective behaviours and risk of fall-related hip fractures: a population-based case-control study. Age Ageing **35:** 491–497.
13. PEEL, N.M., R.J. MCCLURE & J.K. HENDRIKZ. 2007. Psychosocial factors associated with fall-related hip fractures. Age Ageing **36:** 145–151.
14. PEEL, N.M., R.J. MCCLURE & H.P. BARTLETT. 2005. Behavioral determinants of healthy aging. Am. J. Prev. Med. **28:** 298–304.
15. HOSMER, D.W. & S. LEMESHOW. 2000. Applied Logistic Regression. Wiley. New York.
16. DE LAET, C. *et al.* 2005. Body mass index as a predictor of fracture risk: a meta-analysis. Osteoporos. Int. **16:** 1330–1338.
17. JOAKIMSEN, R.M., J.H. MAGNUS & V. FONNEBO. 1997. Physical activity and predisposition for hip fractures: a review. Osteoporos. Int. **7:** 503–513.
18. KANIS, J.A. *et al.* 2005. Alcohol intake as a risk factor for fracture. Osteoporos. Int. **16:** 737–742.
19. KANIS, J.A. *et al.* 2005. Smoking and fracture risk: a meta-analysis. Osteoporos. Int. **16:** 155–162.
20. WOODS, N.F. *et al.* 2005. Frailty: emergence and consequences in women aged 65 and older in the Women's Health Initiative Observational Study. J. Am. Geriatr. Soc. **53:** 1321–1330.
21. HOKBY, A., A. REIMERS & L. LAFLAMME. 2003. Hip fractures among older people: do marital status and type of residence matter? Public Health **117:** 196–201.
22. FAULKNER, K. *et al.* 2003. Is social integration associated with the risk of falling in older community-dwelling women? J. Gerontol. A Biol. Sci. Med. Sci. **58:** M954–M959.
23. KAHANA, E., B. KAHANA & K. KERCHER. 2003. Emerging lifestyles and proactive options for successful ageing. Ageing Int. **28:** 155–180.
24. FORSEN, L. *et al.* 1999. Mental distress and risk of hip fracture. Do broken hearts lead to broken bones? J. Epidemiol. Commun. Health **53:** 343–347.
25. FLETCHER, P.C. & J.P. HIRDES. 2004. Restriction in activity associated with fear of falling among community-based seniors using home care services. Age Ageing **33:** 273–279.
26. HERZOG, A.R., M.B. OFSTEDAL & L.M. WHEELER. 2002. Social engagement and its relationship to health. Clin. Geriatr. Med. **18:** 593–609, ix.

27. Greenfield, E.A. & N.F. Marks. 2004. Formal volunteering as a protective factor for older adults' psychological well-being. J. Gerontol. B Psychol. Sci. Soc. Sci. **59:** S258–264.
28. Zunzunegui, M.V. *et al.* 2003. Social networks, social integration, and social engagement determine cognitive decline in community-dwelling Spanish older adults. J. Gerontol. B Psychol. Sci. Soc. Sci. **58:** S93–S100.
29. Luoh, M.-C. & A.R. Herzog. 2002. Individual consequences of volunteer and paid work in old age: health and mortality. J. Health Soc. Behav. **43:** 490–509.
30. Wilson, R.T. *et al.* 2006. Hip fracture risk among community-dwelling elderly people in the United States: a prospective study of physical, cognitive, and socioeconomic indicators. Am. J. Public Health **96:** 1210–1218.
31. Fries, J.F. 1996. Prevention of osteoporotic fractures: possibilities, the role of exercise, and limitations. Scand. J. Rheumatol. Suppl. **103:** 6–10.
32. Cummings, S.R. 1996. Treatable and untreatable risk factors for hip fracture. Bone **18:** 165s–167s.
33. Lauritzen, J.B., P.A. McNair & B. Lund. 1993. Risk factors for hip fractures: a review. Dan. Med .Bull. **40:** 479–485.
34. Minkler, M., H. Schauffler & K. Clements-Nolle. 2000. Health promotion for older Americans in the 21st century. Am. J. Health Promot. **14:** 371–379.
35. Green, L. & N. Gottlieb. 1989. Health promotion for the aging population: approaches to extending life expectancy. *In* Health Care for an Aging Society. S. Andreopoulos & J.R. Hogness, Eds.: 139–154. Churchill Livingstone. New York.
36. Vita, A.J. *et al.* 1998. Aging, health risks, and cumulative disability. N. Engl. J. Med. **338:** 1035–1041.
37. Anetzberger, G.J. 2002. Community resources to promote successful aging. Clin. Geriatr. Med. **18:** 611–625.
38. Cannuscio, C., J. Block & I. Kawachi. 2003. Social capital and successful aging: the role of senior housing. Ann. Intern. Med. **139:** 395–399.
39. Fried, L.P. *et al.* 1997. Building communities that promote successful aging. West. J. Med. **167:** 216–219.
40. Cattan, M. *et al.* 2005. Preventing social isolation and loneliness among older people: a systematic review of health promotion interventions. Ageing Soc. **25:** 41–67.
41. Bartlett, H.P. & N.M. Peel. 2005. Healthy ageing in the community. *In* Ageing and Place. Perspectives, Policy, Practice. G.J. Andrews & D.R. Phillips, Eds.: 98–109. Routledge. New York.
42. Tinetti, M.E. *et al.* 1995. Shared risk factors for falls, incontinence, and functional dependence: unifying the approach to geriatric syndromes. JAMA **273:** 1348–1353.

Building an Infrastructure to Prevent Falls in Older Californians

The Fall Prevention Center of Excellence

DEBRA J. ROSE,[a] GRETCHEN E. ALKEMA,[b] IN HEE CHOI,[b] CHRISTY M. NISHITA,[b] AND JON PYNOOS[b]

[a]*Fall Prevention Center of Excellence, California State University, Fullerton, California, USA*

[b]*Fall Prevention Center of Excellence, Andrus Gerontology Center, University of Southern California, Los Angeles, California, USA*

ABSTRACT: The Fall Prevention Center of Excellence (Center), a consortium of federal, state, and private organizations, was established in 2005 to guide the implementation of a statewide initiative to prevent falls among older Californians. The process began with the convening of a representative group of recognized leaders in California's health and human services in 2003. This group engaged in a 2-day strategic planning process that culminated in the development of the California Blueprint for Fall Prevention. The overarching goal of the Blueprint is to build a statewide infrastructure for fall prevention services and programs that will serve as a model for the rest of the country. The specific goals of the Center are to establish fall prevention as a key public health priority in California; create, test, and evaluate effective and sustainable fall prevention programs; and build a comprehensive and sustainable fall prevention system in California. To accomplish these goals, the Center is currently engaged in developing and disseminating fall prevention tools and informational resources directed at the needs of both consumer and professional audiences; linking organizations involved in fall prevention while increasing awareness of fall prevention as an important public health issue; and helping communities build their capacity to effectively address falls in older adults through the delivery of integrated fall prevention services and "best practice" programs.

KEYWORDS: falls; older adults; Fall Prevention Center of Excellence; California Blueprint for Fall Prevention

Address for correspondence: Debra J. Rose, Ph.D., Fall Prevention Center of Excellence, KHS 245, California State University, Fullerton, CA 92834. Voice: 714-278-5846; fax: 714-278-5666. drose@fullerton.edu

Ann. N.Y. Acad. Sci. 1114: 170–179 (2007). © 2007 New York Academy of Sciences.
doi: 10.1196/annals.1396.035

INTRODUCTION

California is not only one of the fastest growing states within the United States of America, but it also has an older adult population that is expected to increase at more than double the rate of the total population. In 2005, there were over 3.7 million Californians age 65 and over, and it is projected that this number will increase to more than 8.3 million by 2030.[1] An inevitable consequence of this rapid growth, however, is the need to expand services and programs required by an aging population. The need for fall prevention services and programs, in particular, is underscored by the knowledge that one-third of older Californians fall each year, with approximately 1.3 million older adults suffering injuries from falls, such as hip fractures.[2] Fall-related injuries among older adults also lead to an estimated 213,000 visits to the emergency room and more than 60,000 hospitalizations.[2] Moreover, fall injuries cost California approximately $3.5 billion each year, with hospitalizations averaging $30,000 per admission.[3] Given the significant demographic transition and the epidemiology of falls, there is a pressing need to build a stronger infrastructure for preventing falls among older Californians.

THE CALIFORNIA BLUEPRINT FOR FALL PREVENTION

Although the state of California has legislation addressing the needs of older Californians (Senior Wellness Act of 2001) and a number of state-initiated programs that focus on assisting older Californians to "age-successfully-in-place," the state lacked a comprehensive plan that would coordinate and advance efforts to prevent falls among older Californians. In April 2002, the Archstone Foundation, based in Long Beach, California, invited key stakeholders to be part of a planning committee to strategize the development of a statewide fall prevention blueprint. On February 5–6, 2003, more than 140 leaders of key stakeholder groups from throughout California convened in Sacramento. Representatives reflected a variety of perspectives including those of aging services and programs; public health; legislatures at the local, regional and statewide level; voluntary organizations; academic faculty in geriatric medicine and gerontology; and the healthcare industry. In preparation for the 2-day conference, a white paper titled "Preventing Falls in Older Californians: State of the Art"[4] was prepared by experts in fall prevention and distributed to all attendees. A consensus statement of principles important for establishing a statewide program was also prepared by the conference planning committee and used to guide the strategic planning process (see APPENDIX A). The 2-day conference resulted in the development of the California Blueprint for Fall Prevention, which called for improved communication and coordination between existing programs and services; training of professionals in community-based practice; prevention efforts that address the racial and cultural diversity in California; stable funding for fall prevention; and greater public awareness and education. High on the agenda of attendees was the creation of an entity that would help coordinate activities and provide state-of-the-art information about effective fall prevention strategies.

DEVELOPMENT AND MISSION OF THE FALL PREVENTION CENTER OF EXCELLENCE

The Fall Prevention Center of Excellence (hereafter the Center), as a public–private partnership, was established in January 2005 to ensure the independence, safety, and well-being of older adults through fall prevention efforts by providing leadership, creating new knowledge, improving practice, and developing sustainable fall prevention programs at the individual, program, and systems levels. As a consortium of federal, state, and private organizations, the Center consists of the following four organizations, each of which has a strong background in applied, clinical, and health services research: (1) the Andrus Gerontology Center at the University of Southern California; (2) California State University, Fullerton's Center for Successful Aging; (3) the Veterans Administration Greater Los Angeles Healthcare System Geriatric Research, Education and Clinical Center, along with University of California, Los Angeles; and (4) the California Department of Public Health, State and Local Injury Control. Funded by the Archstone Foundation, the Center has been built on the individual expertise of its partners and their years of collaborations through the planning and coordination of the statewide conference, publications, as well as presentations at professional conferences. The programs encompassed by the Center are shown in FIGURE 1.

With its interdisciplinary team, the Center serves a broad audience including older adults and their families, professionals, program administrators, and policymakers in fall prevention initiatives by employing an evidence-based multifactorial fall prevention approach that includes risk assessment and management, exercise, and environmental modification to reduce falls among older adults. The Center not only provides basic information about fall prevention programs to older adults and their families, but also provides "state-of-the-art" information on fall prevention research to professionals.

The Center's overarching goals are as follows: (1) establish fall prevention as a key public health priority in California; (2) create, test, and evaluate effective and sustainable fall prevention programs; and (3) build a comprehensive and sustainable fall prevention system in California. The first goal will be achieved by (*a*) creating and disseminating information to professionals and consumers, (*b*) utilizing falls surveillance data to provide profiles of falls in California, and (*c*) developing a public awareness/educational campaign for fall prevention that supports grassroots efforts. The second goal will be achieved by (*a*) increasing the knowledge, skills and abilities of existing and future service providers, (*b*) identifying existing fall prevention programs in California, (*c*) implementing evidence-based, multifactorial model programs, and (*d*) developing and evaluating local capacity-building projects. The third goal will be achieved by (*a*) facilitating "stakeholder" organizations to adopt fall prevention as a major goal, (*b*) building and sustaining a statewide fall prevention network among stakeholders, and (*c*) promoting policies that support the development of improved programs and greater resources.

CURRENT EFFORTS AND ACCOMPLISHMENTS

In order to help achieve the Center's overarching goals, the Center has utilized surveillance data to profile falls in California because such information will enhance its ability to (1) understand the breadth and depth of older adult falls in California; (2) identify the characteristics of fallers; and (3) target populations at risk for falls, all of which support the development of fall prevention programs. However, existing data sources about falls often lack reliable, standardized reporting systems and core information (e.g., circumstances of the falls). Given the pressing need for accurate and comprehensive data on falls, the Center prepared an issue brief that emphasizes the importance of (*a*) developing standardized, statewide reporting requirements, and (*b*) improving the documentation of falls.[5] The Center was also involved in creating supplemental screening questions on falls among older adults for the 2007 California Health Interview Survey to gain a better understanding of this problem.

The Center has activated local-level capacity to improve fall prevention efforts in three key ways: Center product development and dissemination, the StopFalls Network, and the California Senior Fall Prevention Projects. The Center has developed various fall prevention tools and informational resources geared toward the needs of both consumer and professional audiences. Examples include a falls home assessment, "Fall Prevention Tips," and "Catch Yourself" information briefs about safety at home and in the community, and technical assistance briefs to help agencies with integrating fall prevention elements into existing programs and developing effective coalitions. In addition, these and many other materials are hosted on the Center website (www.stopfalls.org) and have been disseminated at national conferences and presentations.

The StopFalls Network is a statewide coalition focused on developing stronger linkages among organizations involved in fall prevention and increasing awareness of fall prevention as a public health issue. In addition to older adult and caregiver members, the StopFalls Network currently has 135 organizational members with the heaviest representation from health and social service providers, state and local government agencies, Area Agencies on

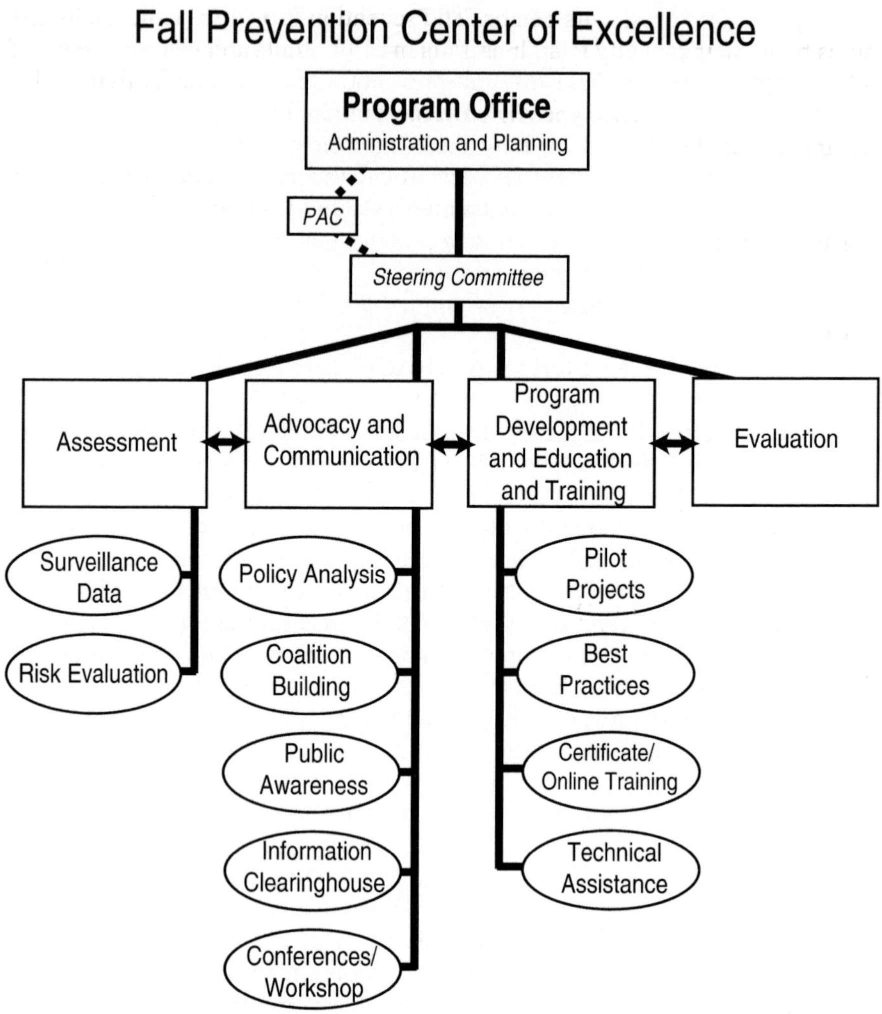

FIGURE 1. Intrastructure of the Center. PAC-Program Advisory Committee.

Aging, hospitals/medical centers, and institutions of higher education. Network members interact through regular teleconferences and electronic information exchange via e-mail and the Center website. The network currently has six working groups, the largest of which focuses on high-risk/frail elders. Other committees include shared resources, communications, policy/advocacy, reimbursement, and continuing education. The continuing education committee has developed a directory of available courses and the policy committee has examined recent California and national legislation, in addition to policy efforts of other states to guide future efforts.

The goal of the California Senior Fall Prevention Projects is to help communities build their capacity to address falls in older adults and deliver integrated fall prevention services that promote good health and independent living. This initiative has funded two types of projects: efforts by community agencies to expand current fall prevention activities with an additional element of medical management, physical activity, and/or home modification, and local coalition development targeted to communities with limited experience in implementing fall prevention activities or those with activities in place that sought to strengthen collaboration among programs. Six Program Expansion and five Coalition Development grants were funded in different areas of the state by the Archstone Foundation in April 2006 for an 18-month period. Several types of entities received Program Expansion grants including adult day care, community multi-service providers, care management, an Area Agency on Aging, and a large hospital. Two of the six Program Expansion programs implemented services that represented all elements of the multifactorial approach to fall prevention. Preliminary findings suggest improved participant physical activity outcomes, participant and provider satisfaction with intervention approach, and increasing organizational and community responsiveness to fall prevention efforts. To date, Coalition grantees have completed community needs assessments and development of a community-wide strategic plan to address falls among older adults, and they are poised to conduct local fall prevention activities, such as a community forum to coalesce community interest in the strategic planning process.

In addition to assisting other organizations build capacity and/or develop cohesive community coalitions centered around the issue of fall prevention, the Center is currently implementing and evaluating three different fall prevention program models in six community sites in Los Angeles and Orange counties. The InSTEP (Increasing Stability through Evaluation and Practice) program includes the core fall prevention components (i.e., medical risk assessment/modification, exercise, and home modification) delivered at different levels of intensity. At the highest intensity, participants will receive a full medical risk assessment (from a medical practitioner), engage in a group-based balance and mobility class two times per week for 12 weeks (taught by a specially trained exercise specialist), and receive an in-home assessment (by an occupational therapist) and recommendations for modification to enhance the home's safety. In contrast, at the lowest intensity, participants will complete a self-assessment of medical risk, which is subsequently reviewed by senior center staff, participate in an exercise class of their choice two times per week for 12 weeks or a home exercise program, and complete a self-assessment of their home's safety with indirect referral to local home modification organizations (resource-listing handout). In addition, half of the programs will test the effectiveness of adding a behavior change intervention (proven effective in the management of certain chronic diseases) to enhance retention and long-term adherence to fall prevention activities.

This initiative will demonstrate how to organize and operate sustainable and effective multifactorial fall prevention programs. Important outcomes to be compared across the three types of model program include fall incidence rates, changes in functional performance, physical activity levels, perceived quality of life, fall and exercise self-efficacy, as well as knowledge and attitudes about falls and physical activity. On the basis of program outcomes, the Center will then produce an array of program materials (e.g., "how-to" manuals, training modules for paraprofessionals and staff, valid and easy-to-use screening and assessment protocols) that will allow health, social service, and other community-based organizations in California and nationwide to replicate and modify these programs to best suit their target audience as well as available resources and capabilities.

NEXT STEPS

The building of a California infrastructure for fall prevention has proven to be a complex and challenging venture. It involves changing attitudes of sometimes skeptical policymakers, agency staff, and older adults themselves about the seriousness of falls and demonstrating that they can be significantly reduced. In some cases, existing programs can be manipulated so that they more effectively address falls. But developing multifactorial programs is difficult because of the fragmentation of the medical, physical activity, and environmental sectors, each of which has its own eligibility requirements, priorities, and procedures. Although there is an increasing array of fall prevention programs that organizations can consider adopting, few have been evaluated for their effectiveness in reducing falls. Providers, who already claim that they are overworked, have yet to be convinced that they should include falls questions when conducting physical exams. In turn, policymakers have been slow to provide adequate reimbursement incentives for the extra time that it takes to do more comprehensive exams for older adults at moderate to high risk of falling. Data about falls continue to be inadequate, with large gaps and widespread underreporting. Fall prevention also competes with many other areas for attention in a time of chronic underfunding of health promotion, exemplified by a gubernatorial veto of a program on osteoporosis and the passage of a fall prevention bill by the state legislature without appropriations.

The Center will continue to address such challenges through its multi-faceted strategy of information, educational, technical assistance, demonstration, and coalition-building activities. The following are examples of some of the activities in which the Center will engage in order to effectively address fall prevention issues at both the state and national levels. At the state level, the Center has been involved in a Senior Falls Workgroup, which is part of a larger Injury Prevention Planning Group convened by the Epidemiology and Prevention for Injury Control Branch of the California Department of Public Health. The Center will help develop recommendations for a state-level strategic framework called Safe California, which can guide California injury prevention efforts for 2007–2010. In addition, given the fact that a statewide network of stakeholders is critical for sharing information, improving practice, and advocating for policies that support fall prevention, the Center is planning to host an invitational statewide summit—involving leaders of key organizations—in December 2007. The goals of this summit are to assess progress since the 2003 state conference on fall prevention, share the latest developments in the field, learn from the experiences of other states, and set an agenda for the next 3 years of activities. At the national level, the Center has expanded its efforts beyond California by participating in the National Council on Aging's Falls Free Initiative that has developed the National Action Plan Blueprint on Fall Prevention,[6] identified promising practices in home modification fall prevention programs, and advocated, along with the Home Safety Council, for passage of the Keeping Seniors Safe from Falls Act of 2006.

In conclusion, the Center will continue to improve the quality of life among older adults through fall prevention efforts by establishing fall prevention as a key public health priority in California, by creating and implementing state-of-the-art, sustainable fall prevention programs, as well as by building a comprehensive fall prevention system in California through the development of a statewide network of stakeholders to share information, improve practices, and advocate for policies that support fall prevention. Given the aging of the population and the endemic nature of falls among older adults, an effective fall prevention infrastructure is critical not only on the local level, but the national and international level as well. The Center's efforts and strategies can serve as a model for the promotion of healthy aging, ensuring that older adults remain active and engaged members of our society.

REFERENCES

1. U.S. Census Bureau. 2006. 2005 American Community Survey data profile highlights. Retrieved March 25, 2007, from http://factfinder.census.gov/servlet/ACSSAFFFacts?_event=Search&geo_id=01000US&_geoContext=01000US&_street=&_county=&_cityTown=&_state=04000US06&_zip=&_lang=en&_sse=on&ActiveGeoDiv=geoSelect&_useEV=&pctxt=fph&pgsl=010&_submenuId=factsheet_1&ds_name=ACS_2005_SAFF&_ci_nbr=null&qr_name=null&reg=null%3Anull&_keyword=&_industry=
2. Alberson, B. February 2003 revised October 2004. Fall prevention issues in California. *In* Creating a California Blueprint for Fall Prevention: Proceedings of a Statewide Conference. Archstone Foundation: 19. Long Beach, CA.
3. Yanochko, P. & B. Hanna. 2006. Multi-factorial Fall Risk Assessment and Intervention for Community Dwelling Seniors: The Role of Home Health Agencies. Caring Choices. Chico, CA.

4. ARCHSTONE FOUNDATION. January 2003 revised October, 2004. Preventing Falls in Older Californians: State of the Art. Archstone Foundation. Long Beach, CA.
5. INJURY SURVEILLANCE WORKGROUP ON FALLS. 2006. Consensus Recommendations for Surveillance of Falls and Fall-Related Injuries. State and Territorial Injury Prevention Directors Association. Atlanta, GA.
6. National Council on Aging. 2005. Falls Free: Promoting a National Fall Prevention Action Plan: National Action Plan. Retrieved March 25, 2007 from http://www.healthyagingprograms.com/content.asp?sectionid=698ElementID=220.

APPENDIX A: CONSENSUS STATEMENT:

Important Principles for Establishing a Statewide Fall Prevention Program

1. Fall prevention should occur in the home, community, and healthcare settings through collaborative efforts of community-based organizations, healthcare organizations, researchers, and policymakers.
2. A long-term commitment to reducing injurious falls involves changing provider behavior, as well as systematic changes in organizations, coalition-building, and follow-up activities to promote new practices.
3. Fall prevention strategies also require changes in behavior of older adults. Communications and marketing are key strategies to increase public awareness.
4. A statewide prevention agenda should address the race/ethnic and cultural diversity of California populations, as well as differences based on region or rural–urban communities.
5. Research should be translated into practice so that appropriate risk assessment occurs at each level of care (medical, rehabilitation, community) and appropriate interventions are tailored to individual needs. Policy and regulations should support basic competencies to assess risk and deliver primary and secondary prevention strategies.
6. A research agenda for fall prevention should be supported.
7. The overall costs of injurious falls by older adults and prevention strategies to health and human services should be recognized. Reducing healthcare costs from injurious falls will involve shifting resources to preventive activities. Some of this resource allocation will be cost-neutral, entail better use of existing resources, or require additional fiscal infusion.
8. Coordination and commitment across systems is necessary to realize savings from preventing injurious falls in older adults. The cost center that realizes savings may differ from the center that needs to invest in prevention or intervention activities. For instance, healthcare costs may be reduced by home modification activities performed by human service agencies.

9. Best practices have been established in multifactorial risk assessment and appropriate interventions; many effective interventions currently exist. Increasing universal access to these effective interventions requires a three-fold approach: recognizing services that are currently available; improving access and matching of need to services; and identifying additional needed resources.
10. There are additional ways to mobilize current resources toward improving fall prevention statewide. For instance, currently available training funds might support education of senior fitness practitioners.

Age-Related Changes in Strength and Somatosensation during Midlife

Rationale for Targeted Preventive Intervention Programs

NANCY L. LOW CHOY, SANDRA G. BRAUER, AND JENNIFER C. NITZ

The University of Queensland, Brisbane, Australia

ABSTRACT: Age-related changes in strength and somatosensation have a negative impact on balance with advanced age. Knowledge of the decades of life when strength and somatosensation show initial and subsequent reductions could inform balance assessment and targeted preventive intervention. We report a cross-sectional study investigating strength and somatosensation of 316 healthy women aged 20–80 years. Demographics, health profile, and activity level were recorded. Strength of quadriceps, hip abductors and adductors, and lower limb somatosensation (tactile acuity, vibration threshold, and joint position error) were measured. Significant age-related changes in strength and somatosensory function were identified, even when demographic variables of activity level, weight, number of reported conditions, medications used, and history of falls were included as covariates. Tested muscles achieved peak force in women in their 40s and then trending reductions presented for those in their 50s and 60s, with significant reductions by the 70s. Tactile acuity, vibration sensitivity, and joint position sense for a non–weight-bearing toe-matching task were significantly reduced by the 40s or 50s, with further reductions by either the 60s or 70s. For the weight-bearing replication task, joint position error was significantly increased by the 60s. A main effect of activity level and body weight was identified for strength measures, and weight also had a main effect for most of the somatosensory modalities. These findings can inform health professionals to use knowledge of early and any subsequent reductions in muscle strength and somatosensation, along with effects of demographics, to develop targeted, innovative programs across the mid-life, a preclinical change period for balance, to promote healthier aging.

KEYWORDS: leg muscle strength; somatosensation; aging women; targeted intervention; balance; mobility; falling or falls

Address for correspondence: Nancy Low Choy, Division of Physiotherapy, SHRS, The University of Queensland, St. Lucia 4072, Brisbane, Australia. Voice: 61 7 33652275; fax: 61 7 33651662. n.lowchoy@uq.edu.au

Ann. N.Y. Acad. Sci. 1114: 180–193 (2007). © 2007 New York Academy of Sciences. doi: 10.1196/annals.1396.014

INTRODUCTION

Age-related changes in strength and somatosensation are two important elements of balance control that have a demonstrated negative impact on balance and mobility in older age.[1–4] As community-ambulant older women continue to have a high risk of falling,[4,5] there is a need for further investigation of sensorimotor factors that could influence early changes in postural stability and functional capacity in women. With improved knowledge of initial and subsequent change periods for strength and somatosensation, new initiatives could be developed to minimize decline in these parameters, and thus balance, before adverse impacts on functional activity and participation occur.

Age-related changes in quadriceps strength have been identified, with women showing adverse changes earlier than men.[6] For women, quadriceps strength is at a peak in younger decades (20s/30s), remains stable for a period before gradually reducing in the 60s, and shows accelerated reduction in advanced age.[6–8] This pattern has also been demonstrated in studies which only include subjects aged 45–75.[9,10] Less specific age-related changes are known for other lower limb muscles although age-decade reference values for many lower limb muscles have been developed.[11]

Mediolateral stability is known to be reduced between 40 and 60 years.[12] Softer surfaces, narrow bases,[3,13] and stepping and walking tasks[14] are particularly demonstrative of reduced stability in midlife. Thus, the strength of muscles associated with the mediolateral plane is important to consider along with measures of quadriceps muscle strength. Given the link between hip abductor muscles and reduced lateral stability in older women,[5] delayed lateral stepping responses to external displacement between young and older subjects,[15] and reduced hip abductor activation in elders with a history of falls,[16] the importance of including tests of hip abductor/hip adductor muscle strength as well as strength of the quadriceps is evident.

Somatosensory function is a major contributor to balance control[17] and studies have demonstrated differences between young and older subjects for a range of modalities. These include decreased somaesthetic sensitivity,[18] raised thresholds for vibratory stimuli,[19] and increased joint position error.[1,4,20–22] Further investigation of these modalities is required as there is less specific information about the type and extent of somatosensory changes that could occur during midlife, a period associated with preclinical changes in balance.

When studies of strength and somatosensory function have included subjects from the mid-decades,[1,3,4] subject numbers have been low (10–30 subjects).[1,3] Although these studies revealed that peripheral sensation and quadriceps strength were important for balance on a soft surface, the age-decades when early changes in quadriceps strength or somtaosensory function occurred were not specified. Thus, a study is required that identifies when initial and/subsequent changes occur in hip muscle strength (controlling the mediolateral plane), quadriceps strength (anterior–posterior plane), and in the various somatosensory modalities.

Reduced activity level is known to have a negative impact on weight and health-related quality of life of middle-aged people,[23] and increased activity level has also been shown to positively affect health, muscle strength, and balance performance in people over 65 years of age.[24–26] Hence, the effects of physical activity,[26] as well as factors known to increase risk of falls, were monitored for all participants in our study.[27–29]

Hence, a study was undertaken using healthy women aged 20- to 80-years old to identify the age-decades when strength and somatosensory function showed initial and subsequent reductions, while controlling for activity level and other demographic or health factors.

METHODS

A cross-sectional design was used to investigate strength and somatosensory function of women aged 20–80 years.

Subjects

Subjects were recruited through the electoral roll and community advertisements and available from the research register at a Centre for Women's Health Research located at a major metropolitan hospital in Brisbane. Subjects were required to be aged between 20 and 80 years, independently mobile in the community without using walking aids, have normal visual acuity or wear corrective lenses, and have no diagnosed disorders commonly associated with balance problems (these included diabetes or peripheral vascular disease; neurological or vestibular diseases; recent musculoskeletal injuries or arthridities requiring active management; and depression or other neuropsychological disorders requiring prescribed medication).[27–29] A total of 463 women met the inclusion criteria and were invited to participate in this study. Sixty-nine percent of these women volunteered (320 subjects) and represented five age groups: 20–39 ($n=64$); 40–49 ($n=84$); 50–59 ($n=83$); 60–69 ($n=44$); and 70–79 ($n=45$).

Procedure

All participants gave their informed consent, as required by the Medical Research Ethics Committee of the University of Queensland. One test session was attended. A structured interview yielded age, the number of co-morbidities and current medications,[29] the number of falls for the previous 12 months,[30] and habitual activity level.[26] Height (cm) and weight (kg) were measured and lower limb dominance for stance and skill was recorded. Lower limb strength and somatosensory function then were measured in random order.

Measures

A spring-gauge was use to measure muscle strength of the quadriceps and hip abductor and hip adductor muscles for both lower limbs.[31] The quadriceps muscle was measured by placing the subject in a sitting position with the pelvis stabilized by a strap. A spring-gauge was attached above the ankle of the test limb, 30 cm below the tibial tuberosity and perpendicular to an attachment. Three maximal contractions of knee extension were recorded and the average torque (Nm) calculated. To measure hip abduction strength, the subject was in a supine position, lying with the non–test leg flexed to enable access to the test leg. One strap from the spring-gauge was attached to the thigh (30 cm from the greater trochanter) and the other to an attachment perpendicular to the line of pull of the spring-gauge. The test-leg hip was in a neutral position and a strap stabilized the pelvis during three maximal contractions into abduction. The average torque (Nm) for the hip abductors was calculated. To measure hip adductor strength, the subject was positioned supine with a spring-gauge attached to the test leg at the same point as for hip abduction. The non–test leg was positioned 20 degrees into hip abduction with the test leg in a neutral position and perpendicular to the line of pull of the spring-gauge. The pelvis was stabilized during three maximal contractions into adduction and then the average torque (Nm) was calculated.

Measures of somatosensory function included tactile acuity,[31,32] vibration sensitivity,[31] and joint position error using a toe-matching task[31] and a modified re-positioning test in weighting-bearing developed by Stillman and colleagues.[33] To measure tactile acuity of the lower limb, nylon monofilaments were applied in random order (with the subject's eyes closed) to each lateral malleolus[31] and to three points on the sole of each foot: the center of the heel, the big toe pad, and the first metatarsal head.[32] A 5.07 monofilament (mm) and then larger or smaller monofilaments were used as required to determine the threshold for tactile acuity for each site. The size of the smallest monofilament detected was recorded.

Lower-limb vibration sensitivity was measured using an electronic unit that generated a 200-Hz vibration of varying amplitude.[31] Threshold perception was recorded (microns of motion) for three trials and an average score calculated for each lower limb. An open-chain toe-matching task using established protocol[31] provided a measure of lower-limb joint position error in non–weight-bearing (sitting). The difference in toe position (degrees) was recorded and the mean for three trials calculated. A closed-chain knee-bend replication task in bilateral stance was used to calculate the joint position error in weight-bearing.[33] In this test, each subject practiced bending the knees (five times) to a rod centered in front of the patella, and then replicated the position without the cue (twice). The final cued knee-bend position and each non-cued position was photographed using a Web Camera (Sony Digital Camera [DSC P50 Cyber Shot]; Sony Australia, Sydney, NSW, Australia) and the three photographs were stored on a computer. An angle-calculator program with established reliability[34] was used to calculate the knee joint angles from the images stored on computer. Markers on the greater trochanter, midline of the knee joint, and lateral malleolus of each subject were identified and registered by using the computer mouse. A computer software program then calculated the knee angles for the cued and non-cued positions. The difference (degrees) between both knee angles was calculated to form the joint position error in weight-bearing.

Analyses

All data analyses were performed using SPSS for windows, Rel. 11.0.1. (SPSS Inc., Chicago, IL). Inspection of data lead to three women being removed from the study on the basis of their weight (over 100 kg) and one due to a recent episode of dizziness. Descriptive analyses of demographics for each decade were undertaken (total $n = 316$). A multivariate analysis was used to establish main effects of age-decade and demographics for strength and somatosensory measures ($P < 0.05$). A series of univariate analyses was then undertaken for the strength and somatosensory measures with the model including age category as the main factor and covariates of height, weight, activity-level, number of prescribed medications, co-morbidities, and history of falls. The least-significant difference between age categories was accepted to control for Type II errors although the P-value was lowered to $P < 0.01$ to protect against Type I errors. Thus the P-values between 0.05 and 0.01 are reported as strong trending findings when the differences between age categories are considered. Age-decade associations were explored between demographics and both strength and somatosensory measures using Pearson's product–moment correlations ($P < 0.05$).

RESULTS

Demographic information of the participants is presented in TABLE 1. The multivariate analyses yielded a main effect of age group for all strength and somatosensory measures investigated ($P < 0.05$) as well as main effects for activity level and weight (TABLE 2). There were no trending or significant main effects of reported conditions and use of prescribed medications for the strength and somatosensory measures of participating women ($P > 0.05$), while history of falls reported yielded a main effect on joint position error for

TABLE 1. Age-decade analysis of demographic information for participating women

	20/30s ($n = 62$)	40s ($n = 72$)	50s ($n = 80$)	60s ($n = 53$)	70s ($n = 50$)
Height[a] (cm)	166.2 (6.3)	163.7 (5.8)	163.4 (5.3)	163.5 (6.4)	158.4 (6.5)
Weight[a] (kg)	63.2 (9.3)	66.5 (17.2)	71.3 (14.9)	69.5 (12.2)	66.4 (12.1)
Dominance for skill (R)[b]	58 (94)	72 (100)	71 (89)	50 (94)	47 (94)
Number of medications[a]	0.7 (0.7)	0.5 (1.0)	1.5 (2.0)	2.7 (2.3)	2.7 (2.4)
Number of conditions[a]	1.3 (1.52)	2.3 (1.90)	3.9 (3.12)	4.9 (3.1)	4.9 (2.9)
Activity level[c]: 1–6	4 (2–5)	3 (2–5)	3 (2–5)	3 (2–5)	2 (1–3)
History of falls[d]	-	1 (1)	5 (1–5)	3 (1–3)	1 (2)

[a]Mean (standard deviation).
[b]Dominance: number (%) preferring the R lower limb for skill.
[c]Median score (range).
[d]Number of women with a history of falls (range in the number of falls reported by these women).

the weight-bearing knee-bend replication task ($F = 7.78$; $P < 0.01$). The significant age-decade associations between relevant demographics and strength or somatosensation are presented in TABLES 3 AND 4, respectively.

The series of univariate analyses for strength and somatosensory measures yielded significant age-related changes ($P < 0.01$), even with adjustments for the six demographic covariates. FIGURE 1 reveals the age-related changes in muscle strength with peak force identified in the 40s for all but one muscle group tested (hip adductors, stance limb). Compared to the strength measures for women in their 40s, the quadriceps revealed trending reductions in strength across the 50s and 60s, and a significant reduction in strength by the 70s. For the hip muscles, the reduction in strength was not apparent in the 50s, but trending reductions emerged for the hip abductors in the 60s, and the strength of both hip abductors and hip adductors was significantly reduced by the 70s.

The initial and subsequent significant reductions in somatosensory function are illustrated in FIGURE 2 ($P < 0.01$). Peak acuity of all sensory modalities tested was determined for those in the youngest age group, with an initial significant reduction for most aspects of somatosensation presenting by the 40s or 50s (tactile acuity, vibration sensitivity, and joint position sense using the toe-matching task) and further reductions in the acuity of these modalities by the 60s or 70s. A significant increase in error for the weight-bearing knee-bend replication task was identified by the 60s.

DISCUSSION

This study confirmed the gradual reduction in strength identified for the quadriceps across the 50s and 60s in earlier studies and the more critical reduction in strength identified in the 70s.[6–10] The reduction in hip muscles also showed trending changes in the 60s with significant reductions in strength

TABLE 2. Main effects of demographics and activity level on sensorimotor function

	Age group	Height (cm)	Weight (kg)	Activity level level (1–6)
Strength				
Quadriceps				
Stance	3.13*	0.83	9.72**	7.28**
Skill	3.56**	1.07	5.10*	7.50**
Hip abductors				
Stance	5.55***	3.29 $P = 0.071$	8.70**	5.81*
Skill	3.93**	5.16*	8.10**	3.10 $P = 0.079$
Hip abductors				
Stance	3.80**	17.01***	4.11*	3.82 $P = 0.052$
Skill	3.97**	9.35**	4.49*	9.84*
Somatosensation				
Vibration sense				
Stance	14.56***	0.06	3.36 $P = 0.068$	0.08
Skill	24.75***	0.06	12.79***	0.16
Toe-matching	5.01**	3.80 $P = 0.052$	0.00	0.63
Knee-bend replication	3.37*	0.14	4.77*	0.03
Tactile acuity (ankle)				
Stance	14.94***	4.98*	0.67	0.09
Skill	8.32***	5.21*	0.06	0.40
Tactile acuity (heel)				
Stance	9.09***	1.25	24.54***	0.03
Skill	10.48***	0.09	16.99***	0.22
Tactile acuity (big-toe pad)				
Stance	14.09***	0.15	6.58*	6.58*
Skill	13.96***	1.34	3.97*	
Tactile acuity (1st metatarsal)				
Stance	14.83***	3.59 $P = 0.059$	17.19***	0.47
Skill	17.12***	6.01*	9.76**	0.06

*Significant at $P < 0.05$; **significant at $P < 0.01$; ***significant at $P < 0.001$.

for those in their 70s. This means that it is critical to target those in the 50- to 70-year age period pre-emptively with interventions before the detrimental reductions in strength of advanced age occur. Our findings support the need for strengthening programs to target the muscles that control the mediolateral plane of movement (hip abductors and hip adductors) in addition to those in the anterior–posterior plane (quadriceps). It is likely that this principle would also apply to the rotational plane and to other lower limb muscles not included in our study as the importance of muscles about the ankle to functional performances in older women has been demonstrated.[35]

Most of the somatosensory modalities tested more clearly showed an earlier significant reduction in acuity in midlife as well as reduced acuity in older decades. Reduced tactile acuity, vibration sensitivity, and joint position sense when the non–weight-bearing toe-matching task was used were evident from

TABLE 3. Significant age-decade associations between weight (kg), physical activity level, and strength of lower limb muscles

	Demographic	Quadriceps strength (Nm)		Hip abductor strength (Nm)		Hip adductor strength (Nm)	
Age: decades	Variables	Stance	Skill	Stance	Skill	Stance	Skill
40s	Weight	–	–	0.266*	0.250*	–	–
	Physical activity	0.203 $P = 0.066$	0.214 $P = 0.072$	–	–	–	–
50s	Physical activity	0.259*	0.280*	0.276*	–	0.198 $P = 0.076$	0.337**
60s	Weight	–	–	–	0.252 $P = 0.072$	0.323*	–
	Physical activity	–	–	0.321*	0.309*	–	–
70s	Weight	0.560***	0.519***	0.466**	0.385**	0.390**	0.415**

*Significant at $P < 0.05$; **significant at $P < 0.01$; ***significant at $P < 0.001$.

TABLE 4. Significant age-decade associations between body weight (kg), reported number of falls, and somatosensory acuity

Age:	Demographic	Tactile acuity (forefoot)		Vibration threshold (Hz)		Joint repositioning error (JPE)	
Decades	Variables	Stance	Skill	Stance	Skill	NWB task	WB task
20s/30s	Weight	0.288*	0.338***	0.318*	0.404**	–	–
40s	Weight	0.422***	0.364**	–	0.382**	–	–
	History of falls	–	–	–	0.347**	–	–
50s	Weight	0.432***	0.330**	–	0.328**	−0.275*	−0.408***
	History of falls	–	–	–	–	0.220*	0.258*
60s	Weight	0.498**	0.329*	0.268 $P = 0.079$	–	–	–
70s	History of falls	–	–	–	–	−0.241 $P = 0.091$	–

*Significant at $P < 0.05$; **significant at $P < 0.01$; ***significant at $P < 0.001$.

the 40s/50s with subsequent periods of reduced acuity by the 60s/70s. The negative association of weight with these somatosensory modalities (20s–60s) supports the need to control weight so that the age-related changes occurring in the somatosenses are minimized. This is particularly important in the midlife change period as somatosensation reflects the reductions in postural stability identified in other studies for balancing on softer and/or narrower surfaces[3,13] or while undertaking functional balance and mobility tasks.[14] It is likely that the reduction in somatosensation would influence both anticipatory and ongoing postural adjustments during movement associated with postural stability and functional capacity,[2] thus explaining the reduced efficiency of balance and mobility tasks identified in midlife and again in older decades.[14] Our findings suggest that pre-emptive interventions need to employ exercises and/or activities that activate the muscle and joint receptors underpinning these somatosensory modalities to ensure that the proprioceptive capacity of muscle is optimally maintained across the midlife period.

The weight-bearing knee-bend replication task revealed trending increases in error for those in their 50s and a significant increase in error for women in their 60s compared to young women. The reduced capacity to replicate a position in standing (increased joint position error) could underpin the increase in postural sway for those over 60 when proprioception is challenged by removing vision.[3,13] These findings support a rationale for using functional weight-bearing exercises that include practice of balance tasks with and without the use of vision to help improve position sense in weight-bearing. This form of context-specific training that integrates a sensory challenge with functional strength training was one element of a program delivered to re-train balance and function with fallers over 65.[36] This form of exercise could readily be adapted and be more challenging if delivered as part of a preventive intervention program for a younger and/or more-able group of elders.

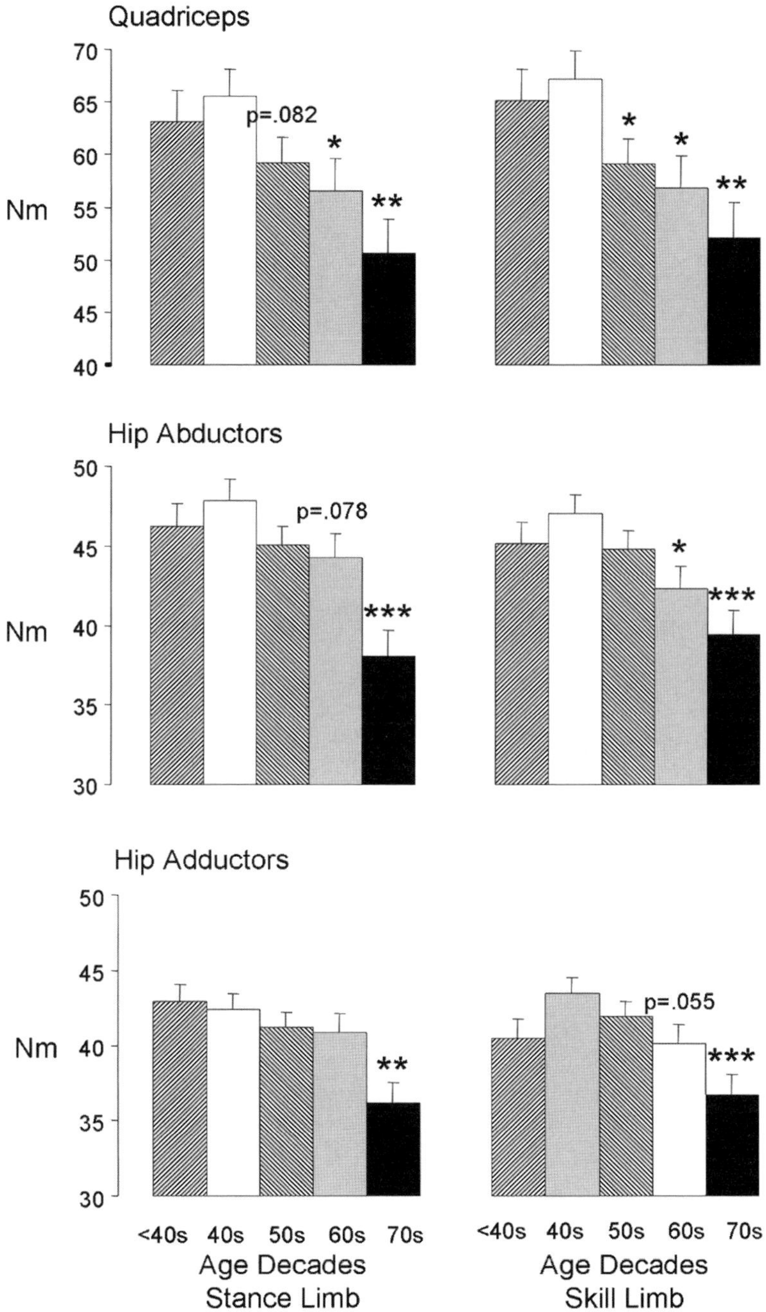

FIGURE 1. Age-related changes in lower limb muscle strength compared to women in their 40s. *Trending reduction $P < 0.05$; **significant $P < 0.01$; ***$P < 0.001$.

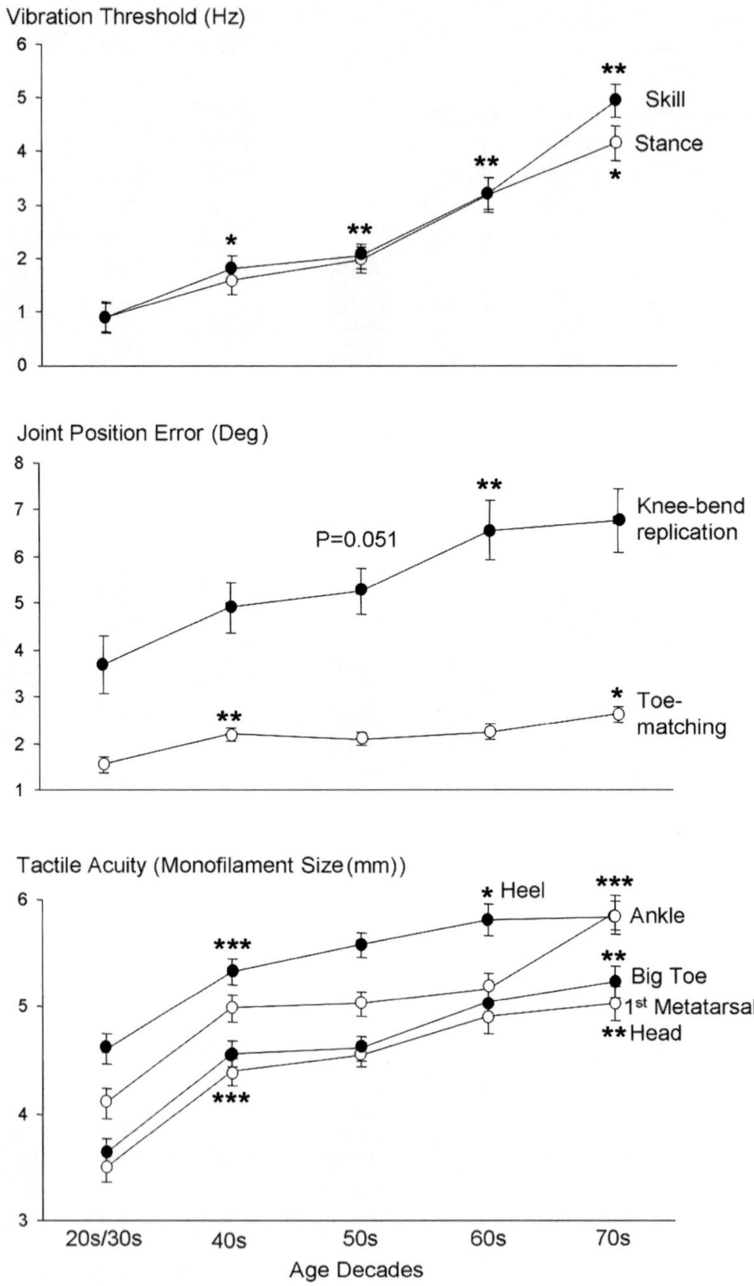

FIGURE 2. Age-related changes in somatosensory function compared to women in their 20s/30s: *Trending finding $P < 0.05$; **significant finding $P < 0.01$; ***significant finding $P < 0.001$.

It was surprising to find that the error for this weight-bearing replication task was greater than the error identified using the non–weight-bearing toe-matching task, particularly as young adults had demonstrated less error during the development phase of this knee-bend replication test.[33] It is likely that the multiple factors involved in maintaining balance during the test when undertaken without hand support, particularly for the older age groups, could account for the increased error when replicating the knee-bend position. Thus, further development of this weight-bearing test is recommended so that the additional sources of error that could confound outcomes for this test are minimized. This step is particularly important as a main effect of falls history was determined for this specific test of joint position error. It was notable that it was women with a history of falls in their 50s (as well as those in their 70s) who yielded an association with this weight-bearing test, and thus the need to target earlier age decades to identify those with a history of falls and provide targeted pre-emptive interventions.

The finding that activity level had a main effect on muscle strength was not surprising, but the effect of activity level for the muscles controlling the mediolateral plane was less consistent than the strong effect for the quadriceps muscle. Increased physical activity, however, had a positive association with all strength measures for women in their 50s and 60s. Thus, while increasing frequency and intensity of physical activity is important,[25–28] consideration needs to be given to the capacity of each activity to target the mediolateral hip muscles if maximal benefit of the physical activity is to be gained.

The findings of this study can inform clinicians of the age-related changes in the sensory and motor systems, and this knowledge could be used to develop and deliver innovative programs across the midlife, a preclinical change period for balance, as well as in older age groups to foster healthier aging.

ACKNOWLEDGMENTS

We wish to acknowledge the Australian Menopause Society Equipment Grant and the funding received through the Royal Brisbane and Women's Hospital Research Foundation to assist with research undertaken at the Betty Byrne Henderson Centre. Funding for a research assistant and staff assistance with recruitment and appointments is appreciated.

[The authors declare no conflict of interest with any organization supporting this research.]

REFERENCES

1. Hurley, M., J. Rees & D. Newham. 1998. Quadriceps function, proprioceptive acuity and functional performance in healthy young, middle-aged and elderly subjects. Age Ageing **27:** 55–57.

2. LIN, S. & M. WOOLLACOTT. 2005. Association between sensorimotor function and functional and reactive balance control in the elderly. Age Ageing **34:** 358–363.
3. LORD, S. & J. WARD. 1994. Age-associated differences in sensori-motor function and balance in community dwelling women. Age Ageing **23:** 452–468.
4. LORD, S., D. LLOYD & S. LI. 1996. Sensori-motor function, gait patterns and falls in community dwelling women. Age Ageing **25:** 292–299.
5. LORD, S., M. ROGERS, A. HOWLAND & R. FITZPATRICK. 1999. Lateral stability, sensorimotor function and falls in older people. J. Am. Geriatr. Soc. **47:** 1077–1081.
6. MURRAY, M., E. DUTHIE, S. GAMBERT, *et al.* 1985. Age related difference in knee muscle strength in normal women. J. Gerontol. **40:** 275–280.
7. HAKKINEN, K. & A. HAKKINEN. 1991. Muscle cross-sectional area, force productions and relaxation characteristics in women at different ages. Eur. J. Appl. Physiol. **62:** 414–419.
8. LINDLE, R., J. METTER & N. LYNCH. 1997. Age and gender comparisons of muscle strength in 654 women and men aged 20 to 93 years. J. Appl. Physiol. **83:** 1581–1587.
9. FRONTERA, W., V. HUGHES, J. LUTZ & W. EVANS. 1991. A cross-sectional study of muscle strength and mass in 45–78 year old men and women. J. Appl. Physiol. **71:** 644–650.
10. SUNNERHAGEN, K., M. HEDBERG, G. HENNING, *et al.* 2000. Muscle performance in an urban population sample of 40 to 79 year old men and women. Scand. J. Rehabil. Med. **32:** 159–167.
11. BOHANNON, R. 1997. Reference values for extremity muscle strength obtained by hand-held dynamometers from adults aged 20 to 79 years. Arch. Phys. Med. Rehabil. **78:** 26–32.
12. NITZ, J., N. LOW CHOY & R. ISLES. 2003. Medio-lateral postural stability in community dwelling women over 40 years of age. Clin. Rehabil. **17:** 725–727.
13. LOW CHOY, N., S. BRAUER & J. NITZ. 2003. Changes in postural stability in women aged 20 to 80 years. J. Gerontol. **58:** M525–M530.
14. ISLES, R., N. LOW CHOY, M. STEER & J. NITZ. 2004. Normal values of balance tests in women aged 20–80. J. Am. Geriatr. Soc. **52:** 1367–1372.
15. MAKI, B., M. EDMONSTONE & W. MCILROY. 2000. Age-related differences in laterally directed compensatory stepping behavior. J. Gerontol. Med. Sci. **55A:** M270–M277.
16. BRAUER, S., Y. BURNS & P. GALLEY. 2000. A prospective study of laboratory and clinical measures of postural stability to predict community dwelling fallers. J. Gerontol. **55A:** M469–M476.
17. WOOLLACOTT, M. 2000. Systems contributing to balance disorders in older adults. J. Gerontol. **55A:** M424–M428.
18. KENSHALO, D. 1986. Somesthetic sensitivity in young and elderly humans. J. Gerontol. **41:** 732–742.
19. PERRET, E. & F. REGALI. 1970. Age and the perceptual thresholds for vibratory stimuli. Eur. Neurol. **4:** 65–76.
20. SKINNER, H., R. BARRACH & S. COOK. 1994. Age-related decline in proprioception. Clin. Orthop. Rel. Res. **184:** 208–211.
21. KAPLAN, F., J. NIXON, M. REITZ, *et al.* 1985. Age-related changes in proprioception and sensation of joint position. Acta Orthop. Scand. **56:** 72–74.
22. PETRELLA, R., P. LATTANZIO & M. NELSON. 1997. Effect of age and activity on knee joint proprioception. Arch. Phys. Med. Rehabil. **76:** 235–241.

23. LITTMAN, A., A. KRISTAL & E. WHILE. 2005. Effects of physical activity intensity, frequency and activity type on 10-y weight changes in middle-aged men and women. Int. J. Obesity **29:** 524–533.
24. BUCHNER, D., S. BERESFORD, E. LARSON, *et al.* 1992. Effects of physical activity on health status in older adults. II. Intervention studies. Annu. Rev. Pub. Health **13:** 469–488.
25. HUGHES, V., W. FRONTERA & M. WOOD. 2001. Longitudinal muscle strength changes in older adults: influence of muscle mass, physical activity and health. J. Gerontol. **56A:** 209–217.
26. HIRVENSALO, M., T. RANTANEN & E. HEIKKINEN. 2000. Mobility difficulties and physical activity as predictors of mortality and loss of independence in the community-living older population. J. Am. Geriatr. Soc. **48:** 493–498.
27. CAMPBELL, A., M. BORRE & G. SPEARS. 1989. Risk factors for falls in community-based prospective study of people 70 years and over. J. Gerontol. **44:** M112–M117.
28. LEIPZIG, M., R. CUMMING & M. TINNETTI. 1999. Drugs and falls in older people: a systematic review and meta-analysis: I. Psychotropic Drugs. J. Am. Geriatr. Soc. **47:** 30–39.
29. TINETTI, M., M. SPEECHLEY & S. GINTER. 1988. Risk factors for falls among elderly persons living in the community. N. Engl. J. Med. **319:** 1701–1707.
30. PEEL, N., C. CARTWRIGHT & M STEINBERG. 1998. Monitoring slips, trips and falls in the older community: preliminary results. Health Promotion J. Aust. **8:** 148–150.
31. LORD, S. & R. CLARK. 1996. Simple physiological and clinical tests for the accurate prediction of falling in older people. J. Gerontol. **42:** 199–203.
32. BELL-KROTOSKI, J. & E. TOMANCIK. 1987. The repeatability of testing with Semmes-Weinstein monofilaments. J. Hand Surg. America Assoc. **12:** 155–161.
33. HOLEWSKI, J. J., R. M. STRESS, P. M. GRAF, *et al.* 1988. Aesthesiometry: quantification of cutaneous pressure sensation in diabetic peripheral neuropathy. J. Rehabil. Res. Dev. **25:** 1–10.
34. RUSSELL, T., G. JULL & R. WOOTTON. 2003. Can the Internet be used as a medium to evaluate knee angle? Man Ther. **8:** 242–246.
35. SUZUKI, T., J. BEAN & R. FIELDING. 2001. Muscle power of the ankle flexors predicts functional performance in community-dwelling older women. J. Am. Geriatr. Soc. **49:** 1161–1167.
36. NITZ, J. & N. LOW CHOY. 2004. The efficacy of a balance strategy training circuit for community based elders who have had a fall. Age Ageing **33:** 1–6.

Living Well until You Die

Quality of Care and Quality of Life in Palliative and Dementia Care

NEIL SMALL

School of Health Studies, University of Bradford, Bradford, United Kingdom

ABSTRACT: The incidence and prevalence of dementia across the world is increasing, carrying crucial implications for the discourse about healthy aging and longevity. For such a discourse to progress, it must engage with the challenge of dementia. If we live into our 80s, then every fifth one of us will have dementia. Of the four who do not, another one is likely to be sharing their life with someone who does. Existing care regimes have failed to consistently act upon best practice in dementia care. That best practice can be pursued via mutual learning between palliative care and person-centered dementia care. Even with cognitive disability and life-limiting illness, it is possible to live well, and it should be an imperative that people be supported to have a better quality of life than they do now. Implementing evidence-based best practice is not enough—it is also necessary to engage with social attitudes toward dementia. Otherwise there is the danger that a significant proportion of older people will be marginalized from a newly emerging ontology of old age.

KEYWORDS: palliative care; dementia; quality of life

THE IRONY OF LONGEVITY

The increasing prevalence of dementia creates challenges for societies, for health and social care professionals, and for individuals and families. It also creates challenges for the development of a discourse of healthy aging. An understanding of healthy aging encompasses a commitment to enhance the quality of a person's life as he or she ages. If we ask, "What can we do to maximize quality of life for a person with dementia?" two areas of challenge are revealed. The first exists at the level of social attitudes, whereas the second is evident in professional practice. Both cognitive impairment and terminal

Address for correspondence: Neil Small, B.Sc. (Econ.), M.S.W., Ph.D., Professor of Health Research, School of Health Studies, University of Bradford, 25 Trinity Road, Bradford BD5 0BB, UK. Voice: +44 (0)1274236456; fax: +44 (0)1274 236458.
N.A.Small@bradford.ac.uk
www.bradford.ac.uk/acad/health/research/pcg

Ann. N.Y. Acad. Sci. 1114: 194–203 (2007).
doi: 10.1196/annals.1396.019

illness have resulted in social exclusion. A characteristic response to the former has been to assume that a loss of self has occured.[1] Responses to the latter have included the sequestration of the dying person to a liminal world of the "not yet dead."[2] When both cognitive impairment and terminal illness co-exist, there are powerful barriers to considering that person's health and social well-being as a priority. If there are barriers to encouraging healthy aging for people with dementia that are consequent on prevailing social attitudes, there are also barriers to physical and mental well-being because of the prevalence of bad practice and the paucity of care based on evidence of best practice. This paper addresses each of these areas of challenge and argues that much can be done to help people with dementia live well until they die. This can be achieved, in large part, through promoting a hybrid construction combining palliative care and person-centered dementia care.

THE EPIDEMIOLOGY OF DEMENTIA

Dementia is poised to become the dread disease of the developed world in the 21st century. Different eras throw up a disease that is particularly dreaded. That disease serves as the nemesis to the shape taken by the perennial hubristic assumption that man is in control of the world we live in, that we are as gods.[3] Hubris appears to be always present, even if its characteristic features change. When we thought we could shape the world via industrialization and urbanization, our dread disease was one of proximity, tuberculosis. When we presumed that reason, and its standard-bearer science, would solve all problems, our dread disease was cancer, one of blind, meaningless replication. Now the times are characterized by individuals defining their own truth, believing that we are who we are because of our capacity to choose and to reason. Consequently, loss of intellectual capacity characterizes our current dread disease—dementia.[4] Dread engenders fear and awe, reactions that stop you in your tracks or make you run away—not reactions that prompt close or creative engagement.

But dread diseases are dread because they are both ontologically and epidemiologically challenging. Worldwide 24 million people now live with dementia and this number is forecast to rise to 81 million by 2040. Between 2001 and 2040 numbers in developed countries are expected to rise by 100%, while those in India, China, South Asia, and the western Pacific are forecast to rise by 300%.[5] Central to the explanation of this increase is an overall increase in life expectancy, more evident in some countries, but also a world trend. Figures from the United Kingdom underline the connection between increased age and numbers with dementia. The prevalence rate increases from 1 in 1000 for 40- to 65-year-olds to 1 in 5 for people over 80 years old.[6] Dementia presents in a number of ways—each has different patterns of onset and differing impacts on functioning. Alzheimer's disease and vascular dementia are the two most often observed, accounting for approximately 55% and 20% of all dementias, respectively.

TABLE 1. Changing epidemiology of dementia in the United States, United Kingdom, and Australia

United States: Estimates from the year 2000 were that 4.5 million people were living with Alzheimer's dementia. This figure represents a doubling of the recorded incidence since 1980.[29] It is predicted that by 2050 this figure will rise to between 11.3 and 16 million. Approximately 1.1 million of these people live with moderate to severe dementia and it is estimated that by 2015 this will rise to 2.8 million.
United Kingdom: Dementia affects 750,000 people.[6] Eighteen thousand people affected by dementia are less than 65 years old. The estimates are that by 2010 there will be 870,000 people with dementia.
Australia: Population estimates published in 1998 identified 134,000 people with dementia. In the population aged 65–84 years, 5% of people were affected, and for people over 85 years, this rose to 25%.[30] By 2002 overall numbers had risen to 162,000.[31] It is projected by 2040 there will be 500,000 people with dementia and by then dementia is expected to be the primary cause of major chronic illness in Australia.

While there is a strong trend evident worldwide, there is some variation in incidence and prevalence figures within countries. Dementia is a syndrome, variously diagnosed, and mediated by social norms. Certain presentations are now more likely to be named as dementia. In the past, similar characteristics may have been identified using a different term, *senility*, for example, or an explanation from a different epistemology may have been sought,[7,8] for example, from that of mysticism or spirituality. TABLE 1 provides more detail on the changing epidemiology of dementia in the United States, the United Kingdom, and Australia.

People die with dementia in one of three ways[9]: They reach the end of their life with dementia, but die from another condition. Others with dementia reach the end of their life with a complex mix of dementia and other medical conditions which together lead to their death. Finally, some people live until their dementia is very advanced and die from the complications that this brings.[10]

QUALITY OF CARE FOR PEOPLE WITH DEMENTIA

Patterns of care for people with dementia vary within and between countries, but they are most often characterized by complex combinations of health and social care, sometimes domiciliary and sometimes institutional. Considerable proportions of care are provided by family and lay caregivers. Consequently, growing numbers of the elderly, including the very elderly, rely on other elderly persons for care.

In this section I will concentrate on care at the end of life and will argue that there are widespread inadequacies in this care. These can be summarized as arising from too much intervention or too little.

Too Much

There has been a widespread identification of the imposition of unnecessary interventionist regimes during hospitalizations including tube feeding, laboratory tests, the use of restraints, and intravenous medications. The result is not good-quality care, and there is evidence of pressure ulcers, constipation, shortness of breath, and pain in such patients. One study found that 51% of patients with advanced dementia received a new feeding tube during a terminal hospitalization compared with 11% of patients with metastatic cancer. There is no evidence base for the efficacy of tube-feeding in this population.[11]

Too Little

While it is evident that there can be interventions that are unnecessary and not evidence-based, there is also a story of too little care, as evidenced in:

(1) poor pain control;
(2) dehydration and malnutrition;
(3) emotional and social neglect;
(4) absence of spiritual care; and
(5) lack of support for family caregivers.[4]

If people with dementia are to live as well as they can until they die then the sorts of failings of care regimes identified above have to be addressed. This requires not only that what is known about best practice be implemented, but also that more fundamental problems about perceptions of people with dementia and the pessimistic assessments of the possibilities for improving their lives have to be engaged with. In the next section I will introduce two ways of approaching care that have the potential to change assumptions and approaches to care.

PALLIATIVE CARE AND PERSON-CENTERED CARE

The modern hospice movement and specialist palliative care developed first in the United Kingdom and then spread across the world in the late 1960s and 1970s and at an increased speed in the 1980s.[12] Hospice and specialist palliative care are now present in more than a hundred countries, with some countries seeing very large numbers of initiatives falling within the parameters of this treatment approach. (For example, in the United Kingdom in 2003–2004 there were 58,000 admissions a year to palliative care units: in the United States there are 3139 hospice and palliative care services and there are 280 in Australia.[13])

TABLE 2 offers a summary of the key constituents of a palliative care approach. It also summarizes person-centered dementia care, which I shall consider below.

TABLE 2. Definition and characterization of palliative and person-centered care

Palliative care	Person-centered care
Palliative care is an approach that improves the quality of life of patients and their families facing the problems associated with life-threatening illness, through the prevention and relief of suffering by means of early identification and impeccable assessment and treatment of pain and other problems—physical, psychosocial, and spiritual.	Person-centered care is an approach to caring for people with dementia that affirms continuing "personhood" and that, in so doing, challenges a previously prevalent "malignant social psychology."
Palliative care:	Person-centered care:
• Provides relief from pain and other distressing symptoms	• Affirms the value of a person and his or her life regardless of disability
• Affirms life and regards dying as a normal process	• Recognizes that people actively cope with dementia
• Intends neither to hasten nor postpone death	• Seeks to minimize the impact of cognitive, functional, and behavioral deficits on a person's quality of life
• Integrates the psychological and spiritual aspects of patient care	• Addresses physical, emotional, spiritual, and social aspects of the person's life
• Offers a support system to help patients live as actively as possible until death	• Recognizes the role of life history and personality
• Offers a support system to help the family cope during the patient's illness and in their own bereavement	• Insists that each person be treated as an individual
• Uses a team approach to address the needs of patients and their families, including bereavement counseling if indicated	• Recognizes that family and staff play a key role in maintaining personhood and quality of life
• Will enhance quality of life, and may also positively influence the course of illness	• Recognizes that family and care staff need support.[33]
• Is applicable early in the course of illness, in conjunction with other therapies that are intended to prolong life, such as chemotherapy or radiation therapy, and includes those investigations needed to understand better and manage distressing clinical complications.[32]	

When people are dying with dementia there is a gap they can fall into: Traditional geriatric care, care of the elderly, and old-age psychiatry, despite making considerable advances in some areas, have not always given optimal relief in dementia. Palliative care units have often not known how to respond to cognitive and behavioral problems.[14,15] Palliative care has had difficulties when prognosis is uncertain and when dying is potentially protracted.[16] Specialist palliative care teams and hospices are underused by older people and by people with non–cancer diagnoses, including dementia.[17] Indeed the majority of palliative care teams in the United Kingdom, even by the mid 1990s, did not include dementia as falling within their remit.[18] In the United States, admission of people with dementia into hospice programs has been promoted since the early 1980s.[19]

Person-centered dementia care emerged in the 1980s and 1990s. Its antecedents were similar to those of palliative care, namely a critical stance to current care regimes and a belief that if the challenge of the condition could be reframed, then therapeutic opportunities would be revealed. Specifically, Kitwood argued that dementia could be understood as evidencing a dialectical interplay between neurobiology and the social environment.[20] Care regimes could be modified such that they did not just undertake custodial containment, but could be used to improve well-being. For example, it became possible to identify behaviors that, if understood as attempts at communication rather than as symptomatic of neuropsychiatric disease, could be responded to in ways that would meet emotional or social needs.[21,22] By 2004 it was possible to record the wide international impact of this approach[23] and to identify considerable research support for its efficacy.[24]

Similar trends are evident in hospice and palliative care and in person-centered dementia care. These are summarized in TABLE 3.

LEARNING FROM EACH OTHER—COMBINING APPROACHES

In this article I have argued that we have to address both social attitudes and practical failures if we are to act in a way that improves the quality of life of people with dementia. Practically, we now have more treatment possibilities. These have come from a growing knowledge of brain function, including an ability to focus intervention on the least-affected areas of the brain and from an ability to enhance memory and perception.[25] Successful interventions have been developed to improve cognition and enhance functioning in the present.[26,27] The most commonly identified causes of death for people with dementia include pneumonia, cardiovascular events, pulmonary embolism, and dysphagia.[10] There is also a high incidence of falls, the consequence of which can precipitate decline and death.[28] All of these causes of death or physical decline can be managed in a way that ameliorates their impact. Palliative care regimes that plan for the totality of the person's needs are the best way to achieve this. There is much therefore to counter therapeutic pessimism. All these advances, both attitudinal and practice-based, can be furthered by a closer liaison between palliative care and dementia care. What each approach has to offer the other is summarized in TABLE 4.

CONCLUSIONS

This article began by suggesting that both the existence of dementia and its prevalence pose a challenge for the advancement of healthy aging. It has

TABLE 3. Some similarities between palliative care and person-centered dementia care

Aspects of the approach	Palliative care	Person-centered dementia care
V's Cartesian dualism	Better understanding of pain as being multi-faceted—physical, emotional, social, and spiritual pain	Better understanding of the brain and the way that the environment interacts in ways that can be either positive or negative
Critical of existing services	Poor quality: the dying experienced abandonment, in particular by the medical profession.	Poor quality, a prevalent therapeutic nihilism, or a lack of a clear sense about what is best to do. The result is that too little help or too much inappropriate help was provided
Knowledge comes from the patient	People became subjects in their own dying—that is, care was shaped to meet their particular needs.	Listening to what the person is saying or discerning what their actions are telling us; developing approaches to facilitate this
Context of the whole life	Caregiver's burden and bereavement care become legitimate foci for palliative care services.	Caregivers and professional staff need to be supported as well as the person with dementia. This person with dementia retains a self even if that self appears to change between the person who was, is, and will be.

been argued that there is much we can do to help people with dementia live well until they die. To achieve this we have to implement best practice in terms of quality of care. This requires changes in service organizations and in the training, attitudes, and practice of health and social care professionals. Such changes would be better facilitated if modes of care could embrace the experiences of modern palliative care which, at the level of social attitudes and self-perceptions as well as at the level of evidence-based care provision, has shown that it is possible to enhance living until you die.

A concern with healthy aging is a social construct that can be misused in that it is vulnerable to a juxtaposition of descriptive and normative meanings. Here the shift would be captured by putting together "healthy" and "successful" aging. To successfully age invokes an ethos of achievement. To construct for oneself a healthy aging may be a valuable counterweight to deficit models of aging, the assumption that aging is a period of decline, but it risks disenfranchising those for whom autonomy and self-aware agency have ceased to

TABLE 4. Learning from each other

What palliative care could offer people with dementia	What dementia care could offer palliative care
Inhibition of unnecessary interventions	Expertise across many diseases
Use of techniques to improve care	Ability to respond to multiple pathologies
Inclusion of more consultation with patients and caregivers	Development of programs for long-term engagement with patient
Adaptation of bereavement care for the particular circumstances of dementia	Engagement of patients who have no verbal abilities
	Work with cognitive and behavioral problems

be a key feature of their lives. If we see healthy aging as achievement, as a manifestation of a modernist imperative to "make oneself," then we risk creating another barrier separating the person with dementia from the society they live within.

The challenge of promoting health aging is one of how to expand the span of healthy living to the end of the life span, whenever that is. To include people with dementia we need to develop an epistemology that incorporates evidence-based practice within a person-centered care paradigm. To redress the dangers of social isolation and either therapeutic neglect or nihilism, we need to question what our dread of dementia might indicate about the way we see ourselves and the way we conceptualize the society we live within.[34]

ACKNOWLEDGMENTS

Much of my thinking about dementia and palliative care has evolved from working with Professor Murna Downs (University of Bradford) and Dr. Katherine Froggatt (Lancaster University).

REFERENCES

1. Kitwood, T. 1997. Dementia Reconsidered: the Person Comes First. Open University Press. Buckingham, UK.
2. Glaser, B. & A. Strauss. 1965. Awareness of Dying. Aldine. Chicago, IL.
3. Patterson, J.T. 1987. The Dread Disease. Harvard University Press. Cambridge, MA.
4. Small, N., M. Downs & K. Froggatt. 2006. Improving end-of-life care for people with dementia: the benefits of combining UK approaches to palliative care and dementia care. *In* Care-Giving in Dementia. Research and Implications, Vol. 4. B.M.L. Meisen & G.M.M. Jones, Eds.: 365–392. Brunner-Routledge. London.
5. Ferri, C.P., M. Prince, M. Brayne, *et al.* 2005. Global prevalence of dementia: A Delphi consensus study. The Lancet **366.9503:** 2112–2117.

6. ALZHEIMER'S SOCIETY. 2006. Policy positions: demography Available at: www. alzheimers.org.uk/News_and_Campaigns/Policy_Watch/demography.htm. Accessed 15 March 2006.
7. FOX, P. 1989. From senility to Alzheimer's disease: the rise of the Alzheimer's disease movement. Milbank Q. **67.1:** 58–102.
8. FOX, P. 2000. The role of the concept of Alzheimer's disease in the development of the Alzheimer's Association in the United States. *In* Concepts of Alzheimer's Disease: biological, Clinical and Cultural Perspectives. P.J. Whitehouse, K. Maurer & J.F. Ballenger, Eds.: 209–233. The Johns Hopkins University Press. London.
9. COX, S. & A. COOK. 2002. Caring for people with dementia at the end of life. *In* Palliative Care for Older People in Care Homes. J. Hockley & D. Clark, Eds.: 86–103. Open University Press. Buckingham, UK.
10. JOLLEY, D. & D. BAXTER. 1997. Life expectation in organic brain disease. Advances in Psychiatric Treatment **3:** 211–218.
11. MITCHELL, S.L., D.K. KIELY & M.B. HAMEL. 2004. Dying with advanced dementia in the nursing home. Arch. Intern. Med. **164:** 321–326.
12. SMALL, N. 2000. The modern hospice movement: "bright lights sparkling" or "a bit of heaven for the few"? *In* Oral History, Health and Welfare. P. Thompson & J. Walmsley, Eds.: 288–308. Routledge. London.
13. INTERNATIONAL OBSERVATORY ON END OF LIFE IN CARE. 2006. www.eolcv-observatory.net.
14. LUCHINS, D.J. & P. HANRAHAN. 1993. What is appropriate health care for end-stage dementia? J. Am. Geriatr. Soc. **4:** 25–30.
15. VOLICER, L., A. COLLARD, A. HURLEY, *et al.* 1994. Impact of special care unit for patients with advanced Alzheimer's disease on patients' discomfort and costs. J. Am. Geriatr. Soc. **42:** 597–603.
16. HANRAHAN, P. & D. J. LUCHINS. 1995. Access to hospice programs in end-stage dementia: a national survey of hospice programs. J. Am. Geriatr. Soc. **43.1:** 56–59.
17. ADDINGTON-HALL, J. & I. HIGGINSON. 2001. Palliative Care for Non-Cancer Patients. Oxford University Press. Oxford, UK.
18. LLOYD-WILLIAMS, M. 1996. An audit of palliative care in dementia. Eur. J. Cancer Care **5:** 53–55.
19. VOLICER, L. 1986. Need for hospice approach to treatment of patients with advanced progressive dementia. J. Am. Geriatr. Soc. **34:** 655–658.
20. KITWOOD, T. 1990. The dialectics of dementia: with particular reference to Alzheimer's disease. Ageing Soc. **10:** 177–196.
21. RADER, J. *et al.* 1985. How to decrease wandering, a form of agenda behaviour. Geriatr. Nurs. **6:** 196–199.
22. STOKES, G. 2000. Challenging Behaviour in Dementia: a Person-Centred Approach. Speechmark Publishing. Bicester, UK.
23. BROOKER, D. 2004. What is person-centred care for people with dementia? Rev. Clin. Gerontol. **13:** 212–222.
24. BROOKER, D. 2006. Dementia care mapping: a review of the research literature. Gerontologist **45:** 11–18.
25. DIAMOND, M. 1988. Enriching Heredity: the Impact of the Environment on the Brain. Free Press. New York.
26. WOODS, R.T. & P.G. BRITTON. 1985. Clinical Psychology with the Elderly. Croom Helm. London.

27. DRÖES, R.M. 1997. Psychosocial treatments for demented patients. *In* Care-Giving in Dementia, Vol. 2. B.M.L. Miesen & G.M.M. Jones, Eds.: 127–148. Routledge. London.
28. HERMANS, D., J. LISAERDE & E. TRIAU. 1989. Sense and non-sense of a technological health care model in terminally ill demented patients. The first international conference on the palliative care of the elderly. J. Palliat. Care **5:** 39–42.
29. HERBERT, L., P. SCHERR, J. BENIAS, *et al.* 2003. Alzheimer's disease in the US population: prevalence estimates using the 2000 census. Arch. Neurol. **60:** 1119–1122.
30. AUSTRALIAN INSTITUTE OF HEALTH AND WELFARE. 1998. Australia's Health 1998: the Sixth Biennial Health Report of the Australian Institute of Health and Welfare. AIHW. Canberra.
31. GILES, L., I. CAMERON & M. CROTTY. 2003. Disability in older Australians: projections for 2006–13. Med. J. Aust. **179.3:** 130–133.
32. WORLD HEALTH ORGANISATION. 1996. Cancer Pain Relief, 2nd ed. WHO. Geneva (www.who.int/hiv/topics/palliative/PalliativeCare).
33. KITWOOD, T. 1997. The concept of personhood and its relevance to a new culture of dementia care. *In* Care-Giving in Dementia, Vol. 2. B.M.L. Miesen & G.M.M. Jones, Eds.: 3–13. Routledge. London.
34. SMALL, N.A., K. FROGGATT & M. DOWNS. 2007. Living and Dying with Dementia. Dialogues about Palliative Care. Oxford University Press. Oxford.

Complementary and Alternative Medicine Use by Older Australians

ANTHONY L. ZHANG,[a] CHARLIE C. L. XUE,[a] VIVIAN LIN,[b] AND DAVID F. STORY[a]

[a]*Division of Chinese Medicine, School of Health Sciences, World Health Organization Collaborating Centre for Traditional Medicine, RMIT University, Melbourne, Victoria, Australia*

[b]*School of Public Health, La Trobe University, Melbourne, Victoria, Australia*

ABSTRACT: Complementary and alternative medicine (CAM) use by Australians is substantial and increasing, but little is known about its use by the elderly. We here present the findings for the elderly cohort in our recently conducted national survey on CAM use by adult Australians. In May and June 2005, computer-assisted telephone interviews, using random-digit telephone dialing, were employed to gather data on CAM use in the last 12 months. Of 1067 adult participants interviewed, 178 were 65 or older. More than half of these (57.8%; 95% CI, 50.7%–64.9%) had used at least one of 17 common forms of CAM and 60.4% of the CAM users had consulted CAM practitioners. Clinical nutrition, chiropractic, massage therapy, meditation, and herbal medicine were the most common forms of CAM used by the elderly. A higher proportion of the elderly had always used both CAM and conventional medical treatments (37.9%) than had those aged 18–34 (15.7%) and 35–64 (26.9%). Elderly CAM users (60.2%) were more likely than younger users to discuss their use with their doctors. Of those who did not do so, 24.1% were not asked by their doctors and 16.0% considered that their doctor would disapprove. In conclusion, we found that a substantial proportion of older Australians use CAM. The elderly are also more likely than younger adults to discuss their use of CAM with their doctors, but doctors need to play a more active role in initiating such communication.

KEYWORDS: complementary and alternative medicine; CAM use; older Australians; population-based survey

INTRODUCTION

As for most developed counties, the population of Australia is aging rapidly. Currently, 13% of Australians are over the age of 65, and this figure will be

Address for correspondence: Dr. Anthony Zhang, Ph.D., Division of Chinese Medicine, School of Health Sciences, RMIT University, P.O. Box 71, Bundoora, VIC 3083, Australia. Voice: +613 9925 7758; fax: +613 9925 7178.
tony.zhang@rmit.edu.au

Ann. N.Y. Acad. Sci. 1114: 204–215 (2007). © 2007 New York Academy of Sciences. doi: 10.1196/annals.1396.032

more than double by 2051.[1] Age-related chronic diseases, such as arthritis and chronic obstructive pulmonary disease, are often not effectively managed by conventional therapies either because of their inability to produce sustained relief or because of concerns about potentially serious adverse effects. Thus, it is likely that the increasing use of complementary and alternative medicine (CAM) therapies in Australia and comparable countries can, at least, be partially explained by limitations of available conventional treatments, particularly those for chronic diseases. On the basis of findings of previous population surveys, in Australia and elsewhere, it is also likely that many individuals seek CAM therapies for general health enhancement and improved well-being.[2,3]

In addition to the reports from Australia, increasing CAM use is also being reported in other developed countries, including the United States, the U.K., and Canada.[4–6] It has been concluded from the findings of these studies[2,4–6] as well as from systematic reviews[3,7,8] that older people are less likely to use CAM than are younger adults. However, a study conducted in the United States demonstrated that in 2006 CAM was used by more than a half (54%) of Americans aged 65 and older.[9] The popularity of CAM among elderly Australians is not clear. Various Australian regional studies of CAM use[2,10] as well as studies on specific elderly subgroups, such as females,[11] have reported disparate findings, with estimates of prevalence of CAM use by older Australians ranging from 15% to 37%.

The increasing use of CAM in the Australian population and the possibility that the elderly, particularly those with chronic diseases, might be choosing to use CAM instead of or in conjunction with conventional therapies are strong reasons to investigate CAM use among the older population. In addition to ascertaining prevalence of use by the elderly, we considered that such a study should also investigate the reasons why the elderly might choose to use CAM and important issues in regard to concurrent use of CAM and conventional medical therapies.

METHODS

Between May and June 2005, we conducted a population-based survey on CAM use by adult Australians. Age was considered to be an important factor with regard to CAM use and, accordingly, we sampled for three age ranges: 18–34, 35–64, and 65 and above. Particular efforts were made to avoid common weaknesses of population surveys by ensuring a representative sample (in addition to age, our sample was representative of the Australian population in respect to gender and state of residence), appropriate choice of the CAM therapies surveyed, and the use of valid sampling methods.[12] Sampling was by random-digit telephone dialing and the data sought were obtained by computer-assisted interviews. The study investigated the use of what we established were the 17 most common forms of CAM used in Australia. The findings provide a national profile of CAM use in Australia, with the elderly being a particular focus.

Approval to conduct the study was obtained from the RMIT University Human Research Ethics Committee. The detailed methodology has been described elsewhere and is summarized below.[12]

Survey Method

The survey was administered to 1067 participants proportionally recruited from the six Australian states and two territories. The sample was also representative in respect of gender and age, based on Australian Bureau of Statistics data.[13] To ensure random recruitment within identified households, the member of the household with the next birthday was selected for interview. No financial incentive was offered. Interviews were conducted in English with an average duration of 13.5 minutes.

At the start of the interview, participants were asked whether they had used one or more of the 17 selected forms of CAM in the preceding 12 months. If so, they were then asked questions about their CAM use, such as visits to CAM practitioners, their rationale for choosing to use CAM (including referral), perceived benefits and risks of CAM and conventional medical treatments, concurrent use of CAM and conventional treatments, and communication with their medical doctor about their CAM use.

Statistical Analysis

All percentages were weighted to adjust for demographic variations in the sample (participant's state of residence, gender, and broad age ranges). Chi-square tests were used to compare proportional differences. The Statistical Package for Social Sciences (SPSS, Chicago, IL) for Windows, version 13.0, was used for data analysis. Probability values (P) less than 0.05 were considered to indicate significant differences between groups.

RESULTS

Characteristics of the Survey Participants

The overall characteristics of the population sample were comparable to relevant population profiles with regard to age range, gender, self-reported health status, and employment and education status, as reported by the Australian Bureau of Statistics in 2005.[13,14]

Of the 1067 completed interviews, three participants did not provide their ages and were excluded from data analysis. Of the 1064 participants, 178

TABLE 1. Participant demographics

Characteristic	Age <65 (n = 886)		Age ≥65 (n = 178)	
	n	%	n	%
Gender				
Female	442	49.9	101	56.7
Male	444	50.1	77	43.3
Region				
NSW	286	32.3	66	37.1
VIC	216	24.4	46	25.8
QLD	179	20.2	27	15.2
SA	72	8.1	15	8.4
WA	80	9.0	16	9.0
TAS/NT/ACT	53	6.0	8	4.5
Country of birth				
Australia	206	23.4	44	24.9
Overseas	674	76.6	133	75.1
Education[a]				
High school or less	364	41.3	99	56.2
Postsecondary education	517	58.7	77	43.8
Employment[a]				
Employed	661	75.8	29	16.4
Not in labor force	211	24.2	148	83.6
Annual household income (A$)[a]				
<20,000	98	12.4	65	48.9
20,000–40,000	155	19.6	40	30.1
>40,000	538	68.0	28	21.1
Had private health insurance in past 12 months				
Yes	385	43.8	83	46.6
No	495	56.3	95	53.4
Self-reported health status[a]				
Excellent/very good/good	766	86.6	134	75.3
Fair/poor	119	13.4	44	24.7

[a]$P < 0.05$ comparing those aged 65 and older with those less than age 65.

(16.7%) were aged 65 or older. In comparison with younger participants (TABLE 1), the elderly cohort had higher proportions of females, of individuals with only fair or poor self-reported health status, and of individuals not in the workforce. These differences were consistent with data published by the Australian Bureau of Statistics ($P > 0.05$ for each comparison).[13,14] In addition, the elderly cohort had smaller proportions with postsecondary education and household income over $40,000 per annum.

Use of CAM Therapies and Visit to CAM Practitioners

Of those aged 65 or older, 57.8% (95% CI, 50.7%–64.9%) had used at least one of 17 forms of CAM in the preceding 12-month period. The prevalence of

TABLE 2. Use of complementary and alternative medicine in the last 12 months

Type of CAM	Age <65 % (SE)	Age ≥65 % (SE)	*P*-value[a]
Clinical nutrition	48.2 (1.7)	35.4 (3.5)	0.001
Chiropractic	16.1 (1.2)	15.7 (2.7)	0.900
Western massage therapy	30.0 (1.5)	13.9 (2.6)	<0.0001
Meditation	18.2 (1.3)	13.6 (2.5)	0.134
Western herbal medicine	17.2 (1.3)	12.3 (2.4)	0.104
Acupuncture	9.3 (1.0)	9.2 (2.1)	0.990
Yoga	12.9 (1.1)	7.9 (2.0)	0.057
Naturopathy	11.4 (1.1)	7.5 (1.9)	0.124
Osteopathy	4.3 (0.7)	5.9 (1.7)	0.335
Reflexology	3.7 (0.6)	5.7 (1.7)	0.199
Energy healing	7.3 (0.9)	5.0 (1.6)	0.268
Homeopathy	6.3 (0.8)	4.6 (1.5)	0.370
Aromatherapy	18.6 (1.3)	4.2 (1.5)	<0.0001
Qigong, martial art, and tai chi	6.5 (0.8)	2.9 (1.2)	0.057
Chinese herbal medicine	7.9 (0.9)	2.6 (1.2)	0.010
Chinese therapeutic massage	5.6 (0.8)	2.4 (1.1)	0.067
Chinese medicine dietary therapy	2.6 (0.5)	1.1 (0.8)	0.230
At least 1 of 17 forms of CAM	71.4 (1.5)	57.8 (3.6)	0.0003

[a]*P*-value comparing those aged 65 and older with those less than age 65.
SE = standard error.

CAM use by elderly participants was considerably lower than that for younger adults in the study (71.4%, $P < 0.05$). Most of the elderly CAM users (60.4%) had visited a CAM practitioner at least once during the 12-month period, that is, more than one-third (34.9%, 95% CI: 28.0%–41.8%) of the elderly participants had visited CAM practitioners. TABLE 2 shows the 17 forms of CAM therapy included in this study relative to the proportions of participants under 65 years of age and of those 65 or older that had used them.

The CAM therapies most commonly used by the elderly were clinical nutrition, herbal medicine, chiropractic, massage therapy, and meditation. However, compared to younger participants, considerably smaller proportions of elderly participants had used clinical nutrition, Western massage therapy, aromatherapy, and Chinese herbal medicine ($P < 0.05$ in each case [TABLE 2]).

Participants were asked to estimate the number of visits they had made to acupuncturists, Chinese or Western herbal medicine practitioners, Chinese or Western therapeutic massage therapists, chiropractors, osteopaths, homeopaths, naturopaths, and aromatherapists. Nearly one in three older Australians (29.3%) had visited a practitioner on at least one occasion in the 12-month period. The number of visits to all 10 types of practitioners by elderly participants, over the 12-month period, amounted to 10.4 visits/person. On the basis of a national figure of 2.67 million people of 65 years or older,[13] it is estimated that 8.14 million CAM visits were made annually by elderly Australians. In comparison, 38.8% of the younger participants in the study visited CAM practitioners annually, which, on a population basis, is 61 million visits in a year.

TABLE 3. Characteristics of elderly CAM users compared to CAM users younger than 65

Characteristic	Aged <65 % (SE)	Aged ≥65 % (SE)	*P*-value[a]
Gender			
Female	77.0 (2.0)	63.1 (4.8)	0.004
Male	65.9 (2.2)	51.4 (5.5)	0.012
Region			
NSW	73.4 (2.6)	66.9 (5.9)	0.293
VIC	72.4 (3.0)	58.4 (7.3)	0.059
QLD	73.2 (3.4)	60.0 (8.5)	0.127
SA	64.1 (5.9)	47.4 (12.1)	0.218
WA	69.9 (4.9)	48.9 (12.5)	0.098
TAS/NT/ACT	58.0 (8.2)	8.1 (9.6)	0.013
Country of birth			
Australia	72.6 (1.7)	58.2 (4.2)	0.001
Overseas	67.2 (3.3)	58.2 (7.4)	0.254
Education			
High school or less	64.4 (2.5)	51.7 (5.0)	0.021
Postsecondary education	76.4 (1.9)	65.8 (5.3)	0.040
Employment			
Employed	72.6 (1.7)	73.1 (8.0)	0.954
Not in labor force	68.5 (3.2)	55.1 (4.0)	0.009
Annual household income (A$)			
<20,000	66.2 (4.8)	51.7 (6.3)	0.069
20,000–40,000	66.5 (3.8)	61.8 (7.1)	0.546
>40,000	72.9 (1.9)	71.6 (8.5)	0.881
Had private health insurance in past 12 months			
Yes	76.0 (1.9)	54.3 (5.0)	<0.0001
No	65.5 (2.4)	62.0 (5.3)	0.538
Self-reported health status			
Excellent/very good/good	71.0 (1.6)	55.2 (4.2)	0.0002
Fair/poor	73.6 (4.1)	66.1 (7.1)	0.346

[a]*P*-value comparing those aged 65 and older with those less than age 65.
SE = standard error

Characteristics of Older Australian CAM Users

For the elderly participants in our study, CAM use was more common among females, those with postsecondary education, those with higher household incomes, those without private health insurance, those who considered their health to be only fair or poor, and those currently employed, than among those not in each of the specified categories (TABLE 3).

Use of Conventional Medical Services Vs. Complementary Medicine

A higher proportion of the elderly participants in the study had visited a general medical practitioner (GP) and/or a medical specialist (92.8% over the 12-month period) than those under 65 (78.4%, $P < 0.05$). Moreover, a smaller proportion of elderly participants who visited medical doctors also visited CAM practitioners (56.2%) than did younger participants who visited a medical doctor (74.9%, $P < 0.05$). However, multivariate analyses did not indicate that there was a correlation between the use of medical doctor by the elderly and either CAM use ($P > 0.05$) or visits to CAM practitioners ($P > 0.05$).

Higher proportions of elderly than of younger participants in the survey visited their medical doctors for health checkups, immunization, or to obtain prescriptions (32.8% in total, compared to 23.8% among younger adults, $P < 0.05$), for cardiac conditions (21.0% versus 7.8%, $P < 0.05$) and hypertension (15.7% versus 2.4%, $P < 0.05$). Among the elderly participants, there was no significant association between visits to medical doctors for a specific medical condition and the use of CAM ($P > 0.05$).

The proportions of elderly and younger CAM users with certain opinions of either CAM or conventional treatments were comparable, notably confidence that Western medicine would improve or cure their conditions, satisfaction with medical services, and the appreciation that there were potential risks associated with CAM and Western medicine. However, compared with younger CAM users, smaller proportions of elderly users considered that CAM provides a more "holistic" approach to health than conventional medical treatments (58.9% versus 78.7%, respectively, $P < 0.05$) and that CAM offers a more "natural" therapeutic approach than conventional medical treatments (65.5% versus 83.0%, respectively, $P < 0.05$).

Approximately six out of 10 (59.5%) of the elderly participants visited a medical doctor only, that is, they did not visit CAM practitioners. This is a significantly higher proportion than for younger participants (39.1%, $P < 0.05$). However, more than one-third of elderly CAM users (37.9%) had always used both CAM and conventional medicine in the preceding 12 months. This is also a significantly higher proportion than that for younger CAM users in the study (26.9% for those aged 35–64 and 15.7% for those aged 18–34, $P < 0.05$).

Communication with Medical Doctors Concerning CAM Use

Among elderly CAM users, 90.6% had also visited a GP in the 12-month period. Of the CAM users who had also visited a medical doctor, more than half (60.2%) stated that they always discussed their CAM use with their medical doctors. This is a significantly higher proportion than that for CAM users who had also visited medical doctors aged 35–64 (49.4%) and for those aged 18–34 (40.2%, $P < 0.05$). Of the users who indicated that they discussed their use of CAM with their medical doctors, 67.7% considered it important to do so and 21.1% were actually advised by their doctors to use some form of CAM.

In respect of the elderly CAM users who had visited a GP but did not discuss their use of CAM with their doctors, 43.6% did not consider it necessary to do so, 24.1% did not because they were not specifically asked about it by their doctors, and 16.0% considered that their doctors would disapprove of their use of CAM.

Previous and Future Use of CAM

Among the elderly participants, nearly three-quarters (73.8%) had used at least one form of CAM at some time in their lives and nearly two-thirds (65.9%) indicated that they would consider using some form of CAM in the future. Both these proportions were lower than those for younger participants, being 86.2% and 88.5%, respectively.

DISCUSSION

Although the findings of our study indicate that the number of the elderly Australian population (65 years and older) using CAM is less than that of younger adults, it is nevertheless appreciable. Thus, 57.8% of the elderly participants in our national survey had used at least one of 17 common forms of CAM in the 12 months preceding the survey and, of these, 64.9% had visited a CAM practitioner in the 12-month period. The prevalence of CAM use by the elderly compares with a prevalence of 71.4% for younger adults. Of the 17 forms of CAM included in the survey, clinical nutrition, chiropractic, Western massage therapy, meditation and Western herbal medicine were the most commonly used by the elderly cohort. Although clinical nutrition and Western massage therapy were two of the most popular complementary therapies used by the elderly, they were used to a significantly greater extent by the younger participants. Similarly, aromatherapy and Chinese herbal medicine were significantly more popular among younger participants.

Some correlations between socio-demographic characteristics and CAM use suggested by the study outcomes are worthy of note. Thus, CAM use overall would appear to be more common among adult females than males, irrespective of age. Perhaps not unexpectedly, for both the elderly and younger participants, CAM use was more prevalent among those with higher levels of education, and those from households with higher incomes. Also, there was a higher proportion of elderly CAM users without private health insurance coverage than those with insurance coverage. The reverse was the case for younger CAM users.

We did not investigate the specific health-related reasons that individuals elected to use CAM therapies, but differences between elderly and younger participants in regard to the relationship between self-reported heath status and CAM suggest possible different rationales for using CAM between the older and younger cohorts in the study. The proportion of elderly participants in our survey sample of 1064 that considered their health was poor or only fair was almost twice that for the younger cohorts, presumably reflecting the higher incidences of chronic diseases in the elderly. It is therefore of interest that a higher proportion of the elderly actually using CAM considered that their health was fair/poor rather than good/excellent since this was not the case for younger CAM users, for whom there was no substantial difference in self-reported health status. Thus, it appears that health status, whether perceived or confirmed, is a determinant of CAM use in the elderly, but not in younger individuals.

Consistent with findings of previous surveys on CAM use,[15,16] the lack of effective communication between patients and their medical doctors about the patient's use of CAM is of concern. Most forms of CAM therapy, whether substance-based or not, may produce adverse effects and/or interfere with other treatments. Of particular concern with substance-based CAM therapies is the possibility of serious adverse interactions involving conventional therapeutic drugs and even surgical or manipulative treatments. It is likely that elderly CAM users may be more susceptible to such problems on account of their age, health status, and relatively high use of medication. For the 12-month period investigated, we found that more than one-third of the elderly CAM users also had conventional medical treatment; this was a considerably higher proportion than for younger CAM users. Although our findings indicate that elderly CAM users are more likely to discuss their use of CAM with their medical doctors than are younger users, some 40% of elderly CAM users did not do so. The situation appears to be considerably worse in the United States, where more than half (57%) of CAM users over 65 did not disclose their use of at least one of their CAM therapies to their doctors[17] and about three-quarters (77%) of users over 50 did not disclose their use.[9]

Of those in our study not disclosing CAM use to their doctors, almost one-quarter indicated that they were not asked about possible CAM use by their doctors, whereas about 16% thought that their doctors would disapprove of such therapies. Reinforcing the conclusions of previous studies,[15,16] it is obvious that there is a need for improved communication between medical practitioners and their patients in regard to CAM, with doctors are taking a more active and perhaps less judgmental role in initiating such discussions.[18,19] It is possible that some medical practitioners do not ask their patients about their use of CAM because of their lack of appreciation of the popularity of CAM in the community and perhaps because they themselves have a poor understanding of the complementary/alternative therapies.[20,21] In regard to the latter, it is worth noting that many medical schools, particularly in the United States, include introductory material on the main CAM therapies in their curricula.[22]

Considerably more research on quality, safety and efficacy on commonly used forms of CAM therapies is urgently needed to provide a basis for evidence-based communication between CAM users and the healthcare professions (CAM and conventional) concerning CAM use. It is worth noting that the

Australian government, both state and federal, has taken initiatives to develop and implement regulatory frameworks for a number of CAM therapies, such as chiropractic, osteopathy, acupuncture, and Chinese herbal medicine. Also, in line with recommendation of an expert committee established by the federal government to examine the use of complementary medicine in the Australian healthcare system, funding is now being provided through national agencies to support research on CAM therapies.[23]

The prevalence of elderly CAM users in Australia is consistent with the finding of a recent study by the National Center for Complementary and Alternative Medicine (NCCAM), which reported CAM use by 54% of Americans aged 65 and older.[9] Our study demonstrated that the use of CAM by older Australians is broadly comparable to its use by Australian adults overall. We found that more than 60% of the older CAM users in Australia (more than one-third of all older Australians) visited CAM practitioners, with an average of more than 10 visits annually. That is, more than eight million visits to CAM practitioners were made by older Australians over a 12-month period. These findings are also similar to those reported for the elderly American population.[17]

In conclusion, the present study offers useful insights into the high level of use of CAM and CAM practitioners by older Australians. There is a relatively high concurrent use of CAM and Western medicine by both the elderly and younger adults. Although it seems that Australians are more likely to discuss their use of CAM with their doctors than are their U.S. counterparts, but the number that fails to do so remains a concern. More concerted efforts are required from medical and CAM practitioners as well as CAM users to improve such communication and to thereby to bring about safer and more rationale use of CAM. Because of the relatively small sample size in this study and exclusion of the institutionalized elderly[24] (7% of older Australians), further studies on CAM use by elderly Australians are needed.

ACKNOWLEDGMENTS

This study was funded by the RMIT University, the Sydney Institute of Traditional Chinese Medicine, the Chiropractors' Association of Australia, and the Australian Acupuncture & Chinese Medicine Association.

[The authors declare that they have no competing financial interests.]

REFERENCES

1. Australian Bureau of Statistics. 2005. Australian Demographic Statistics: June 2005. [Access available from www.abs.gov.au].
2. MacLennan, A.H., S.P. Myers & A.W. Taylor. 2006. The continuing use of complementary and alternative medicine in South Australia: costs and beliefs in 2004. Med. J. Aust. **184:** 27–31.

3. ERNST, E. 2000. Prevalence of use of complementary/alternative medicine: a systematic review. Bull. World Health Organ. **78:** 252–257.
4. THOMAS, K. & P. COLEMAN. 2004. Use of complementary or alternative medicine in a general population in Great Britain. Results from the National Omnibus Survey. J. Public Health **26:** 152–157.
5. EISENBERG, D.M. *et al.* 1998. Trends in alternative medicine use in the United States, 1990–1997: results of a follow-up national survey. JAMA **280:** 1569–1575.
6. RAMSAY, C., M. WALKER & J. ALEXANDER. 1999. Alternative medicine in Canada: use and public attitudes. Public Policy Sources **21:** 1–31.
7. WOOTTON, J.C. & A. SPARBER. 2001. Surveys of complementary and alternative medicine: Part I. General trends and demographic groups. J. Altern. Complement. Med. **7:** 195–208.
8. HARRIS, P. & R. REES. 2000. The prevalence of complementary and alternative medicine use among the general population: a systematic review of the literature. Complement. Ther. Med. **8:** 88–96.
9. NATIONAL CENTER FOR COMPLEMENTARY AND ALTERNATIVE MEDICINE. 2007. Complementary and Alternative Medicine: What People 50 and Older are Using and Discussing with Their Physicians. NCCAM. Washington, DC.
10. KERMODE, S., S. MYERS & L. RAMSAY. 1998. Natural and complementary therapy utilisation on the north coast of NSW. Aust. J. Holist. Nurs. **5:** 7–13.
11. ADAMS, J. *et al.* 2003. The profile of women who consult alternative health practitioners in Australia. Med. J. Aust. **179:** 297–300.
12. XUE, C.C.L. *et al.* 2007. Complementary and alternative medicine use in Australia: results of a national population based survey in 2005. *J. Altern. Complement. Med.* **13:** In press.
13. AUSTRALIAN BUREAU OF STATISTICS. 2005. Population by Age and Sex, Australian States and Territories: June 2005. [Access available from www.abs.gov.au].
14. AUSTRALIAN INSTITUTE OF HEALTH AND WELFARE. 2004. Australia's Health 2004. AIHW Cat. No. AUS 44. Canberra.
15. CROCK, R.D. *et al.* 1999. Confronting the communication gap between conventional and alternative medicine: a survey of physicians' attitudes. Altern. Ther. Health Med. **5:** 61–66.
16. FLANNERY, M.A. *et al.* 2006. Communication about complementary and alternative medicine: perspectives of primary care clinicians. Altern. Ther. Health Med. **12:** 56–63.
17. FOSTER, D.F. *et al.* 2000. Alternative medicine use in older Americans. J. Am. Geriatrics Soc. **48:** 1560–1565.
18. EISENBERG, D.M. 1997. Advising patients who seek alternative medical therapies. Ann. Intern. Med. **127:** 61–69.
19. HALLER, C.A. 2001. We should always ask our patients about unconventional treatments. West. J. Med. **175:** 164.
20. MILDEN, S.P. & D. STOKOLS. 2004. Physicians' attitudes and practices regarding complementary and alternative medicine. Behav. Med. **30:** 73–82.
21. CORBIN WINSLOW, L. & H. SHAPIRO. 2002. Physicians want education about complementary and alternative medicine to enhance communication with their patients. Arch. Intern. Med. **162:** 1176–1181.
22. WETZEL, M.S., D.M. EISENBERG & T.J. KAPTCHUK. 1998. Courses involving complementary and alternative medicine at US medical schools. JAMA **280:** 784–787.

23. National Health and Medical Research Council. 2007. Complementary and Alternative Medicine Special Call for Research Applications. [Access available from www.nhmrc.gov.au/funding/types/granttype/strategic/cam.htm].
24. Australian Bureau of Statistics. 2001. 2001 Census of Population and Housing: Aging in Australia. [Access available from www.abs.gov.au].

'Anti-Aging Medicine' and the Cultural Context of Aging in Australia

Preliminary Findings from Ongoing Research on Users and Providers of 'Anti-Aging Medicine' in Australia

BEATRIZ CARDONA

Centre for Cultural Research, University of Western Sydney, Penrith, New South Wales, Australia

ABSTRACT: This paper explores the management of the aging body within the anti-aging discourse and its implications on notions of "successful" and "healthy" aging policies. By looking at some of the preliminary findings of our current study of 'anti-aging medicine' in Australia, including interviews conducted with stakeholders in the anti-aging debate, this study explores some recurrent values and perceptions regarding 'anti-aging medicine,' the renegotiation of boundaries between illness and health, and the social, cultural, and economic forces shaping understandings and practices around aging and decisions to use anti-aging technologies.

KEYWORDS: 'anti-aging medicine;' healthy aging; positive aging

BACKGROUND INFORMATION

"The war on anti-aging medicine,"[1] with leading gerontologists and anti-aging practitioners and marketers fighting over the meaning of old age[2] and the scientific validity and legitimacy of existing anti-aging therapies to alter the aging process[3] and prolong life, has so far excluded the views and perceptions from the consumers of these technologies. Despite evidence of a huge public interest in and willingness to pay for interventions that delay at least the appearance of aging,[4] there is no current information available on the reasons why people pursue these practices. Common assumptions on why people use 'anti-aging medicine,' as pointed out by Lucke and Hall,[4] include fears of disability

Address for correspondence: Beatriz Cardona, Centre for Cultural Research, University of Western Sydney, Locked bag 1797, Penrith South DC, NSW 1797 Australia. Voice: +61 2 9685 9600; fax: +61 2 9685 9610.
b.cardona@uws.edu.au

**Ann. N.Y. Acad. Sci. 1114: 216–229 (2007). © 2007 New York Academy of Sciences.
doi: 10.1196/annals.1396.000**

and death and the pursuit of health in old age. The present study explores these beliefs along with the social discourse and structural transformations that give rise to the construction of the aging self using these narratives.

Definitions of what aging means in the 21st century, explains Katz,[5] include terms such as "positive aging," "healthy aging," and other terms from a new discourse promoting stereotypes of aging as a period of independence, choice, mobility, and autonomy. These new images of aging contrast sharply with past perceptions of aging as a period of decline, decrepitude, and dependency. According to Katz this change in perception is the result of efforts by gerontologists in the 1960s to address an ageist discourse that fostered inequality and victimized human differences.[5] Notions of positive aging, however, are closely linked to anti-aging values and the spread of an anti-aging culture, which, according to Katz, reinvents notions of maturity and growing older without aging. The process of growing older unaffected by aging assumes the possibility that the physical and physiological signs of aging, such as the wearing out of the body, reduction in mobility, and disability associated with age can be treated to compress morbidity and improve the quality of life. Part of the controversy surrounding this anti-aging discourse is not generated by the potential of scientific intervention to achieve such compression of morbidity and improvement in the quality of life of elder people, but rather because it is characterized by an implicit ageism and a focus on consumerism.

In Australia public health campaigns and health-promotion strategies, such as the National Strategy for an Aging Australia, approach the issue of aging by promoting strategies that prioritize physical health and aim to enhance the ability of aging individuals to manage their health needs. Campaigns promoting the benefits of exercise and healthy dietary regimes have been part of this strategy. Less attention has been given to the physical, social, and economic environments in which elderly people live and the relevance of this environment for health outcomes.[6] The cultural context informing some of these health-oriented discourses represents health outcomes as phenomena that lie in the hands of individuals, requiring appropriate lifestyle modifications.[7] Age has become a concern to a wider public and, as the inefficiencies of old age become costlier and costlier in the public mind, the body itself becomes a matter of increasing concern.[8] Preventive medicine and aspirational sciences that seek to prevent age-related illnesses exist alongside governmental strategies that disinvest in long-term health care and financial support for an aging population, while directing aging individuals toward developing their own strategies for the governance of old age. Choosing 'anti-aging medicine,' this study will argue, represents one of many strategies that both constrain and facilitate identity in old age. Users of these controversial technologies seek, with varying degrees of success, to resolve the contradictions and challenges posed by aging in contemporary society.

MATERIALS AND METHODS

This study proceeded using the following means:

(1) Twenty five in-depth qualitative interviews were held in Sydney, Melbourne, and Brisbane with both users and providers of 'anti-aging medicine' as well as other key stakeholders in the anti-aging debates.
(2) A variety of texts relevant to the discourse, including anti-aging literature, biogerontological research magazines, media channels, and popular magazines, were critically analyzed.
(3) Social theories of globalization were used to understand the interplay between the global and local structures that shape the expansion of the anti-aging market.

PRELIMINARY ISSUES EMERGING FROM THE INTERVIEWS

This paper presents only a partial and summarized view of some of the emerging issues. A final report with a comprehensive analysis of 'anti-aging medicine' in Australia through the views of consumers and providers and other key players in the anti-aging debate will be made available in future.

Preliminary issues emerging from the interviews showed:

- conflicting definitions and understandings of 'anti-aging medicine';
- shifting interpretations of social discourses on aging: anti-aging and "healthy aging"; and
- deception, self-deception, or self-assertion by the anti-aging consumer.

Conflicting Definitions and Understandings of 'Anti-Aging Medicine'

The providers of 'anti-aging medicine' in this study had a background in medicine, having first worked as general practitioners before migrating to 'anti-aging medicine.' The decision to specialize in this area was based on positive personal experiences with 'anti-aging medicine' along with financial incentives. Different positions regarding the meaning and objectives of 'anti-aging medicine' were enunciated; some providers saw anti-aging as part of complementary and alternative medicine (CAM), while others reinforced its "scientific" and mainstream biomedical character:

> I work the natural way, stimulating the creation of hormones using natural products. It is dangerous and I don't recommend the use of synthetic hormones. [anti-aging practitioner]

> I don't consider 'anti-aging medicine' as alternative. I consider more things like naturopaths and homeopaths as alternatives. But 'anti-aging medicine'

> is proper medicine. It is preventive medicine using products such as human growth hormone. [anti-aging practitioner]

The issue of what is "legitimate" 'anti-aging medicine' was evident, as various and conflicting positions were adopted by practitioners working in different fields. Although these physicians referred to themselves as "anti-aging doctors," there was an absence of consensus as to what 'anti-aging medicine' means, the risks and benefits associated with some products, and the legalities of prescribing drugs such as human growth hormone (HGH) for off-label uses.

Disparities in knowledge and source of information for the use and prescription of products such as hormone treatments for anti-aging purposes emerged, with some practitioners relying more heavily than others on pharmaceutical companies' guidelines to inform their practices, while others accessed resources and information from various medical journals and research. Personal experiences with anti-aging treatments by practitioners themselves also provided important clinical evidence used to inform their practices.[9]

> My patients look at me and they can't believe I am 74. I tell them my secret, hormones, human growth hormone. I have been using it for the last 10 years and as you can see I don't look my age at all. When people see this they want to try the same treatments for themselves.

Central to current debates on 'anti-aging medicine' by leading gerontologists and anti-aging supporters concerns the issue of legitimacy as well as the perception of misuse, exaggeration, and/or extrapolation of biogerontological research for commercial interests. More importantly there is disagreement among biogerontologists as to whether there are any anti-aging therapies currently available to halt, slow, or reverse the aging process. From the point of view advanced by Olshansky *et al.*,[3] which supports a definition of aging based on the biochemistry of cellular metabolism, 'anti-aging medicine' does not currently exist, as no interventions yet exist capable of modulating the aging process itself at the molecular level.

Anti-aging researchers, such as Livesey and Raffaele,[10] challenge this strict molecular definition of aging and point to a clinical-organismal definition to argue that there are many therapies scientifically proven to at least slow, and in some cases reverse, the changes in the function and structure of aging human bodies. Among therapies cited by these scientist–entrepreneurs are hormones, such as HGH, DHEA, melatonin, and nutritional supplements, that contain synthetic and natural anti-oxidants which are argued to slow down the aging process by preventing or delaying physiological decline and aiding lost functional abilities to be regained.

The use of these therapies in private clinics and for commercial interests, rather than in hospitals under the control and supervision of geriatricians and for medically defined age-related illnesses, reflects tensions and power issues between institutions and individuals for the control and management of knowledge about bodies, illness, and aging. Critics of 'anti-aging medicine' point to the risks involved in poorly supervised practices, the financial costs for consumers, and false expectations that the "fountain of youth" can be found in hormone treatments. This criticism is rebuked by both anti-aging practitioners and consumers in this study by pointing to clinical evidence and by questioning the motives behind such attacks as well as challenging perceptions of consumers as vulnerable and unable to make informed decisions about the products they purchase to satisfy a need not yet met by orthodox medicine.

> I think you could do yourself a lot of damage if you acted irresponsibly with what's out there, no question. I think even in terms of a lot of medicine that previously was on prescription is now over the counter. . .so, yes, it's certainly a case of buyer beware. [Ronald, aged 51]

> I get information in the internet before buying and so far nothing has made me sick or unwell. . . . on the contrary I feel better so how you feel is important, you can judge for yourself. [Galia, aged 49]

The expanding influence of 'anti-aging medicine' threatens the status and funding of biogerontological research[11] and challenges the dominance of institutional discourses and management of knowledge about aging bodies. In his analysis of what is at stake in "the war on anti-aging medicine," Vincent[1] concludes that the field of gerontology is genuinely faced with a crisis provoked by unprecedented understanding of the biological mechanisms of aging.[1] He argues that it is a cultural crisis brought about by the belief in the infallibility of science to cure aging and death.

Through this study it also becomes evident that this "crisis" gerontology is facing resonates with the cultural construction of aging as *decline*. The dominant decline ideology promoted by science stigmatizes aging and old age, making it possible for society to create another type of crisis: the midlife crisis, prompting aging individuals to seek remedies to halt the advent of this loss associated with later life. Positioning aging as a demographic, social, and economic "crisis" in the 21st century has added further impetus to negative perceptions of the role and contribution of aging populations to social and economic development. It has also prompted the adoption of global and local political rationales which, along with new scientific concepts of temporal biological processes, are all parts of the contingent social conditions that can produce anti-aging and enhancement technologies directed against aging.[12] The extent to which gerontology engages with these crucial issues, including the impact of narratives of decline and loss in later life and the belief science has the answer to these challenges, will be an important factor in determining how 'anti-aging medicine' develops in the new century.

Shifting Interpretations of Social Discourses on Aging: Anti-Aging and "Healthy Aging"

Crucial to an understanding of the motivations and forces behind the decisions to consume anti-aging technologies and therapies lie social, cultural, and economic narratives advanced by institutional agencies and disciplines governing the lives of aging populations. Modern constructions of aging advanced by medical institutions, Third Age advocates, global consumer markets, and national and international policies on aging position the aging process as one characterized by both risks and opportunities. The role of the state, as Phillipson points out,[13] is crucial in the construction of old age, as evident in its allocation of resources for the aged and policies, such as enforced retirement and welfare entitlements. However in the 1970s the state began to retreat from its acquired obligations toward the elderly, including rationalization of health and community care services to the more recent restructuring of welfare entitlements, personal financing of old age, and retirement age, leading to what Phillipson calls a "collapse of consensus" about what to expect in old age.

These challenges to personal and collective identity have been further exacerbated by the dramatic expansion and global influence of biomedicine and its attempts to fill in the vacuum left by the retreat of the state in guaranteeing consensus about old-age rights, expectations, and possibilities. The "biomedicalization of aging," as first referred by Estes and Binney,[14] attempts to construct aging as a medical problem to be alleviated in part by medical intervention. Despite the array of evidence of the importance of social, economic, and behavioral factors in explaining the experiences of old age, the biomedical model is becoming the "institutionalized thought structure," influencing research, policymaking, and views about aging and science itself.[7]

The most important implications of a biomedical model of old age are the construction of old age as pathological and undesirable and biomedicine as a form of social control of the elderly through medical definition, management, and treatment.[7] The following extracts provide example of how consumers of 'anti-aging medicine' have incorporated mainstream discourses about aging promoted by medical knowledge and social policy directions to justify their decisions to consume anti-aging products and therapies.

> What's the good of living in your 80s if you're going to be in a nursing home? 'Anti-aging medicine's' really the insurance policy. [John, aged 53]

> I just don't want to end up sick and debilitated and being looked after by people. Plus, I've got a very strong sense of responsibility and being personally responsible for what I do and not wanting to impose on other people. [Sharon, aged 43]

The power of official discourses on aging as illness and decay, and new disciplines of self-care and self-funding produce new identities and behaviors for individuals, as evidenced in the respondents' concerns with 'anti-aging medicine' as an insurance against the risks of financial or care "burden" on others. Significant also in this analysis is the understanding of the respondents' life-long experiences structured around social class, economic opportunities, and cultural preferences. Health-related behaviors are influenced by social determinants such as education, income, occupation, life events, and social networks,[15] calling for a more nuanced social and structural analysis of the informants' life-course.

Most of the users of 'anti-aging medicine' in this study identified themselves as financially well off and belonging to upper-middle-class backgrounds. They all had tertiary education and 90% were employed in business and professional sectors. Their perceptions of what health and healthy means coincide with previous research in Australia, including the Health Status of Older People (HSOP) project, conducted in 1996,[16] showing that older people place great importance on keeping physically and socially active. However, in contrast to the HSOP study, anti-aging consumers do not see "absence of disease" as an unlikely ideal, but, on the contrary, a real possibility to be achieved through lyfestyle and consumption choices. This higher expectation also contrasted sharply with an East Sydney Area Health Service study of elderly people from economic and socially disadvantaged backgrounds in the Eastern Sydney area which found that elderly people were more willing to endorse a view of aging as an unremitting decremental process with expectable stages of decline.[17]

This perception of an inevitable aging-related decline along with a belief that good health can largely be attributed to "good fortune" positions them in sharp contrast with anti-aging consumers in this study. Such contrast in views and expectations is a clear indication that the social and economic conditions of people's lives significantly influence the nature and meaning of their practices. The structured dependency model of age-related inequality,[18] if applied to anti-aging consumers in their privileged social and economic position, suggests that the material forces and access to resources and consumer-culture information inform their attitudes, beliefs, and expectations and consumer choices about what is possible in old age. An important analysis, which is beyond the scope of this paper, concerns thus the tendency to homogenize the experiences of aging individuals under health-promotion narratives and Third Age ideologies, often oblivious to enduring socioeconomic differentials in the life-course[19,20] and the response-ability[7] to achieve "successful" and healthy aging.

Active, healthy and productive aging, as pointed out by Biggs, has taken on a "moral significance that mediates our everyday understanding of what it is to age well" (p. 96).[21] The opportunities for identity-building offered by social phenomena including consumer lifestyles, plastic surgery, and retirement communities offers an almost inexhaustible array of identity alternatives that exist, however, within the very specific and limited confinements of the youthful body. Some of the main concerns of gerontologists[5,22] with 'anti-aging medicine' and values lie therefore in the fact they are market-driven, and they are becoming the dominant discourse shaping the views of old age. Part of this increasing dominance in shaping current ageist culture, argues Haber, is the result of a current global concern with the increase in the numbers of older people and the impact that this increase has on the local and global economy.

This new type of aging govermentality,[23] based on complex and contradictory models positioning aging as both decline but also opportunity, shifts the burden of the state's responsibility toward its aging populations to *individuals* through narratives of personal empowerment, thus reconceptualizing individuals as autonomous actors who can choose behaviors, practices, and most importantly, products.[12] Aging individuals are being encouraged, through coercive as well as persuasive global and local policies and campaigns, such as healthy aging, to become self-managing and to internalize the values of independence, productivity, and self-care in later life. In turn, commercial anti-aging narratives and ideologies are part of an array of strategies available to mainly affluent aging individuals to maximize these individual self-managing regimes and chances to accomplish and exceed.

> What I am doing now is about "healthy aging". . .using anti-aging, eating the right food, exercising and doing everything I can to remain healthy for as long as I can. [Galia, aged 49]

> I try to medicate every day, and I do a lot of metaphysical work around stuff—listening to my body. If my body says something's going wrong, I get it checked out fairly quickly. Lots of testing, but testing for specific things, like specific markers to do with cardiovascular disease and cancer. Like, your hormone levels, for example—your estrogen and progesterone. [Sharon, aged 43]

Anti-aging discourses and consumer markets seek to provide a particular version of the self-managing techniques individuals require in order to achieve 'successful aging' and the dangers and consequences of failing to implement this type of care. Websites, such as those of the American Academy of 'anti-aging medicine', capitalize on global discourses on a demographic aging population crisis, the risk of disability, burden, dependency and loss of mobility. Alongside this marketing of fear, anti-aging markets offer hope and the possibility not only of growing old unaffected by illness, but also of reverting to a younger self or no aging at all. Consumers of 'anti-aging medicine' in this study believed their anti-aging products was allowing them to control their aging process and avert all the dangers associated with old age, including loss of vitality, look, and symbolic power. The tendency to explain their youthful appearance and vitality as evidence that 'anti-aging medicine' works, revealed a lack of awareness and uderstanding of the complex interplay of factors determining how a person ages, including genetic, environmental, socio-economic, and cultural factors.

> You look at one person who is 60 on anti-aging [therapy] versus someone who is not, there's like ten years difference between the two. . . .I've got one friend and she's 55 and she looks 35 and another friend who's 73 and she looks 53. They have the energy that normally you wouldn't have at that age. [Diane, aged 63]

> In fact, people were quite surprised at my age. They're quite shocked when they find how old I am, because I can do things that men of 45 are actually stopping doing. I look younger than them and my energy levels are higher in some ways. [Edward, aged 59]

The goal of reverting to a younger self, expressed in the above comments, seeks more than the attainment of health in later life—it also proposes new benchmarks of what a person should look and act like at certain age and the desirability of subjecting the body to technoscience and anti-aging treatments in order to alter the "normal" to new values that then position the person to have an advantage over others the same age. This type of enhancement—one that creates a "cultural complicity"[24] by which the choices to adopt certain enhancement technologies are created by cultural forces and commercial interests and consumer cultures—ignores the long-term consequences for society: new standards of normality, success, and moral virtue around the body's ability to remain ageless.

References to interventions to improve the self and alter the mood also make a biological reductionist assumption that all behavior, interactions, and physiological functions are related to neuronal structures. This perspective erases the social, creating a subjectivity reliant on assistance—psychopharmaceutical drugs to make a person happier, or more "authentic" or less depressed—and ignores the social, economic, and cultural forces that may be at play. As Rose explains,

> This certainly raises important questions about how we configure the boundaries of the normal and the pathological, the treatable and the acceptable. It does indeed raise questions about the kinds of humans we want to be and the role of the market in this reshaping of ourselves as "neurochemical selves" (p. 3).[25]

Terms such as "health promotion," "healthy living," and "healthy aging" highlight the mandate for work and attention toward attaining and maintaining health. The shift in cultural expectations of constant self-discipline risk abatement and surveillance as personal responsibility produces a paradox in health: health is both *more* biomedicalized through surveillance and risk technologies done by the individual at home and *less* medicalized as the key responsibility shifts from the professional physician to reliance upon collaboration with the individual patient/user/consumer.[26] As the following extract indicates, consumers of 'anti-aging medicine' extend their surveillance mechanisms to include risk containment of anti-aging products which could paradoxically make them seriously ill while promising to make them "better than well."

> I'm concerned, for example, that DHEA might increase prostate cancer, the potential, and testosterone might do the same, so I vigorously check my PSA. I've had a biopsy on my prostate, for any slight indication of trouble, it was a very painful biopsy, and it proved negative. [Edward, aged 59]

> I have a blood test every three months, just to make sure there are no side effects from the hormone treatment and I see X. . . probably every three weeks and let him know anything that I feel. But apart from that, I guess it's like any medical research. Someone says that Lipitor is very good for cholesterol. I haven't got time to get the knowledge of those people who say Lipitor is good, to actually take it. So it's like . . . I take Panadol if I have a severe pain. I don't go on the internet and study every research document on Panadol and what might be the side effects, or cough syrups. . .or coffee! And the other thing is, life is full of all these things that aren't properly explained anyway. [John, aged 53]

Deception, Self-Deception, or Self-Assertion: The Anti-Aging Consumer

Aged populations are constructed differently, empowered at times and disempowered at others, to reinforce specific ideologies and practices by various institutions and governing bodies. A recent report from the United States General Accounting Office (GAO) on anti-aging health products[27] highlighted the potential physical and economic harm to senior citizens from these products. The GAO report argues that it is *senior citizens* who are most at risk from anti-aging "cure all" claims.[27] The report claims that 4 of 10 senior citizens use herbal dietary supplements, arguing that these older persons are susceptible to claims of cures for chronic diseases and conditions, such as cancer, memory loss, arthritis, and atherosclerosis. The portrayal of senior citizens by regulatory bodies as vulnerable to fraud and quackery contrasts sharply with images of *silver surfers* and *bay boomers,* actively pursuing lifestyle and health opportunities, advanced by media and market aging research.[28] In these politics of representation ageism manifests itself under the guise of either negative or positive images, obscuring the material and cultural realities—both positive and negative—that shape the health choices and experiences in later life.

Representations of anti-aging consumers as vulnerable and in need of in need of protection can sometimes be contrasted with anti-aging users in this study who, while acknowledging the possibility of therapies not working, also reaffirmed their critical judgment, awareness of possible side effects from the treatments, and willingness to take risks in the hope of maximizing health outcomes and enhancing specific physical and personality traits in accordance with cultural and consumer expectations. This finding, however, does not exclude the possibility that users are being misled and/or misinformed regarding the actual benefits of the products being consumed. Moreover this is a warning on how the problem of illness and health bears on the constitution of the self as an ethical subject, the sense of what we do with our freedom, and the extent to which we engage with attributions of self responsibility.[29]

> I saw something on the internet which I thought was going to help me. I read all about it and I thought this is it and I asked the doctor to give it to me. Well the doctor gave it to me and it was all right, but the thing is what I didn't realize was that when I researched that product I thought I had all of the answers but there were other products were better which I wasn't aware of. [Harold, aged 66]

The positioning of many anti-aging consumers in this study as in "partnerships" with health providers and in a more democratic power relationship is indicative of how new consumer ideologies in health provision are contextualized in the anti-aging doctor/consumer relationship as well as how the self is asserting itself against mainstream medical indifference.[30] The new framework of providing consumer-oriented and individualized medical treatment in contexts dealing with healthy, relatively well-informed aging individuals, opens up new interactions and power dynamics as well as new avenues to "problematize" health and exercise control over aging individuals.

Individual agency in the form of product knowledge and consumer-oriented anti-aging practices, along with proactive engagement in body management and surveillance against a background of constrained possibilities of growing old outside the framework of agelessness, are illustrative of a Foucaldian model of how the identities of older people are both facilitated and constrained by biotechnologies. In this paradox the self is caught between "an ontological battle of learning about the self and health needs and having their 'needs' decided by others" (p. 3).[29] The following comments are illustrative of this paradox, particularly in relation to the manner in which users of anti-aging therapies interpret their "choices" as a transgression of normative decline ideologies and a self-assertion against mainstream medical indifference.

> [It would be good] if doctors were given long enough time to work with people rather than treating their diseases [and doing] something more interesting than what they're doing now, like creating incentives. People are not "incentivized" by their clinical condition, they are "incentivized" by other things, like weight, image and body appearance. [Sharon, aged 43]

> Well, I am not silly, I know they [anti-aging products] don't do everything they say. . . .I know some drugs are lifesaving and some others cause great harm, but you can also look at the medical profession and be very critical of some of the drugs they prescribe to people. So my main concern is with the validity of conventional medicine to give answers to the problems we face

> today and to provide a plan of action that looks at preventing the diseases that sooner or later will confront us. [Leslie, aged 61]

Cultural reductionist models of senior citizens as either "vulnerable" or "empowered" render complex human experience and interactions into caricatures and exclude the likelihood of these traits mutually co-existing, along with a myriad of other traits, experiences, and possibilities. As it currently stands, such constructions tell us more about how disciplinary apparatuses and sites of domination construct the self to produce reductive social and historical analyses[23] with vested interests in promoting particular versions of what later life is or should be about. Protecting consumers against fraudulent claims and potentially dangerous practices is an important task. However, given the perspectival nature of knowledge, a multifaceted approach is needed that is capable of incorporating existing tensions between competing institutions telling different versions or "truths" about aging, health, and illness and the skepticism of an increasing number of aging individuals about the ability of governments and mainstream medical practices to address the challenges of growing old in modern Western societies.

CONCLUSION

> On one level I think it is a problem that as a society we feel the need to have to do these [anti-aging] procedures. It's all about our currency of people, you know. "Am I worthwhile, am I worthy?"—all that sort of crap. "As I get older, I need to be looking good each day in my job..." I think it's tragic that we have those attitudes, but they are alive and well and prevalent, and see, I am an opportunist, and I'm a chameleon, so I do it because I think, well, it helps me get what I want out of my life because I live in a society. I haven't yet opted out. [Sharon, aged 43]

I find this quote from one of our anti-aging consumers to encompass some of the anxieties and tensions confronted as their aging self is recreated within a particular social discourse. Sharon's comments also point to the need to look at anti-aging technologies and products through an examination of the relations among individuals, technologies, consumer markets, political ideologies to control aging populations, and scientific concepts of what is natural and what can be "cured" or enhanced.

Attacks on 'anti-aging medicine' because of its pseudoscientific character and unregulated practices, I believe, is not going to resonate with many disillusioned and anxious consumers who view mainstream medical practice as poorly equipped to deal with the cultural context of aging in modern Western societies.

The point is therefore not about the increasing success of 'anti-aging medicine' and pseudoscience in enticing consumers into their practices, but rather the urgent need for social gerontologists, researchers, and philosophers "to determine the conditions and contexts in which an individual's adaptation to aging is either facilitated or limited."[31] This assessment should take into account the role of consumer cultures and anti-aging ideologies in an era of aggressive global capitalism and reduced governmental commitment towards the financial and health provision of its aging populations.

ACKNOWLEDGMENTS

I thank Dr. Brett Neilson, chief supervisor for my Ph.D. thesis, for his advice and support in the writing of this paper.

REFERENCES

1. VINCENT, J. 2003. What is at stake in the 'war on anti-aging medicine'? Aging & Society **23**.
2. SAGE CROSSROADS. 2004. 'Is Aging a Disease?' Live debate, January 22, Harry R. Moody vs. Arthur Caplan.
3. OLSHANSKY, J., L. HAYFLICK & B. CARNES. 2002. Position Statement on Human Aging. The Journals of Gerontology, Series A: Biological Sciences and Medical Sciences, August, Vol. 57A.
4. LUCKE, J. & W. HALL. 2005. Who wants to live forever? EMBO Rep. **6:** 98–102.
5. KATZ, S. 2001/2002. Growing older without aging? Positive aging, anti-ageism, and anti-aging. Generations **25:** 27.
6. COTA. 2004. Healthy Aging: The Big Picture. Fifty Plus News, http:/www.fiftyplusnews.com/cota/march00.shtml.
7. MINKLER, M. & C.L. ESTES. 1991. Critical Perspectives on Aging: the Political and Moral Economy of Growing Old. Baywood Publishing Co. Amityville, New York.
8. GILLEARD, C. & P. HIGGS. 1998. Aging and the limiting conditions of the body. Sociological Research on Line **3**. "/s/sro/srosro.html" Sociological Research Online.
9. MALTERUD, K. 2006. The social construction of clinical knowledge: the context of culture and discourse. J. Eval. Clin. Pract. **2:** 248–256.
10. ANTI-AGING MEDICINE ASSOCIATES OF MANHATTAN. 2002. Corporate website http://www.antiagingny.com (accessed May 20, 2005).
11. BINSTOCK, R.H. 2004. Anti-aging medicine and research: a realm of conflict and profound societal implications. J. Gerontol. **59A:** 523–533.
12. HOGLE, L. 2005. Enhancement technologies and the body. Annu. Rev. Anthropol. **34:** 695–716.
13. PHILLIPSON. 1998. Reconstructing Old Age. Sage, London.
14. ESTES, C.L. & E.A. BINNEY. 1991. The biomedicalization of aging: dangers and dilemmas. *In* Critical Perspectives on Aging: the Political and Moral Economy of Growing Old. Meredith Minkler & Carroll L. Estes, Eds.: 10. Baywood Publishing Co. Amityville, New York.

15. Kending, H. 1996. Understanding health promotion for older people: sociological contributions. *In* Sociology of Aging. V. Minichiello, N. Chappell, H. Kending & A. Walker, Eds.: 13. International Sociological Association. Melbourne, Australia.
16. Kending, H., H. Helme, K. Teshura, *et al.* 1996. Health status of older people: data report, Melbourne: Victoria Health Promotion Foundation, Lincoln Gerontology Center and National Ageing Institute. 80 pages.
17. Sheriff, J.N. & L. Chenoweth. 2006. Promoting healthy ageing for those over 65 with health check log: a pilot study. Australas. J. Ageing **25**(1): 46–49.
18. Fennell, G., C. Phillipson & H. Evers. 1988. The Sociology of Old Age. Open University Press. Philadelphia.
19. Alcock, P. 1996. Social Policy in Britain: themes and Issues. Macmillan. London.
20. Walker, A. 1990. Poverty and inequality in old age. *In* Aging in Society: an Introduction to Social Gerontology. J. Bond & P. Coleman, Eds. Sage. London.
21. Biggs, S. & S. Daatland. 2004. Ageing and diversity: a critical introduction. *In* Ageing and Diversity: multiple pathways and cultural migrations. S. Biggs & S. Daatland (Eds.): 1–12. Policy Press. Bristol.
22. Haber, C. 2004. Life Extension and History: the Continual Search for the Fountain of Youth. The Journal of Gerontology, June, **59A**.
23. Kellner, B. & S. Best. 2005. Foucault and the critique of modernity. *In* Postmodern Theory: Critical Interrogations. Macmillian Press, Ltd. London .
24. Elliot, C. 1998. What is Wrong with Enhancement Technologies? Centre for Bioethics, University of Minnesota.
25. Rose, N. 2005. Will Biomedicine Transform Society? Background briefing, Radio National, 13th March.
26. Clarke, A., L. Mamo, J. Fishman, *et al.* 2003. Biomedicalization: technoscientific transformations of health, illness and U.S. biomedicine. Am. Sociol. Rev. **68:** 34.
27. U.S. General Accounting Office. 2001. Report at http:/www.senate.gov/~aging/hr73rpt.pdf.
28. Katz, S. 1990. 'Busy bodies: activity, ageing and the management of everyday life.' Journal of Ageing Studies, June, vol. 14, issue 2.
29. Powell, J. 2001. Rethinking Structure and Agency: Bio-ethics, Aging and Technologies of the Self. Centre for Social Science. John Mores University. Liverpool, UK.
30. Arney, W. & B. Bergen. 1984. Medicine and the Management of Living. University of Chicago Press. Chicago, IL.
31. Katz, S. 2005. Cultural Aging: lifecourse, Lifestyles and Senior Worlds. Broadview Press. Ontario, Canada.

Are We Getting Healthier As We Grow Older?

Implications for Babyboomer Labor Force Participation

DEBORAH J. SCHOFIELD, MEGAN E. PASSEY, ARUL EARNEST, IAN C. GLOOR, AND RUPENDRA SHRESTHA

Northern Rivers University Department of Rural Health, School of Public Health, Faculty of Medicine, University of Sydney, Lismore, NSW, Australia

ABSTRACT: The Intergenerational Report (IGR) released by the Department of the Treasury of the Commonwealth of Australia in 2002 highlighted pressures that in the future would threaten the sustainability of the Australian government's budget balance. These pressures result from the growing needs of an aging population and labor shortages that will limit economic growth and taxation revenue. The IGR has become a driving force in planning government policy. The Treasurer has recently said that "the whole economic agenda of the government at the moment is drawn from the IGR." In response, the Prime Minster and Treasurer have promoted deferred or gradual retirement as part of the solution. However, about 50% of men and 20% of women retire early as a result of ill health, indicating that poor health is potentially a limiter of economic growth. This paper reports lower labor force participation among persons with poorer health and that these persons move out of the labor force at a faster rate as they age. A range of measures suggests some decline in health in the pre-retirement age group (those aged from 40 to 64 years). This indicates that better health may be a facilitator of greater labor force participation in the baby boomer cohort. However, there is evidence that improving economic conditions in Australia leading to low unemployment has created an environment more favorable to the employment of older workers with health problems as there has been a rise in labor force participation in these groups, and measures to prevent chronic disease may further increase the employment prospects.

KEYWORDS: **labor force participation; aging; health trends; health**

Address for correspondence: Dr. Deborah Schofield, Associate Professor and Director of Research, Northern Rivers University Department of Rural Health, School of Public Health, Faculty of Medicine, University of Sydney, P.O. Box 3074, Lismore NSW 2480 Australia. Voice: 61 2 6620 2601; fax: 61 2 6620 7270.

deborahsch@ncahs.health.nsw.gov.au

Ann. N.Y. Acad. Sci. 1114: 230–240 (2007).
doi: 10.1196/annals.1396.002

BACKGROUND

The Intergenerational Report (IGR) released by the Treasurer of the Commonwealth of Australia in 2002[1] highlighted future pressures that would threaten the sustainability of the Australian government's budget balance because of the growing needs of an aging population and subsequent spending on health and related services[2] as well as labor shortages which will limit economic growth and tax revenue. The IGR has become a driving force in planning government policy and the Treasurer has recently said that "the whole economic agenda of the government at the moment is drawn from the IGR."[3]

Australia has labor force participation rates of about 75% for men aged 55–59 years and 45% for men aged 60–64 years. Older women's labor force participation is lower, at about 50% for women aged 55–59 years and 25% for women aged 60–64 years.[4] Labor force participation rates are relatively low among older Australians. Australia falls within the bottom third for OECD countries, with an effective retirement age of older men being about 58 years in 2000.[5]

In response, the Prime Minster and Treasurer have promoted deferred or gradual retirement as a method for increasing labor force participation.[6] For example, the Treasurer recently said "if we are going to make this economy sustainable, we have got to . . . encourage older people to maintain a connection with the workforce. . . . With people living longer and in better health they have the capacity to maintain a connection with the workforce."[7]

In Australia there has been increasing life expectancy in all age groups.[8] Nonetheless, 50% of men and 20% of women retire early as a result of their own ill health.[9] However, there has been little evidence in Australia about whether the general health of the older working-age population is improving along with a capacity for increased labor force participation. If this improvement does not occur, then the future supply of labor may not increase as the government information has foreshadowed, potentially leading to greater labor market shortages.

A number of international papers, including numerous American studies, have reported trend declines in disability rates, although these have tended to have been for persons beyond traditional retirement age.[10,11] There have also been American studies that have concluded that disability is rising, although previously this has been seen to be driven by aging of the population and a rise in reported disability among young people.[12] However, a recent report by the National [U.S.] Bureau of Economic Research suggests that Americans aged 50–55 years report poorer health and greater disability than the previous age cohort, with rises in diabetes, obesity, and psychiatric problems in this age group.[13] Similarly, another study of obesity and disability trends indicated that while disability has been reported as declining, if current trends in obesity continue, disability rates will increase by 1% per year more in the 50- to 69-year age group than if there were no further weight gain.[14]

TABLE 1. National health survey data in persons aged 40–64 years (1995, 2001, and 2005)

		1995		2001		2005	
	Age cohort	N	W	N	W	N	W
Males	40–44	2020	661,458	952	714,797	968	753,376
	45–49	1869	637,055	799	658,190	891	706,858
	50–54	1467	497,376	745	640,445	779	644,431
	55–59	1188	410,751	568	499,396	756	604,375
	60–64	975	348,382	483	400,268	653	460,043
Females	40–44	2102	663,476	1083	739,541	1063	764,177
	45–49	1884	618,412	907	677,491	970	719,148
	50–54	1432	478,295	757	640,398	869	664,388
	55–59	1157	396,405	616	491,644	848	602,258
	60–64	996	352,445	599	405,840	688	453,548

NOTE: N is the unweighted number of records; W is the weighted number of persons represented by the records

In Australia, too, the prevalence of some health conditions (such as deaths from coronary heart disease) has declined,[15] while that of others (such as diabetes) has risen.[8,16] Similarly, there has been a decrease in the prevalence of some health risk factors (such as smoking) and a rise in others (such as obesity).[10] There has been mixed evidence about trends in disability. The Australian Institute of Health and Welfare (AIHW) reported rising age-adjusted national rates of disability between 1981 and 1998.[8] However, the AIHW later concluded that rising age-adjusted rates of disability were attributable to changes in the Australian Bureau of Statistics Disability, Ageing and Carers surveys. Nonetheless, it was concluded that there had not been a decrease in age-adjusted rates of disability.[17]

In this paper, a range of health measures will be used to determine whether older workers in the baby boomer cohort are becoming healthier, and the implications for policy promoting longer employment will be discussed.

METHODS

The 1995, 2001, and 2005 National Health Surveys conducted by the Australian Bureau of Statistics were used to examine trends in various measures of health status for the population aged 40–64 years. Information about the number of records and the weighted number of individuals those records represented in the sample for each year of the survey is provided by age and sex in TABLE 1.

Measures of health used in the National Health Surveys included self-reported health status, whether days were spent away from work (or study) in the previous 2 weeks, the number of long-term conditions, and specific reported long-term health conditions (such as diabetes and rheumatism). Some

TABLE 2. Labor force participation by self-assessed health status in persons aged 40–64 years (2005)

Sex	Self-assessed health status	Actual participation
Male	Excellent	93
	Very good	92
	Good	86
	Fair	64
	Poor	29
	All	83
Female	Excellent	76
	Very good	76
	Good	64
	Fair	43
	Poor	24
	All	66

health status variables were not comparable across all years and for these only the comparable data sets were used.

Health status measures were analyzed in relation to measures of labor force participation. This was done on an age and sex basis for individual years and also for age cohorts. Participation in the labor force was defined as being either employed or looking for work. Not participating in the labor force was defined as being out of the labor force and not seeking employment.

Data analysis was undertaken using SAS version 9.1 (SAS Institute Inc., Cary, NC).

RESULTS

Labor Force Participation and Self-Report Health Status

The data from the National Health Surveys were first used to determine the relationship between labor force participation and a broad health indicator, self-reported health status. Using data from the 2005 National Health Survey, it was found that labor force participation was lower when health is poorer (TABLE 2). Labor force participation is more than 90% for males in excellent to very good health and about 75% for females. However labor force participation declines as health status declines, with male participation 29% for males and 24% for females in poor health.

Overall, for Australian men aged 40–64 years, there has been little change in health status over the 10 years from 1995 to 2005. However, within this group there has been decreases in some age groups. There was a decrease in thc proportion of men in excellent or very good health in the 40- to 44- and the 50- to 54-year age groups (from 59.8 to 54.4% and from 50.6 to 49.2%, respectively). Conversely there has been an increase in the 55- to

TABLE 3. Cohort analysis of proportion of the population participating in the workforce by self-assessed health status, by sex, aged 40–64 years (1995, 2001, 2005)

Sex	Health status	Age	1995	2001	2005
Male	Very good	40–44	0.97	0.97	0.98
		45–49	0.98	0.96	0.98
		50–54	0.95	0.96	0.95
		55–59	0.85	0.83	0.88
		60–64	0.62	0.59	0.70
	Fair	40–44	0.84	0.80	0.82
		45–49	0.80	0.88	0.76
		50–54	0.75	0.71	0.74
		55–59	0.47	0.51	0.58
		60–64	0.39	0.45	0.34
Female	Very good	40–44	0.85	0.80	0.82
		45–49	0.82	0.82	0.87
		50–54	0.70	0.75	0.83
		55–59	0.53	0.56	0.67
		60–64	0.19	0.34	0.36
	Fair	40–44	0.58	0.73	0.79
		45–49	0.51	0.59	0.51
		50–54	0.44	0.49	0.48
		55–59	0.30	0.43	0.35
		60–64	0.12	0.13	0.18

64-year-old groups (45.0–49.1% and 37.2–43.7% for those aged 55–59 and 60–64, respectively).

For women in the 40- to 64-year age group, there was a slight rise in the proportion of women reporting excellent or very good health (from 54.1 to 55.3%) and rises of up to 4.5%age points in the 5-year age groups, with an exception of a small decline in the 50- to 54-year age group (from 58.1 to 57.7%).

As expected, workforce participation of 5-year cohorts declined with age, as can be seen by following the shaded boxes across and down in TABLE 3. However the decline was much more rapid for those in fair health than for those in very good health. For example, of men in very good health aged 50–54 in 1995, 95% were in the workforce, falling to 70% 10 years later when they were aged 60–64 years—a decrease of about 25%. By contrast, the labor force participation of men of the same age but in fair health fell from 75 to 34%, a decrease of 55%. For younger men in very good health aged 40–44 years, labor force participation declined within 10 years from 97 to 95%—a decrease of only 2%—compared with men of the same age in fair health, whose labor force participation decreased from 84 to 74%—a fall of 12%.

There was a small decline in labor force participation for women 40–44 years old in 1995 and in very good health over the 10 years to 2001. For older women aged 50–54 in 1995 who were in very good health, there was a decrease of about 50% over the same period. The decrease in labor force participation for

TABLE 4. People reporting days away from work (or study) in the last 2 weeks by ages from 40 to 64 (1995, 2001, and 2005)

	1995		2001		2005	
Age	Male (%)	Female (%)	Male (%)	Female (%)	Male (%)	Female (%)
40–44	8.0	8.3	8.3	10.4	8.7	9.9
45–49	6.0	8.8	9.6	11.7	8.1	10.2
50–54	7.1	8.2	7.4	9.7	7.3	11.0
55–59	10.8	9.7	9.7	9.0	6.7	8.5
60–64	7.5	7.5	11.6	6.7	7.4	6.4
All	7.6	8.6	8.9	10.2	7.8	9.8

women in fair health over the 10-year period was somewhat greater than for men, with the decline being 17% for women aged 40–44 years in 1995, and 59% for women aged 50–54 in 1995.

Between 1995 and 2005, labor force participation has remained relatively unchanged for men aged from 40 to 54 years and in good health, although there was with an increase in the older age groups. For men in fair health, there were both increases and decreases in the different age groups. However, women in fair health have experienced increases in labor force participation in most age groups. There were also increases for women in the older age groups who were in very good health. This suggests that as the labor market improves and unemployment declines, as it has in Australia, that even groups who previously had relatively low rates of employment may benefit from new job opportunities.

Persons Reporting Days Away from Work

More women reported days away from work (or study) due to ill health in the 2 weeks prior to the 2005 survey than did men in all but the 60–64 age group, where 6.4% of women aged 60–64 reported days away from work compared with 7.4% for men. Somewhat surprisingly, days away from work generally declined with age, with those aged 55–64 reporting the lowest proportion of days away from work for both men and women (TABLE 4).

The proportion of reported days away from work in the 2 weeks prior to the survey increased between 1995 and 2005. For men aged 40–64 years, this increase was from 7.6–7.8% reporting days away from work, and for women 8.6–9.8%. The youngest age groups reported the largest increases, with those aged 45–49 reporting an increase from 6.0 to 8.1% for men and 8.8–10.2% for women. However, there were decreases among the older age groups from 10.8 to 6.7% for men and from 9.7 to 8.5% for women aged 55–59 years, for example.

TABLE 5. Long-term conditions and labor force participation, by sex, persons 40–64 years, 2005

Number of long-term conditions	Labor force participation (%)
0	86
1	86
2	81
3	77
4	74
5 or more	59

Long-Term Health Conditions

Labor force participation decreased as the number of long-term conditions reported increased in 2005 (TABLE 5). However there was no difference between one and no long-term conditions. This is because relatively minor conditions, such as astigmatism and psorisis, are included. As a result, there is a relatively large proportion of respondents who report one or more long-term conditions, with the probability of having multiple long-term conditions increasing with age. For example, 12.5% of men and 20.4% of women aged 40–44 reported five or more long-term conditions, rising to 42.6% for men and 48.8% for women in the 60- to 64-year age group.

There has been an increase in the proportion of persons aged 40–64 years reporting five or more long-term conditions (from 25 to 32%) and in all age groups between 1995 and 2005, with the increase ranging from of about 5 to 40% (TABLE 6).

TABLE 7 shows the labor force participation relative to specific conditions for 2005. The conditions associated with the lowest rate of labor force participation in the 40- to 64-year age group was diabetes mellitus type 1 with only 43% labor force participation, followed by rheumatism, and alcohol- and drug dependency.

Of these long-term conditions, there has been an increase in reported prevalence of the group of circulatory conditions including atherosclerosis, heart disease, and stroke (from 4 to 8%) and mental health conditions excluding drug and alcohol dependence (from 4 to 12%).

Persons with more than one long-term condition had lower labor force participation. For example, those with mental health conditions (excluding drug and alcohol dependence) and digestive disorders, labor force participation was less than 30%, and there has been an increase those reporting having these two conditions—from 0.4 to 0.8% between 1995 and 2005.

TABLE 6. Proportion of persons with five or more long-term conditions, by 5-year age group and sex in persons aged 40–64 years (1995, 2001, and 2005)

Sex	Age	1995	2001	2005
Male	40–44	0.11	0.11	0.12
	45–49	0.15	0.20	0.18
	50–54	0.22	0.27	0.22
	55–59	0.29	0.36	0.34
	60–64	0.35	0.40	0.43
Female	40–44	0.15	0.17	0.20
	45–49	0.20	0.27	0.27
	50–54	0.26	0.36	0.33
	55–59	0.36	0.38	0.39
	60–64	0.40	0.39	0.49

TABLE 7. Long-term conditions and labor force participation in persons 40–64 years (2005)

ICD9 Code	Labor force participation (%)
Diabetes mellitus type 1	43
Rheumatism	45
Alcohol and drug dependency	47
Nervous system disorders	51
Urinary system disorders	55
Neoplasms excluding skin cancer	55
Circulatory conditions	55
Digestive system including stomach ulcer	57
Mental conditions, excluding drug and alcohol dependence	59
Congenital anomalies	61
Other digestive system disorders	61
Endocrine/nutririon/metabolism/immunity (excluding diabetes type 1)	62

DISCUSSSION

While it was not unexpected to find that labor force participation was lower for people reporting poor health, the numerous measures pointing to some decline in health status, particularly for the 40- to 54-year age group, was somewhat unexpected, given the optimism of Australian research of about a decade ago concerning compression of morbidity (where illness is experienced at later ages as longevity increases).[18–22] However, the recent report by the Productivity Commission on the economic impacts of an aging Australia suggests that the weight of evidence is now that better health among older people is not going to reduce health expenditure.[10] This view is formed largely as a result of the rise of chronic diseases, such as diabetes, and associated risk factors, such as obesity, offsetting decreases in the prevalence of diseases, such as cancer.

The lower labor force participation among people with worse health status and greater number of long-term conditions identified in our study highlights the importance of increasing efforts to reduce risk factors for these conditions, as this is likely to have a beneficial effect in reducing disability. For example, in a longitudinal study, it was found that smoking, body mass index, and exercise patterns in midlife and late adulthood are predictors of subsequent disability. Not only do persons with better health habits survive longer, but in such persons, disability is postponed and compressed into fewer years at the end of life.[21] Additionally, recent declines in some risk factors such as smoking, with an associated decline in mortality from coronary heart disease, lung cancer, chronic lung disease, and strokes,[22] illustrate the health benefits that can be gained from comprehensive approaches to risk factor modification.

Reductions in cardiovascular disease mortality over the last 30 years have largely been due to reductions in smoking, serum cholesterol, and diastolic blood pressure, resulting from a range of public health and clinical interventions designed to decrease the incidence of disease and to improve treatment of acute events.[8,22] However, over the same period, there has been an increase in overweight and obesity, a trend that is continuing and that has been associated with increasing diabetes and musculoskeletal conditions.[8,10,22] In the long-term the higher prevalence of both diabetes and obesity is likely to be manifest in increases in cardiovascular disease, thus potentially undermining gains to date from reductions in other risk factors. The increased prevalence of circulatory conditions identified in our study may be an early indication of this phenomenon in addition to a reflection of the higher survival rate of those with cardiovascular disease. In addition to the obvious health benefits that could be achieved, the potential to increase labor force participation provides further compelling evidence for mounting comprehensive public health interventions targeting obesity.

Although there is some evidence that ill health is not generally decreasing for the pre-retirement age group, there is some room for optimism, as there is evidence that improving economic conditions in Australia leading to low unemployment have created an environment more favorable to the employment of older workers with health problems, and there has been a rise in labor force participation in these groups. If economic improvements were coupled with advances in prevention of chronic, work-limiting conditions in middle and later working life, then greater labor force participation of older workers may be achievable.

REFERENCES

1. Commonwealth of Australia. 2002. Intergenerational Report 2002-03. Budget Paper No. 5. Department of the Treasury. Canberra, ACT, Australia.

2. Schofield, D.J. & G.P. Rothman 2007. Projections of commonwealth health expenditure in Australia's first Intergenerational Report. *In* Modelling our Future. A. Gupta & A.M. Harding. Eds.: 149–168. Elsevier. Amsterdam, the Netherlands.
3. Davis, M. 2006. Future shock: Costello's tax warning. Australian Financial Review, p 8. 11 April.
4. Australian Institute of Health and Welfare. 2002. Older Australia at a Glance. AIHW. Canberra.
5. The Organisation for Economic Co-Operation and Development. 2004. Labour Force Participation of Older Workers. OECD Economics Department paper, May 2004. OECD. Paris. http://www.oecd.org/dataoecd/25/4/31743847.pdf.
6. The Prime Minister J. W. Howard MP. 2003. Address to the Symposium on Mature Age Employment. Sydney, 27 August. www.pm.gov.au/news/speeches/speech462.html.
7. The Treasurer, P. Costello. 2005. Doorstop Interview Toorak Place Retirement Village. Melbourne. 24 October. http://www.treasurer.gov.au/tsr/content/transcripts/2005/143.asp.
8. Australian Institute of Health and Welfare. 2006. Australia's Health. AIHW. Canberra.
9. Australian Bureau of Statistics. 1998. Retirement and Retirement Intentions Survey. ABS. Canberra.
10. Manton, K.G. *et al.* 1997. Chronic disability trends in elderly United States populations: 1982–1994. Proc. Natl. Acad. Sci. USA **94:** 2593–2598.
11. Crimmins, E.M. *et al.* 1997. Further evidence on recent trends in the prevalence and incidence of disability among older Americans from two sources: the LSOA and the NHIS. J. Geront. Ser. B: Psych. Sci. Soc. Sci. **52:** S59–S71.
12. Kaye, H.S. *et al.* 1996. Trends in Disability Rates in the United States, 1970–1994. Disability Statistics Abstract 17. Disability Statistics Center, University of California, San Fransisco, CA. http://www.barrierfreechoices.com/studies/Trends%20in%20Disability%20Rates%20in%20the%20United%20States%201970-1994.pdf.
13. Soldo, B.J. *et al.* 2007. Cross-cohort Differences in Health on the Verge of Retirement. Working Paper 12762. National Bureau of Economic Research. Cambridge, MA. http://www.nber.org/papers/w12762.
14. Roland, S. *et al.* 2004. Increasing obesity rates and disability trends. Health Affairs **23:** 199–205.
15. Productivity Comission. 2005. Economic Implications of an Ageing Australia. Research Report. Productivity Comission. Canberra.
16. Australian Institute of Health and Welfare. 2001. Australian Health Trends. AIHW. Canberra.
17. Australian Institute of Health and Welfare. 2003. Australia's Welfare. AIHW. Canberra.
18. Mathers, C. 1999. International Trends in Health Expectancies: do They Provide Evidence for Expansion or Compression of Morbidity. Occasional Papers Series No. 4. Department of Health and Aged Care. Canberra.
19. Goss, J. *et al.* 1994. Economic Perspective on the Health Impact of the Ageing of the Australian Population in the 21st Century. Australian Institute of Health and Welfare. Canberra.
20. Eckerman, S. 1992. Projected Health Expenditure and Ageing: revised Methodologies. Australian Institute of Health and Welfare. Canberra.

21. VITA, A.J. *et al.* 1998. Aging, health risks, and cumulative disability. N. Engl. J. Med. **339:** 481–482.
22. ABELSON, P. & TAYLOR, R. 2001. Programs to reduce tobacco consumption. *In* Returns on Investment in Public Health: an Epidemiological and Economic Analysis prepared for the Department of Health and Ageing. P. Abelson, Ed.: 15–32. Department of Health and Ageing. Canberra.

Independence, Well-being, and Social Participation in an Aging Population

FIONA ALPASS,[a] ANDY TOWERS,[a] CHRISTINE STEPHENS,[a]
ELJON FITZGERALD,[b] BRENDAN STEVENSON,[b] AND JUDITH DAVEY[c]

[a]*School of Psychology, Massey University, Palmerston North, New Zealand*

[b]*School of Māori Studies, Massey University, Palmerston North, New Zealand*

[c]*New Zealand Institute for Research on Ageing, Victoria University, Wellington, New Zealand*

ABSTRACT: In 2005 the Health, Work, and Retirement (HWR) Longitudinal Study was established at Massey University in order to identify the later-midlife factors that lay the basis for community participation, independence, and health in later life. Information was collected via postal questionnaire on physical and mental health, psychosocial factors, work and retirement attitudes, and socioeconomic and demographic indicators of a sample of 6662 community-dwelling adults aged 55–70 years. This report provides an overview of these results, and highlights the impact that the transition from work to retirement has on the health and retirement adjustment of older people with respect to independence, well-being, and social participation.

KEYWORDS: retirement adjustment; work status; general health; mental health

INTRODUCTION

This article provides an overview of the establishment and initial results of a research project funded by the Health Research Council of New Zealand to investigate the health of older adults in the transition from work to retirement. This project, the Health, Work, and Retirement Longitudinal Study, aims to follow the choices and life circumstances of midlife New Zealand adults to determine the factors that predict positive health outcomes in old age.

Address for correspondence: Dr. Fiona Alpass, School of Psychology, Massey University, Private Bag 11222, Palmerston North, New Zealand. Voice: (0064) 6 3569099; ext.: 2071; fax: (0064) 6 3505673.
F.M.Alpass@massey.ac.nz

Ann. N.Y. Acad. Sci. 1114: 241–250 (2007).
doi: 10.1196/annals.1396.009

RATIONALE FOR THE STUDY

In view of the increased access to food, medicines, and health care that have reduced worldwide mortality rates and increased life expectancy, many developed countries face the prospect of a growing old-age population. Globally, the proportion of individuals aged 60 and over has tripled in the past 50 years, and this number is set to more than triple once more in the next 50 years.[1] In New Zealand, specifically, 20% of the population will be aged 65 and over by the year 2026 and this proportion is projected to rise further by 2051.[2] As a result many older adults will experience what has become known as a "Third Age," a phase of life between exit from the workforce and the "Fourth Age," a period of dependency.

Concern is rising at the potential ramifications of this population transformation, as a burgeoning older-adult population increases already existing pressures on medical and social security systems. In New Zealand the labor market growth rate is projected to slow and then become negative from the 2020s as baby boomers retire,[3] so there will be fewer workers to support the old-age population leading to an increasing "old age" dependency ratio (those aged 65+ years as a proportion of those in the working-age population). The current ratio of 5.5 workers to each person over 65 in 2001 is expected to more than halve over the next five decades.[4] One response to the projected demands of an aging population is to encourage people to postpone retirement and stay longer in the workforce, thereby prolonging economic activity and contribution. Another way of coping with the costs of an aging population is to ensure that those in retirement can remain independent. There are two key determinants central to post-retirement retention in the workforce *and* maintenance of independence—health status and health changes.[5–8]

Research shows that normal aging involves changes in our physical and mental health.[9,10] However, many health conditions that present as we age, such as musculoskeletal or cardiovascular illness, actually have their antecedents in early midlife,[11] and it is increasingly evident that such health deficits are not necessarily the result of aging itself. Alpass and Towers[12] found that retirees reported poorer health than workers, but showed that these differences were not explained by age. Furthermore, for employed people at midlife, job satisfaction and hours worked were positively related to health, and those who felt social pressure to retire reported poorer physical and mental health.[13] This suggests that either retirement is not good for one's health, or that poorer health may increase the likelihood of retirement, and that the transition from work to retirement may be seen as part of many life events experienced during midlife which have an impact on well-being in retirement. However, there is little evidence on the relationship of health to workforce participation and positive aging in retirement in New Zealand. Thus, the need for a broad-based retirement and health research initiative is clear.

ESTABLISHING THE HEALTH, WORK, AND RETIREMENT LONGITUDINAL STUDY

To assess how the transition from work to retirement has an impact on older people's independence and well-being in their retirement years, the Health Research Council of New Zealand helped establish the Health, Work, and Retirement (HWR) Longitudinal Study at Massey University. The HWR study is collecting and analyzing information on physical and mental health, psychosocial factors, work and retirement status and attitudes, and socioeconomic and demographic status over time as older individuals make the transition from work to retirement. It aims to test the hypothesis that psychosocial variables at midlife predict health after retirement, and to specify the determinants of health and well-being (including participation and independence) in old age. The proposed research is based on understandings that good health is integrally related (as antecedent, outcome, and component) to independent living and community participation.

The HWR study is currently funded for 3 years (two waves of data collection) and each wave involves three studies.

1. The first study is a postal survey of a representative sample of New Zealanders aged 55–70 years, with a specific oversampling of Māori (the indigenous people of New Zealand).
2. The second study involves face-to-face interviews with a representative sample of Māori in the 55- to 70-year age group.
3. The third study involves face-to-face interviews with a representative subsample from the general population.

The postal survey has already been completed and both sets of face-to-face interviews are under way.

METHODOLOGY

Participants

A sample of New Zealanders aged 55–70 was randomly selected from the New Zealand electoral roll, a compulsory voting register of New Zealand adults over 18 years of age that currently covers 98% of the adult population of the country. For the purpose of oversampling Māori, the grand sample is divided into two specific subsamples: the general sample and the Māori sample. The general sample was randomly selected from all adults aged 55- to 70 on the electoral roll, and when this was completed the Māori sample was then randomly selected from all 55- to 70-year-old Māori voters on the roll using the Māori descent indicator included in the roll.

Postal Questionnaire

The postal questionnaire included questions on the following:

1. Individual and household demographics (e.g., gender, ethnicity, income levels)
2. Work participation and retirement factors (e.g., work status, job satisfaction, career commitment, attitudes toward retirement, retirement planning)
3. Social participation and independence factors (e.g., volunteerism, leisure orientation, social support)
4. Health factors (e.g., SF-36, a medical complaint checklist)

All questionnaires included a consent form for participants to complete if they wished to continue in the longitudinal study.

Survey Methodology

The postal survey was modeled on the Tailored Design Method described by Dillman,[14] and consisted of a five-stage posting schedule. Participants were sent an initial letter of introduction telling about the study and of their random selection. One week later the questionnaire, information sheet about the study, and a free-post return envelope was sent. After 2 more weeks, a reminder postcard was sent thanking responders and requesting nonresponders to participate. Three weeks later a replacement questionnaire (identical to the original) was sent to all nonrespondents to further encourage participation. Five weeks later, another reminder postcard was sent to the remaining nonrespondents further urging them to participate.

RETURN RATES AND SAMPLE CHARACTERISTICS

Return Rates

The initial questionnaire was sent to 13,045 persons: this included 5264 for the general sample and 7781 for the Māori sample. After excluding 518 participants who could not complete the questionnaire (e.g., they had moved from their listed address, they were hospitalized, or had died), this potential grand sample was reduced to 12,527. Overall 6662 questionnaires were returned (3108 for the general sample, and 3554 for the Māori sample) representing an overall response rate of 53%. Furthermore, 3135 (47%) of the respondents in the first data-wave consented to participate in the longitudinal study.

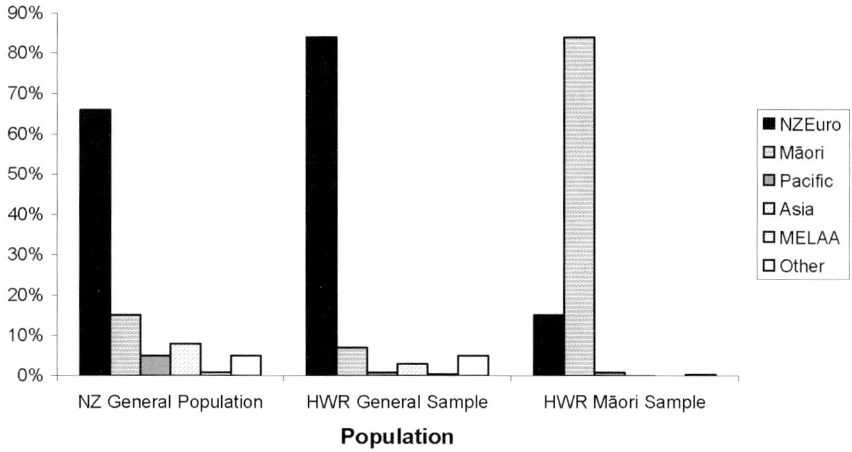

FIGURE 1. A comparison of the ethnic composition of the New Zealand general population and the HWR general and Māori samples.

Sample Characteristics

Approximately 43% of the grand sample work full-time, 21% work part-time, 2% are unemployed, 21% are completely retired, and the remainder are homemakers or students or their work status is unknown. As a consequence of specific oversampling of Māori, 48% of the grand sample are Māori, 47% New Zealand European, 1% Pacific, 1% Asian, and 3% are MELAA (Middle Eastern/Latin American/African) or of other ethnic descent. When broken down by subsample, the general sample mirrors the ethnic composition of the New Zealand general population, aside from a slight increase in "other" ethnicities (FIG. 1), while the Māori sample has a surprising number of "Māori" identifying as non-Māori. This may be explained by recent ethnicity calculation changes by the New Zealand electoral commission, which used to classify nonresponders to ethnicity classifications as "New Zealand European," but now classes these nonresponders as 'Māori' in order to capture potential Māori nonresponders.

FIGURE 2 shows that the HWR sample has a slightly higher female-to-male ratio compared to our reference population, and that the distribution of participants across age bands closely approximates that of the New Zealand older-adult population.

FIGURE 3 shows that the SF-36 mental and general health subscale base rates of the HWR sample, while slightly lower in both comparisons, still closely approximate the New Zealand norms in their range and their sex differences. However, it must be noted that the New Zealand SF-36 normative scores in this graph represent the norms for the population of 15+ years rather than the older population, so this may account for the slightly poorer SF-36 scores for the

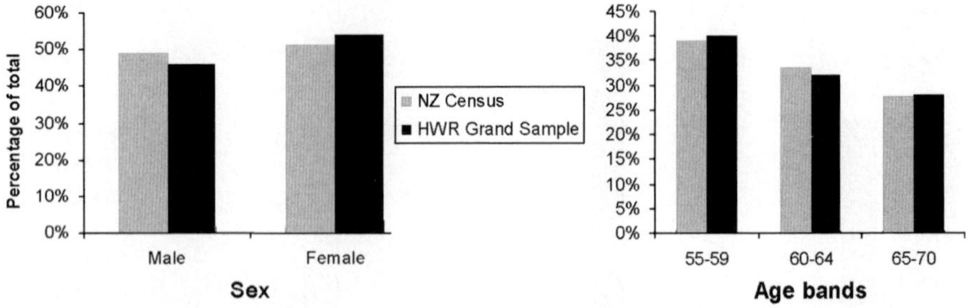

FIGURE 2. A comparison of sex and age distributions between the HWR grand sample and the New Zealand older-adult population.

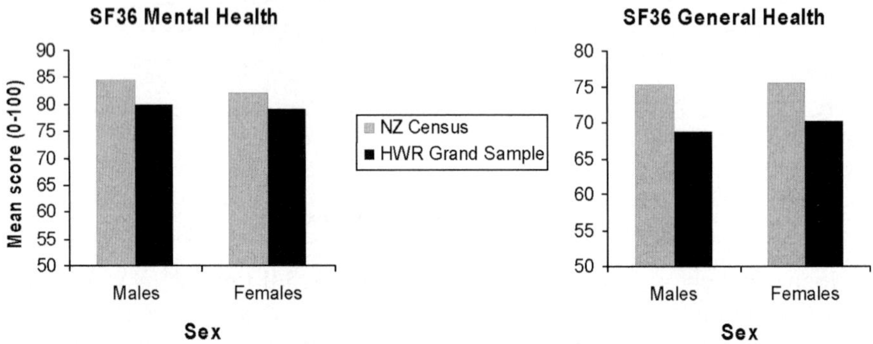

FIGURE 3. A comparison of the SF-36 mental and general health subscale scores between the HWR grand sample and the New Zealand 15+ population norms.

HWR sample in comparison to the New Zealand norms. To this end, other than the slightly high female-to-male ratio in the current sample, our grand sample provides a suitable proxy for the New Zealand 55- to 70-year-old population.

INITIAL FINDINGS

The Relationship between Workforce Status and Health

The survey required participants to note their *current* work status and the work status they would *prefer* to be in. This enabled us to find participants with *matched* work statuses (i.e., those currently classified in the status they also prefer to be in) and those with a positive or negative mismatch (i.e., those preferring less and more work, respectively). These groups were then compared on levels of mental and general health in order to ascertain whether work-status mismatch was associated with worse health. FIGURE 4 shows that a general downward health trend is evident as work status transitions from levels of matched

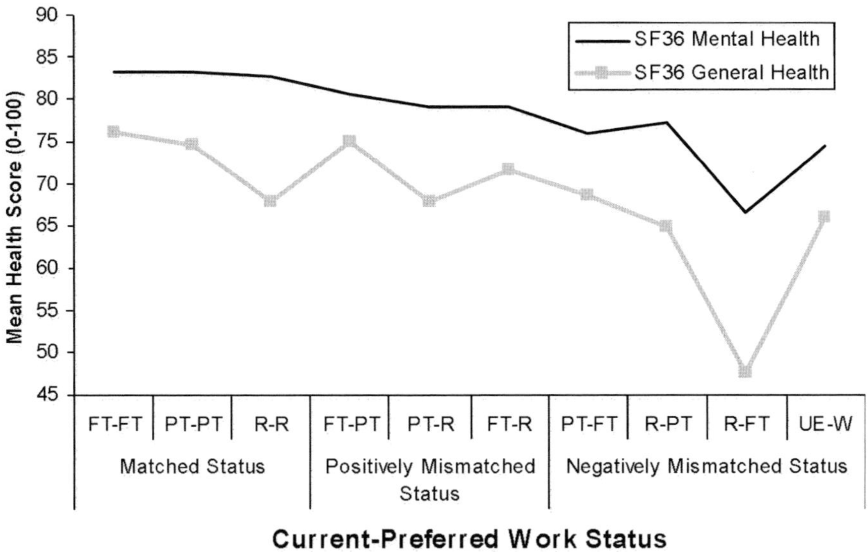

FIGURE 4. A comparison of SF-36 mental and general health scores for participants with matched and mismatched work statuses. Work status key: FT = full-time; PT = part-time; R = retired; UE = unemployed and seeking work; W = full or part-time work.

through positively and negatively mismatched levels. One-way ANOVAs reveal that there were statistically significant differences between these three groups on mental and general health scores, $F(2, 3552) = 52.86$, $P < 0.001$, and $F(2, 3522) = 37.98$, $P < 0.001$, respectively. Specifically, Scheffe's post-hoc analysis revealed that the matched and positively mismatched group had significantly better mental *and* general health than the negatively mismatched group ($P < 0.001$), and matched individuals also had significantly better mental health (but not general health) than the positively mismatched group ($P < 0.001$).

Contributors to Retirement Adjustment

A comparison of the levels of retirement adjustment for those retirees who had matched and mismatched retirement statuses shows that mismatch is associated with poorer retirement adjustment. FIGURE 5 shows that those who are currently retired and prefer to be retired have significantly better retirement adjustment scores than those who have retired but would like some part-time employment, and this group in turn had significantly better adjustment levels than those retirees preferring full-time employment.

Income and general health have always been perceived as important contributors to retirement adjustment.[15,16] However, as illustrated in TABLE 1, although

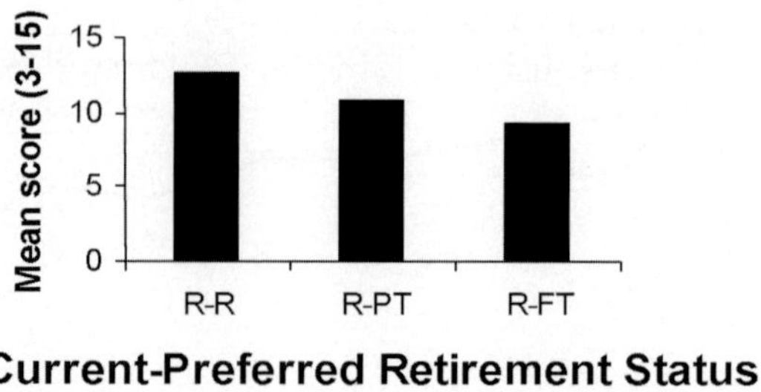

FIGURE 5. A comparison of the retirement adjustment levels of retirees with matched and mismatched work statuses. Retirement status key: R = retired; FT = full-time work; PT = part-time work.

TABLE 1. Statistical weights showing bivariate and multivariate associations between retirement adjustment and income, retirement planning, social integration and health factors

	Statistical weights showing association with retirement adjustment	
	Bivariate correlation coefficient (r)	Mutliple regression coefficient (β)
Income	0.08*	0.04
Retirement planning	0.12**	0.06
Number of health problems	−0.17**	−0.04
Social integration	0.25**	0.17**
General health	0.28**	0.10
Mental health	0.35**	0.22**

NOTE: $^{*}P < 0.05$, $^{**}P < 0.001$

bivariate correlations support their relationship with adjustment score, neither factor remains a significant predictor of retirement adjustment when subjected to multivariate regression analysis. Instead, regression analysis reveals that social integration and mental health levels may be significant predictors of retirement adjustment.

However, these results are only cross-sectional in nature and, therefore we cannot pinpoint cause and effect, but we will follow these relationships with interest as longitudinal data build.

THE FUTURE

With anticipated funding from the New Zealand Foundation for Research, Science and Technology (FoRST), we intend to broaden the scope of the current study and extend the duration to the year 2012 at least. The HWR study will combine with an existing family-focused study (Enhancing Wellbeing in an Ageing Society; EWAS) to broaden the longitudinal participant base to almost

5000 New Zealanders aged 40–85 years. These participants will be assessed in two waves over 5 years, using existing postal surveys, introducing focus groups, and enhancing the current face-to-face interviewing by introducing Computer-Assisted Personal Interview (CAPI) technology for an expanded subsample.

The current focus on health will be broadened to include the following objectives:

1. Economic participation: Exploring economic security across the life span.
2. Social participation: Understanding the basis for meaningful participation and integration in all aspects of society.
3. Intergenerational transfer: Patterns and complexities of intergenerational relationships and their contribution to positive aging.
4. Resilience and health: Social, psychological, economic, and environmental determinants of healthy aging.

REFERENCES

1. UNITED NATIONS. 2002. World Population Ageing: 1950–2050. DESA. New York.
2. STATISTICS NEW ZEALAND. 2004. Older New Zealanders —65 and Beyond. Statistics New Zealand. Wellington, NZ.
3. STEPHENSON, J. & G.M. SCOBIE. 2002. The Economics of Population Ageing. New Zealand Treasury. Wellington, NZ.
4. STATISTICS NEW ZEALAND. 1998. New Zealand Labour Force Projections 1996 (Base)–2051. Statistics New Zealand. Wellington, NZ.
5. DAVEY, J.A. & J. CORNWALL. 2003. Maximising the Potential of Older Workers. New Zealand Institute for Research on Ageing. Wellington, NZ.
6. DORFMAN, L.T. 1995. Health conditions and perceived quality of life in retirement. Health. Soc. Work **20:** 192–199.
7. HANSSON, R.O., P.D. DEKOEKKOEK, W.M. NEECE, *et al.* 1997. Successful aging at work: annual Review, 1992–1996: the older worker and transitions to retirement. J. Vocat. Behav. **51:** 202–233.
8. SAMARTINO, F.J. 1987. The effect of health on retirement. Soc. Secur. Bull. **50:** 31–47.
9. ALPASS, F., N. LONG, N. PACHANA, *et al.* 2001. Correlates of hearing handicap in a veteran sample. Aust. J. Rehab. Counsel. **7:** 38–42.
10. Masoro, E.J. & S.N. Austad (Eds.). 2001. Handbook of the Biology of Aging. Academic Press. San Diego, CA.
11. ILMARINEN, J. 1994. Physical requirements associated with the work of aging workers in the European Union. Exp. Aging Res. **28:** 7–23.
12. ALPASS, F. & A. TOWERS. 2004. Predicting health disparities among older workers and retirees. Poster session presentation at the New Zealand Psychological Society Annual Conference "Psychologies for Aoteoroa New Zealand", August, 2004, Te Papa, Wellington, New Zealand.
13. ALPASS, F., C. STEPHENS, J. DAVEY, *et al.* 2005. A study of the health of older adults in the transition from work to retirement. Paper presented at the *38th National*

Conference of the Australian Association of Gerontology: exploring the meaning of ageing through research policy and practice, Gold Coast, November.

14. DILLMAN, D.A. 2000. Mail and Internet Surveys: the Tailored Design Method, 2nd ed. John Wiley & Sons. New York.
15. GALL, T.L., D.R. EVANS & J. HOWARD. 1997. The retirement adjustment process: changes in the well-being of male retirees across time. J. Gerontol. B Psychol. Sci. Soc. Sci. **52:** 110–117.
16. KIM, J.E. & P. MOEN. 2002. Retirement transitions, gender, and psychological well-being: a life-course, ecological model. J. Gerontol. B. Psychol. Sci. Soc. Sci. **57B:** 212–222.

"Grey Nomads" in Australia

Are They a Good Model for Successful Aging and Health?

PAUL F. D. HIGGS[a] AND FRANCES QUIRK[b]

[a]*Centre for Behavioural & Social Sciences in Medicine, Division of Medicine, University College London, London, England, United Kingdom*

[b]*School of Medicine and Department of Psychology, James Cook University, Townsville, Queensland, Australia*

ABSTRACT: Lifestyle factors have been identified as being very important in determining health in later life. Nutrition, exercise, and social environment all interact to promote, or to limit, opportunities for an active and healthy post-working life. Not only are rates of chronic illness and disability reduced through the promotion of healthy lifestyles, but also quality of life is maintained through the compression of morbidity. Governments in Australia, as in the European Union and North America, have highlighted the importance of behavioral change in health promotion strategies with the aim of having an impact on the health-related lifestyles of their populations. This paper examines the example of a group of older Australians, the "grey nomads," who may present opportunities for examining health-related lifestyle changes. The term *grey nomad* refers to a portion of the older population in Australia who choose to use their later years and retirement as opportunities for travel and leisure, mainly within the confines of the Australian continent. As such, they are similar to groups in North America, such as the "snow birds," who travel to the southern United States to escape the colder winters of more northerly latitudes. Similar seasonal migrations occur from Northern to Southern Europe. What all share in common is an active culture/lifestyle of attempting to "age successfully." Grey nomads also participate in the creation of what can be termed *postmodern communities*, where they and other regular travelers may develop a sense of community feeling with others who are also regularly returning to the same spot year after year. Social support is highly predictive of health outcomes and such mobile communities may prove a positive factor in promoting good health. In this paper we examine whether the "grey nomads" represent a good model for improving health-related lifestyles in later life.

Author for correspondence: Frances Quirk, School of Medicine and Department of Psychology, James Cook University, Townsville, Qld, 4811, Australia. Voice: +61 7 4781 4253; fax: +61 7 4781 5117.

frances.quirk@jcu.edu.au

Ann. N.Y. Acad. Sci. 1114: 251–257 (2007). © 2007 New York Academy of Sciences.

doi: 10.1196/annals.1396.030

Keywords: grey nomads; successful aging; third age

Australia, in common with most nations in the industrialized world, has an aging population. Internationally this is a consequence of both increased longevity, better health, and the demographic consequences of the baby boom of the early postwar years in the middle of the 20th century.[1] Australia has participated in these changes as well as having many of its own features connected to various waves of migration. In Australia the growth of the population over retirement age has also occurred in all states and territories, although it has not been shared by all groups of Australians, particularly Aboriginal Australians and the Torres Straits Islanders.[2] One particular concern is about whether this increase in the numbers and proportions of older people in various national populations will inevitably lead to higher levels of disability and dependency. This simple correlation has been challenged by many writers who instead propose the emergence of a "compression of morbidity," in which increased longevity does not lead to a greater period of life spent in ill health; instead, such conditions are concentrated in a relatively short period prior to death. What this leads to is a disconnection between increasing numbers of older people and higher levels of illness and disability with a concomitant growth in demand for health and social services. What this also points to is the potential for people in later life to maintain good health for longer.[3,4]

While these facts are well documented and have been the subject of considerable policy debate and activity, what has not received as much attention has been the cultural shifts that have been occurring in later life as the baby boom generation ages.[5] These changes will have particular bearing on the pursuit of healthy aging in the coming decades because of the way in which dispositions and expectations set up at earlier ages will work themselves out as these cohorts age. In this paper we focus on the Australian phenomenon of the *grey nomad*, a term used to describe some older Australians' leisure retirement activities and in particular their tendency to spend relatively long periods of time traveling the length and breadth of the country. While this might be a particularly Australian cultural activity, it has considerable international resonance with other countries such as the United States and Canada, some of whose older population also experiences later-life travelling.[6] In order to understand the significance of the grey nomad for the topic of healthy aging it is necessary to study it within the changing cultural context of old age.

The categorization of what constitutes old age has always been a social process dependent on cultural norms as much as physiological signs. This indeterminacy has led to particular difficulties once the concept of mass retirement and pension entitlement was introduced as a mainstay of social policy in the early 20th century.[7] Using chronological age as a marker for entry into a formalized category of old age sits uneasily with both individual physiological change and the social role (or lack of role) that the aged are expected to occupy.

Setting ages for retirement at 70, 65, or 60 has blurred the idea of old age into one of later life, which is not so easily identified by reference to incapacity, frailty, or life expectancy. In the final decades of the 20th century, retirement emerged as a distinct life stage which in some countries stretched to a length of 20 years after formal employment.[8] What has been described as the "selling" of retirement has been so successful that for groups an active choice of early retirement was desired and meant that many older people left the workforce at ages well below the age at which states set their retirement ages. It must also be acknowledged that many older workers found that redundancy merged into retirement as their occupations disappeared in the economic restructuring that many countries experienced in the 1980s and 1990s. While retirement became a distinct life stage in the second half of the 20th century, its cultural coordinates had not changed greatly since those that had operated in the mid-century, where the expectation of what has been termed the *institutionalized life course* laid down a normative trajectory of childhood, adulthood, middle age, and old age, with its corresponding activities of education, work, child-rearing, and retirement.[9] In such a schema, dependent upon a modernity based upon industrialization, there were expectations of time-appropriateness and linear progression. Education and training occurred in early life and seniority was built into organizations and remuneration. In a similar fashion the vast majority of children were born to married young women, and the two-parent nuclear family was seen as an uncontroversial policy objective. This pattern of life, it has been argued, has become "deinstitutionalized."[10] The notion of training for a "job for life" has become anachronistic as retraining and continuing professional development have made it likely that employees of all ages are as likely to be involved in education or in "life-long learning." Similarly, notions of seniority and length of service–based salaries have been replaced by more specific activity– and outcome–focused forms of remuneration. Family life has been transformed in many countries by relatively high rates of lone parenthood, recombinant family structures, and the options offered by biomedical fertility treatments. Retirement has not been unaffected by these changes as deinstitutionalization has made possible a greater flexibility of roles in later life than was ever conceptualized by the midcentury theorists of role theory and disengagement. The notion of age norms and age-appropriate activities has been rendered problematic by societies which are struggling with them too. Certainly, optimistic commentators such as Peter Laslett[11] have identified later life as a Third Age, a potential "crown of life" that can be used for personal self-enrichment and which is not a residual category of social policy. The engagement by recent cohorts of retirees with these opportunities can be evidenced not only among the grey nomads but also in the retirement communities of Florida and Arizona, along the Mediterranean coastline of Spain, and in the boom in tourism and travel for those who are over 60.[12] However, this is not just a cohort effect, it is also a generational cultural phenomenon connected to the very different experiences of those growing up in the post-war economic boom with those who preceded them. Mass affluence was not just a backdrop; it was also the basis for the development of cultural styles and a focus on identity and difference that came into full effect in the 1960s. Combining the de-institutionalization of the life course with the generational transformation wrought by the baby boom generations situates contemporary later life in a new location, one which implies a new set of questions to be asked about the achievement of a healthy and successful later life.

LIFESTYLE FACTORS

An important dimension of the transformation of the experiences of later life has been the increasing cultural significance of lifestyle. This is now seen as more of a project built around choices than as a combination of social styles connected to status and economic position.[13] Emerging out of the growth of affluent consumer societies, the focus on differentiation and "style of life" has become important to both health policy agendas and to the lives of those growing older and entering retirement. The implied passivity of old age as conceived in the late-19th and for most of the 20th century is not a model that is attractive to cohorts who have grown up in societies where circumstances are not just given, but can be changed. While not all may be able to successfully take advantage of the opportunities available, the possibility of making positive lifestyle choices has been culturally legitimated. The promotion of health and well-being in later life is seen as having its foundations in the behaviors and attitudes laid down at earlier points in the life course, as well as being capable of revision at later points. Lifestyle change is one of the main objectives of health policy in countries, such as Australia, Canada, and the United States.[1] Taking exercise, eating the right foods, and maintaining positive well-being are social factors that are seen to be beneficial to healthy aging. The cultural transformation of later life has for many retirees also been accompanied by an affluence, when viewed in comparison with the past circumstances of retirees. In Australia the "over 55s" have 25% of the country's disposable income and nearly 40% of its total wealth.[14] As a consequence, many older Australians are in a position to engage in the construction of post-work lifestyles in which healthy choices are part of an aspirational healthy "lifestyle."

SUCCESSFUL AGING AND THE THIRD AGE

The idea that later life is not just a residual stage of life overshadowed by decline and disability has been popularized in the concept of the "Third Age," advanced by the British historian Peter Laslett.[11] In his work, Laslett makes the case that the period marked out by the relinquishing of work and family responsibilities can be regarded as the "crown of life" when older people cannot only experience a positive form of life, but can also use it as time of self-realization. Projects and ambitions unrealizable at earlier points of the life course, because of time and other pressures, are now capable of being achieved. While this may be viewed as embodying the experiences of particularly affluent retirees, it would appear that it resonates much more widely in the ideas of "productive aging." It is also apparent in the cultural underpinning of the grey nomads and in their U.S. and Canadian counterparts, the "snow birds." Both groups use the flexible opportunities available in retirement to travel or seasonally migrate to better climates. The phenomenon of Canadians traveling to Florida and to other warmer parts of the United States during the bitterly cold Northern winter has been the focus of a number of ethnographic studies.[12] One of the features of these studies is the description of the alternative community structures that emerge and function over a limited time period and that seem to be as sustained as those who live in more geographically permanent circumstances.[12] While there has been relatively little social research carried out on Australia's grey nomads, their activities are seen to be on par with those in North America. In particular, outdoor and active pursuits are popular among older Australian visitors to Queensland.[6] All of this might bear out particular generational motivations as well as reflect patterns of wealth and income; however, the emergence of a lifestyle associated with the activities of the grey nomads suggests that it needs to be studied in relation to whether it is one model for successful aging.

A NEW MODEL FOR AGING?

One of the important considerations in understanding the potential of such new cultural forms of aging is that, in part, it is not determined by the simple connection between later life, illness, and disability. This is not to suggest that these issues are not important in relation to this group, but rather it is to point out that the cultural possibilities of leading a grey nomad lifestyle are not curtailed by poor health and early mortality. Neither does it seem to be the case that there is some form of health selection occurring, positively selecting the healthiest of the older population.[15] Grey nomads have a health profile similar to the rest of the older population and this means that they have similar needs and conditions. In one study in which the most prevalent chronic disease was hypertension, it was not the fact of chronic disease that was preventing travel. However, such a profile could overburden areas that attract such visitors. In one popular location 13% of visitors were aged 65 or over and more than three-quarters came from other states in Australia. While this might seem to place additional burdens on health services in places popular with visitors, it also suggests that many grey nomads are taking the invocation "adding life to years" seriously. They are using the cultural and economic resources built up over the second half of the 20th century and exploring different ways and lifestyles of aging. These lifestyles may privilege activity, health, and vitality, but there is no evidence that these older persons are more likely to have fully integrated them into their day-to-day behavior than are other cohorts. However, the desire to successfully age should not be regarded as unimportant, given that it is the key to spending more time in the Third Age.

SOME TENTATIVE CONCLUSIONS

Grey nomads are a phenomenon of the changing nature of later life and the way in which social change and generational culture has rewritten the script of old age. Healthy aging is an important objective for both individuals and societies if they are to achieve beneficial outcomes. In this light, not only do we need to have a better understanding of the psychological and social factors that contribute to the broadest definition of health, but we also need to know how healthy aging processes connect to the changing circumstances of social life in different circumstances. In other words we need to study developments, such as the grey nomads, because of the way in which these unexpected phenomena may shed light on the important objective of improving life and health for all. In this, as always, more research is needed.

REFERENCES

1. World Health Organization. 2002. Active Aging. A Policy Framework. World Health Organization. Geneva.
2. Australian Bureau of Statistics. 2004. Australian Social Trends 2004. Ausinfo. Canberra ACT.
3. Crimmins, E.M. 2004. Trends in the health of the elderly. Annu. Rev. Publ. Health **25:** 79–98.
4. Fries, J.F. 2003. Measuring and monitoring success in compressing morbidity. Ann. Intern. Med. **139:** 469–503.
5. Quine, S. & S. Carter. 2006. Australian baby boomers' expectations and plans for their old age. Aust. J. Aging **25:** 3–8.
6. Mings, R. 1997. Tracking 'snowbirds' in Australia: winter sun seekers in far north Queensland. Australian Geographical Studies **35:** 168–182.
7. Baldwin, P. 1990. The Politics of Social Solidarity. Cambridge University Press. Cambridge, UK.
8. Gilleard, C. & P. Higgs. 2000. Cultures of Aging: self, Citizen and the Body. Prentice Hall. Harlow.
9. Kohli, M. *et al.* 2001. Time for Retirement. Cambridge University Press. Cambridge.
10. Gilleard, C. & P. Higgs. 2005. Contexts of Aging: class, Cohort and Community. Polity Press. Cambridge, UK.
11. Laslett, P. 1989. A Fresh Map of Life. Widenfield and Nicholson. London.
12. Katz, S. 2005. Cultural Aging. Broadview Publishers. Peterborough, Ont.

13. Ransome, P. 2005. Work, Consumption and Culture: affluence and Social Change in the Twenty-first century. Sage. London.
14. Commonwealth Department of Health and Aged Care. 2001. Population Aging and the Economy. Ausinfo. Canberra ACT.
15. Tate, J. *et al.* 2006. Grey Nomads: health and health preparation of older travellers in remote Australia. Aust. Fam. Physician **35:** 1–2.

Affordable Rental Housing for Older People in Australia

What Do Older People Want?

COLLEEN MARIA CARTWRIGHT

Aged Services Learning and Research Collaboration, Southern Cross University, Coffs Harbour, Australia

ABSTRACT: This study aims to determine the perspectives of older community members and service providers in relation to design and location of affordable rental housing in a coastal retirement area of NSW, Australia. This was achieved by holding focus groups and interviews with older community members ($n = 21$) and service providers ($n = 43$). All participants reported a shortage of affordable rental housing suitable for older people in the area. The preferred option was for cluster units, possibly with an on-site carer, located in or close to town. Participants agreed that if choice was limited to "one-bedroom with large living area" or "two-bedrooms with small living area," the majority of single older people would choose the former option, provided there was adequate storage. The importance of asking the "right" question was demonstrated by responses to options for one- or two-bedroom units. Further research is needed to identify what baby boomers think will meet their need for future affordable housing.

KEYWORDS: affordable housing; older people; social integration; transport

INTRODUCTION

Working from the principle that affordable, accessible, and appropriate housing contributes to health and well-being, and thus to longevity, a study was conducted in a coastal region of New South Wales (Australia) to identify, or develop, physical and financial models of sustainable affordable rental housing that meet the needs of the older community while achieving commercial returns. This paper addresses the perspectives of older people and service providers in relation to physical models, that is, the design and location of affordable rental housing.

Address for correspondence: Professor Colleen Cartwright, Aged Services Learning and Research Collaboration, Southern Cross University, Hogbin Drive, Coffs Harbour, NSW 2450, Australia. Voice: +61 2-6659-3382; fax: +61 2-6659-3622.

colleen.cartwright@scu.edu.au

Ann. N.Y. Acad. Sci. 1114: 258–266 (2007).
doi: 10.1196/annals.1396.036

Background

Throughout Australia there has been a reduction in the availability of affordable housing, with estimates of households in unaffordable housing in NSW ranging from 8.12% to 10.65%. Although the majority of older people own their own homes, in 2002–2003 approximately 13% of older-person households were living in rental accommodation and almost three-quarters (72%) of older lone-person households was made up of women living alone.[1] In addition, some older homeowners cannot meet increasing home maintenance costs and increased rates (particularly in coastal areas), leading some people to sell their homes to look for more affordable accommodation. Despite receiving Commonwealth Rent Assistance, the level of assistance is still inadequate for those on low incomes to enter the private rental market, and in many cases they are unable to compete with those on higher incomes for low-cost housing.

Other factors having an impact on affordable housing availability include:

- shrinking public housing stock;
- market response, where new development is aimed at the top end of the market;
- gentrification of areas which previously had low-rental stock; and
- loss of caravan parks, which close or convert to upmarket resorts.

The Australian Housing and Urban Research Institute (AHURI) has reported that the current shortage of affordable housing in Australia is likely to worsen:

> The absolute number of people entering old age as renters will increase, and more will be renting in the private market. The ability of older renters to access affordable and stable housing is set to be an even more pressing issue than it is now (p. 2).[2]

The shifting emphasis in aged care from residential services to home and community services means that there will be increasing numbers of frail, and in many cases, very old people in this region requiring affordable accommodation; the majority of these older people will be women living on their own: "Many of these individuals are likely to need housing that incorporates some form of support for at least some period" (p. 3).[2]

The area of study has a fast-growing population and a large part of the growth is older retirees. Communities in the study area currently have 22–28% of their population aged 60 or above, compared to 17% for Australia and NSW; by 2016 they will all have over 30% aged 60 or above, with over 40% in two communities. Demand for affordable rental housing in the area is likely to increase, while supply is shrinking.

Definitions

The following definitions were used in the study:

Affordable housing: cost of housing does not exceed 30% of income for people in the bottom 40% of income earners; and the cost of housing leaves sufficient income for other basic needs, such as food, clothing, transport, medical care, and education. However, for low-income households, most State Housing Authorities set the affordability level at 25% of household income.

Older people: the earliest age at which someone could claim an age pension in Australia has been 60 for women (until recently) and 65 for men. This research used 60 as a starting point.

METHODOLOGY

Ethics approval was obtained from the Human Research Ethics Committee of Southern Cross University.

An Advisory Committee was established which included a representative from Department of Housing, the Enterprise and Training Company, Ltd, the NSW Council on the Ageing, and Coffs Harbour Future of Ageing Committee.

Participants

Older community members (OCMs) ($n = 21$) were aged 60 or above and include people recently retired and/or starting to live on an age pension. Participants were recruited through community organizations and the research team's networks; 15 were female and 6 male; ages ranged from 60 to 84 years; 8 were widowed, 6 were married, 4 were separated/divorced, and 3 were single. All but one were in receipt of age-related pensions.

Service providers (SPs) ($n = 43$) were recruited from government and non-government housing and community care organizations; 30 were female and 13 male; ages ranged from less than 40 to over 60.

Questionnaire Development

A semi-structured questionnaire was developed from issues identified in the literature and piloted with three older people and one service provider.

Process

The study area contained five main communities. One focus group was conducted with older community members in each community and an interview was held with one individual who was unable to attend a focus groups. Six focus groups were held with service providers (one in each community plus an additional focus group for service providers whose work encompassed the whole region). Individual interviews were held with four senior staff members of government housing organizations. With the permission of participants, interviews and focus groups were tape-recorded. Tapes were transcribed and responses to each question summarized.

RESULTS

Following is a list of the questions asked of both older community members and service providers. Responses have been summarized, with direct quotes used to highlight issues raised.

Q1: Are you aware of any problems in your community in relation to affordable housing for older people?

All participants reported a shortage of appropriate affordable houses/units (i.e., $135 to $160 per week) from private landlords or from the main provider of public housing in the region. In some areas, high numbers of people moving up from southern areas for the lifestyle has pushed prices higher; in other areas topographic problems, such as hills or flood-prone land, limits the availability of suitable land. In addition, some areas lack adequate infrastructure, such as footpaths or ramps, or have few ground-floor units, so older people cannot physically access the housing.

One elderly woman was going to be evicted because the landlord wanted to sell the units; it took more than 12 months for service providers to find alternative accommodation for her, which caused her great stress.

> SP: *The landlord was getting more and more angry with her because she couldn't move out but she had nowhere to go. Because she was 75% blind, she couldn't leave [the town] because she'd been there for so long and was in her 80s, and couldn't set herself up in a new suburb with no support.*

Other issues identified by participants included:

- size of units: there were very few affordable two-bedroom (or large one-bedroom) units available. Most one-bedroom units were too small "and claustrophobic."
- safety and accessibility: many units had no rails in bathrooms, with showers over baths, doorways not wide enough for wheelchairs, and design problems.
- location: distance from town and no access to suitable public transport created problems.

> OCM: *If you don't have a car you are up the creek. There is no public transport for medical appointments, etc. I pay $38 return to community transport for my chemotherapy and taxis are too expensive.*

Q2: Would you [your clients] prefer a stand alone house, a granny flat (small unit in back garden of relatives' home), a semi-detached unit (two on one block), or a cluster (e.g., six units, set among other houses).

While welcoming choice, the majority of participants favored cluster units set among other houses. Benefits of clusters included having other older people around to talk to and for security; and being more convenient for homecare and other services. Despite this, and although it was not part of the questions asked, a number of older community members commented that they do not want to live in retirement villages. There were strong feelings about not being stigmatized and integration was also seen as a way of retaining community and avoiding ghettos:

> OCM: *If you have elderly people as neighbors and you are going up town, then you see if they want something too. When you are in the normal community, the little kids come out for a walk or are on their scooters and that sort of thing. Places like [retirement villages] can be out of touch with real life.*

Q3: What do you think about shared housing with people of a similar age?

Several people had heard of the Abbeyfield model (8–10 en-suite rooms in a large house, with a caretaker and some meal provision), but most did not like the model. Choice of food, security of tenure, and being able to choose who you live with were thought to be important aspects of independent living.

> OCM: *It sounds very much like a mix between a hostel and a boarding house to me. What happens when you are no longer able to clean your own room and do your washing . . . does that mean you're out? What about family coming to stay?*

Q4: What about high-rise accommodation if the complex had a lift?

The majority of participants thought that this would be a reasonable option, provided there was good security, including a secure car park. Some older participants said that they feel more secure the higher they are, "away from where people can get at you." However, concern was expressed about what would happen if there was a fire and the elevator was out of order, as a number of participants said that they would not be able to manage the stairs, especially if the lights were out.

> SP: *A lot of older people are terrified of lifts and won't go in on their own anyway, so they probably wouldn't go out either.*

Q5: If you needed some support to stay living in the community, would you prefer a cluster of units with a carer's unit included or a carer who could come when called (e.g., with a medi-alert or something similar)?

Participants were divided between the two options. Service providers expressed concern about the cost to residents of having a carer on site and there was also concern that having a carer on site may make the residents feel as if they are already in residential care, when most would be trying to retain their independence.

Participants were asked to consider the model where the carer receives free rent in return for providing some care to the other residents:

> SP: *The caretaker would have to be trained and have the right attitude. If they are receiving free rent it would not exactly be a volunteer but the position is not like a paid staff worker. They would be living there and it would also be their home. Just how much help could they reasonably be asked to provide?*

Having adequate qualifications or training (e.g., CPR) and understanding the medical problems of their clients was considered essential. One option suggested was that if a group of older people were all eligible for a carer to receive a carer's pension, this funding could be pooled and provide a wage for an on-site or on-call carer.

At the suggestion of a member of the Advisory Committee the next question was asked in two sections:

Q6 (a): Thinking now about design: Would you prefer a one-bedroom unit for less rent or a two-bedroom unit, but paying a little more rent?

Almost all participants said that they [their clients] would prefer two-bedroom accommodation, even if it meant slightly higher rent.

> OCM: *For out of season clothes, art and craft materials and equipment, and to have room for family and visitors when they come to stay.*

> SP: *Older people are getting into computers and all that as well, they need somewhere for their music and stereos.*

One service provider was concerned that older people are often admitted to hospital when all they need is additional care at home, because there is no room in the unit for another person to sleep over.

Q6 (b): If the overall size of the unit had to remain the same, would you prefer two bedrooms with a small living space or one bedroom with bigger living space?

This question resulted in many of the older participants changing their preference. The majority said that they would prefer one bedroom, with a big enough living space that it could be screened off as sleeping space for guests. Others added that a living area should be big enough to have friends over for a meal, or just for the comfort of everyday activities.

> OCM: *In my current living space, if I have more than one person in for a cup of tea it's too crowded. After all, I spend more time in the living room than the bedroom.*

> SP: *The loungeroom is so small you couldn't fit a couch and a table, you had to make a decision between a couch or table & chairs.*

Most service providers also thought that their single clients would choose one bedroom with a larger living space, provided they had adequate storage. For couples, two bedrooms was seen as mandatory by almost everyone.

> OCM: *For a couple you would need the extra room just to have space from each other. For reading or computer or crafts, like my wife does calligraphy, so a second room is needed not so much as a bedroom but for all the other things.*

There was general agreement on the need for good lockable storage as well as space to do things.

> OCM: *Even big cupboards in the laundry would be helpful, to store my photos and things from a lifetime of memories.*

Some participants thought that a storage shed could be set up to recharge batteries on powered scooters or wheelchairs and "could double as the men's working shed as well." However, others thought that if medical equipment had to be stored, it would be essential to have a second bedroom. It was also noted that in the future there will be a need for space for computers and other technology (and for some older people that need already exists).

Whichever option was preferred, all participants agreed on the need for adaptable housing and good design, with wide doorways and entry ways plus ramps and rails.

> SP: *If you have a wheelchair, obviously you are going to need a bit more space to get around the coffee table and have a normal life. A walking frame, you need to get around.*

One of the service providers also noted that one-bedroom, larger living space units cost less to build, which would allow providers of public housing to maximize the number of premises available.

Q7: If you needed to choose between a quiet area that was a little bit out from the center of town, perhaps needing a bus or car trip to get into town, or somewhere closer to facilities but a bit more noisy, which would you choose?

Most of the older community members lived close to, or in, town and the majority, particularly those who no longer drove their own cars, said that it suited them well. Being within walking distance of the shops, doctor, and other services enhanced independence, as well as allowing people to socialize, for example, to have a coffee or lunch out with a friend.

> OCM: *I am right in town and I love being close to town. It gets a bit noisy but not too bad and you get used to it. It's alive.*

A small number of the older participants did not want to be in town but acknowledged that "when you stop driving you need to have transport organized or you are stuck." Throughout the discussions reliable transport was consistently raised as an important aspect of affordable housing.

Service providers agreed that most older people would want to be in or close to town, in order to easily access medical and other services, and community amenities, and not have to rely on public transport, or community transport, which was already struggling to meet the increasing needs in the region.

Other reasons given for the target group wanting to be located centrally included:

> SP: *Electric scooters are on the increase, and people don't want to pay for taxis. Scooters provide independence and recreation.*

There was general agreement that, when new accommodation options were being considered, transport should be included with planning. However, others felt that older people would be more likely to walk to shops or for a coffee if they were in town but might not do that if they had to catch a bus. For people on walkers or in wheelchairs, public transport does not provide a real option and community transport limits them to specific times.

Finally, one respondent urged that "we do not continue making the errors of the past."

> SP: *By not planning where we are putting people, what areas we will have what housing in, we have in the past developed mini ghettos and major social problems. Older people, people with disabilities, the disadvantaged need to be integrated into the general community. We need to keep that high on our agenda."*

DISCUSSION

The study highlighted the growing shortage of affordable, accessible, safe, and appropriate housing for older people in coastal communities in New South Wales. As many other coastal areas of Australia are facing similar population changes, it is highly likely that this problem is widespread in Australia. Lack of appropriate housing choices can have a serious negative impact on the health and longevity of older people.

In terms of housing design and location, the favored option was for small clusters of units, in or close to towns, and integrated into the community. The importance of asking the "right" question, and not operating from assumptions or pre-conceived ides, was demonstrated by responses to options for one- or two-bedroom units, with size of the living space and adequate storage being of paramount importance. Lack of adequate public transport was identified as a major issue, not just for those living outside of the main towns, but also for older people who have to travel to the larger centers to access medical care.

In addition, given government policies designed to keep people in the community and out of residential aged-care facilities, there is a need for creative thinking to develop new ways of paying carers to support older people in the community, perhaps with some negotiated accommodation options for the carers. It may also be possible to develop short training courses, which could equip such carers with a range of basic support skills, including CPR and dispute mediation.

While this research investigated the preferred options of older community members and their service providers, further research is needed to identify what the baby boomers think will meet their need for future affordable housing.

ACKNOWLEDGMENTS

The study was funded by a grant from the NSW Department of Housing and the Enterprise and Training Company of Coffs Harbour, Ltd.

REFERENCES

1. Australian Bureau of Statistics. Australian Social Trends 2004; Chapter 7—Housing, Part 3—Homelessness. ABS Canberra, Australia, 2004.
2. AHURI Research & Policy Bulletin. Housing futures in an ageing Australia. Issue 43: May 2004.

Retirement Planning and Expectations of Australian Babyboomers

Are They Ready to Retire?

WENDY HUNTER, WEI WANG, AND ANTHONY WORSLEY

Deakin University, School of Exercise and Nutrition Sciences, Burwood, Victoria, Australia

ABSTRACT: The "babyboomer" generation, those who were born between 1946 and 1965, account for over 30% of the population in Australia, and it is no surprise that the government is concerned about the financial implications for future healthcare costs. While many babyboomers are more financially secure than previous generations have been on account of compulsory superannuation contributions made over the past 15 years, there are still some who are financially vulnerable and expect to rely on government pensions and welfare assistance. Changes to family structures and job security also mean that those who are less financially secure will need social support. Using an ecological framework model, we explored the retirement expectations and experiences of some Australian babyboomers through focus groups and individual interviews to identify key issues and their plans to address these issues. Four main themes are reported in this paper: retirement attitudes and expectations, finances, health, and food. The results suggested that for many persons retirement equated freedom. Little future planning was undertaken for retirement other than contributing to superannuation schemes, and expectations were reported in general terms, such as wanting to remain independent and healthy and to have time to socialize and travel.

KEYWORDS: babyboomers; retirement; expectations; health; food

INTRODUCTION

The "babyboomer generation" is a term given to those born from 1946 up to 1965 as a result of an extended period of high birth rates after the end of World War II.[1] As in many western countries babyboomers account for nearly a third of the Australian population and a third of this group were born overseas.[1] There

Address for correspondence: Wendy Hunter, Deakin University, School of Exercise and Nutrition Sciences, 221 Burwood Highway, Burwood, Victoria, 3125, Australia. Voice: +613 9244 3799; fax: +613 9244 6017.
wendy.hunter@deakin.edu.au

**Ann. N.Y. Acad. Sci. 1114: 267–278 (2007). © 2007 New York Academy of Sciences.
doi: 10.1196/annals.1396.024**

are some differences between the babyboomer generation and past generations, including an increase in life expectancy, types of illnesses now experienced by babyboomers, family structures, job (in)security, and financial security.[2] For previous generations, families have been relatively close-knit structures, with families often living within the same suburb or town, but today many babyboomers not only live in different cities from those of their children, but they are often in different countries, thus creating a greater reliance on formal support networks, such as welfare services, rather than relying on the informal support of the family.[2]

Babyboomers have faced mass retrenchment and redundancy during their lifetimes and cannot expect to have the same job for the remainder of their working lives.[3] Many will have to gain a new skill or upgrade their skills to retain their employment prospects; however some companies are reluctant to provide training opportunities for older staff, believing that it is not cost-effective and that the older staff members are likely to find it difficult to learn new skills.[4] If made redundant, older people appear to experience difficulties in obtaining employment, with many organizations holding stereotypical attitudes toward the capabilities and productivity of older workers.[5,6] Job insecurity has a considerable impact on the financial security, health, and quality of life of people as they age, particularly for low-income workers.[4,7] Although the babyboomers are the first generation who have benefited from the compulsory contributions to superannuation funds,[4] the financial support offered by superannuation depends on the length of time the person has contributed, the amount he or she has contributed, and the type of scheme; therefore, some babyboomers, especially women, will still be retiring with minimal private income from superannuation and will depend on the state pension.[8]

Patrickson and Ranzijn suggest in 2004 that only a third of unemployed people aged between 55 and 70 would like to return to the workforce, with the majority contented to regard themselves as retired.[4]

Some health professionals and government representatives suggest there will be an increased demand on health and welfare services as the babyboomer generation will move into high-risk years for health problems, such as cancer.[9] As people age and become increasingly dependent, this will have implications for the provision of formal support, such as government welfare services, and holds planning implications for all levels of government.[2,10]

THE ECOLOGICAL FRAMEWORK

Looking at these issues in isolation can place misleading emphasis on any one issue, and therefore we have investigated the retirement plans and expectations of babyboomers in relation to health and nutrition and, through the use of a conceptual framework, this paper will show the interrelationship between some of the individual and environmental factors that influence babyboomers.

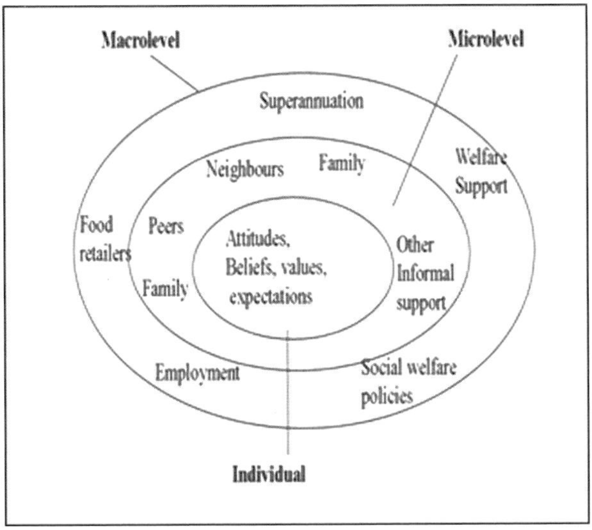

FIGURE 1. Bronfenbrenner's ecological model for retirement planning.

The ecological framework model has been commonly used in family studies and more recently has been applied to food, nutrition, and health studies.[11–13] The framework proposes that successful adaptation to changing circumstances will depend on the individual's adaptive range, which is generall governed by his or her stage of development.[14]

Bronfenbrenner's model looked at several levels of determinants that influence people, including microlevel and macrolevel determinants.[14] Microlevel determinants include the individual's immediate setting of family, peers, friends and so on, while at the macrolevel are the rules and regulations that influence the choices and behaviors of the individual as well as cultural, ethnic, political, and religious norms.[15] Thus, the framework enables complex issues, such as retirement planning and expectations for health and nutrition, to be explored systematically (FIG. 1).

EXPERIMENTAL METHODS

Participants

Data were collected through five focus groups and 12 face-to-face interviews; participants were aged between 41 and 70 years of age. For the focus groups, purposeful sampling was used by contacting relevant organizations in the Melbourne Metropolitan area that are associated with services for middle-aged and older people. Interview participants were recruited through a variety of methods, but the most predominant method was "snowballing," which

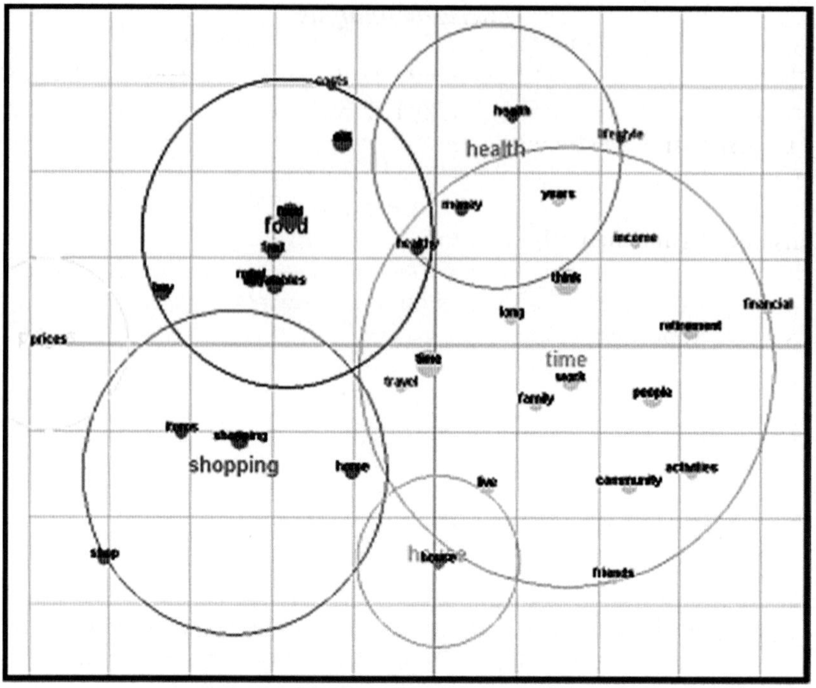

FIGURE 2. Concepts identified in the analysis of the interviews using Leximancer software.

involves participants nominating others to participate in the study. The snowball sampling method is a non-probability sampling technique.[16]

Interviews and focus group session were audio-recorded and transcribed by a professional transcription company. On receipt of the transcripts, the audio files were listened to and compared with the transcriptions to ensure accuracy; alterations were made where necessary.

Focus Group Analysis

The transcripts were coded using a qualitative software program, MAXQDA v2.01 software (VERBI Software, Consult. Sozialforschung. GmbH, Marburg, Germany), and analysis was conducted using content analysis.[17] Level-one coding was based on the themes of the interview/focus group questions, which had been developed based on published literature in the area. From the participants' responses to these questions, two more levels of coding were applied. A definition was created for each of the codes to ensure that the researchers were consistent in their application of the coding.

Interview Analysis

The range of issues which respondents thought important was identified from a series of interview transcripts by Leximancer (Version 2.2) (Leximancer Pty Ltd, Brisbane, Australia) that is based on content analysis. Leximancer is a data-mining and visualization tool that analyzes text and displays the extracted information visually via a "concept map." The concept map displays main concepts and their relationships with other concepts (FIG. 2). Leximancer quantifies and displays the conceptual structure of the texts.[18] After attaching the interview transcript files to the program, Leximancer performs the following steps: text pre-processing and concept identification, concept editing and learning, cassification and indexing, and mapping and statistics.[18]

The map of concepts relating to "retirement" used for the interview data is shown in FIGURE 2. Each word on the map indicated a concept related to retirement raised by the participants. Each concept has a colored circle (the small dots at the center of each word), indicating the occurrences of the concept relative to others that the participants mentioned in the interviews. The size of these dots indicates the centrality of the concepts in the text. The big circles indicate the groupings of concepts that can be aggregated to form themes (e.g., food and health).

RESULTS AND DISCUSSION

In total 44 people aged between 41 and 70 years of age participated in either the interviews or focus groups (TABLES 1 and 2 show the social profile of participants). The sample included 29 pre-retirees, who considered themselves to be were between 1 and 15 years away from retirement, and nine recent retirees, that is, people who had retired within the previous 5 years.

The themes that will be discussed in this paper are:

(1) retirement attitudes and expectations;
(2) finance and superannuation;
(3) health; and
(4) food and nutrition.

While "time" was a major theme for the interviews (FIG. 2) and was raised consistently in the focus groups, it was discussed in the context of the other issues in TABLE 3.

Retirement Attitudes and Expectations

Freedom

Both pre-retirees and recent retirees were generally positive about retirement and the general consensus was that that this stage of their lifecourse equates

TABLE 1. Focus group participant profile ($n = 32$)

Gender	Men	12	Women	20	–	–
Age range	41–54	12	55–64	13	Over 65	7
Marital status	Single/widowed	8	Married/Cohabiting	24	–	–
Education	Secondary/ trade school	16	Tertiary/ post-graduate study	16	–	–
Accommodation	House	28	Unit	4	–	–
Home ownership[a]	Own	27	Rent	2	–	–
Garden	Available	30	Not available	2	–	–
Car availability	Available	32	–		–	–
Expected retirement[b]	<10 years	14	>10 years	6	Already retired	9
Annual household income[c]	<$35,000	11	$35,000–$50,000	11	>$50,000	8

[a] Three participants declined to answer this question.
[b] Three participants declined to answer this question.
[c] Two participants declined to answer this question.

to freedom. Participants felt that it would give them time to do the things they want to do rather than what they have to do:

> *Friends, activities, social occasions and* [I] *still enjoy the bliss of not having a dopey boss on your back.* (FG3, male).

This positive attitude toward retirement was also found in the recent study conducted by the Australian Pensions Insurance Agency (APIA),[8] whereas in the United States there appears to be growing anxiety about retirement among the babyboomer generation.[19]

One concern for some participants was that they would have to work longer than originally planned because of changes to government legislation, the decline in the birth rate, and the perception that "so many young people are on social welfare," resulting in insufficient funds to provide them with a pension. This view was predominant among low-income participants.

Adaptation

Pre-retirees felt that they would adapt easily because they expected to have the time to undertake physical and social activities, such as golf, swimming, or going to the gym. Important microlevel factors that were expected to provide satisfaction in retirement included relationships with family, and many of the participants looked forward to spending time with their families and particularly to helping to look after grandchildren:

> *We are also a group that is so used to be flat chat* [very busy] *7 days a week, 24 hours a day that when we retire and everything stops, are we going to sit back and think what the hell now.* (FG2, female)

On the other hand, recent retirees who had adapted to their changing circumstances successfully, talked about the need to create a meaningful structure for

TABLE 2. Interview participant profile ($n = 12$)

Gender	Men	7	Women	5
Age range	<60 years	8	>60 years	4
Marital status	Single/widow	6	Married/cohabiting	6
Education	Secondary or trade	7	Tertiary/post-graduate	5
Accommodation	House	7	Flat/unit	5
Home ownership[a]	Own	7	Rent	3
Garden	Available	11	Not available	1
Car availability	Available	10	Not available	2
Expected retirement[b]	<10 years	5	>10 years	4
Annual household income	<$35,000	7	>$35,000	5

[a]Two participants declined to answer this question.
[b]Three participants were already retired.

themselves and adjust to their new circumstances and acknowledge that this took time.

One participant summed up the feelings of the majority of the men:

> *I also have this feeling of satisfaction with having fulfilled my debt to society by having worked and put into the society that we live in, and now it is my turn to relax and enjoy...*" (FG4, male)

These men seemed to adapt more readily to their new role(s); they were proactive in joining clubs and other activities and, except for the financial benefit, did not miss paid employment.

However, a few men found adapting to their new circumstances difficult and suggested that retirement could be boring. They also talked about feeling useless and losing social networks:

> *I would love to be partly retired; I would like to be working 2 or 3 days a week. I find that, having worked for some 50 plus years, you miss having a group of people around you...and that is one big thing I am missing in retirement.* (FG4, male)

The APIA study also found that many of their participants would like to have a choice about when they retire; for some this may involve a move into a part-time role rather than ending work completely,[8] although negative attitudes and discriminatory behaviors by employers to older workers suggest that much more needs to be undertaken by governments to enable older workers to have this choice.[4]

Finance and Superannuation

Finance was the main issue that was given consideration for future planning by pre-retirees. However, there still seemed to be some vagueness over how much is likely to be needed:

TABLE 3. Key themes from focus groups and interviews

From focus groups	From interviews
Retirement attitudes and expectations	Retirement attitudes and expectations
Finances and superannuation	Activities/socializing
Health	Food and shopping
Food	House
Accommodation	Health
Activities/socializing	Time

> *I think our biggest thing, everybody if they want to be truthful, the biggest worry of the future is will you have enough finances? It's all about money.* (FG2, female)

Macrolevel factors, such as government policies and employment conditions, influenced the amount of superannuation funds participants had available, while microlevel factors included the family structure of the individual—for example whether they had children. Having sufficient money to live the sort of life they wanted was very important to pre-retirees, particularly for those who wanted to travel or move home when they retire.

Participants believed that it was important to make sure they were financially secure, as this would influence their lifestyle, and some had adopted a range of financial strategies to fund their retirement, including buying an investment property. Other participants, particularly lower-income women, reported having limited superannuation funds, which would need to be supplemented by government support. The APIA study found that over 40% of babyboomers did not have sufficient savings to fund their retirement and believe that there was insufficient time for them make significant growth in their retirement fund.[8] Lack of adequate funds to support retirement expectations was also the major cause of anxiety for United States babyboomers, with 34% citing financial independence as a principal factor for a good quality of life in retirement.[19]

Recent retirees suggested that finance has a considerable influence on the choices they make and that they have had to adapt to living on a restricted income, even if they had "good" superannuation schemes.

> *I think financially it has been much more difficult. Financially we have to make a lot of choices.* (FG3, female)

The majority of participants also commented that they expected to spend most of their money during their retirement and at the microlevel they worried about the potential consequences for their families as it will eliminate any inheritance for their children. These participants were also concerned that their children and grandchildren do not have the same opportunities to own their homes. Conversely, some thought that their adult children should be able to provide for their own future. At the macrolevel they were also concerned that they would have insufficient funds to pay nursing home fees should they be necessary in the future.

Health

Health appeared to be a major concern for participants regardless of socioeconomic status. All of the participants wanted to remain sufficiently fit and healthy to continue with their current lifestyle and put off moving into aged-care facilities for as long as possible. This reflects the findings of other studies as reported by Quine and Carter, who suggest that babyboomers are more likely to use aids, such as cosmetic surgery, complementary medicine, and therapeutic devices, in a quest to defer the effects of aging for as long as possible.[20] There was concern about macrolevel factors, such as access to adequate health services on retirement and a general expectation that their health would deteriorate. At the macrolevel, pre-retirees, however, appeared to have less trust and confidence in the medical profession, believing that they have to take control of their own health. One pre-retiree said:

> *It is lifestyle choices you make now, let alone lifestyle choices you are making when you are retiring. We are putting down the foundations now—that is probably how I view it, is that what I am doing now is laying the foundations for that health—future health, so continue the routine that we've established.* (FG1, female)

Diet and physical activity were identified as a means of maintaining good health during retirement, for example, in preventing or modifying type 2 diabetes or heart disease. The pre-retirees felt that keeping a positive attitude was also very important and worried about becoming depressed or disabled through ill health. Dealing with health problems caused some anxiety among the recent retirees who were trying to cope with deteriorating health.

Food and Nutrition

Participants were asked to consider what they thought their diets would be like in the future and why their foods might change. More than three-quarters of the pre-retirees believed their diets would become healthier when they retired because they would be able to take more time for shopping and preparing food. They would be looking for food that was nutritious and fresh. They recognized the importance of a healthy diet as they were getting older, but didn't want to take the diet too seriously, as explained by one man:

> *Oh well, I still enjoy a meal at McDonalds sometimes, but I don't sort of enjoy it as much as I did when I was younger, and that will probably change over the next 20 years, although I hope I never lose the ability to still enjoy a cream-cake, now and again . . ..* (FG1, male)

Other studies that have looked at current dietary intakes and food choices of babyboomers found that overall babyboomer women have better diets than their children, but they are moving away from dietary health fads and concentrating on the enjoyment of food.[21–23]

The recent retirees appeared to agree with the pre-retirees. They felt they were able to give their diet more consideration now that they were not working. They also reported shopping more frequently so that their food was fresher. Others, however, also reported that changes to their dietary habits had occurred due to major illnesses. In general, recent retirees felt they had a varied diet and as long as they could afford to, they would continue to try new things subject to their affordability, but recent retirees were more price aware than pre-retirees when food shopping, probably because of having to work with a more restricted income.

At the macrolevel, pre-retirees from high socioeconomic backgrounds expected to continue to socialize by going out for meals in coffee shops and restaurants with friends. However, participants from low socioeconomic backgrounds preferred to eat at home, as they felt the cost of food in restaurants did not give good value for money. While eating as a social activity was also important for recent retirees, it wasn't something they could do regularly because of financial constraints. They were more likely to eat at friends' and relatives' houses if they ate out.

CONCLUSIONS

In conclusion, healthy aging is important for individuals, as is the confidence that there are sufficient resources to cope with the expected increased demands of this aging population. This paper presents some of the issues identified by babyboomers that they believe will enhance or lessen their quality of life in retirement. It would seem that most are not ready for retirement, even though they generally seem to view retirement positively; they are specifically concerned about finances and health. Sharing eating occasions with others is considered an important aspect of socializing, but little planning has gone into other aspects of their daily activities and how their needs might change. Through the application of the ecological framework model, it can be seen that the individual's ability to adapt to changing circumstances has considerable influence on his or her ability to adjust to changes in lifestyle after retirement.

Further research needs to be conducted to determine whether these results are representative of the babyboomer generation in general and to identify factors that will help babyboomers to plan for successful aging and retirement.

ACKNOWLEDGMENTS

We would like to acknowledge the support of Sodexho Seniors and Sanitarium Foods. We also acknowledge and thank MECWA and the Greater Dandenong Council for allow us to recruit their staff for focus groups and interviews as well as members of the Lilydale Genealogy group and all interview participants.

REFERENCES

1. AUSTRALIAN BUREAU OF STATISTICS. 2006. Age matters, Babyboomers and the 2001 census. Newsletter, Issue 2, http://www.abs.gov.au/ websitedbs/D3110129.NSF/f128006d2d1e7c10ca2566fd0081ba4b/3cd12b553c337b 4aca256d050000ec99!OpenDocument. Accessed March 29, 2007.
2. HUGO, G. 2003. Australia's ageing population. Aust. Planner **46:** 109–118.
3. PATRICKSON, M. & L. HARTMANN. 1995. Australia's ageing population: Implications for human resource management. Int. J. Manpower **16:** 34–46.
4. PATRICKSON, M. & R. RANZIJN. 2004. Bounded choices in work and retirement in Australia. Employee Relat. **26:** 422–432.
5. VAN DER HEUVAL, A. 1999. Mature aged workers: Are they a disadvantaged group in the labour market? Aust. Bull. Labour **25:** 11–22.
6. TAYLOR, P. & A. WALKER. 1998. Employers and older workers: attitudes and employment practices. Ageing Soc. **18:** 641–658.
7. LORETTO, W. & P. WHITE. 2006. Employers' attitudes, practices and policies towards older workers. Hum. Resour. Manage. J. **16:** 313–330.
8. AUSTRALIAN PENSIONER INSURANCE AGENCY. 2007. Understanding Over 50s: An APIA Research Report. http://www.apia.com.au/pdf/APIA%20Over %2050s.pdf. Accessed March 29, 2007.
9. KENNEDY, B.J., S.A. BUSHHOUSE & A.P. BENDER. 1994. Minnesota population cancer risk. Cancer **73:** 724–730.
10. SLACK-SMITH, D. & M. LAVERTY. 2005. Diversity and the Babyboomers: findings of the Compass Consultation on Australia's Ageing Population. Novartis Pharmaceuticals Australia Pty Ltd. : 8–10.
11. STORY, M., D. NUEMARK-STZAINER & S. FRENCH. 2002. Individual and environmental influences on adolescent eating behaviours. J. Am. Dietetic Assoc. **102:** S40–S51.
12. PETERSON, K.E., G. SORENSON & M. PEARSON, *et al.* 2002. Design of an intervention addressing multiple levels of influence on dietary and activity patterns of low-income postpartum women. Health Educ. Res. **17:** 531–540.
13. ELDER, J.P., L. LYTLE & J.F. SALLIS, *et al.* 2007. A description of the social-ecological framework used in the trial of activity for adolescent girls (TAAG). Health Educ. Res. **22:** 155–165.
14. WHITE, J.M. & D.M. KLEIN. 2005. Family Theories. Sage Publications. Thousand Oaks, CA.
15. REIFSNIDER, E., M. GALLAGHER & B. FORGIONE. 2005. Using ecological models in research on health disparities. J. Prof. Nurs. **21:** 216–222.

16. McGee, T.R. 2007. Sociological investigations. *In* Public Sociology: an Introduction to Australian Society. J. Germov & M. Poole, Eds.: 77. Allen & Unwin. Crows Nest.
17. Mayring, P. 2000. Qualitative content cnalysis. Forum Qualitative Social Research **1:** 2, http://www.qualitative-research.net/fqs/fqs-eng.htm. Accessed March 29, 2007.
18. Smith, A.E. 2000. *Machine mapping of document collections: the Leximancer system*. Proceedings of the Fifth Australasian Document Computing, Sunshine Coast, Australia, December 1, 2000, http://www.leximancer.com/documents/adcs2000.pdf.
19. Michaelson, D. 2005. The MetLife Survey of American Attitudes Toward Retirement: what's Changed? Findings from a National Study. MetLife Mature Market Institute: 6–7.
20. Quine, S. & S. Carter. 2006. Australian baby boomer's plans and expectations for their old age. Australas. J. Ageing **25:** 3–8.
21. Dobson, A., G. Mishra & W. Brown, *et al.* 1997. Food habits of young and middle-aged women living outside the capital cities of Australia. Aust. N. Z. J. Public Health **21:** 711–715.
22. Ball, K., G.D. Mishra & C.W. Thane, *et al.* 2004. How well do Australian women comply with dietary guidelines? Public Health Nutr. **7:** 443–452.
23. MacKay, H. 1997. Generations. Pan MacMillan. Sydney, Australia.

Age Identity, Age Perceptions, and Health

Evidence from the English Longitudinal Study of Ageing

PANAYOTES DEMAKAKOS,[a] EDLIRA GJONCA,[a] AND JAMES NAZROO[b]

[a]*Department of Epidemiology and Public Health, University College London, London, United Kingdom*

[b]*Department of Sociology, University of Manchester, Manchester, United Kingdom*

ABSTRACT: This study explores the association between age identity, perceptions of age, and health. It uses data from the first and second waves of the English Longitudinal Study of Ageing (ELSA). The sample consists of 8780 men and women. Four independent variables were used: perceptions of when middle age ends and old age starts, and the ages respondents would like to be and felt they were (desired and self-perceived ages). The outcomes variables were: self-assessed health, self-reported limiting long-standing illness or disability, hypertension, and diabetes. Statistical analysis included descriptive, bivariate, and multivariate tests. A series of logistic regression models were estimated, which were adjusted for age, sex, marital/cohabitation status, and socioeconomic position. The analysis showed that self-perceived age and perceived chronological end of middle age related to all four health outcomes over and above other covariates. In contrast, desired age did not relate to any of the outcomes, while perceived start of old age related only to self-reported health dimensions. These findings indicate the importance of age identity and age perceptions for health in old age.

KEYWORDS: aging; age identity; age perceptions; self-perceptions; health; ELSA

INTRODUCTION

As well as chronological age, aging needs to be understood in terms of individual perceptions and experiences. Age perceptions and age self-perceptions

Address for correspondence: Dr. Panayotes Demakakos, Department of Epidemiology and Public Health, University College London, 1–19 Torrington Place, London, WC1E 6BT, UK. Voice: +44-20-7679-1712; fax: +44-20-7813-0242.
p.demakakos@ucl.ac.uk

**Ann. N.Y. Acad. Sci. 1114: 279–287 (2007). © 2007 New York Academy of Sciences.
doi: 10.1196/annals.1396.021**

(age identity) are in part shaped by dynamic interactions with social and cultural systems that frame our understandings of age and aging in the context of aging populations and changing experiences of chronological age. Research has challenged the notion that chronological age dominates the aging process by showing that there are many dimensions of age relevant to it.[1] In contemporary Western societies the majority of older people (*a*) perceive themselves younger than their chronological age,[2–4] (*b*) think that middle age ends well after 60,[5] and (*c*) would like to be much younger than their chronological age.[3] Also, there is accumulating evidence that age identity and age perceptions relate to a wide array of health outcomes, including mortality, functional status, morbidity, and self-perceived health.[2,4,6–11]

This study aims to explore whether age identity and age perceptions relate to self-reported and clinically measured health outcomes in a national sample of older adults in England. It focuses on the following dimensions of age identity and age perceptions: (*a*) "self-perceived age," which is the age our respondents felt they were; (*b*) "desired age," which is the age our respondents would like to be; (*c*) "perceived chronological end of middle age"; and (*d*) "perceived chronological start of old age." Self-perceived age is expected to measure one's evaluation of one's present status and situation against one's representations of age and aging, which form one's personal aging model.[12] Desired age is expected to reflect values of age in contemporary society and how far respondents have come to terms with their aging. Perceived boundaries of middle and old age are expected to reflect one's personal model aging[12] and internalization of societal and cultural norms.

METHODS

Data

The data for this study come from the first two waves of the English Longitudinal Study of Ageing (ELSA). The first wave took place in 2002–2003 and the second in 2004–2005. This paper involves 3950 men and 4830 women aged 50 years old or older (born on or before February 29, 1952) who were residents of England and took part in both waves of the study. The technical details for both waves can be found elsewhere.[13]

Measures

The questions on the perceived chronological boundaries of middle and old age were the following: "Please could you tell us at what age you consider old age to start?" and "We would also like you to tell us at what age you consider middle age to end?" The questions on self-perceived and desired ages were as follows: "How old do you feel you are?" and "How old would you like to be?" Replies to all these questions were initially recorded in the form of single years of age and were subsequently recoded. The perceived boundaries of middle and old age were coded in 5-year intervals, except all cases reporting values $\leq$50 or $>$80, which were grouped together. The self-perceived age variable had three categories: those feeling as old as their chronological age, those feeling older than it, and those feeling younger than it. Desired age was a dichotomy with two categories: those who want to be as old as their chronological age versus those who would like to be younger than it.

Four health variables were used as dichotomized outcome variables: self-perceived health, self-reported long-standing illness or disability, hypertension, and diabetes. The self-perceived health question was as follows: "Would you say your health is . . . excellent/very good/good/fair/poor" and responses were dichotomized; those who reported having excellent or very good health against all others. The self-reported long-standing illness or disability question read as follows: "Do you have any long-standing illness, disability or infirmity? By long-standing [is meant] anything that has troubled you over a period of time, or that is likely to affect you over a period of time" and possible responses to it were yes or no. Prevalent diabetes was measured either as self-reported diagnosis of diabetes by a doctor in the past or as a raised glycated hemoglobin (HbA1c) level ($\geq$6%).[14] Similarly, prevalent hypertension was measured either as self-reported diagnosis of hypertension by a doctor in the past or as mean value of three successive measurements of systolic and diastolic hypertension $\geq$140/90 mmHg.

Chronological age, sex, marital/cohabitation status, and wealth were employed as covariates. Chronological age was used in 5-year intervals; all those aged 80+ years were combined into the oldest age group. Marital/cohabitation status was treated as a dichotomy: those living with their spouse or partner against all others. Wealth was measured at the benefit-unit level (benefit unit is either a couple or a single person with any dependent children they may have) and reflected on the value of all financial assets possessed by the benefit unit against owed debts. All data were from the second wave of the ELSA, except for the two variables on perceived boundaries of middle and old age, which were from the first wave of ELSA.

Analysis

The analysis was descriptive, bivariate, and multivariate. It included a brief description of the sample characteristics followed by a breakdown of the age-adjusted prevalence rates of all four health outcomes by the four independent variables. The multivariate analysis included the estimation of logistic regression models for all four outcomes, which were adjusted for age, sex, marital/cohabitation status, and wealth in two steps. Analysis did not include any cases with missing values and no weights were used. The level of statistical significance used was $P \leq 0.05$.

RESULTS

Three-fourths of the sample, 73.4% ($n = 5311$), perceived themselves to be younger than their chronological age, while 18.1% ($n = 1309$) were feeling as old as their chronological age and 8.6% ($n = 619$), a comparatively large number of respondents, reported feeling older than their chronological age. As regards their desired age, the vast majority, 91.4% ($n = 6306$), reported wanting to be younger than their chronological age, while only 8.6% ($n = 591$) would not like to be younger than their chronological age. With respect to the perceived chronological end of middle age, 12.5% ($n = 1007$) reported that middle age ended at or before 50; 35.7% ($n = 2865$) that it ended between 51 and 60 years of age; 19.4% ($n = 1551$) that it ended between 61–65; while a large part of our sample, 32.4% ($n = 2602$), believed that middle age ended well after the retirement age limit of 65. As regards perceived chronological start of old age, 13.3% ($n = 1056$) reported that old age started at or before 60; 12.9% ($n = 1025$) that it started between 61 and 65; 33.8% ($n = 2690$) that it started at some point within the post-retirement period, between 66 and 70; and 40% ($n = 3177$) said that old age began after 70. The bivariate analysis showed that self-perceived age and perceived chronological limits of middle and old age related significantly ($P \leq 0.05$) to age-adjusted prevalence rates of all four health outcomes (self-perceived health, self-reported long-standing illness or disability, hypertension, and diabetes). On the contrary, desired age did not relate significantly to any of the age-adjusted prevalence rates, except for self-perceived health.

The multivariate analysis (TABLE 1) showed that perceived chronological end of middle age and self-perceived age were significant correlates of all selected health outcomes. Perceived chronological end of middle age remained significantly related to all health outcomes regardless of the adjustment for the major sociodemographic confounders (age, sex, and marital/cohabitation status) and socioeconomic status (wealth). Self-perceived age was shown to be by and large the most important correlate of all four health outcomes. Its relationships with them were of a dose–response type. Respondents who were feeling younger than their chronological age had a significantly lower chance of reporting ill health on any of the health dimensions compared to the other two categories. Similarly respondents who were feeling as old as their chronological age had better health than those who were feeling older than the chronological age. The only exception to this dose–response rule was with the association between self-perceived age and diabetes, where the difference between those feeling younger than their chronological age and those feeling as old as their chronological age was not statistically significant.

TABLE 1. Logistic regression analyses for the associations of self-perceptions about age with health outcomes[a,b]

	Poor, fair, or good self-perceived health	Long-standing illness or disability	Diagnosed or measured hypertension ≥140/90 mmHg	Diagnosed diabetes HbA1c ≥ 6.0
Desired Age				
Model I[c]				
Desired age < chronological age	1.00	1.00	1.00	1.00
Desired age = chronological age	1.22 (1.03 to 1.44)*	1.01 (0.85 to 1.20)	0.99 (0.82 to 1.21)	1.11 (0.81 to 1.53)
Model II[d]				
Desired age < chronological age	1.00	1.00	1.00	1.00
Desired age = chronological age	1.17 (0.98 to 1.40)	0.97 (0.82 to 1.16)	0.96 (0.79 to 1.17)	1.07 (0.78 to 1.49)
Self-perceived Age				
Model I[c]				
Self-perceived < chronological age	1.00	1.00	1.00	1.00
Self-perceived > chronological age	0.20 (0.16 to 0.25)**	0.31 (0.25 to 0.37)**	0.60 (0.49 to 0.74)**	0.40 (0.31 to 0.52)**
Self-perceived = chronological age	0.48 (0.42 to 0.54)**	0.61 (0.54 to 0.70)**	0.77 (0.66 to 0.89)**	0.92 (0.74 to 1.15)
Model II[d]				
Self-perceived < chronological age	1.00	1.00	1.00	1.00
Self-perceived > chronological age	0.22 (0.18 to 0.28)**	0.34 (0.28 to 0.41)**	0.63 (0.51 to 0.78)**	0.45 (0.34 to 0.58)**
Self-perceived = chronological age	0.48 (0.42 to 0.55)**	0.62 (0.54 to 0.70)**	0.76 (0.66 to 0.89)**	0.92 (0.74 to 1.16)
Perceived Chronological Start of Old Age				
Model I[c]	1.24 (1.20 to 1.28)**	1.12 (1.08 to 1.15)**	1.04 (1.00 to 1.07)*	1.09 (1.03 to 1.15)**
Model II[d]	1.20 (1.16 to 1.24)**	1.09 (1.06 to 1.12)**	1.02 (0.98 to 1.05)	1.06 (1.01 to 1.12)*
Perceived Chronological End of Middle Age				
Model I[c]	1.19 (1.16 to 1.22)**	1.08 (1.05 to 1.11)**	1.07 (1.04 to 1.10)**	1.13 (1.07 to 1.19)**
Model II[d]	1.15 (1.12 to 1.19)**	1.05 (1.02 to 1.08)**	1.05 (1.02 to 1.09)**	1.10 (1.05 to 1.16)**

[a]For all analyses presented in this paper the latest versions of the archived ELSA data sets were used.
[b]Results are given in the form of odds ratios (95% confidence intervals).
[c]Model I is adjusted for age, sex, and marital/cohabitation status.
[d]Model II is adjusted for age, sex, marital/cohabitation status, and wealth.
$^{**}P \leq 0.01$; $^{*}P \leq 0.05$.

It is worth noting that adjusting for wealth did not attenuate the strength of any of the associations between self-perceived age and the four health outcomes.

Moreover, the multivariate analysis showed that perception of the chronological start of old age only related significantly to self-perceived health and self-reported long-standing illness or disability, but not to hypertension or diabetes; any statistically significant association between perceived chronological start of old age and hypertension or diabetes was a byproduct of an interaction between the former and chronological age. Finally, desired age did not relate to any of the health outcomes after adjustment for wealth.

DISCUSSION

In a national sample of older people we found that those who were feeling younger than their chronological age, or believed that middle age ended later in life, reported better health and suffered significantly less from hypertension and diabetes than those who either did not feel younger than their chronological age or believed that middle age ended earlier in life. We also found that those who preferred to be younger than their chronological age did not have worse health than those who preferred to be of the same age as their chronological age. Finally, we found that those who believed old age to start later in life reported better self-perceived health and less self-reported long-standing illness or disability than those who, on the contrary, believed old age to start earlier in life.

Our analysis showed that self-perceived age not only relates to all four health outcomes over and above chronological age, sex, marital/cohabitation status, and wealth, but also that this age identity variable had the strongest association with the health outcomes. This finding is in accordance with the little longitudinal evidence available, which shows the significant predictive value of self-perceived age for mortality and survival in older samples[8] and with an increasing number of cross-sectional studies on the associations between self-perceived age and various dimensions of health.[2,4,11,15] It is also a finding that should be considered in relation to longitudinal studies showing that self-perceived life expectancy, which is a concept close to that of self-perceived age, is predictive of mortality in older samples.[7,9]

Our analysis also suggested that the perceived chronological boundaries of middle and old age were related to health, but not in a uniform way. Both variables were associated with the self-reported health measures, but only perceived chronological end of middle age was associated with hypertension and diabetes, both a combination of direct measurement and self-reported diagnosis. A plausible explanation for this difference might be that perceived chronological end of middle age captures the transition from middle to old age as experienced by the individual better than perceived chronological start of old age. These findings are similar to those of the only previous study that we found on this topic. This study showed associations between perceived chronological end of middle age and a wide array of cardiovascular and functional health outcomes, including mortality from coronary heart disease, in a 7-year follow-up.[6]

There is more than one explanation for why age identity and perceptions of age relate to health in such a generic and strong manner. One plausible explanation is that such age-related variables reflect a self-assessment of one's own age based on a series of judgments about one's own condition and therefore are expected to relate to health simply because they accurately epitomize one's health, functional, socioeconomic, and psychological status. Such an explanation relates to a possibility that variables such as self-perceived age might associate with health because they encapsulate complex information about the individuals themselves and also about their social and cultural environments.[8] In relation to this explanatory approach, it should be mentioned that recent research has shown that age identity and perceptions of age are associated with mortality and other health outcomes independently of self-perceived health and prior health conditions.[6,7,9]

Another explanation draws on the suggestions of a recent paper that it is psychosocial characteristics, such as self-efficacy, sense of control, personal commitment, and social engagement that relate to longevity.[16] Considered together with the horizon hypothesis,[17] which argues that self-perceived longevity increases one's sense of control, this leads to the possibility that age identity and age perceptions might relate to health outcomes through psychosocial pathways.[18,19] It might be that a younger age identity and positive age perceptions catalyze an increase in psychosocial resources that can benefit later health and extend life.

Finally, the lack of any association between desired age and health may be because the former is a construct that reflects aging stereotypes and the ageism bias of contemporary Western societies, rather than individual experiences of coming to terms with one's own age, which is likely to have been more relevant for individual health.

STRENGTHS AND LIMITATIONS

The most important limitation of our analysis is that it is effectively cross-sectional. Thus, it does not provide clues on the direction of causality. It is possible that instead of the age variables influencing health outcomes, health has an impact on them. The paper's most important strength is that that it uses a range of age-related variables and thus gives a comprehensive account of how age perceptions and age identity might relate to health. To the best of our knowledge this is the first time that perceived chronological start of old age has been used as a correlate of health. A further strength is that it uses not only self-reported, but also clinical health data. The use of such data might provide useful insights into the pathogenetic pathways to which age perceptions and age identity might relate. To the best of our knowledge, no other study has focused on associations between diagnosed or clinically measured hypertension or diabetes and age identity and perceptions. Finally, the use of wealth as a covariate in the regression analysis adds strength to our results, as this is a socioeconomic marker that, we think, provides a very accurate contemporary account of one's accumulated socioeconomic position.

ACKNOWLEDGMENTS

The data in this study were made available through the UK Data Archive. ELSA was developed by a team of researchers based at University College, London, the Institute for Fiscal Studies, and the National Centre for Social Research. The funding for ELSA is provided by the U.S. National Institute of Aging, and a consortium of UK government departments coordinated by the Office for National Statistics. James Nazroo is grateful to the ESRC, whose funding of the project "Inequalities in health in an aging population: patterns, causes and consequences" (RES-000-23-0590) supported his work on this paper. The developers and funders of ELSA and the Archive do not bear any responsibility for the analyses or interpretations presented here.

REFERENCES

1. SETTERSTEN, R.A. & K.U. MAYER. 1997. The measurement of age, age structuring, and the life course. Annu. Rev. Sociol. **23:** 233–261.
2. LOGAN, J.R., R. WARD & G. SPITZE. 1992. As old as you feel: age identity in middle and later life. Soc. Forces **71:** 451–467.
3. UOTINEN, V., T. RANTANEN, T. SUUTAMA & I. RUOPPILA. 2006. Change in subjective age among older people over an eight-year follow-up: "Getting older and feeling younger?" Exp. Ageing Res. **32:** 381–393.
4. WESTERHOF, G.J. & A.E. BARRETT. 2005. Age identity and subjective well-being: a comparison of the United States and Germany. J. Gerontol. B Psychol. Sci. Soc. Sci. **60:** S129–S136.
5. LACHMAN, M.E. 2004. Development in midlife. Annu. Rev. Psychol. **55:** 305–331.
6. KUPER, H. & M. MARMOT. 2003. Intimations of mortality: perceived age of leaving middle age as a predictor of future health outcomes within the Whitehall II study. Age Ageing **32:** 178–184.
7. SIEGEL, M., E.H. BRADLEY & S.V. KASL. 2003. Self-rated life expectancy as a predictor of mortality: evidence from the HRS and AHEAD surveys. Gerontology **49:** 265–271.
8. UOTINEN, V., T. RANTANEN & T. SUUTAMA. 2005. Perceived age as a predictor of old age mortality: a 13-year prospective study. Age Aging **34:** 368–372.

9. VAN DOORN, C. & S.V. KASL. 1998. Can parental longevity and self-rated life expectancy predict mortality among older persons? Results from an Australian cohort. J. Gerontol. B Psychol. Sci. Soc. Sci. **53:** S28–S34.
10. KNOLL, N., N. RIECKMANN, U. SCHOLZ & R. SCHWARZER. 2004. Predictors of subjective age before and after cataract surgery: conscientiousness makes a difference. Psychol. Aging **19:** 676–688.
11. WESTERHOF, G.J., A.E. BARRETT & N. STEVERINK. 2003. Forever young? A comparison of age identities in the United States and Germany. Res. Aging **25:** 366–383.
12. FURSTENBERG, A.L. 2002. Trajectories of aging: imagined pathways in later life. Int. J. Aging Hum. Dev. **55:** 1–24.
13. CHESHIRE, H., K. COX, C. LESSOF & R. TAYLOR. 2006. Methodology. *In* Retirement, Health and Relationships of the Older Population in England: The 2004 English Longitudinal Study of Aging. J. Banks, E. Breeze, C. Lessof & J. Nazroo, Eds.: 367–383. The Institute of Fiscal Studies. London.
14. AMERICAN DIABETES ASSOCIATION. 2006. Standards of medical care in diabetes–2006. Diabetes Care **29:** S4–S42.
15. BARRETT, A.E. 2003. Socioeconomic status and age identity: the role of dimensions of health in the subjective construction of age. J. Gerontol. B Psychol. Sci. Soc. Sci. **58:** S101–S109.
16. FRY, P.S. & D. L. DEBATS. 2006. Sources of life strengths as predictors of late-life mortality and survivorship. Int. J. Aging Hum. Dev. **62:** 303–334.
17. MIROWSKY, J. 1997. Age, subjective life expectancy, and the sense of control: the horizon hypothesis. J. Gerontol. B Psychol. Sci. Soc. Sci. **52:** S125–S134.
18. MARTIKAINEN, P., M. BARTLEY & E. LAHELMA. 2002. Psychosocial determinants of health in social epidemiology. Int. J. Epidemiol. **31:** 1091–1093.
19. MARMOT, M. 2006. Introduction. *In* Social Determinants of Health. M. Marmot & R.G. Wilkinson, Eds.: 1–5. Oxford University Press. Oxford, UK.

Community Attitudes to the Regulation of Life Extension

MAIR UNDERWOOD,[a] HELEN BARTLETT,[a] AND WAYNE HALL[b]

[a]*Australasian Centre on Ageing, The University of Queensland, Brisbane, QLD, Australia*

[b]*School of Population Health, The University of Queensland, Brisbane, QLD, Australia*

ABSTRACT: Technologies purported to extend human life are already being marketed widely, and are being used by community members, despite a lack of evidence on their efficacy or safety: in fact, the use of some putative anti-aging technologies (e.g., human growth hormone) is illegal. Existing regulation is proving to be ineffective, especially in the face of Internet sales. Further advances in the field of life extension are a distinct possibility, exacerbating the need for a policy response. This paper presents the preliminary results of a study of community attitudes to life extension, with a focus on attitudes to the control and availability of strong life-extending technologies.

KEYWORDS: **longevity; public opinion; health policy; qualitative research**

BACKGROUND

Life extension has been divided by some into what has been termed *weak* and *strong* forms. Weak life extension refers to an increased average life expectancy, with the maximum human life span remaining unchanged.[1] Strong life extension refers to the more ambitious goal of dramatically increasing life expectancy and also increasing maximum life span.[1] This paper is primarily concerned with strong life extension.

While no strong life-extension intervention has yet been proven to have an impact on the fundamental aging process,[2] numerous claims have been made. One existing technology that purports to extend life is human growth hormone (hGH). It is marketed as an anti-aging technology despite the fact that there is little evidence to support this claim,[3,4] and there is even evidence of an elevated risk of cancer associated with its use.[5] The use of hGH as

Author for correspondence: Professor Helen Bartlett, Director, Australasian Centre on Ageing, The University of Queensland, Brisbane, QLD 4072, Australia. Voice: +61 7 3346 9080; fax: +61 7 3365 7503.

h.bartlett@uq.edu.au

Ann. N.Y. Acad. Sci. 1114: 288–299 (2007). © **2007 New York Academy of Sciences.**
doi: 10.1196/annals.1396.033

an anti-aging therapy is against the law in Australia (as it is in America and other countries). Dehydroepiandrosterone (DHEA) is also marketed as an anti-aging technology on Australian websites despite limited evidence regarding its efficacy,[6] evidence of the risks entailed in its use (such as an increased risk of cancer[7]), and the illegality of this use of DHEA.

Calorie restriction is the only intervention that has consistently been shown to slow the rate of aging in model organisms,[8] but the transferability of these results to humans has been questioned.[9] Indeed, there is limited evidence regarding its efficacy in a human context. Nonetheless, people across the world have voluntarily placed themselves on calorie-restricted diets in an effort to extend their lives. Numerous authors have expressed concerns about the potential health ramifications of unsupervised caloric restriction, especially among older people.[8,10,11]

The marketing of technologies through the internet has created a need for greater regulation of potentially life-extending technologies. "To date, there are no indications that market forces are weeding out risky, ineffective, economically harmful, and fraudulent anti-aging interventions—no sign, that is, of free market regulation" (p. 305).[12]

When Will Life Extension be Possible?

While there are currently no interventions that have been proven to have an impact on the maximum length of the human life span, biogerontologists are cautiously optimistic about the success of human life extension on the basis of research involving model organisms.[13] While there are no guarantees that extraordinary human longevity can be achieved, Louria argues that it is no longer a question of *whether* life expectancy will increase to 100 or 120 years, but *when*.[14] The optimism of the scientific community is demonstrated by the fact that 60 demographers, gerontologists and aging researchers made estimates of the life expectancy of a child born in the year 2100 that averaged 292 years (with estimates up to 5000 being made).[15]

Will There be a Demand for Life Extension?

Assumptions have been freely made about community attitudes towards life extension. Some have assumed a high,[16] or even universal,[2] demand for life extension technologies; others have suggested that life extension would be inherently undesirable.[17,18] None of these claims has any empirical basis because there has been very little research into community attitudes to life extension. Calnan, Montaner, and Horne asked 1187 British community members aged 18 or over about their attitudes towards a hypothetical scenario in which gene therapy was used to slow the aging process and extend average life expectancy to 100 years. They found that 37% of participants thought that such a technology should be banned, 5% thought it should be available at high cost, and only 30% thought it should be freely available.[19]

A British survey of 1003 people aged 16 or over, also examined people's attitudes to the possibility of living until 100.[20] This study examined personal interest in the prospect, and in particular what sacrifices participants would be willing to make in order to reach the age of 100. They found that the vast majority of participants would not sacrifice their independence (90%), health and vitality (93%), or the company of friends and family (94%). The most likely sacrifices identified were travel (42%), sex (40%) and being able to eat and drink whatever they wanted.

In one of the few in-depth analyses of community attitudes to life extension, Lucke, Ryan, and Hall conducted structured interviews with 31 Australian community members aged 50 years or over.[21] Respondents were ambivalent about research into life extension. Some were enthusiastic, but others were opposed to such research. These findings have been confirmed and expanded upon in the current study described in this paper.

While research on community attitudes to life extension is quite limited, the demand for existing products that purport to affect the aging process (such as hGH and DHEA) would imply a considerable demand for such technologies. Market researchers currently estimate the anti-aging and life enhancement market at US$36 billion, and predict that it will rise to US$49 billion by 2012, and more than US$100 billion by 2020.[22]

The Need for an Anticipatory Policy Response

If human life extension does become possible in the near future, as some argue it will, and there is some evidence of public demand for such technologies, then we must prepare for this eventuality. "The history of biomedical science shows how unexpectedly progress can catch the scientific community and society unawares by accomplishing the 'impossible'" (p. 29).[23] When modifying features that have been universal constants of human experience, like aging, the environment at risk is the social one.[23] Therefore, "it is not too early to consider contingent policy responses" (p. 180).[21]

The voice of the lay public has been curiously missing from debates over the social consequences of human life extension.[24] This paper represents one of the few efforts to seek lay opinions to contribute to the debate over appropriate policy responses to life extension.

METHODS

This paper builds on the work of Lucke, Ryan, and Hall[21] in aiming to answer the following questions:

1. How do members of the public understand the possibilities for increasing life expectancy?
2. How likely is the general public to take up life-extending interventions?
3. How do key stakeholders view the possibility of life-extension interventions and to what extent do their views agree or disagree?
4. What are the research and policy priorities for Australia regarding life extension?

This paper reports the preliminary findings from the first phase of a 3-year study approved by the University of Queensland's Research Ethics Committee, and funded by the Australia Research Council. This qualitative phase scoped the issues of importance to community members using focus groups and in-depth interviews. A quantitative phase is to follow involving a telephone survey that will also explore individual characteristics that influence attitudes.

The aim of the first phase is to explore community attitudes to life-extension technologies. Using a grounded theory approach, data collection was guided by a small set of open-ended questions that were posed to focus groups of community members. The results of the focus groups then informed the construction of an interview schedule used during semi-structured interviews with individual participants.

To ensure that the results were well grounded in the data, interviews were recorded and transcribed verbatim. The data were then entered into NVivo 2 (1999 QSR International Pty Ltd., Doncaster, Victoria, Australia), a qualitative data analysis program that aids the coding and retrieval process. The data were coded and examined for similarities which enabled the grouping of categories, followed by the identification of emerging themes.

SAMPLING

Convenience sampling was used to recruit community members via the print media, posters placed around the university campus, and the 50+ Registry of the Australasian Centre on Ageing at the University of Queensland. Furthermore, different groups (based on their occupations, affiliations or personal beliefs) of individuals were targeted on the basis of propositions about their diversity of opinions and behavior, in accordance with theoretical sampling.

The results presented here are based on six focus groups (averaging 11 individuals per group, total $n = 64$) and 17 individual interviews. Three focus groups with community members were conducted: two with older community members (50+ years) and one with younger community members (aged 18–31 years). An additional three focus groups were conducted with specific groups: a student evangelical Christian discussion group, a Baptist Bible discussion group, and a group of self-identified "transhumanists" (individuals who advocate the use of new medical technologies, such as life extension and reproductive cloning, to transcend human limitations). Focus groups lasted between 1 and 1.5 hours each.

Individual interviews, lasting 1.25 hours on average, were conducted with 10 women and seven men aged 50–81. Four of these participants were targeted because of their occupations (a funeral director, a pro-euthanasia activist, a pro-life activist, and an academic with expertise in the area of death and dying). All interviews were taped with the consent of the participants.

RESULTS

This paper presents key aspects of the preliminary results of the first, qualitative stage of the project with a focus on community attitudes towards the regulation and distribution of future life-extension technologies. The findings are organized around two main themes: control of life extension and availability of life extension. Participant details are included after each quote. Quotes from general community members are followed by the participant's age and gender, whereas quotes from targeted interviews are followed by a brief descriptor (e.g., "pro-life"). Quotes from focus groups are followed by a general description of the group (e.g., "Christian" or "younger community").

The Availability of Life Extension

Community members were most concerned about restrictions on the availability of life-extension technologies should they become available. Quite different opinions were expressed on this matter.

Universal Availability and Individual Choice

Many felt that life extension technologies should be available to all:

> *I guess everybody should be able to access it* [50, woman].

> *I think it would have to be available to everyone* [Christian].

> *I mean what right has one human being got above another to have treatment for something, or extension of their life, over someone else? Everybody's got their important valuable contribution to make* [52, woman].

These participants felt that life extension should be freely available and that the individual should make the final decision whether or not to take up the technology.

> *In an ideal world I think it would be based on people's choice, because there's going to be people who don't care* [61, woman].

> [Life extension] *should be an individual choice. . . . Extension of life should not come from the government, it should come from the individual who do that with the full understanding of what's going on, and that would be very fine* [pro-euthanasia].

Some of these participants emphasized the subjective nature of such a decision:

> *It should be the individual person* [that makes the decision]. *Because what to one man is agony, isn't to another* [81, man].

Unequal Access

Some participants saw no problem with unequal access to life extension. For example, one participant argued that equal distribution would not be viable:

> *We would spend so much money and time and resources getting people whose life expectancy is so poor up to a certain level, ours would probably end up going down because we've got to spend all our money and time and effort on that* [transhumanist].

Furthermore, some felt that it should not be left to the individual to decide:

> *I believe in social responsibility so it has to be nutted out* [thought through], *probably ethically by society, to come up with an idea, an idea that has to be run through the political gauntlet, and only then will something responsible be done. But to make it only a matter of individual choice seems to me to be irresponsible, and it smacks of only those people who have the money to pay for it* [academic].

These participants felt that individuals may not consider the broader implications of such as decision:

> *If you leave it up to the masses to make the decision, nine time out of ten it's the wrong decision anyway. It's what they want, right? Whether or not it's good for the community is irrelevant* [69, man].

Furthermore, individuals may not have the capacity to make a decision:

> *It might come down to soundness of mind and body when the decision to extend life is taken. If they're not in sound mind it's not really appropriate* [52, woman].

Factors Determining Availability

While many thought life extension should be available to all, they thought that this was unlikely to happen because access to life extension would be determined by cost and ability to pay:

> *Well I suppose pragmatically that's how it's going to work* [the availability of life extension being determined by affordability], *because that's how the health system works now actually and those that can afford it get it, they get the best treatment now* [61, woman].

Some thought that allowing cost to determine availability was not only sensible, but beneficial to society:

> *I think the smartest thing a government could do would be, if that sort of a bullet* [strong life extension] *was available, to put a hell of a price on it.*
>
> *Why is that?* [interviewer]
>
> *To keep it out of the hands of the masses. Otherwise, the government's going to have to look after them in their old age, aren't they? The government can't afford to do that* [69, man].
>
> *I think that health innovations are always going to have an inequality to them simply because, health costs. . . . we need people to "experiment on," we need to perfect the science to get it cheaper so that it is available for other people. And that's the only way that any health innovation has ever come into existence. The rich people get it first and once the manufacturing process or the service delivery gets cheaper that's when other people get to try it out. . . . And in some ways the rich are our lab rats, because that's how we learned that some breast implants weren't very good, you know?* [laughter] *Now poorer people can afford it and they've got a safer version* [younger community].

While some felt that access to life extension should, or at least would, be determined by its affordability, others mentioned different criteria. Some participants felt that individuals should deserve to extend their life:

> *. . . they would have to deserve to have access to it* [the life extension technology]. *In other words they should have done something to make sure that what they were offering to become part of using this technology, was a body that wasn't a total mess and a brain that wasn't a total mess* [69, man].

Indeed some felt that the opportunity should only go to those who would make a significant contribution to society:

> *Some of these famous professors who keep discovering stuff, or Einstein . . . really outstanding people . . . let them live to two or three hundred. . . . But for the average person who is literally just doing what they can do in their little sphere of people they know and family, there is really not much point* [68, woman].

Problems Arising from Limited Availability

While some participants thought there should be some sort of discrimination when it came to accessing life extension technologies, they acknowledged that this would be problematic.

Participant 1: *Maybe a GP should have to decide whether or not a person's eligible.*

Participant 2: *But then you're discriminating.*

Participant 3: *You are discriminating.*

Participant 4: *Does it come down to if you've got qualifications, if you're smart, you can contribute to society?*

Participant 5: *Your genes are good enough?* [Christians].

I don't know, it's hard. I just don't think you could have it as one blanket response like "everyone's gonna have their life extended." . . . I think it should definitely be discriminatory, but then that's also hard to apply [52, woman].

Participants recognized that the unequal distribution of life-extension technologies would create or exacerbate existing social inequities:

Participant 1: . . . *we're already having enough problems trying to just keep people healthy, particularly people in rural areas and y'know, indigenous communities are not even being able to provide enough quality of health services, and that. . .*

Participant 2: *It seems like the first world gets the benefit of most of our medicine and the wealthy get richer and live longer, but do we share that and make life-extension open to everybody and increase everybody, so people in poverty live in poverty longer, people who are rich get richer for longer and there's a greater gap?* [Christians].

It [strong life extension] *will create divides, and that's another of the things that I completely hate. . . . I think that everyone would resent this, this strong life extension* [pro-euthanasia].

The Control of Life Extension

One of the greatest issues of concern to community members was with the control of future life-extension technologies.

Participant1: *There'd be issues of control no doubt about it.*

Participant 2: *There'd be a black market* [Christians].

You know, who's going to set the guidelines? You know, is it going to be someone like me, somebody like you, who is it going to be? Or people like Philip Nitschke [an Australian euthanasia advocate] *who doesn't have any issue with troubled teenagers accessing euthanasia?* [pro-life].

They recognized a need for control, but acknowledged that this would be difficult to achieve:

. . . *look at the issues they had with RU486* [the morning-after birth control pill], *trying to figure out how to control it. That pill, just the division it caused*

> *within society. . .* [Christian].

Participants recognized the need for a regulatory body if life-extension technologies become available, but many participants did not think that the government should fulfill this role.

> *Who should control these technologies? [interviewer]*

> *. . . I don't think governments should, even though you know, there's an argument to say that we democratically elect governments but you know if the government that is in isn't your choice, you don't feel like you have democratically elected them personally. . . . So I don't know, do we trust governments with that kind of decision? It's a huge decision* [61, woman].

> [I wouldn't want] *that the government should be saying yes for people to live long or not live long. Rather like going to a vet* [81, man].

> Some felt a governmental regulatory body may result in individuals feeling pressured:

> *You don't want Big Brother breathing down your neck, especially if it was a case of, you know, corralling you* [55, woman].

Members of the government were not the only people that participants felt should be excluded from the regulation of life-extension technologies.

> *. . . it* [the regulatory body] *certainly can't be religious leaders, they've got too much bias* [61, woman].

> *I think a regulatory body* [would be needed] *but I wouldn't like to think there were any politicians on it, and I wouldn't like to think there were any Catholics on it. Because really that wouldn't be right. And maybe there shouldn't be any religious people on it either* [76, man].

Some suggested that control should be given to:

> *"the people who develop it"* [78, woman] or the *"experts in the field"* [69, man].

> *Scientists and medical people on the regulatory body. But certainly not appointed by any particular* [political] *party. . . .Scientists, but nothing to do with religion and nothing to do with politics* [76, man].

So, for some community members, a politically and religiously neutral panel of experts was the ideal choice for members of a regulatory body.

DISCUSSION

This preliminary study revealed that community members identified a need to regulate life-extension technologies if they become available. Curiously, community members made little reference to the safety and efficacy of life extension technologies (topics that are of great concern to academics[3]), but this may reflect a lack of technical knowledge, or possibly an optimistic belief that their safety and efficacy can be readily assessed much like pharmaceutical drugs are today.

Community members were most concerned with the control and availability of life-extension technologies. They felt that regulation is necessary to control availability and prevent a black market.

In general, the community members raised similar issues to those debated in the academic literature. For example, like the community members, many authors suggest that potential life-extension technologies should be available to all (or even that life extension is a human right[25]), but several also acknowledge that, at least in the beginning, the availability will be determined by affordability.[26] Participants acknowledged that the current system for the allocation of health resources is frequently unequal, and believed that this would also apply if life-extension technologies became available. Furthermore, they acknowledged the problems that result from the current system, and saw these problems as being exacerbated by the advent of life extension.

Some authors regard the unequal distribution of life-extension technologies as a reason not to pursue the goal of human life extension.[27] Others disagree, with some arguing that ideas of "leveling everyone" by impeding research or access are impractical.[28]

Community members had little faith in government regulation. They felt that the government could not, or should not, regulate life extension. They also felt that people with a religious background should be excluded because of their potential for bias. Some community members had more faith in researchers/scientists regulating life-extension technologies. Perhaps these participants attributed a certain "objectivity" to scientists that politicians and religious leaders were seen as lacking.

This preference for a particular regulatory body has been noted in the literature (it seems often without empirical support). For example, Francis Fukuyama states that "the public holds scientists in much higher regard than politicians, not to mention theologians or philosophers" (p. 6).[17]

Like community members, many academics see a need for the regulation of life-extension technologies. They point to existing problems with the regulation of "anti-aging" technologies, and in particular the nutritional supplement industry in the United States.[12,29,30] When it comes to the regulation of life extension, most academics are primarily concerned with safety and efficacy. While the distribution and availability of life extension are discussed in the academic literature, these discussions rarely mention formal regulation. This paper adds a community perspective to the debate about the regulation of life-extension technologies.

It is not too early to consider the regulation of life-extending technologies. Technologies that purport to slow the human aging process are already being marketed and used (placing considerable health and financial risks on community members), and there is a distinct possibility that new technologies will be made available in the future. Given the level of community concern, the policy response should focus on regulating the availability of life-extension technologies.

REFERENCES

1. Moody, H.R. 2001/2002. Who's afraid of life extension? Generations **25:** 33–37.
2. Olshansky, S.J., L. Hayflick & B.A. Carnes. 2002. No truth to the fountain of youth. Sci. Am. **286:** 92–95.
3. Liu, H. *et al.* 2007. The safety and efficacy of growth hormone in the healthy elderly. Ann. Intern. Med. **146:** 104–115.
4. Perls, T.H., N.R. Reisman & S.J. Olshankšky. 2005. Provision or distribution of growth hormone for "antiaging": clinical and legal issues. JAMA **294:** 2086–2090.
5. Binstock, R.H. 2004. Anti-aging medicine and research: a realm of conflict and profound societal implications. J. Gerontol. **59A:** 523.
6. Horani, M.H. & J.E. Morley. 2004. Hormonal fountains of youth. Clin. Geriatr. Med. **20:** 275–292.
7. Boia, L. 2004. Forever Young: a Cultural History of Longevity. Reaktion Books. London.
8. Dirks, A.J. & C. Leeuwenburgh. 2006. Caloric restriction in humans: potential pitfalls and health concerns. Mech. Aging Devel. **127:** 1–7.
9. Demetrius, L. 2005. Of mice and men. EMBO Rep. **6:** s39–s44.
10. Le Bourg, E. 2005. Calorie restriction to retard aging and increase longevity. Presse Med. **34:** 121–127.
11. Schlienger, J.L., A. Pradignac & F. Grunenberger. 1995. Nutrition of the elderly: a challenge between facts and needs. Horm Res. **43:** 46–51.
12. Mehlman, M.J. *et al.* 2004. Anti-aging medicine: can consumers be better protected? Gerontologist **44:** 304–310.
13. Hunter, P. 2007. Is eternal youth scientifically plausible? EMBO Rep. **8:** 18–20.
14. Louria, D.B. 2005. Extraordinary longevity: individual and societal issues. J. Am. Geriatrics Soc. **53:** s317–s319.
15. Richel, T. 2003. Will human life expectancy quadruple in the next hundred years? Sixty gerontologists say public debate on life extension is necessary. J. Anti Aging Med. **6:** 309–314.
16. Harris, J. 2004. Immortal ethics. Ann. N. Y. Acad. Sci. **1019:** 527–534.
17. Fukuyama, F. 2002. Our Posthuman Future: consequences of the Biotechnology Revolution. Farrar, Straus and Giroux. New York.
18. Kass, L.R. 2002. Life, Liberty and the Defense of Dignity: the Challenge for Bioethics. Encounter Books. San Francisco.
19. Calnan, M., D. Montaner & R. Horne. 2005. How acceptable are innovative health-care technologies? A survey of public beliefs and attitudes in England and Wales. Soc. Sci. Med. **60:** 1937–1948.
20. Ipsos MORI. 2006. Three score years and ... when? www.ipsos-mori.com/polls/2006/bupa.shtml Accessed 7 February 2007.

21. LUCKE, J.C., B. RYAN & W. HALL. 2006. What does the community think about lifespan extension technologies? The need for an empirical base for ethical and policy debates. Australas. J. Aging **25:** 180–184.
22. SPECTRA INTELLIGENCE. 2006. Antiaging and Life Enhancement Pharmacotherapeutics. http://www.piribo.com/publications/therapeutic/antiaging_life_enhancement_pharmacotherapeutics.html Accessed 19th January 2007.
23. JUENGST, E.T. *et al.* 2003. Biogerontology, "anti-aging medicine," and the challenges of human enhancement. Hastings Cent. Rep. **33:** 21–30.
24. LUCKE, J.C. & W. HALL. 2005. Who wants to live forever? EMBO Rep. **6:** 98–102.
25. DE GREY, A. 2005. Life extension, human rights, and the rational refinement of repugnance. J. Med. Ethics **31:** 659–663.
26. HAYFLICK, L. 2000. The future of aging. Nature **408:** 267–269.
27. GLANNON, W. 2002. Reply to Harris. Bioethics **16:** 292–297.
28. SOLOMON, L.D. 2006. The Quest for Human Longevity: science, Business, and Public Policy. Transaction Publishers. New Brunswick, NJ.
29. REISMAN, N.R. 2004. Legal issues associated with the current and future practice of anti-aging medicine. J. Gerontol. A Biol. Sci. Med. Sci. **59:** 674–681.
30. PERLS, T.T. 2004. Anti-aging quackery: human growth hormone and tricks of the trade: more dangerous than ever. J. Gerontol. A Biol. Sci. Med. Sci. **59:** 682–691.

The Elderly in a Period of Transition

Health, Personality, and Social Aspects of Adaptation

IGNAT C. PETROV

Medical University, Sofia, Bulgaria

ABSTRACT: The dramatic changes in overall mode of life experienced since 1989 by the people in Central and Eastern Europe (CEE) provides a good opportunity to study the aging processes and the status of the elderly. In the 1980s, CEE countries were already at an advanced stage of demographic aging of their populations. The first years of the period of transition from totalitarianism to democracy and market economies were marked with an abrupt but varying degree of economic and social deterioration in CEE. This was followed by partial improvement. Earlier, in the late 1960s and 1970s, mortality rates had increased along with a stagnation and decrease of life expectancy in CEE. This contrasted with the steady improvement in the European Union (EU) during the 1970s and 1980s. By 1989, there already was a striking gap in life expectancy between eastern and western Europe, revealing the existence of an East–West health divide. The most important health problems contributing to greater mortality in CEE proved to be cardiovascular and alcohol-related diseases as well as violence and injury. Suicide rates also were found higher among the (elderly) population in the countries of CEE and Russia compared with the West. The gap was most expressed among the men. The reasons for deterioration in health are complex. There is evidence that negative economic and social changes can lead to psychosocial stress and to unhealthy behaviors. The morale of the elderly people in transition was affected and many-sided. The Sofia study described here has revealed positive as well as negative personal attitudes toward the changes in life. Feelings of liberation, hope, and new perspectives were reported along with sentiments of disappointment, pessimism, and fear. In sum, transition processes have run unevenly in different countries and have had differing impacts on individuals. Nevertheless, general trends toward improving the economic, psychosocial, and health aspects of the elderly are recently prevailing.

KEYWORDS: aging; elderly; transition; Eastern Europe

Address for correspondence: Ignat C. Petrov, Boulevard Vitosha 36, Sofia 1000, Bulgaria. Voice: (+359 2)9876877; fax: (+359 2)9874145.
ignatpetrov@yahoo.com

Ann. N.Y. Acad. Sci. 1114: 300–309 (2007).
doi: 10.1196/annals.1396.041

INTRODUCTION: HETEROGENEITY IN AGING

There is a considerable diversity in aging among individuals and social groups, a diversity between countries and world regions, physical environments, ecological, economic, and social conditions, different historical and political periods, and various cultures and traditions.

AGING: A PSYCHOSOCIAL VIEW

Aging is a biological as well as psychic and social phenomenon. Aging is a process of individual development and life-long adaptation. An essential distinguishing feature of the human from the other living beings is that psychic and social development may not coincide with the biological trajectory of aging. In individual cases successful aging could be a time of continued growth, development, and achievement. Some talented personalities have created their best works at the very end of their lives notwithstanding the obvious signs of biological decline. With aging some people develop an expertise called *wisdom*.[1,2] And, for the majority of aging people, the late-life period could also be successful and meaningful in the terms of simply "being in life."

PRECONDITIONS FOR SUCCESSFUL/ACTIVE AGING

In most cases the preconditions for having a successful and active life include relatively good health, economic security, and family and social coherence. The notion of *quality of life* comprises these basic prerequisites for successful aging and contentment. Another useful concept is that of *active aging* as an aim to be followed.

Studies on the Bulgarian rural elderly population revealed that value orientations at the age 70 and over were focused on few vital themes: the person's own health, the family relations in a narrow circle (children and spouse/husband), and economic security.[3,4]

Generally, elderly persons, and especially the very old, are to some extent, more vulnerable to physical and psychosocial stress,[5,6] to which the accumulation and gradual progression of multiple chronic pathology contributes. In some cases, this impairs mobility, an independent mode of life, and/or vision or hearing. Varying degrees of cognitive and psychomotor changes may also appear. All this could evoke feelings of insecurity, anxiety, and/or inferiority in some of the elderly. Bereavement and widowhood are more common in aging, and a dependency on the immediate surroundings may be increased. In the very old, there could be a sense of dissatisfaction at insufficient care and consideration and there is increased worry about maintaining the status achieved during the course of life.[3,4] Intergenerational problems—concretely in the relationships with their children and their families—may arise. The social integration of the elderly is of crucial importance.[7] Urbanization and modernization and the sharp social–economic fluctuations may contribute to a sense of *devaluation* of the elderly regarding his/her role and social status.

The notion of transition has different implications. With respect to the demographic transition we now see increasing proportions of the world population live to experience old age, and the great increase in the world population comes from the rapidly developing countries. Another part of the contemporary world, the countries of Central and Eastern Europe (CEE), have experienced for the last 17 years dramatic political, social, and economic changes within the transition from totalitarianism to democracy and a market economy.

AGING IN TRANSITION IN CEE COUNTRIES

During the 1990s social and economic circumstances deteriorated to varying extents in all CEE countries. Economic crises were followed by more or less marked economic growth. But even in countries that made the most successful transition, economic growth was accompanied by increased income inequalities and poverty levels.[8] The decreased GDP as well as the increased income inequality in CEE transitional countries correlated statistically with changes in mortality and life expectancy.[9] Of course, transition affected mostly the incomes of the elderly people.

I well remember two big economic crises in Bulgaria: in 1990 and again in 1996–1997. Inflation and a real decrease in the incomes of the elderly followed both crises. Real wages, and especially pensions, declined considerably.

LIFE EXPECTANCY IN THE COUNTRIES IN TRANSITION

During the 1950s and 1960s mortality decreased and life expectancy improved throughout much of Europe.

In the late 1960s and 1970s this began to change. The steady improvement in the EU contrasted sharply with the stagnation and decline in Eastern Europe. By 1989, there already was a striking gap in life expectancy between East and West.[10] In western Europe, life expectancy continued to increase steadily from 1980 to 2000. In most CEE countries there was a stagnation, followed by a slow improvement (in Poland, the Czech Republic, Hungary, and Slovakia). Bulgaria and Romania began to show improvements only in the late 1990s.

The evolution was more uneven in the successor countries of the Soviet Union. There was a series of unexpected fluctuations, with declines and slight improvements, the changes in the 1990s broadly mirroring the economic crises experienced.[11,12]

By 1994 in Russia the male life expectancy at birth was 57.2 years and by 2001 it was 58.9 years.[11]

By 2002 in Bulgaria the life expectancy at birth was 68.5 years among men and 75.4 years among women. It increased slightly up to 2004. Bulgarian men live 10 years less, and Bulgarian women 7 years less than their Swedish counterparts.

In 1970 male life expectancy at age 15 was 56 in the western European countries; 55 in the countries of CEE (excluding the Soviet Union); and 52 in the Soviet Union. In 1997 male life expectancy at 15 years was 60 in the western European countries, 54 in the countries of CEE, and 48 in Russia. At 1970 life expectancy at age 15 years was similar in the countries of eastern and western Europe. From 1970, life expectancy at age 15 improved continuously in the EU countries, but not in eastern Europe. By 1990 there was a 4-year gap, which, by 1997, had widened in men to 6 years.[10]

Only 41% of Russian males today can expect to survive to the age of 65 versus 81% for the western European males. The gap in the remaining life expectancy at age 65 increased by five times: from 1 to 5 years. Andreev, McKee, and Shkolnikov wrote in 2003 that the probability of remaining alive and healthy declines faster in the Russian Federation, with the gap between eastern Europe and Russia widening at older ages, with premature death in men, but survival in a poor state of health in women.[12]

Looking at sex differences it is apparent that the men in transition countries have been especially vulnerable. In all industrialized countries men have a lower life expectancy than women do. However, this gap is especially large in the transition countries of CEE and the former Soviet Union FSU. The ratio of men to women is also correlated statistically with GDP per head of population.[9]

MORTALITY AT OLDER AGES

Most of the East European and former Soviet Union countries have experienced rising adult mortality in the period since 1970. This increase accelerated after 1990 and was largely confined to men of working age. The "mortality crisis in the early 1990s" was short-lived in East Germany, Poland, the Czech Republic, and was delayed in Hungary. In Bulgaria and Romania there was a continuing steady deterioration in male's mortality and no improvement in women's until the late 1990s. Death rates at older ages among both men and women in the FSU countries are now about twice those in western Europe and about 50% higher as in CEE. In Bulgaria and Romania death rates over 65 years remained rather high during the 1980s and into much of the 1990s compared to the EU average, with improvement seen in the late 1990s only.[13]

MOST IMPORTANT HEALTH PROBLEMS IN TRANSITION

The gap in life expectancy and mortality rates revealed the existence of a health divide between western countries and the CEE countries. There is diversity in this, but the most important health problems in the CEE countries seem to be cardiovascular and alcohol-related diseases as well as violence and injuries, especially those due to traffic.[14]

Recently the deaths due to cardiovascular disease began to decrease in some of the transition countries (e.g., Poland) but remained high in other countries (e.g., Bulgaria).

The high levels of cardiovascular diseases and of mortality due to cardiovascular diseases in CEE reflect a high level of many risk factors, such as fat and alcohol consumption, smoking, very low levels of fruit and vegetables in the diet, and others. The poverty and the differences in access and quality of health care could also contribute to the greater mortality from cardiovascular disease.[9,10,15]

All CEE countries experienced a short-lived, but important increase in deaths from injuries, especially those due to traffic. This was especially large in the Baltic states where, although death rates have returned toward their previous levels, they are still much higher than in many old EU states. While all causes of injury are more common, the gap is particularly great for homicide and suicide.[16]

SUICIDE AMONG THE ELDERLY IN A PERIOD OF TRANSITION

Several studies, including the study of Pearson and Conwell[17] on populations in 37 countries, showed that the highest rate of suicide in the populations was among the elderly, and especially among men aged 75 and over. In his dissertation "Towards a socio-economic and demographic theory of elderly suicide," Kiemo[7] thoroughly examined data (all from the period around 1995) from 49 countries at various stages of development. One of the findings was that suicide rates were higher among the populations in the countries of CEE and especially those of the FSU compared with their counterparts from the most developed countries. The gap was most expressed among the men. The author assumed that (elderly) men are more vulnerable and that social conditions following on the transition to marked economies have affected the role and status of the elderly adversely. Transition may cause an imbalance in social integration and especially in age integration, which could aggravate the elders' rate of suicide.

DETERMINANTS OF POOR HEALTH IN THE CEE

Bobak *et al.*[10] estimated that of the 6-year gap in the life expectancy at birth between CEE countries and the rest of Europe in the early 1990s, only 15% developed in infancy; 43% originated in the 35- to 64-year-old age group and 23% in those 65 and older. Cardiovascular diseases accounted for 54%, followed by external causes (23%) and respiratory diseases (16%). The contribution of the external causes (injuries and violence) was higher in the countries of the FSU.

The lack of efforts to control common risk factors for chronic disease (blood pressure, smoking, obesity, and cholesterol) probably made a substantial contribution to the gap, mostly through the mortality rise since 1960s. The prevalence of obesity is high in most countries of the region. Nutrition is often poor, with high intakes of saturated fats and low intakes of fresh fruits and vegetables, and leisure-time physical activity is typically low.

The causes of dramatic fluctuations in mortality in the 1990s are not fully understood, but there is a great consensus that changes in health are related to deterioration in social and economic conditions in all CEE countries, such as economic instability, migration, increased unemployment, and income and social inequalities.[15,18] The negative economic changes had can lead to unhealthy behaviors. The frequency of addictions to alcohol or cigarette smoking is much higher in the European countries in transition than in the rest of Europe.[19] High consumption of alcohol and binge drinking also contributed to the changes in mortality, particularly through alcohol poisoning, injuries and violent deaths, and possibly cardiovascular disease.[20]

In some countries in transition, the public health systems were partly decentralized after 1989. At the same time, national economic crises reduced public sector funding. Some of the problems in the 1990s resulted from the disintegration of public health services and a lack of funds. The partly decentralized governments and new legislation often brought about a reduction in control, blurring of responsibilities, and loss of discipline.[15] Collaboration between different sectors and institutions (e.g., in Bulgaria) is inadequate.

PSYCHOSOCIAL PROBLEMS OF THE ELDERLY DURING TRANSITION

It was hypothesized that totalitarian regimes could lead to a "learned helplessness." In those regimes psychosocial resources, such as perceived mastery, optimism, and sense of control, were, on average, low. The worsening in 1990s can also be attributed to psychosocial stress. It can result from acute, transition-related dislocation in the labor market, income inequalities, changes in family life, social disorganization and underdevelopment of the civil society, and lack of adequate public policy response to the needs of the people.[21]

The changing demographic situation left its mark on the status and roles of the elderly. Many of the countries in CEE showed at the end of the 20th century signs of expressed aging of the population. According to the United Nations' data, Bulgaria was already in 2002 in the eighth rank in the world list regarding the aging of its population. The low birth rate and the migration of young people during the 1990s contributed to this. By about 1995 Bulgaria had already the most top-heavy age structure.

It could be supposed that in a situation of economic crisis this age structure might increase some negative societal attitudes toward the elderly, who might be regarded as an economic burden. I have observed many individual cases of the negative attitudes of ageism, and I think that the interest toward the problems of the elderly people has decreased considerably in Bulgaria during the 1990s among institutions, young people, and the whole society. The interest of the ministries and universities in this country to the issues and study of gerontology is low.

The economic and social changes of transition have affected undoubtedly the intergenerational relations and relations in the family. The income, and mostly the income of the elderly, dropped significantly with the inflation. In Bulgaria it was usual during socialism that the elderly parents helped their adult children with their savings. But after the economic crises of the 1990s, it became very difficult for the majority of the elderly people to even pay for their everyday expenses. Now many of them must rely on the financial help of their children.

On the other hand, the relatively preserved family solidarity in countries like Bulgaria is a powerful positive factor in the difficult situation of transition.

Health systems are also in transition. A common goal is the implementation of modern primary care. A few countries have succeeded, but many, including Bulgaria, have not.

The rise of social inequality, the inadequate healthcare system, and the rise of criminality may contribute to the increased feeling of insecurity especially among the elderly people.

What are the feelings and attitudes of the elderly themselves toward the changes in the mode of life after 1989?

Muller[22] studied the social attitudes in East Germany after the unification of the country. He has found that 45% of the East Germans were more satisfied with life in 2000 in comparison with 1989/1990; 30% were more dissatisfied; and 12% found no change.

In the period of 1992–1993 we studied a group in Sofia of autonomous elderly people with active lives.[23–25] The study was directed on their feelings and attitudes in a general way toward the life changes during the transition. The interviews with 61 elderly women and men (mean age, 74 years) and a Self-Evaluation Test used revealed a considerable variability of views, confirming that old age is psychologically heterogeneous. Notwithstanding the richness of diverse views, 70% of the elderly were polarized at their extreme: 35% showing a fully positive and optimistic attitude toward the changes (+), and 35% definitely negative ones (−). The smallest groups had more mixed attitudes, classified as +−; ++−; or +−−. The positive feelings (of relief, joy, hope, etc.) were connected first of all with ideas of liberation and freedom. The negative feelings (of disappointment, pessimism, incertitude, etc.) implied worries about economic difficulties, delinquency, moral and cultural degradation, as well as fears of the restoration of communism, concerns about national unity and the future, and fear of war.[23,25]

The attitudes of the elderly toward the current transition process were influenced by biographical factors (family and friends circle, school, profession, historical events, values and ideas acquired in childhood and youth, life-long experiences) as well by the actual situation and personality structures. Among the 34 personality, health, and social variables calculated in this study, nine correlated significantly at $P < 0.05$ with positive versus negative views on the changes. Favorable factors were living with someone; husband/spouse living; higher level of education; stable personality; higher self-esteem on the scales of happiness, health and intelligence of the Self-Evaluation Test; less attention paid subjectively to the everyday economic difficulties; and also good family relations and with the children. Their children and/or grandchildren influenced the attitudes of the elderly toward the changes, and the elderly did influence them.[24,25]

Comparing the interviews of 1992 with those of 1993, we observed some deepening of negative feelings from 1992 to 1993.[24] Nevertheless, the mentality of these active elderly was rich, holding much valuable in the sphere of family and of social relations.

In 2005 we interviewed again 32 elderly from the same group. The negative views prevailed slightly. But generally, the attitudes were some more balanced, and the extreme + and − were less common. Among all positive feelings and attitudes, the ideas of liberation and freedom continued to be prevalent. More frequently expressed was the satisfaction with the coming entering of the country in the EU. Among the negative attitudes, some fears have been not yet reported (the fears about a restitution of communism, about national unity, and possible war). New subjects of disappointment have appeared (about inadequacy of the reforms in the health and social care systems, about the increased poverty levels and increased inequalities in incomes, about corruption, etc.).

CONCLUSION

The social and economic changes of transition in the CEE countries during the last 17 years have brought along a hope, a sense of liberation, and new perspectives, but also the recognition of a heavy burden for many years on the population of these countries, including the elderly. The process of transition has run unevenly in different countries, but the general trends toward improving the demographic, economic, health, and psychosocial aspects of these countries are recently on the rise. The active participation of the elderly people in the life will be essential.

REFERENCES

1. BALTES, P.B. 1993. The aging mind: potentials and limits. Gerontologist **33:** 580–594.
2. BALTES, P.B. & U.M. STAUDINGER. 2000. Wisdom: a metaheuristic (pragmatic) to orchestrate mind and virtue towards excellence. Am. Psychol. **55:** 122–136.
3. PETROV I.C. 1976. Self-Evaluation of Some Personality Aspects of the Aged by the Dembo-Rubinstein Test. Experimental-Psychological and Clinical-Psychopathologic Investigation [in Bulgarian; summary in English]. Ph.D. dissertation, Meditsinska Akademija, Sofia, Bulgaria.
4. PETROV, I.C. 1975. Medical-psychological problems in old people according to their self-evaluation of happiness [in Bulgarian; summary in English]. Nevrol. psihiat. i nevrohirurg. (Sofia) **14:** 306–313.
5. PETROV, I.C. 1979. Mental Aging [in Bulgarian]. Meditsinska Academia – Center za Nauchna Informatsia po Meditsina i Zdraveopazvane. Sofia, Bulgaria.
6. PETROV, I.C. 1981. General psychopathology of aging [in Bulgarian; summary in English]. *In* Psychology and Psychopathology of Aging. K. Konstantinov, I.C. Petrov & C. Christozov, Eds.: 48–73. Meditsina i Fizkultura. Sofia, Bulgaria.
7. KIEMO K. 2003. Towards A Socio-Economic and Demographic Theory of Elderly Suicide. A Comparison of 49 Countries At Various Stages of Development. PhD Dissertation, Uppsala. Repro Ekonomikum.
8. HUMAN DEVELOPMENT REPORT. 2003. United Nations Development Programme. Oxford University Press. New York/Oxford.
9. MARMOT, M. & M. BOBAK. 2000. International comparators and poverty and health in Europe. Br. Med. J. **321:** 1124–1128.
10. BOBAK, M. & M. MARMOT. 1996. East-west mortality divide and its potential explanations: proposed research agenda. Br. Med. J. **312:** 421–425.
11. WORLD HEALTH ORGANISATION. 2002. The World Health Report 2002. World Health Organisation. Geneva.
12. ANDREEV, E.M., M. MCKEE & V.M. SHKOLNIKOV. 2003. Health expectancy in the Russian Federation: a new perspective on the health divide in Europe. Bull. WHO **81:** 11, 778–788.
13. NOLTE, E., M. MCKEE & A. GILMORE. 2004. Morbidity and mortality in transition countries in the European context. Background paper for the session "Morbidity, Mortality and Reproductive Health: Facing Challenges in Transition Countries" at the European Population Forum 2004. http://www.unece.org/pau/epf/nolte.pdf
14. BOBAK, M., D. BLANE & N. MARMOT. 1998. Social Determinants of Health: their Relevance in the European Context. Review of Health Promotion and Education Online: Verona Initiative. http://www.rhpeo.org/ijhp-articles/e-proceedings/verona/2/index.htm.

15. Bobak, M., M. McCarthy, F. Perlman & M. Marmot. 2004. Modernizing public health. *In* Health Systems in Transition. J. Figueras, M. MiKee, J. Cain & S. Lesof, Eds.: European Observatory on Health Systems and Policies. http:/www.eurasiahealthtransitionconference.org/Health_systems_Modernizing_Health.pdf.
16. McKee, M., L. McLehose & E. Nolte. 2004. Health Policy and European Union Enlargement. Open University Press–McGraw-Hill. Maidenhead, Berkshire, UK.
17. Pearson, L. & Y. Conwell. 1996. Suicide and Aging: international Perspectives. Springer Publishing. New York.
18. Bobak, M. & J. Powles. 2001. Poverty and Non-communicable Diseases in Central and Eastern Europe and the Former Soviet Union. A Report for the Non-Communicable Diseases and Mental Health Cluster, WHO, Geneva. University College London. London.
19. Cockerham, W.C. 1997. The social determinants of the decline of life expectancy in Russia and Eastern Europe: a lifestyle explanation. J. Health Soc. Behav. **38:** 117–130.
20. McKee, M. 1999. Alcohol in Russia. Alcohol and Alcoholism **34:** 824–829.
21. Cornia, G.A. & R. Paniccia. 2000. The transition mortality crisis: evidence, interpretation and policy responses. *In* The Mortality Crisis in Transitional Economies. G.A. Cornia & R. Panicia, Eds. Oxford University Press. Oxford.
22. Muller, O. 2004. Value Systems of the Citizens and Socio-Economic Conditions—Challenges from Democratisation for the EU Enlargement. Policy Paper Series "Democratic Values." Policy Paper 10. East Germany: Democratisation par excellence? http://www.kulsoz.euv-frankfurt-o.de/EU%20Project/papers/Pp10.pdf.
23. Petrov, I.C. 1996. Feelings and attitudes toward the changes during social and economic transition: a study of Sofia autonomous elderly subjects. Rom. J. Gerontol. Geriatr. **17:** 73–82.
24. Petrov, I.C. 1997. Feelings and attitudes toward the changes during social and economic transition. Part II. Evolution of the feelings and attitudes: comparison with health, personality and social variables. Rom. J. Gerontol. Geriatr. **18:** 31–48.
25. Petrov, I.C. 1997. The elderly in a period of transition in Sofia. Int. J.Geriatr. Psychiatry (London) **12:** 773.

The Human Mosaic

Cultural Beliefs and Health Professions Training

ROSELLEN M. ROSICH

Alaska Geriatric Education Center, University of Alaska Anchorage, Anchorage, Alaska, USA

ABSTRACT: Developing culturally competent healthcare service must incorporate the needs of ethnic elders as well as the types of services provided. The question arises as to how to define optimal ways of teaching cultural diversity to the future healthcare professionals who will be treating older adults of indigenous heritage. Basing its work on an ethnogeriatric approach developed by the Stanford Geriatric Education Center, the Alaska Geriatric Education Center adapted the tool and utilized it with indigenous populations within Alaska. The reformatted approach examined and charted historical events (by cohort analysis) that have influenced Alaska's Native Elder's healthcare belief system over generations. Historical profiles produced from this approach may be used in studying and treating seniors from diverse ethnic backgrounds and provide an educational tool in training gerontology or geriatric healthcare professionals. The approach may also be used for organizing individuals' perceptions of the historical events in their lives within clinical health and mental health settings.

KEYWORDS: **ethnogeriatrics; health beliefs; cohort analysis; historical trauma; training**

Currently very little information is known about aging in Arctic populations and specifically aging among the indigenous people of Alaska (i.e., Inupiat, Yup'ik, Aleuts, Alutiiq, Athabascan, Southeastern Tlingit, Tsimshian, and Haida). Less is known about historical events and their impact on Alaska Natives in older cohorts (55+ years) with respect to health and mental healthcare beliefs (i.e., ethnogeriatrics). The paucity of information available on aging and health among Alaska Native Elders has an impact on training and developing a diverse, culturally competent, and sensitive healthcare professional workforce. In addition, historical events and their impact on the various indigenous people

Address for correspondence: Rosellen M. Rosich, Ph.D., Alaska Geriatric Education Center, University of Alaska Anchorage, 2702 Gambell St., Suite 103, Anchorage, AK 99503. Voice: 907-786-1739; fax: 907-786-4898.

afrmr@uaa.alaska.edu

Ann. N.Y. Acad. Sci. 1114: 310–316 (2007).
doi: 10.1196/annals.1396.029

of Alaska diverge quite extensively according to the region of the state in which the group resides.

Alaska's unique geography and population characteristics, combined with transportation and economic factors, present enormous challenges for delivery of healthcare services for all its elderly inhabitants. It is perhaps not surprising that under these conditions sizeable portions of the state of Alaska (52% of both urban and rural census areas) have been designated as Health Professional Shortage Areas and Medically Underserved Areas, in of which many Alaska Native Elders reside.

As the nation's demographic landscape has shifted, significant health disparities in access to healthcare utilization, quality, and health outcomes continue to be documented for a variety of ethnic groups. However, a number of federal agencies (e.g., The U.S. Department of Health and Human Services, Health Resources and Services Administration) have accumulated evidence stressing that increasing the quality of the health care provided, as well as eliminating barriers to care and health disparities, may be accomplished through an indirect micro-level route that encompasses training and educating a culturally competent and sensitive healthcare professional workforce. The Health Resources and Services Administration, Bureau of Health Professions, funds Geriatric Education Centers (GECs). These GECs are dedicated to improving geriatric education and training, supply, distribution, and quality of healthcare professionals by means of strong community–academic partnerships within the geographic region of the United States in which they are located. A strong focus is placed upon eliminating barriers to care and health disparities, to assure quality of care, and to improve public health and healthcare systems. Educating a culturally sensitive and competent geriatric healthcare work force is a cornerstone of these centers' training programs.[1]

CULTURALLY COMPETENT SERVICE DELIVERY AND ALASKA NATIVES

Developing culturally competent service delivery incorporates an understanding of the needs of ethnic elders and the types of service provided. Culture and historical context (e.g., influences of historical events), such as immigration patterns and discrimination, is reflected in the values, health beliefs, illness behaviors, self-image, degree of trust, and expectations of providers held by older persons and important in training healthcare professionals.[2] Integrating cultural competency training into the academic health and mental health curriculum in order to improve healthcare service delivery is an area of growing interest and one that helps to eliminate structural and cultural barriers to health care.

The term *ethnogerontology* first appeared in the 1970s literature describing cross-cultural aging. In 1987 faculty members of the Stanford Geriatric

Education Center (SGEC) adapted the term *ethnogeriatrics* to focus specifically on health care for elders from different cultures. The SGEC has produced a number of monographs and working papers on the topic of health and mental health care for the older members of the populations of various ethnic groups. It is interesting to note that only two monographs on aging and health have been produced by the SGEC on Alaska Native Elders/American Indians and both have indicated in their introductions a lack of information about aging and health among Eskimo and Aleut populations. The monographs indicate that there are significant gaps in the ethnogeriatric understanding of American Indian/Alaska Native (AI/AN) aging and health issues. Especially lacking are descriptive observations on the continuities and discontinuities in the cultural experiences and expressions of older Indians, Eskimos, and Aleuts, especially within cities.[3] Lack of knowledge about health care and aging among Alaska Native Elders is especially critical given that program planning and implementation requires the development of strategies based upon the nature of various AI/AN populations.[3]

As clearly cited in the SGEC monographs, very little is known regarding the healthcare belief system of Alaska Natives, and even less about the impact that their historical context has had upon their values, health beliefs, illness behaviors, degree of trust, and expectations of providers held by the older person.[2] This is further substantiated by work being done by the newly funded National Resource Center for American Indian, Alaska Native and Native Hawaiian Elders, a center located at the University of Alaska Anchorage and funded by the U.S. Department of Health and Human Services, Administration on Aging, whose main charge, through their Listening to our Elders project, is to collect information about Alaska Native Elders' healthcare beliefs and practices. Because training and education of healthcare professionals is now believed to have an impact on both the quality and accessibility of health care, it is critical to develop informed training about the beliefs implicit in the Alaska Native culture and to identify those historical events (by cohort analysis) that have had an effect upon the indigenous populations' healthcare beliefs and practices.

THE COHORT ANALYSIS APPROACH

The cohort analysis approach yields individual historical profiles for each population that a healthcare professional is working with. A critical notion of this approach deals with the differential impact that an event may have on an individual according to what period of the life span he or she occupied when the event took place. Subsequently, significant events and periods experienced by cohorts of interest during specific age periods (i.e., childhood, adolescence, young adulthood, middle age, and later adulthood) are also generated within this approach. TABLE 1 provides significant life events and

TABLE 1. Cohort experiences of Alaska Natives

1900–1920	1920–1940	1940–1960	1960–1980	1980–Present
The "Great Death" killed 60% of Alaska Native people	Polio and TB epidemics	WWII: Relocations of Aleuts and seizure by Japanese	Vietnam war	Oil boom and pipeline development
Gold rush	Boarding schools required by Bureau of Indian Affairs	Alaska–Canada (Alcan) road built connecting AK to U.S. through Canada	Earthquake and tidal wave	Telephone and TV in every village
Nomadic or seasonal lifestyles predominate	Nomadic and seasonal lifestyles moves into community	Alaskan Statehood	Rise of alcohol abuse and criminal behavior	Village sobriety movement and education
Adaptations to death with survivors giving up some old cultural beliefs		Loss of lands and rights to statehood	North Slope lease auctions	Welfare and anti-poverty programs
Alaskan Brotherhood established in southeast AK			Relocation act	Self-determination
Become U.S. citizens			AK Native claims; settlement act and AK national interest lands; conservation act (loss of land and hunting and fishing rights)	Oil spill and clean-up

Continued

TABLE 1. *Continued.*

1900–1920	1920–1940	1940–1960	1960–1980	1980–Present
			Molly Hootch act to establish schools in all villages	Highest suicide and social-ill rates in U.S.
				Migration from villages to urban areas
				Sexual abuse by catholic priests
Age at Historical Experience				
Current Age Cohort				
Children and adolescents	Young adults and middle-aged	Middle-aged and young-old	Young-old and old	Old
Children	Adolescents and young adult	Young adults and middle-aged	Middle-aged and young-old	Young-old and old
	Children and adolescents	Adolescents and young adults	Young adults and middle-aged	Middle-aged and young-old
	Children	Children and adolescents	Adolescents and young adults	Young adult and middle-aged

experiences of Alaska Natives generated by the Alaska Geriatric Education Center (AKGEC).

This is just an initial attempt by the AKGEC to understand the historical context and cohort experiences of Alaska Natives. It is important to remember when using this approach that there is great diversity in experiences and perception of those experiences, both across the Alaska Native cultural groupings as well as within each group. Historical events and their impact on the various indigenous people of Alaska diverge quite extensively according to the region of the state in which the group resides. For instance, World War II had a much larger impact on the Aleut people residing on the Aleutian Islands off Alaska, in comparison to other groups, because of their forced relocation by the United States government and seizure of two islands (Attu and Kiska) by the Japanese.

Regardless of the need to consider events specific to individual groups nested within the larger ethnic group, this tool is still useful in training healthcare professionals. Significant events such as the "great death," a term used by Harold Napoleon in his book *Yuuyaraq, The Way of the Human Being*, to describe the waves of death that swept over indigenous people and villages is thought to be a contributing factor to what is referred to as "historical trauma" within the Alaska Native population.[4] The historical trauma response has been defined today as "...a cumulative emotional and psychological wounding, over the lifespan and across generations, emanating from massive group trauma experiences. The historical trauma response (HTR) is the constellation of features in reaction to this trauma."[5] Educational policies set by the Bureau of Indian Affairs separated Alaska Natives from their families and sent them out of state to attend school; although TABLE 1 shows the socialization of Alaska Natives to Western ways as occurring in the 1920s, it actually began in the late 1870s.[6] Thus, Alaska Native cohorts who were children when they experienced separation from others through death or school, and are now Elders or seniors being relocated to assistive living facilities, may exhibit historical trauma-like symptoms. A healthcare professional who is unaware of this history may jump to the wrong conclusion and not provide the most optimal or culturally competent care for that particular person.

CONCLUSION

The cohort analysis is a tool that provides insight into the impact the historical context might have on individuals' health beliefs. Preliminary research indicates that historical events such as the "great death" and boarding-school experiences for Alaska Natives may be reflected within their values, health beliefs, illness behaviors, self-image, and degree of trust of the health professional as well as their expectations of their providers. Further investigation into significant historical differences between the various Alaska Native groups needs to be performed by the AKGEC. However, preliminary evidence indicates that the cohort analysis approach is useful as an educational aid in training healthcare professionals in gerontology and geriatrics and in training a culturally competent and sensitive healthcare workforce.

ACKNOWLEDGMENTS

Thanks are extended to Elizabeth "Cookie" Rose and Patrick Frank for their guidance and assistance in employing the cohort analysis approach and identification of the historical events which may have had an impact on the earlier parts of the lives of these Alaska Elders. A special thank you goes to the Stanford Geriatric Education Center for use of the cohort analysis tool and their Visiting Scholar Ethnogeriatrics Program, of which this paper is an outgrowth. The Alaska Geriatric Education Center is supported by a grant from the U.S. Department of Health and Human Services, Health Resources and Services Administration, Bureau of Health Professions.

REFERENCES

1. U.S. Department of Health and Human Services, Health Resources and Services Administration. 2001. First annual report to the Secretary Department of Health and Human Services and to the Congress: Review and recommendations on interdisciplinary, community-based linkages title VII, part D Public Health Service Act. National Advisory Committee on Interdisciplinary Community Based linkages. Accessed September 10, 2007 from http//bhpr.hrsa.gov/interdisciplinary/gec.html.
2. Yeo, G., N. Hikoyeda, M. McBride, *et al.* 1998. Cohort Analysis as a Tool in Ethnogeriatrics: historical Profiles of Elders from Eight Ethnic Populations in the United States. 2nd ed. Working Paper #12, Stanford Geriatric Education Center, Stanford Medical School, Stanford University, Palo Alto, CA.
3. McCabe, M. & J. Cuellar. 1994. Aging and Health: American Indian/Alaska Native Elders. 2nd ed. Working Paper #6, Stanford Geriatric Education Center, Stanford Medical School, Stanford University. Palo Alto, CA.
4. Napolean, H. 1996. Yuuyaraq: the Way of the Human Being. Alaska Native Knowledge Network. Fairbanks, AK.
5. Brave Heart, M.Y.H. 2004. The historical trauma response among natives and its relationship to substance abuse. *In* Healing and Mental Health for Native Americans: speaking in Red. E. Nebelkopf & M. Phillips, Eds.: 1–15. Altamira Press. Walnut Creek, CA.
6. La Belle, J. 2005. Boarding School: historical Trauma among Alaska's Native People. Working Paper, National Resource Center for American Indian, Alaska Native, and Native Hawaiian Elders, University of Alaska Anchorage, Anchorage, AK. Accessed September 10, 2007 from http://elders.uaa.alaska.edu/Publications/yr2-2 boarding-school.pdf.

Lifestyle of the Elderly in Rural and Urban Malaysia

DORIS PADMINI SELVARATNAM AND POO BEE TIN

Faculty of Economics and Business, National University of Malaysia, Selangor, Malaysia

ABSTRACT: Malaysia is steadily progressing toward an aging population demographic pattern. While aging is a natural process, its impact can be painful individually as well as for the nation. Individually there is a loss of a paying job after retirement, loss of physical and mental fitness, and also occasionally the loss of social integration due to lack of mobility. For a nation, an aging population means a growing dependency ratio, a greater need of care, and more medical facilities for this age group. This article looks at the various economic and social implications of the aging population in Malaysia in general, and in the rural and urban setting specifically. The paper focuses on a research sample of 132 (66 rural, 66 urban) elderly persons. The findings suggest that the demographic patterns of the elderly vary from the rural to the urban setting, with differing issues that need to be addressed to alleviate problems encountered related to loneliness, lack of financial stability, and emotional strain. Policy suggestion will be geared toward providing a solution to problems at hand as well as aiding the working group members to prepare and sustain a comfortable livelihood for the aged in their later years.

KEYWORDS: lifestyle; healthy aging; longevity; social network; economic and social implication

INTRODUCTION

The lifestyle of the elderly is often not studied because of the age of the subject. Researchers are often captivated by the notion that the younger generation is more productive and contributes to the economic well-being of a nation. However, we cannot ignore the fact that the elderly lifestyle has an impact on the younger generation, too. Healthy lifestyle practiced by the elderly will help to ensure that they will have a healthy living arrangement. Consequently, a healthy elderly generation would place a smaller burden on the younger generation in terms of financial, medical, and physical needs.

Address for correspondence: Doris Padmini Selvaratnam, Ph.D., Senior Lecturer, School of Economics, Faculty of Economics and Business, National University of Malaysia, 43600 Bangi, Selangor, Malaysia.
pegasus@pkrisc.cc.ukm.my

Ann. N.Y. Acad. Sci. 1114: 317–325 (2007).
doi: 10.1196/annals.1396.025

The pattern of aging population prevalent in the developed countries is now beginning to be evident in Malaysia. Malaysia's demographic profile is also progressing toward a graying society. The elderly population in Malaysia is expected to increase from 1,032,300 people (5.9%) in 1991 to 3,439,600 people (9.9%) by the year 2020.[1] The number of elderly will increase higher among the Chinese (16.6%) and Indians (11.6%) compared to the Bumiputera (8.0%) in the year 2020, compared to 1991 with 7.6%, 5.4% and 5.4%, respectively. The relative high percentage of elderly among the Chinese could be due to the healthier food consumption, taking supplements, and the pattern of exercises followed and practiced by them compared to those of the Bumiputera (ethnic Malays and other indigenous groups) and Indians, who favor eating food rich in coconut milk and sugar and are less likely to exercise.

The total dependency ratio of the elderly in Malaysia is expected to increase from 10.2% (1991) to 12.1% (2010) and is projected to further increase to 16.5% (2020). The majority of the elderly people tend to reside in rural areas,[1] possibly because the elderly return to their home villages after their retirement.

DEFINITIONS

The Elderly

The Nutrition of Elderly report (1992) by the committee on Medical Aspects of Food Policy (COMA) cited by Blades[2] defines elderly people as those over the age of 65 years. It also defines the following age groups: the "young elderly," aged from 65 to 74 years, and the "older elderly," those 75 years of age or older. The World Health Organization (WHO) cited by Blades[2] classifies those aged 60–74 years as elderly, those 75–89 years as old, and those over 90 years as very old. Malaysia has also adopted this definition. Nevertheless, in Malaysia, the formal retirement age is 56, and people over this age are generally referred as elderly. For the research, the definition of the elderly as those aged 55 and above is used for the total elderly sample collected in the Intensified Research Priority Area (IRPA) project.

Lifestyle and Social Capital

Lifestyle of an individual is often studied in the context of comparing the level of nutritional intake, health, neighborliness, exercise and recreational activities, family relations, and work or source of income. For the purpose of this study the researchers focused on the aspect of networking among the elderly, their relatives, friends, and the community members who live around them. In general, social capital can be referred to institutions, networking relationship, and norms that shape the number and quality of social interactions (in terms of quantity and quality) in a society.[3] Social capital is acknowledged as new field of specialization and as the fourth capital after physical capital, financial capital, and human capital as the engine of economic growth.[4–6] Social capital's element of networking is an important aspect in the study of the elderly as social exclusion can be detrimental to one's health in the long run. Being connected can give an elderly person the reassurance of company and a listening ear and also ensures that the person is well cared for in his or her golden age.

RESEARCH OBJECTIVES

The research aimed to assess whether the increased material wealth has led to a decline in social capital among the elderly. The specific objectives of the research are, first, to identify rural and urban economic and social implications of aging; second, to identify lifestyle patterns in rural and urban settings; and, finally, to recommend healthy aging programs for Malaysian society.

REVIEW OF THE LITERATURE

Lifestyle

Arendt argues that there is no relation between income and social relationships and the activities of the elderly.[7] Most of the elderly people prefer to spend their time among friends or children and grandchildren at home. Forging social relationships is important at old age to reduce loneliness and feeling of being marginalized by society. Sometimes the unfriendly environment and high perception in fear of crime can lead the elderly to spend most of the time inside their house,[8] putting them further at risk of being isolated and marginalized. Studies show that an elderly person's income is merely sufficient for daily expenses.[9]

While various studies on the elderly and lifestyle focus on health, nutritional intake, exercise, poverty, and consumption patterns, current research emphasizes social networking among the elderly. This research looked into the elderly person's involvement in various social organizations, associations, religious activities, and local activities.

Social Capital

Social capital has a dimension that has been ignored in economic development efforts to enhance social well-being.[10] A greater emphasis is often given to physical capital, human capital, and financial capital. The social capital of a society includes the institutions, the relationships, and the attitudes and values that govern interactions among people and contribute to economic and social development. Social capital accrues through the capabilities engendered by the presence of trust in a community.[11] The existing social networks can be strengthened as an effective social control system.[12] Nevertheless, the measurement of social capital should include both formal and informal institutions like the family.[13]

METHODOLOGY

This research was based on field surveys. Research questionnaires were administered during interview with respondents at study locations. A stratified random sampling based on age, ethnicity, and strata (rural, suburban, and urban) was used to obtain the samples. Research samples consist of individuals at the community level and well represent the population of Malaysia. There was no marital status or gender discrimination. The total elderly sample size was 132 from the total of 986 samples collected in 2005 for the wider research project funded by Intensified Research Priority Area, Ministry of Science, Technology and Innovation, Malaysia (MOSTI) (05-02-02-10036 EAR).

RESEARCH FINDINGS

This section will discuss the research findings based on simple frequency counts, percentage, and cross-tabulations of data obtained from the field survey.

Demographic Characteristics of Respondents

Of the 132 respondents, 51 were female respondents and 81 male. A total of 107 (81.06%) were married, 4 were single, and the remainder of 21 respondents were either widows or widowers; 50% were from an urban area, while the remaining 50% were from a rural area. A majority of 58.3% had a primary school educational background, 13.6% had a lower secondary education (SRP or PMR), while 11.4% did not have any formal education. A total of 45.5% still had paid jobs to sustain their living, while 54.5% did not have a paid job anymore.

A majority of the sample group (56.8%) had an income less than RM691, 16.8% RM691 to RM1000, 9.8% RM1001 to RM2000, 5.3% in the income bracket of RM2001 to RM3000, and 11.3% had an income bracket of more than RM3001. A total of 58.1% of the respondents had 1–5 household size, 38% with 6–10 household size, and the balance 11–15 household size.

Economic and Social Implications

Economic Implication

After the age of 56 years, which is the formal retirement age, many Malaysians still continue to actively participate in paid employment. The reasons for this vary from sustaining their lifestyle before retirement, to maintaining their livelihood above the Poverty Line Index (PLI), or just to occupy themselves.

Among the economic implications of aging are lower growth rate and decline in per-capita income.[14] The majority of elderly Malaysians lose their main source of income once they retire.[15] The Ninth Malaysia Plan notes that the elderly are among the most vulnerable members in society and are part of the hard-core poor in society. The Household Expenditure Investigations Report 98/99[16] shows that the overall monthly expenditure of Malaysian urban and rural elderly households (where the householder is above 65 years old) is RM1518 and RM908, respectively. Nevertheless, the expenditure for food, clothing, and rent alone is RM531 and RM874, respectively, for the rural and urban elderly Malaysian households. This suggests that maybe the PLI needs to be scrutinized to ensure that the elderly are adequately provided for.

The country's PLI is RM398 (2005) compared to RM272 (1977) for Peninsular Malaysia, RM503 (2005) compared to RM352 (1977) for Sabah, and RM482 (2005) compared to RM304 (1977) for Sarawak. The overall PLI for Malaysia has increased to RM691 (2005) compared to RM294 in 1977 (RM9). A total of 56.82% of the respondents live below the PLI and their burden will be exacerbated in the light of rising concern of increasing medical costs and general cost of living.

There are more working elderly in the rural area ($n = 34$) compared to the urban area ($n = 26$). This could be attributed to the lifestyle of the rural community in Malaysia, who own estates and plantations. Having retired from formal jobs, they "return to the land." In view of the total of 49 elderly rural respondents in our study having income less then RM691, it is not surprising why the majority of the elderly in the rural area are still involved in paid employment. On the other hand, only 26 elderly urban respondents have incomes less than RM691. More elderly people in the urban areas are living with their children today due to income inadequacy and on account of the government's campaign to encourage elderly folks to be cared for by their children. Furthermore, tax incentives to buy special equipment and medicine is another encouraging factor.

Social Implications

Around 31 respondents had lived in their neighborhood for 21–30 years, another 28 lived there for less than 10 years, and only 20 respondents have lived for more than 40 years in their neighborhood. The majority of the respondents, regardless of whether they were in urban or rural areas, said price was the overriding factor in choosing their residential place. This was followed by facilities and location factors. A very small sample group chose type of neighbors and ethnic composition, heritage, and aesthetic values as their top priority in selecting a neighborhood. The respondents were further asked their opinion on the level of trustworthiness in the neighborhood. Regardless of the residential area, the majority ($n = 121$) of the elderly responded "no" or "not sure." Only a total of 8.3% elderly respondents (10.6% in urban and 6.1% in rural area) believed that there will be a high degree of trustworthiness in society in the future. A majority of the elderly rural and urban respondents ($n = 86$) said that it is difficult to trust others, while 10 and 36 respondents said it is either very easy or just easy to trust others.

Lifestyle Patterns in Rural and Urban Settings

Living in isolation can enhance the dangers of marginalization from the mainstream of economic and social activities. Loneliness can also be a cause of desolation and lead to depression. While a majority of 104 elderly respondents are staying with someone, 21.21% of the respondents are staying alone. Nevertheless, as will be seen just below, although they live alone there are many opportunities for them to be active in their respective communities.

This research has attempted to investigate the lifestyle of the elderly based on their networking in various organizations. This is seen as an important factor because elderly people also need to stay connected with people around them. Involvement and participation in various associations, religious groups, and local activities like *rukun tetangga* (neighborhood watch group) or *gotong-royong* (neighborhood voluntary teamwork) can help the elderly to stay active in their respective communities.

Religious activity participation was similar for the rural ($n = 15$) and urban ($n = 16$) locations; this is a poor count for both categories, possibly because the majority are Muslims, who pray at home. There was a higher participation in organization, association, or political party activities among the rural elderly ($n = 29$) compared to the urban elderly ($n = 16$), possibly because of the close proximity in the rural area. Meanwhile, the urban elderly played a bigger role ($n = 9$) in *rukun tetangga*, compared to the rural elderly ($n = 6$).

Healthy Aging Programs

Currently the government has had in place a National Social Welfare Policy since 1992, a National Policy for the Elderly since 1995, and a guideline for Daily Care for the Elderly. The government has also launched an Older

Persons Action Plan since 1999 to help ensure that the elderly experience a happy, meaningful, and integrated life and are actively participating in the local community activities.

Among the programs that the researchers would like to recommend are: First, creation of job opportunities for the elderly. Paid jobs are always a plus as they provide income to sustain the elderly person's livelihood. Such jobs include being a tour guide or a historical events narrator, among other jobs that are not taxing to the elderly person's physical ability while still tapping on their mental capability. Unpaid jobs, on the other hand, are welcomed by those who seek to engage in activities that help to pass their time without expecting remuneration. This is suitable for the more affluent elderly members. And jobs like taking care of the young help to inculcate social, moral, and religious values.

Second, social exclusion can become hazardous to a society. Social inclusion is vital to ensure that the elderly not only have longevity, but also have a socially healthy lifestyle. Social activities need to be geared specially for the elderly. *Rukun tetangga* is a grass-roots security system which Malaysia has implemented since the 1960s. Because of the improved public safety and security, this program has become less popular. The Prime Minister recently called for a reactivation of this activity because its purpose is not only getting together to ensure the security of the neighborhood, but also to get to know all the members in that area.

Third, the elderly need regular exercise programs. Although various social organizations offer exercise activities like the *tai chi*, *kwantong,* yoga, and meditation, the elderly need to be more actively involved. When the elderly gather in groups they will feel more relaxed and feel free to participate in these activities compared to their inclination to do so if they are living alone or in a strange neighborhood.

Fourth, discounts are widely available for domestic travel by land and air to selected foreign destinations during promotional periods. But these travelers are small in number. More attention needs to be given to the aspect of affordable medication and health care for the elderly. Although since 1979 the government has given a tax exemption of RM1000 for children who take care of their elderly parents, the elderly persons also need to have special discount cards to purchase medications.

Malaysia has a diverse source of traditional medicines. More research and development needs to be done to tap these resources to ensure that alternative medicine is available at a more affordable price. Some of these medicines can help eliminate problems of arthritis, osteoporosis, dementia, and diabetes, which can ensure a more healthy longevity.

Elderly housing villages or communities is another option that the government could use to provide integrated care and nursing for the elderly while they are socially included in the variety of religious and other social activities available in the community. Being in one area, but still enjoying living in individual units, will definitely provide a good environment for them. Health officials will also find it easier to see to their needs. Children who are busy with dual careers will also feel reassured that their parents are receiving the best medical attention here. But they must also be responsible to visit their parents often and take them on family outings, too.

Health-check home visit is an initiative that Malaysian Red Crescent Society and the Gerontological Association have started that helps to keep the elderly linked to health needs and socially active. Care provided by family members, community, and the nation will help elderly people enjoy a meaningful life.[17] In addition, the younger generation, recognizing the future importance of aging gracefully and of having a comfortable income to sustain a balanced lifestyle, must become aware of the importance in investing in good health and health insurance as well as saving to retire comfortably, taking into consideration the prospect of rising inflation and a rising cost of living.

CONCLUSIONS

The research undertaken showed that the majority of the elderly in rural and urban Malaysia are socially connected. These older persons realize the importance of family ties and neighborly connections. Staying linked to people near and far is essential to ensure that the elderly are not marginalized, isolated, or otherwise socially excluded. The majority of the respondents are living with someone. A total of 28 respondents live alone. Nevertheless, all the respondents do stay connected with their children, relatives, friends, or neighbors.

The urban elderly respondents are better off in terms of economic wealth, while the rural elderly respondents still have to be formally employed to maintain their livelihood. Nevertheless, the level of social capital is higher in the rural community, with higher participation in political organizations, religious activities, and the neighborhood watch group called *rukun tetangga*.

The researchers would like to conclude that the research could be further strengthened with a nationwide study (a larger sample or a nationwide survey) comparing the urban and rural elderly Malaysians, and the age, ethnicity, and gender of the elderly. A comparative study of the younger and older generation will also provide important details about the lifestyle pattern between the two groups. Cross-country analysis is also suggested to gain information on indigenous knowledge of healthy lifestyles for all members of society.

ACKNOWLEDGMENTS

The article is based on a wider research project funded by Intensified Research Priority Area (IRPA) and a grant (05-02-02-10036 EAR) from the Ministry of Science, Technology and Innovation, Malaysia (MOSTI).

REFERENCES

1. PALA, J. 2005. Aliran Penuaan Penduduk di Malaysia. Siri Monograph No. 1. Department of Statistics Malaysia. Kuala Lumpur.
2. BLADES, M. 2002. Nutrition and the elderly in residential care. Nutr. Food Sci. **32:** 174–179.
3. THE WORLD BANK. 1999. What is Social Capital? *Poverty*Net http://www.worldbank.org/poverty/scapital/whatsc.htm Accessed June 12, 2006.
4. WOOLCOCK, M. & D. NARAYAN. 2000. Social capital : implications for development theory, research & policy. World Bank Res. Observ. **15:** 225–249.
5. COLEMAN, J. 1988. Social capital in the creation of human capital. Am. J. Sociol. **94:** S95–S120.
6. WOOLCOOK, M. 1998. Social capital and economic development: towards a theoretical synthesis and policy framework. Theory Soc. **27:** 151–208.
7. ARENDT, J.N. 2005. Income and outcomes for elderly: do the poor have the poorer life. Soc. Indic. Res. **70:** 327–347.
8. TOM, S. & R. O'TOOLE. 2005. Crime-free neighbourhood. Reasononline. Feb. Issue. http://www.reason.com/0502/fe.st.crime.sthml. Accessed August 15, 2006.
9. CHENG, Y.H., I. CHI, K.W. BOEY, *et al.* 2002. Self-rated economic condition and the health of the elderly persons in Hong Kong. Soc. Sci. Med. **55:** 1415–1424.
10. NARAYAN, D. 1999. Bonds and Bridges. Social Capital and Poverty. The World Bank Policy Research Working Paper No. 2167. Washington, DC.
11. FUKUYAMA, F. 1995. Trust: the Social Virtues and the Creation of Prosperity. Hamish Hamilton. London.
12. GOVERNMENT OF MALAYSIA. 2003. The National Social Policy. National Printing Press. Kuala Lumpur.
13. SPELLERBERG, A. 2001. Framework for the Measurement of Social Capital in New Zealand. Social Capital Program Team. Statistics New Zealand.
14. SELVARATNAM, D.P. 2004. Implication of aging population for Malaysia: economic perspective. Paper presented at the National Conference on Population Aging in the Developing World: Bridging Research, Policy and Practice. Organised by the Institute of Gerontology, Department of Social Welfare Malaysia, United Nations Population Fund and Gerontological Association of Malaysia, 5–6th October 2004 Summit Hotel, Subang Jaya.
15. ISMAIL, N. & D. GABDA. 2001. Social security scheme and the elderly. *In* Social Security for the Elderly. Abdul Aziz Jemain, Lukman Z. Mohamad, Wan Norsiah Mohamed, Eds.: 77–89. Prentice Hall. Kuala Lumpur.
16. DEPARTMENT OF STATISTICS MALAYSIA. 1999. Household Expenditure Investigation Report 98/99. Department of Statistics Malaysia. Kuala Lumpur.
17. AZIZ, R.A. 2000. Changing values and the family elderly care in Malaysia. *In* Gerontological Approaches to Care of the Aged in the 21st Century. Proceedings of the 6th Asia/Oceania Regional Congress of Gerontology, Seoul. Sung-Jae *et al.*, Eds. Seoul Organizing Committee for the 6th Asia/Oceania Regional Congress for Gerontology.

Healthy Aging for Caregivers

What Are Their Needs?

TINA WU[a] AND KAREN LO[b]

[a]*Department of Social Work, Shih-Chien University, Taipei, Taiwan*

[b]*Simon Fraser University, Biology Department, Vancouver, BC, Canada*

ABSTRACT: The aim of this study was to design a culture-sensitive scale to measure caregiver's burden for the purpose of identifying their needs in Taiwan. In this study, four factors were identified as caregiver's top priority needs: "personal factors," "relation factors," "financial factors," and "reward factors." Definitions for each factor are also discussed. The total variances explained by the four factors are 34%, 13%, 9%, and 8%, respectively. The results show caregivers in Taiwan place more emphasis on their physical and emotional interaction with the care-receivers than is found in other cultures. Policy recommendations inferred from these findings are made. It is necessary for the government or organizations to provide services to fulfill caregivers' needs to promote a healthy aging process of the caregivers themselves. Participants who cared for their elderly parents or relatives were selected randomly from the island of Taiwan. The total selected sample size of caregivers is 2524, who were caring for 3504 persons. Data were analyzed by univariate, bivariate, and multivariate methods. Maximum Likelihood Extraction Factors Analysis and LISREL for Structural Equation Modeling methods were used to confirm and make final the structure of the scale. Recommendations and suggestions are provided for further studies on culturally sensitive projects.

KEYWORDS: caregiver; needs assessment; confirmatory factor analysis; LISREL; structural equation modeling method

INTRODUCTION

Webster's dictionary defines "healthy" as "a state of enjoying health and vigor of body, mind, or spirit," while the definition of aging is "showing the effects or the characteristics of increasing age." When these two concepts are put together, healthy aging could be the state of "enjoying health and vigor

Address for correspondence: Dr. Tina Wu, Associate Professor, Department of Social Work, Shih-Chien University, # 70, Tah-Chih Street, Taipei, Taiwan. Voice: 886-2-25381111: ext.: 6915; fax: 886-2-25338622.

twu@mail.usc.edu.tw

Ann. N.Y. Acad. Sci. 1114: 326–336 (2007). © 2007 New York Academy of Sciences.
doi: 10.1196/annals.1396.017

of body, mind, or spirit while experiencing the effects of increasing age." In Taiwan, as the elderly population continues to grow, healthy aging issues have risen to become the top priority of social/health policy concerns. There are not only issues for individual elderly, but there are also issues for the family members and friends who are caring for them. In the paper, we want to identify prominent need factors of family caregivers as well as to address and discuss issues and services that would improve the caregiver's health and well-being.

The World Health Organization defined health as "a state of complete physical, mental, and social wellbeing, and not merely the absence of disease" (p. 16).[1] The WHO recognized that psychosocial well-being is an important component of health; however, health remained as an abstract concept without objective operational definition. For many years, it has been too difficult for the field to accomplish the mission for defining health in a comprehensive and specific way. Recently, there is a trend to define health in terms of "quality of life." In this definition, health is "the opportunity to make choices and to gain satisfaction from living despite functional limitations." It is implied that health is a dynamic process of interaction between families/communities/organizations and individuals. Therefore, one's health involves freedom of choice, interaction among formal and informal organizations (e.g., caregivers), and individual characteristics. Everyone involved in caregiving triad network needs to take deliberate action to make the changes necessary for people in the circle toward healthy aging. Health is no longer a personal matter in the process; rather, it is the result of network interactions.

Traditionally, aging has been viewed as the process of continual decline in health. In the process of caregiving, the aging process occurs at the same time to both caregivers and care-recipients. In most aging research[2] the elderly care-receivers were usually portrayed negatively. People tend to focus on the physical and mental losses of those being cared for, as well as on decreases in physical attractiveness. Thus, we see that the aging process has a negative impact on the recipients of care. However, very little attention has been paid to the *caregiver's* aging process; instead, researchers have mainly focused on the caregiver's *burden*. If caregivers and caretakers are aging together as a unit, can we apply this aging process view universally and hold it to be true of the aging process of the caregivers as well? What would be the outcome when individuals have the *combined* experience of the aging process and the burden of caregiving?

To date, most family caregiving research has portrayed caregiving as a negative life experience for caregivers.[3] This particular conceptualization has been studied extensively and has been professionally termed *caregiver's burden*. However, the concept of caregiving should involve both positive and negative aspects. Human interaction has its complexity, especially with family members, relatives, or friends. Research focusing on both aspects could provide a better understanding of the caregiving experience and help clinicians and practitioners to work more effectively with the family caregiver.[4]

Many studies have explored factors for the caregiver's well-being. Schulz *et al.*[5] reviewed past research and found factors, such as income, spouse, self-rated health, perceived stress and burden, life satisfaction, self-esteem/self-mastery, and caretaker problem behaviors, consistently influenced the caregiver's psychiatric morbidity. After empirical testing, their study showed physical morbidity was associated with caregiver's depression, social support, the care-recipient's problem behaviors, and cognitive impairment. Although demographics are not generally studied by health care/social services, public-funded service interventions to improve the caregiver's well-being may be developed for such other factors as the caregiver's perceptions regarding caregiving, which is related to how they view the process and the outcome of their caregiving. Usually, caregivers view the concept as a "burden," which tends to have negative effect on their own well-being as well as on their quality of life. Numerous studies have examined the relationship of caregiver burden with such variables of well-being as depression or physical health, and have shown that caregiver burden is a significant factor on both variables.[6–9]

"Caregiver's burden" is a subjective judgment, involving the process of caregiving appraisal, which is defined by Lawton *et al.*[10] as "all cognitive and affective appraisals and re-appraisals of the potential stressor and the efficacy of one's coping efforts." After years of research, many researchers have claimed that caregiving is perceived as both a negative and a positive experience by caregivers. Positive appraisals of care have been generally operationalized as "satisfactions," "uplift," "reward," or "mastery," whereas negative appraisals are reviewed as "anxiety," "tiredness," "breathlessness," "fatigue," "emotional distress," and "depression," for example.[11–14]

Furthermore, Sales[15] has pointed out that family burden and quality of life should include both subjective and objective burden appraisal of caregiving. Objective burden maybe the direct tasks of care (e.g., helping with activities of daily living [ADL], supervising), indirect tasks, length of time taken in dealing with the emotional needs of the caretaker, or effects of caregiving on other aspects of life (family interaction, family routine, leisure, work/employment, mental health, physical health, social network, others outside household, children, financial consequences). As to the subjective burden, it is related to personal reactions to caregiving, such as feeling distressed, stigmatized, worrying, shameful, and guilty of paying attention to one's needs first instead of care-recipient's needs.

In this study, critical aspects of caregiving are investigated and identified. Positive and negative appraisals of care refer to any cognitive and affective evaluation of one's caregiving experience that describes caregiving as good (or bad) and comfortable (or uncomfortable). The framework of the research integrates positive/negative appraisal of four caregiving domains based on review of the literature and results of qualitative interviews with study participants. The four domains are: (1) satisfaction/dissatisfaction with the attachment or relationship with the specific care-recipient (*relationship satisfaction*); (2)

consequential gain/loss from the caregiving experience (*consequential gain*); (3) confidence/fear in taking care of the care-recipients (*role confidence*); and (4) satisfaction/dissatisfaction with fulfilling the socially normatively assigned role (*normative fulfillment*).

By definition, positive relationship satisfaction is a positive appraisal of the attachment relationship with the particular care-recipient. Role confidence is the sense that a caregiver perceives whether she/he is doing a good job in caregiving and whether she/he is the best person to care for the care-receiver. Consequential gain from a caregiving experience includes perceptions of reward incurred as a result of the caregiving experience (e.g., one's insightful view of life and death, closer family ties, personal growth, or any new meaning generated in the caregiver's life). Normative fulfillment through the caregiver role means that caregivers feel good about caregiving because they are considered as virtuous citizens of the society. Any behavior or thinking on the contrary to the positive appraisal defined from above would be considered as negative sides of each appraisal.

In Eastern culture, filial piety has been long recognized as an honorable deed, although social change is gradually taking a toll on this concept. In Taiwanese culture, generally speaking, people respect the one who is taking care of the disabled or the elderly in the family. The society considers it a virtue and honors them verbally and non-verbally. Although the terms "filial piety" and "cause and effect of filial piety" are not well-known in Western culture, the concept of filial piety is not wholly unfamiliar, for it is included in the idea of kindness. The moral of goodness being rewarded by goodness is also illustrated in Western fairytales, such as Cinderella and Snow White.

In conclusion, the findings of the study will help the government and professionals in the field of aging come together to materialize a plan for caregivers. Their healthy aging would in turn help their care-recipients have a better quality of life. As a result of this reciprocity, aging will become not only a natural, but also rewarding process.

METHODS

Subjects

The subjects for this research project were family caregivers providing home-based care for elderly relatives aged 65 or above. Participating caregivers were randomly selected from a government household registry of those who were receiving home-care services at the time of research. However, some participants were in poor physical conditions and could not take part in the research. All together 2524 caregivers responded to the face-to-face questionnaire interview, a response rate of 82.2%. Their family members were recruited as the validating party for the research. The total number for participating care

TABLE 1. Descriptive statistics of participating care-recipients ($n = 3504$)

Variables	Range	Mean/Median	SD	Coding algorithm
Age	65–104	77.3	7.5	Chronological age
Sex	1–2	2	0.5	1: male; 2: female
Education	1–7	1	1.5	1: illiterate; 2: literate 3: elementary school; 4: junior high; 5: senior high; 6: college; 7: graduate study and above
Marital status	1–5	3	1.1	1: living with spouse; 2: not living with spouse; 3: spouse deceased; 4: divorced; 5: never married
Income	1–13	3	2.5	1: NT$10,000; 2: 10,000–19,999 3: 20,000–29,999; 4: 30,000–39,999 5: 40,000–49,999; 6: 50,000–59,999 7: 60,000–69,999; 8: 70,000–79,999 9: 80,000–89,999; 10: 90,000–99,999 11: 100,000–199,999; 12: 200,000–299,999 13: 300,000 and above.

recipients was 3504, a response rate of 82.5%. Since some of the caregivers were caring for more than one family member, the caregivers' number is less than that of the participating care-recipients. The exact care-recipients' age range is 2–104 years; however, only persons over 65 were selected and included in the research design.

The following provides details of the study.

Questionnaire

A questionnaire was developed for the study paying closer attention to Taiwanese family culture values with regard to filial piety in order to discern the caregivers' true appraisal of their caregiving needs, stress, subjective quality of life, as well as other variables including that of individuality, which is a variable valued in Western culture. TABLES 1 and 2 describe both the care-recipients and caregivers.

Originally, the questionnaire design gathered several scales listed in the aging research literature, such as the SF-12, CED-D in Chinese elderly, Mini-Mental State Examination (MMSE), and the Zarit Caregiver Burden (ZBI). These researchers had conceptualized burden centralized around factors of "time spent caring for the care-recipient," "financial costs," "physical disability,"

TABLE 2. Descriptive statistics of participating caregivers ($n = 2524$)

Variables	Range	Mean/Median	SD	Coding algorithm
Age	25–105	68.0	15.3	Chronological age
Sex	1–2	2	0.5	1: male; 2: female
Education	1–7	4	1.5	1: illiterate; 2: literate 3: elementary; 4: junior high; 5: senior high; 6: college; 7: graduate and above
Marital status	1–5	1	1.3	1: living with spouse; 2: not living with spouse; 3: spouse deceased; 4: divorced; 5: never married
Income	1–13	3	2.5	1: NT$10,000; 2: 10,000–19,999 3: 20,000–29,999; 4: 30,000–39,999 5: 40,000–49,999; 6: 50,000–59,999 7: 60,000–69,999; 8: 70,000–79,999 9: 80,000–89,999; 10: 90,000–99,999 11: 100,000–199,999; 12: 200,000–299,999 13: 300,000 and above.
Self-rated health	1–5	3	0.9	1: excellent; 2: good; 3: fair; 4: not good; 5: poor
No. of years caregiving		10.2	11.6	Actual number of years

and "emotional distress and psychological functioning." Although all scales are useful in assessing a specific type of burden, broader measures that can assess culture-sensitive domains of burden should be developed, especially from both subjective and objective perspectives of burden. Consequently, positive and negative domains regarding caregiving are included in the construction of the scale. In addition, culture-sensitive wording was taken into consideration in constructing the sentences, say in avoiding words that would trigger "social standards" judgment.

Sampling and Data Collection

The items of the Taiwanese Culture-Sensitive Subjective Appraisal Scale (TCSSAS) are shown in FIGURE 1.

The TCSSAS was originally designed as a pilot project for a government service intervention program. As well, the data collection process was initiated to serve the same function. All potential care-recipients were identified or screened by government officials or case managers in their own communities. All of them were diagnosed as having "functional disabilities," with more than one ADL and/or IADL limitations. Their caregivers were then recruited to

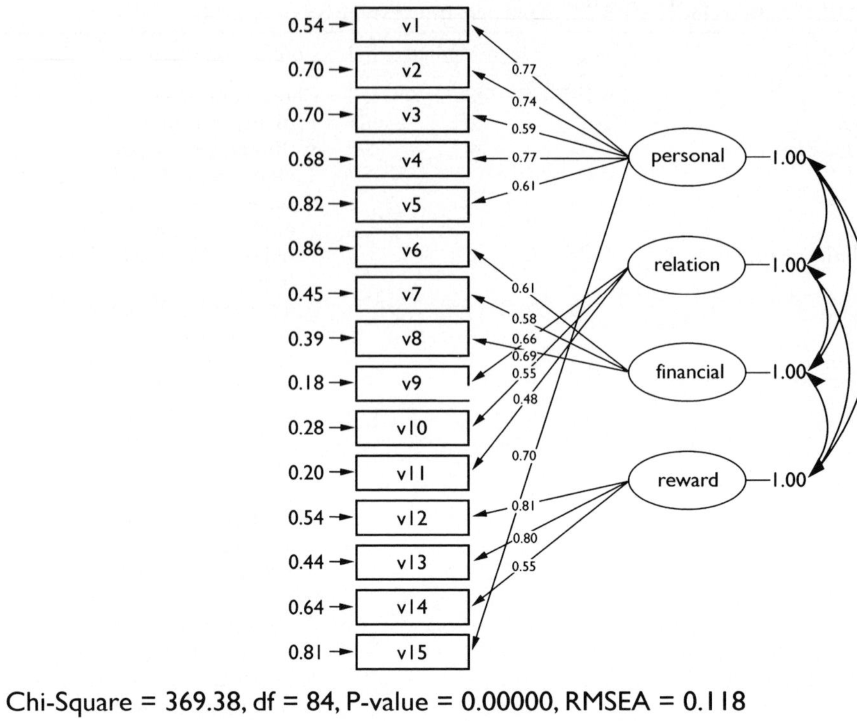

FIGURE 1. Structural equation modeling of TCSSAS.

participate in the project afterwards. Informed consent was obtained from all participating caregivers. All together, there were 4247 care-recipients; however, only 3504 agreed to be part of the investigation for a response rate of 82.5%. Reasons for not wanting to participate were: "not at home," "moved away," "institutionalized," "passed away," and "refused to talk to the interviewers." Of the 2524 caregivers, reasons for those who refused to participate were: "no time," "foreign-aide—cannot communicate in local language (2%)," and "paid-care" for a positive response rate of 82.2%.

Trained interviewers distributed and read the questionnaires to the caregivers, who were then given enough time to complete the questionnaire items. After the interview, demographic information was transferred from the official case record to the back of the questionnaire. Average time spent for each questionnaire/interview is about 30 minutes.

DATA ANALYSES

The purpose of the research is to use the designed instrument to detect the needs of the participating caregivers. Following the examination of descriptive

TABLE 3. The Taiwanese Culture-Sensitive Subjective Appraisal Scale (TCSSAS)

Items	Scale ranking
1. Feeling tired on account of caregiving	0–3 (0: never; 1: seldom; 2: occasionally; 3: often)
2. Feeling one's health is depreciating because of caregiving	0–3
3. The care-recipient makes me angry	0–3
4. Feeling distressed because of care-recipient's poor conditions	0–3
5. Feeling sorry for not doing enough for care-recipient	0–3
6. Feeling segregated from relative and friends because of caregiving	0–3
7. Feeling trapped at home (cannot travel) because of caregiving	0–3
8. Feeling the need to watch over care-recipient all the time	0–3
9. Feeling no time for self because of caregiving	0–3
10. Feeling strained financially (cannot afford the involved caregiving cost) on account of caregiving	0–3
11. Feeling had to live one's life under a tight budget due to caregiving	0–3
12. Feeling relationship with family members is strained due to caregiving	0–3
13. Feeling incapable of taking care of other family members on account of caregiving	0–3
14. Feeling no positive emotional support from other family members on account of caregiving	0–3
15. Feeling disturbed on account of caregiving (e.g., family life needs re-adjustment to accommodate care-recipient's needs)	0–3
16. Feeling one's importance through caregiving	0–3
17. Feeling being respected and well-valued on account of caregiving	0–3
18. Feeling peace in one's heart and fulfilled in one's duty in life through caregiving	0–3
19. Feeling ashamed of care-recipient's behavior	0–3
20. Feeling one no longer has the ability to care for the care-recipient	0–3
21. Feeling limited in having a future or in planning for one because of caregiving.	0–3
22. Feeling arrangement of one's everyday life constrained by need to center on care-recpient's needs.	0–3

statistics, bivariate and multivariate analyses were adopted to detect the relationship between items among the TCSSAS scale. In multivariate analysis, the reliability and validity of TCSSAS were examined, following by principal and confirmatory factor analyzes.

Furthermore, in order to finalize the structure of factor analyses, latent variable analysis or structural equation modeling method were conducted to confirm the variance explained by each domain and item. Previous caregiver burden research shows that burdens of families associated with a care-recipient's illness can take many forms. Thus, it is reasonable to assume that the concept of caregiver's appraisal should be viewed with multidimensional factors. Data were analyzed with SPSS12.0 and LISREL Structural Equation modeling 8.51.

RESULTS AND DISCUSSION

Originally, the instrument had 22 items, but after careful testing and analysis, the items were reduced to 15. In the process, the total variance was accounted for, cumulative rotation Sums of Squared Loadings, or total variance explained by the model changed from 42.87% to 52.1%. In summary, the smaller number of items has better ability to explain the model. An interesting finding is that after deleting all four items related to the *individuality* value (i.e., a value stressed in Western society), the instrument had a better power in the total variance explained. The total variances explained by the four factors are 34%, 13%, 9%, and 8%, respectively (i.e., personal, relational, financial, and reward factors). The results show caregivers in Taiwan place much more emphasis on their physical and emotional interaction with the care-receivers than is found in other cultures.

As to the validity and reliability of the scale, the criteria validity check among items in the same dimension shows a high correlation coefficient of 0.563, while the reliability Cronbach alpha value is 0.86 for the final 15 items. In conclusion, the results confirm the scale is reliable and valid.

What kinds of services should be offered to caregivers in order to lessen their burden in caring for their family members? We wish to propose a triad model for caregiver services instead of a dyad model just between caregivers and caretakers.

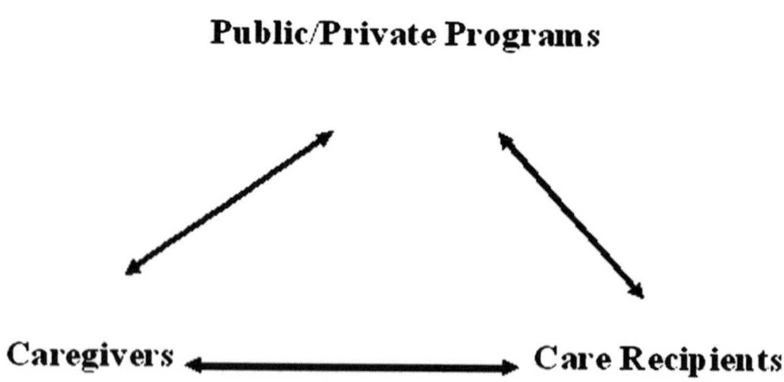

The government should be included in the service paradigm for caregivers. The government should be responsible for integration of service packages for different levels of caregiver's needs. Services, such as companions, escorts, live-in help, provision for light housekeeping and laundry, shopping and meal preparation, medication follow-up, bathing and dressing; registered nurses, licensed practical nurses, certified nurse assistants; trained caregivers, hospice, facility staffing, recreational activities, and insurance/welfare-related information sharing and assistance should be offered.

This study possesses both strengths and weaknesses. The strengths lie in its originality trying to construct a culture-sensitive measurement; however, further testing is required. Taiwanese policymakers should refer to this research as a benchmark for future development for caregiver's services.

ACKNOWLEDGMENTS

We want to thank Dr. Shwu-Chong Wu for providing a raw data set for the study.

REFERENCES

1. World Health Organization (WHO). 1947. Constitution of the World Health Organization. Geneva, p. 16.
2. Palmore, E. B. 2006. Age discrimination: an historical and contemporary analysis. Gerontologist **46:** 848–850.
3. Yamamoto-Mitani, N. *et al.* 2004. Subjective quality of life and positive appraisal of care among Japanese family caregivers of older adults. Quality of Life Res. **13:** 207–221.
4. Kramer, B.J. 1997. Gain in the caregiving experience: Where are we? What next? Gerontologist **37:** 218–232.
5. Schulz, R., A.T. O'Brien, J. Bookwala & K. Fleissner. 1995. Psychiatric and physical morbidity effects of dementia caregiving: prevalence, correlates, and causes. Gerontologist **35:** 771–791.
6. Arai, Y., K. Kudo, T. Hosokawa, *et al.* 1997. Reliability and validity of the Japanese version of the Zarit caregiver burden. Psych. Clin. Neurosci. **51:** 281–287.
7. Gallichio, L., N. Siddiqi, P. Langenberg & M. Baumgarten. 2002. Gender differences in burden and depression among informal caregivers of demented elders in the community. Int. J. Geriatr. Psych. **17:** 154–163.
8. Pruchno, R.A., N.D. Peters & C.J. Burant. 1995. Mental health of co-resident family caregivers: examination of a two-factor model. J. Gerontol. Psychol. Sci. **50B:** P247–P256.
9. Sparks, M.B., C.J. Farran, E. Doner & E. Keane-Hagerty. 1998. Wives, husbands, and daughters of dementia patients: predictors of caregivers mental and physical health. Schol. Inq. Nurs. Pract. **12:** 221–234.
10. Lawton, M.P., M.H. Kleban, M. Moss, *et al.* 1989. Measuring caregiving appraisal. J. Gerontol. Psychol. Sci. **44:** P610–P617.
11. Lawton, M.P., M. Moss, M.H. Kleban, *et al.* 1991. A two-factor model of caregiving appraisal and psychological well-being. J. Gerontol.: Psychol. Sci. **46:** P181–P189.

12. KINNEY, J.M. & M.A.P. STEPHENS. 1989. Hassles and uplifts of giving care to a family member with dementia. Psychol. Aging **4:** 402–408.
13. PRUCHNO, R. 1990. The effects of help patterns on the mental health of spouse caregivers. Res. Aging **12:** 57–71.
14. PICOT, S.J.F., J. YOUNGBLUT & R. ZELLER. 1997. Development and testing of a measure of perceived caregiver rewards in adults. J. Nurs. Meas. **5:** 33–52.
15. SALES, E. 2003. Family burden and quality of life. Quality of Life Res. **12**(Suppl 1): 33–41.

ADDITIONAL RESOURCES

1. COUSINS, R., A.D.M. DAVIES, C. TURNBULL & J.R. PLAYFER. 2002. Assessing caregiving distress: a conceptual analysis and a brief scale. Br. J. Clin. Psychol. **41:** 387–403.
2. LIANG, J., S.C. WU, N.M. KRAUSE, *et al.* 1992. The structure of the mental health inventory among Chinese in Taiwan. Medical Care **30:** 659–675.
3. NORROW-HOWELL, N., E. PROCTOR & P. ROZARIO. 2001. How much is enough? Perspectives of care recipients and professionals on the sufficiency of in-home care. Gerontologist **41:** 6.
4. STRABRIDGE, W.J. 1991. The effect of social factors on adult children caring for older parents. University of Washington, Seattle, USA. Dissertation Abs. Int. **52**(3-A): 1094.

Well for Life: A Way of Life

WENDI A. KRUGER,[a] CATHERINE E. THOMPSON,[a]
ROSEMARY A. McKENZIE,[b] AND LUCIO NACCARELLA[b]

[a]*Aged Care, Victorian State Government Department of Human Services, Melbourne, Victoria, Australia*

[b]*Program Evaluation Unit, The University of Melbourne, Melbourne, Victoria, Australia*

ABSTRACT: Well for Life is an innovative public program in Victoria, Australia that focuses on improving nutrition and increasing physical activity to promote healthy aging. For more than 4 years Aged Care has funded projects for frail older people who live in Public Sector Residential Aged Care (PSRAC) facilities or attend Home and Community Care (HACC) Planned Activity Groups (PAGs). Many stereotypes exist around what frail older people can or cannot do. Well for Life challenges many of these through organizational culture and policy change and workforce development. A person-centered approach to care is adopted that emphasizes improving participants' activities of daily living and encouraging exercise and involvement. An external provider used a participatory evaluation approach to support the first phase of Well for Life. This approach enabled funded agencies to plan, implement, and monitor progress in their projects and assist them to embed Well for Life into policy and practice. The evaluation not only revealed insight into the delivery and adoption of Well for Life, but also assisted management and staff to make Well for Life a reality. Many PSRAC residents and PAG participants have reported significant gains from their involvement in Well for Life. These achievements reflect staff enthusiasm and organizational commitment to Well for Life principles. Evidence supports the benefit of improved nutrition and physical activity for people of any age and Well for Life is adding to the evidence-base for frail older people.

KEYWORDS: older; frail; elderly; fitness; wellness; physical activity; nutrition; well for life; go for your life

BACKGROUND

There has been extensive collaboration between the Victorian State Government Department of Human Services (DHS) and The University of Melbourne in the implementation of the Well for Life program throughout Victoria, Australia.

Address for correspondence: Wendi A. Kruger, Level 12, 50 Lonsdale Street, Melbourne, Victoria 3000, Australia. Voice: 03 9096 7076; fax: 9096 9163.
wendi.kruger@dhs.vic.gov.au

Ann. N.Y. Acad. Sci. 1114: 337–342 (2007).
doi: 10.1196/annals.1396.034

Well for Life is a publicly funded intervention that aims to improve nutrition and increase physical activity among frailer older people. There is evidence that improving nutrition and increasing physical activity has significant health benefits at any age.[1–6] Consequently it is equally important, and maybe more so, for older people to improve their nutrition and increase levels of physical activity to achieve health gains. These health gains in turn can assist an older person to remain functionally able or return to a level of functional ability that may have been lost.

Well for Life was developed with the National Ageing Research Institute (NARI) and the Dietitians Association of Australia (DAA), Victorian Branch, in 2000. It operates under the 'Go for your life,' banner, which is a Victorian State Government initiative to improve physical activity, nutrition, and social involvement for all ages.

In this paper we describe the experiential findings of older people and their involvement in Well for Life. Their stories contribute to the still-developing body of evidence to support the benefits of improved nutrition and increased physical activity among this particular cohort.

WELL FOR LIFE APPROACH

Well for Life emphasizes the need for service providers and organizations to work in partnership. It leads by example in working across aged care and other sectors through team building and the development of resources. It promotes a partnership approach for agencies in developing funding submissions and enhances sustainability beyond the funded period. More than 100 agencies that provide services to residential aged care facilities and community care Planned Activity Groups (PAGs) in Victoria have received Well for Life Incentive Grants since 2003.

Another aspect of the Well for Life approach is to provide opportunities for workforce development, particularly staff training, to improve outcomes and encourage sustainability of activities beyond the funded period. Staff training enhances the capacity of staff to assist in improving the quality of life of older people.

We found many agencies chose to undertake staff training in tai chi to encourage frail older people to be involved in more physical activities. Feedback has shown that tai chi has been especially popular with people with dementia, as the person is able to mirror the moves developed by the instructor. The popularity of this form of physical activity has been such that specific Tai Chi for People with Dementia courses have been developed through funding from the Victorian Government.

CAPACITY BUILDING AND EVALUATION

Well for Life enabled funded agencies to achieve the desired outcomes of improved nutrition and physical activity through Participatory Evaluation. This method combines external evaluation with assistance in *project planning, implementation*, and *monitoring*.

An external provider was engaged to provide this support to the first phase of Well for Life projects as well as undertake an evaluation of project outcomes. It was through this support, provided for the 12-month funding period, that Well for Life participating agencies were assisted with necessary principles and practice to promote a change in organizational culture from one that in many instances operates from a medical model to one that incorporates a more health-promoting person-centered model of care and service.

Further phases of Well for Life have also been provided with support and capacity building to assist in achieving the best possible results for agencies and the older people they serve. The concentration has been on planning and implementation; improved capacity to provide health-promoting interventions; and assistance in developing knowledge and skills to achieve internal project evaluation and quality improvement. Overall the aim is to improve outcomes for frail older people.

The evaluation of Well for Life revealed that many assumptions exist around what older people are able to do. Staff can hold assumptions about older people and sometimes expect them to remain frail and unable to perform many activities. The older person can also be limited in his or her thinking about what they can do. Well for Life has provided the impetus to confront these thoughts and beliefs: challenging those assumptions and encouraging both staff and participants to be actively involved. By engaging in physical activities that assist muscle length and strength, older persons can participate more in their own routines of daily life, such as brushing their hair and grooming.

There is indeed a relationship between becoming more physically able and having the ability to accomplish things that we may tend to take for granted. This then assists the person to feel better emotionally and physically.

EXPERIENTIAL FINDINGS

Many stories emerged of what participating in Well for Life was achieving for people. *Well for Life: A Collection of Stories*[7] was created to give the human perspective to the publication *Well for Life: Highlights from the Evaluation.*[8]

The story collection provided an opportunity for staff and participants involved in Well for Life to share their accounts and provide some examples of how *Well for Life* has assisted them. It has also given rise to people more readily sharing their stories. One such story was of a woman who had been involved in Well for Life and had been prone to falls. She had been participating in exercises in the Well for Life program yet had another fall. Although the physical activities she had been doing did not prevent her from falling, she was able to control how she fell and, importantly, she was able to get up from the floor after she fell, which she had previously been unable to do. This gave her a sense of greater control.

Other stories attest to how staff themselves have benefited from Well for Life by being exposed to a variety of activities, including providing participants with different food choices. Staff involved have also stated that seeing older people in a different light has given them insight into the ability of older people to adapt to more interesting experiences than they thought possible.

In one residential facility, when an unused staff room was set up with an electric organ, a mini pool table that had been in the junk room for some time, and an air hockey table, it surprised the staff that even the women had a go. Staff were amazed when a woman of 90 and another of 84 were both very keen to play. Both women were quite proud of their age and their ability to be involved.

In the same facility an 83-year-old Italian man who has dementia went along to the newly created activity room and saw people with cues. He commented "I could have a go at that!" So another resident, a past pool champion, decided to teach this "new chap" how to play. In the middle of playing, the new chap walked over, picked up the chalk, and chalked the cue. No one had shown him how to do this and no one had done it before him, so he was not copying. Once again the staff and other residents were amazed by what older people could do when given the opportunity. As the story goes, the new chap often wandered up the hall to the room to have a game. Even though he may have needed a bit of encouragement at times, he continued to play and amaze people with his knowledge.

Another story involved a PAG that ran a water program through the local hospital. One woman, who had recently turned 90, attended the water program and returned to the activity group to tell people, a group in their 80s and 90s, how beneficial it was. She then proceeded to "do" all the exercises she had learnt in the water program on dry land, with a microphone in her hand, to demonstrate what she had done in the class. One 84-year-old man in the group, who used a walking frame said, "if that old lady can do that, I can do that too!" He went along to the program. It was learned that until then he had not been in the water for 45 years.

Five months after starting the gym group, a man who felt he had gained a lot from his involvement in Well for Life said that he had a new lease on life. He no longer felt lethargic and could walk faster without experiencing pain or shortness of breath. His cholesterol levels decreased, his posture improved, and his weight decreased. Although he had been unable to completely give up smoking, he did cut down significantly.

One last story was of a novel activity that was conducted at another residential facility. Residents used areas within the facility to walk the distance of the Commonwealth Games Queen's Baton Relay Path, as the games were about to commence in Victoria. One man, who was not very motivated to exercise or walk much as he had suffered a stroke some years ago and was using a wheelchair and walking only with staff supervision, has progressed enormously. He now manages up to 18 kilometers on the exercise bike in one session and frequently requests to go walking with staff. Previously, not a man to converse at length, he now talks shyly, but with pride, in his improvement and contribution to Well for Life. Not only has the program improved his strength and endurance, but being involved has also given him a sense of worth and achievement.

CONCLUSION

Providing opportunities for older people to be involved in their food choices and activities, and for staff to be involved in workforce development and training, combined with a bit of encouragement and persistence with staff and participants, can go a long way.

It helps to break down the barriers created by assumptions and stereotypical attitudes and to embed change from a medical/institutional model to a more person-centered approach within the organizational culture, through support and capacity building.

Well for Life, as a targeted collaborative program aiming to improve nutrition and levels of physical activity among frail older people, has provided the opportunity for this to occur in many public residential aged-care facilities and community care PAGs across Victoria. Many people have had a wonderful time in the process as Well for Life becomes a way of life for older people.

ACKNOWLEDGMENTS

We would like to thank Well for Life staff and participants and The University of Melbourne for their involvement.

REFERENCES

1. Prime Minister's Science Engineering and Innovation Council. 2003. Promoting Healthy Ageing in Australia. Canberra.
2. Gregg, E., M. Pereira & J. Caspersen. 2000. Physical activity, falls, and fractures among older adults: a review of the epidemiologic evidence. J. Am. Geriatr. Soc. **48:** 883–893.
3. Finnucane, D., L.C. Giles, R.T. Withers, *et al.* 1997. Exercise profile and subsequent mortality in an elderly Australian population. Aust. N.Z. J. Publ. Health **21:** 155–158.
4. Simons, R. & R. Andel. 2006. The effects of resistance training and walking on functional fitness in advanced old age. J. Ageing Health **18:** 91–105.
5. Mazzeo, R.S., P. Cavanagh, W. Evans, *et al.* 1998. ACSM Position Stand: exercise and physical activity for older adults. Med. Sci. Sports Exercise **30:** 992–1008.

6. FIATORONE, M.A., E.I. O'NEILL, N.D. RYAN, *et al.* 1994. Exercise training and nutritional supplementation for physical frailty in very elderly people. N. Engl. J. Med. **330:** 1769–1775.
7. VICTORIAN GOVERNMENT DEPARTMENT OF HUMAN SERVICES. 2006. Well for Life: A Collection of Stories. The Department. Melbourne.
8. VICTORIAN GOVERNMENT DEPARTMENT OF HUMAN SERVICES. 2005. Well for Life: Highlights from the Evaluation—Phase One projects. The Department. Melbourne.

A Management-Control System to Assist with the Development, Contracting, and Monitoring of New Services for Older People

You *Can* Always Get What You Want (with Apologies to the Rolling Stones)

STEPHEN JACOBS

Senior Research Fellow, Faculty of Medical and Health Sciences, The University of Auckland, Auckland, New Zealand

ABSTRACT: Implementing community services can be a very complex process where people end up "not seeing the wood for the trees" and then accepting a compromise nobody wants. Neither agency theory nor stewardship theory is particularly robust when explaining the reality that funders and managers of community health service for older people face in managing contracts. There are particular difficulties with goal divergence between policy directives and implementation practices, which result from goals not being well understood and performance requirements being ambiguously defined and infrequently monitored. This leads to accountability issues for public managers, and raises efficiency and effectiveness issues. This paper provides an interim report on a project developing a management-control system to provide a process for health planners and funders to use when developing and contracting services. A major focus is inviting stakeholders into a process of communication that ensures that they have shared understandings of where they want to get to, and what has to be done to get there.

KEYWORDS: management control; older people; performance management; contracting; stewardship; community services; health services; aging

INTRODUCTION

In New Zealand, the Government is the major funder of health services for older people, mostly through allocating resources to 21 District Health Boards

Address for correspondence: Stephen Jacobs, 72 Thompson St., Te Aro, Wellington 6011, New Zealand. Voice: 6443828275.
jacobs.sp@gmail.com

**Ann. N.Y. Acad. Sci. 1114: 343–354 (2007). © 2007 New York Academy of Sciences.
doi: 10.1196/annals.1396.013**

(DHBs) that are charged with providing services that will improve the well-being of their populations. These DHBs then contract their own provider arms, 81 primary health organizations, and a myriad of for-profit and not-for-profit organizations, to deliver services.

Guidance is provided to the DHBs about what they are expected to provide through both strategies—there are more than 35 Government health strategies[1–4]—and a number of guidelines.[5,6] DHB planners and funders of services for older people theoretically know what they are expected to contract for. However, there is often no clarity about how best to develop and implement services that will achieve the required goals or vision.

An example of the difficulties of implementation is care-management services to assist persons with complex and multiple conditions. The effectiveness of the care plan put together after an assessment of a person's needs has been shown to be reduced (understandably) through failure to follow the recommendations by either the person being assessed or the healthcare professionals involved. Often a care plan may be developed but not implemented.[6] However, up to 70% of initiatives designed to improve achievement of desired outcomes through improving management and organizational performance have been shown to fail because of poor design or difficulties in implementation.[7] Learning from evaluation has also been difficult, because results may be delivered late or in a way that practitioners cannot understand or put into practice.[8]

The quality of public services delivered by contractors depends largely on the quality of contract management provided by public managers.[9] However, contract management for health service provision is difficult, a major factor being goal divergence between policy directives and implementation practices as a result of misunderstandings about policy directives and goals, ambiguous definition of performance requirements, and infrequent monitoring. This leads not only to accountability issues for public managers, but also raises efficiency and effectiveness issues. Van Slyke talks of "a complex managerial environment . . . devoid of competition, administrative capacity, and clearly defined and agreed upon contractual goals and outcomes" (p. PNP:D3).[9]

This paper is a progress report on a project undertaken in pursuit of a Ph.D., with the final report due at the end of 2007. The project seeks to develop a systematic approach to developing and implementing new services, that is, to create an instrument that will assist planners and funders in developing and contracting new services. The objective is to enable the development of an action plan for the delivery of services in which all key stakeholders agree:

- on the vision (where they're going)
- with, and are committed to, the objectives of the program
- on the action plan for implementation
- on the critical success factors and how they will be measured
- on a feedback loop, so that the action plan can be reviewed on the basis of evidence.

The instrument will assist planners and funders to build trusting relationships with service providers and other stakeholders on the basis of a shared understanding of goals and success factors, tasks required, task responsibility, and outcomes expected. These can then be written into contracts.

The initial theoretical basis for the instrument was a combination of management control, performance measurement, and management models, and theories about how people communicate with each other. Performance measurement and management theories were used because of the evidence that behavior is influenced by something's being the subject of measurement.[10] The key theories used in the initial development are here discussed.

THEORIES USED IN DEVELOPING THE INSTRUMENT

The V-Model

Based on Juergen Habermas's Theory of Communicative Action, the V-Model[11] postulates that systemic knowledge is holistic understanding of:

a. The norms or expectation of people in a system;
b. The technical system; and
c. The relationship between the two.

Eliciting systemic knowledge is likened to a journey of discovery, with the success of the journey being validated by the commitment of the stakeholders to an agreed-upon process for achieving technical excellence. The V-Model (FIG. 1) provides a process for discussion between stakeholders about their intentions for any program and how they will move from intentions to achieving the outcomes they desire.

The Balanced Scorecard

Influential in health over the last decade,[12–15] the key contribution of the Scorecard model in this project, is the concept that a number of perspectives other than financial are needed. Kaplan and Norton[12,13] proposed four perspectives:

d. Customer—What does the customer want? This has been translated into "Does the service meet the goals and needs of the older person receiving services, and those of their informal carers?" This adjustment has been made because if the person receiving the service is accessing services through the public system, which will usually be the case, then they will not be directly funding that service.
e. Internal business—What must the service excel at?

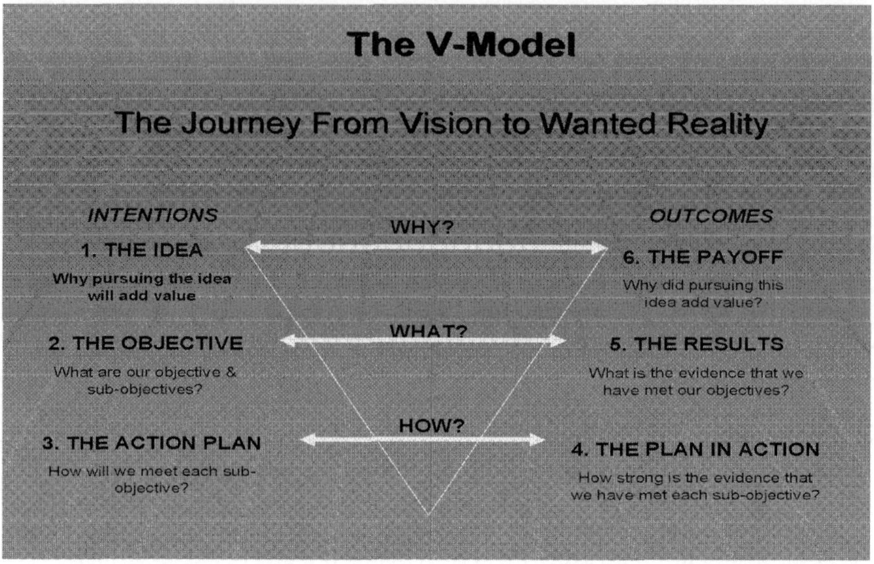

FIGURE 1. The V-Model: the journey from vision to wanted reality.[11]

f. Growth—How can the business continue to improve?
g. Financial—How do funders perceive the service?
h. Community—How does the service have an impact on the health status of the population?[16] This fifth perspective has been added for health services that are publicly funded.

The Performance Prism

The Performance Prism[17] is an approach that raises the concept of the stakeholder's contribution. In health, very few people are passive receivers of services. As shown by the example of adherence (or not) to care plans, it can be seen that the person receiving services contributes positively or negatively to the successful implementation of those services, as has been shown by the impact of adherence or non-adherence to treatment and/or care plans.[6]

The Results-Based Logic Model

Developed in Canada by the Treasury, the Results-Based Logic model[18] puts forward the concepts that:

i. Accounting for results means demonstrating that you have made a difference; and

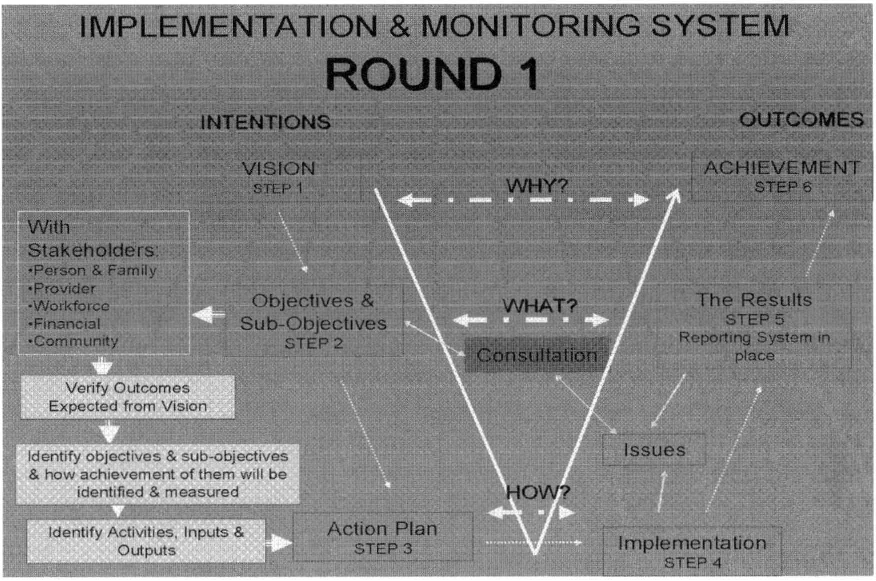

FIGURE 2. A process for engaging with the stakeholders.

j. A common theory of logic links can be used. Evidence of the most effective method of operating is often not available. By using consultation as well as evidence, linkages can be drawn between the set of activities that make up a program and the sequence of outcomes that can be expected to flow from those activities.

This project has involved working with three DHBs, with Capital and Coast DHB (CCDHB) being the first. CCDHB is responsible for health services to approximately 250,000 people in the lower part of the North Island of New Zealand. Seeking to find ways to implement the Health of Older People Strategy within community-based services, CCDHB faced issues of service stability and sustainability, high worker turnover, quality and safety, service gaps, inappropriate existing services, and services that created dependency in patients. CCDHB consulted its population to develop a new agreed-upon approach to integrated home and community care, working within the principles that older people want to have control over their own lives and to maintain and regain health and function as much as possible. The Planning and Funding team was charged with designing the service framework and developing the appropriate services to deliver the new integrated restoration-focused approach. CCDHB anticipated working with both existing and new providers to shift to this new model of operation.

CCDHB's Outcome and Process Key Words	
▪ Strengths based / Active decision making	▪ Flexible
▪ Cultural sensitivity	▪ Ongoing monitoring of client
▪ Restorative (Rehab / Habilitation)	▪ Monitoring mechanisms
▪ Clinical input	▪ Budget management
▪ Outcomes measures	▪ Cost-effective
▪ Enhance social support	▪ Stabilise workforce
▪ Social participation	▪ Provider capacity
	▪ Options
	▪ Responsive

FIGURE 3. Key describing desired outcomes and processes.

CAPABILITIES				
What each stakeholder must do well if the desired outcomes are to be achieved	**What competence and capacity is required?**	**What do they have to do to become competent and capable?**	**What assistance do they need from other stakeholders?**	**What is the timeframe?**
LONGTERM OUTCOME SOUGHT:				
Capability Needed:				
Client				
Informal Carers				
Assessment & Care Planning Service				
Package of Care Provider				
Funder				

Outcome	Capabilities	Indicators (How would you know that the outcome is being achieved or that the capabilities are in place?)	Measurement (How would you measure the outcomes or capabilities?)	Target (What level of achievement means that the service is achieving success in this domain?)

FIGURE 4. The discussion template.

THE INSTRUMENT

From the theoretical base, a process was developed for engaging with the stakeholders within CCDHB to develop an agreed implementation plan to develop the new services. The process is shown in FIGURE 2.

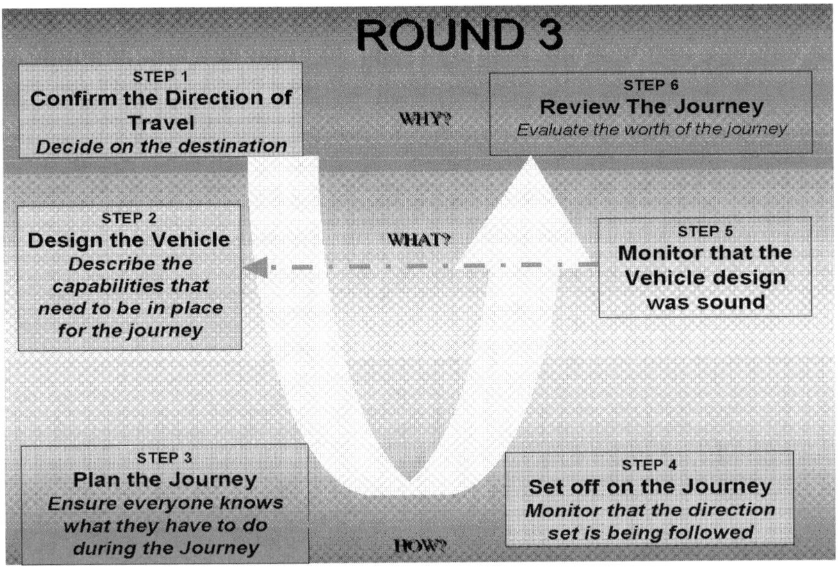

FIGURE 5. The revised V-Model.

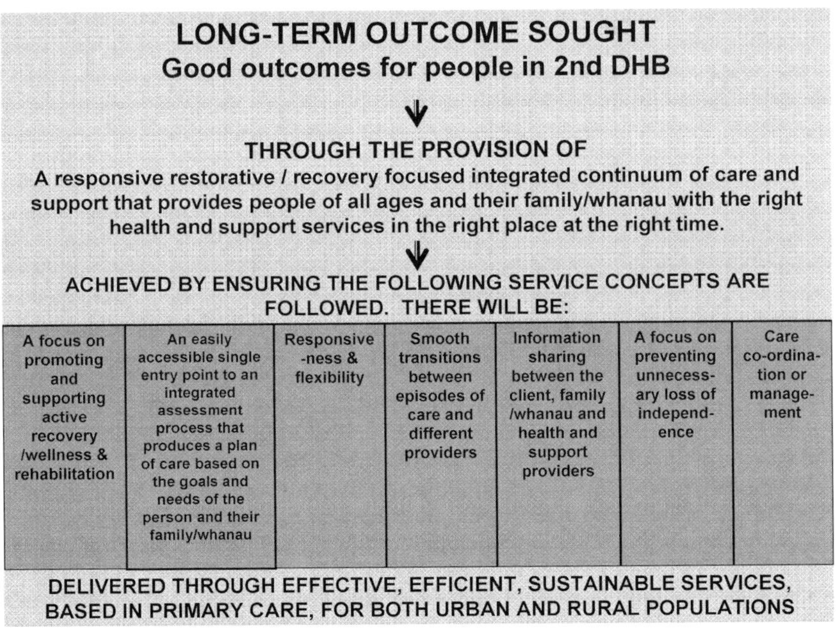

FIGURE 6. The one-page map for the second DHB.

The researcher (S.J.) met with the CCDHB Planning and Funding staff to clarify their understanding of the vision in their consultation document. First, the researcher identified what appeared to be key words in the document referring to outcomes to be achieved and processes to be followed to achieve those outcomes. These are shown in FIGURE 3.

After discussion, these became stated as the following principles:

1. Client *mana* and dignity are preserved.
2. The client receives a service focused on restoring and/or maximizing health and functional ability, with the care and/or service plan developed and delivered in a manner that:

    a. meets their goals, needs and manages their risks, including the goals and needs of their informal/family/*whanau* carers
    b. ensures that the client receives integrated and coordinated services.

3. Social support for the client is enhanced.
4. The services the client receives are solution-focused, flexible, responsive, and provide different options.
5. The client receives a service that is sustainable.

Next these principles were presented to providers of community support services in a format that aimed to assist with discussion about how the principles would be put into action by the providers. The templates shown in FIGURE 4 were used to guide discussion on each principle, seeking agreement about what would actually have to be done, and how success would be measured.

The providers involved were resistant to the process and the approach proved to be very unsatisfactory. It was learned that:

- Stakeholders may interpret the same words differently. Agreement to a written statement may not actually be agreement about what has to happen.
- Providers may think they understand what is required, and may feel they are already doing it. While they may expect only a minor tweaking of what they already doing, this is in actuality a *major* issue for which a significant shift in provider culture is required.
- Providers want to show they understand, even when they don't, especially when contracts and future funding are involved.
- Talking about measurement and linking the achievement of outcomes to payment for services at an early stage in the development process causes caution and anxiety.
- Academics can use terms that every day providers don't understand easily.
- The templates used had a sound theoretical base, but they did not work as a mechanism for assisting communication. People felt overwhelmed and threatened by a plethora of empty boxes that they were asked to fill in for each principle.

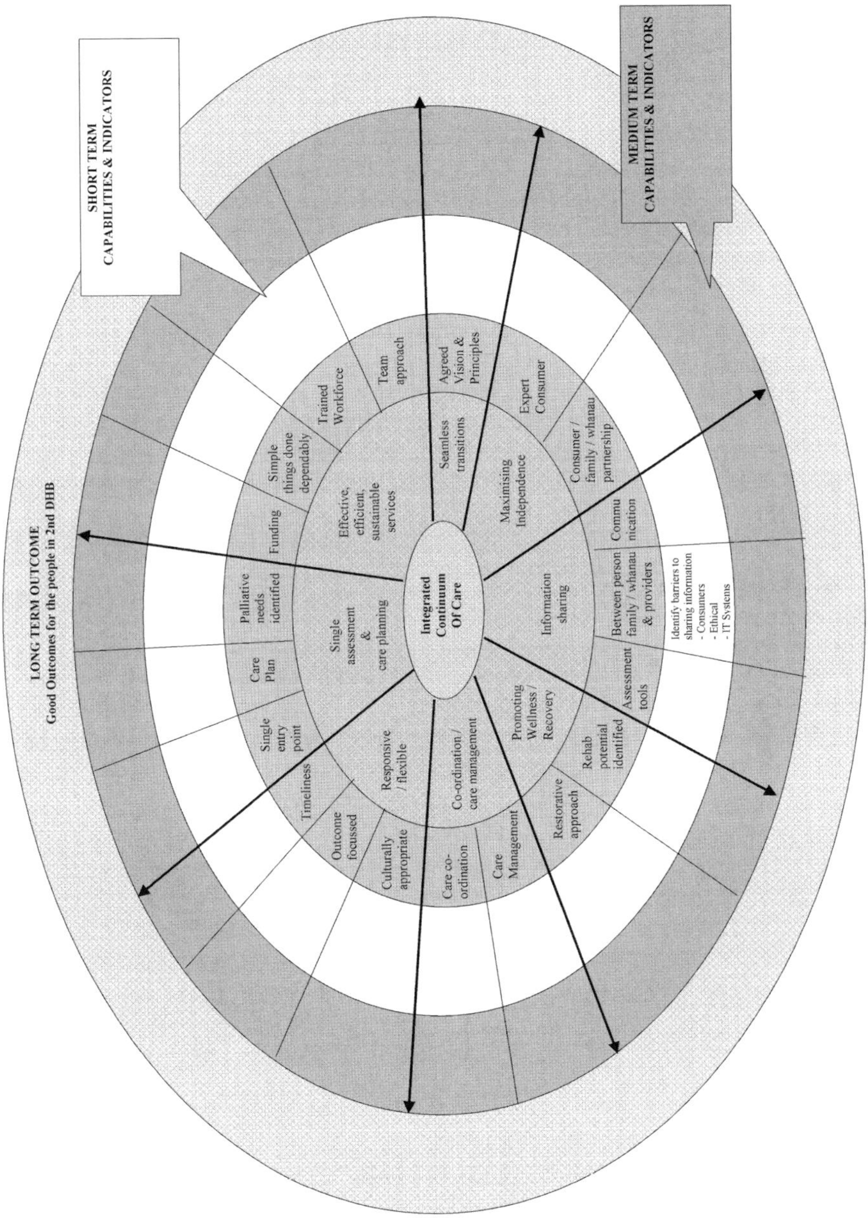

FIGURE 7. The target diagram for the second DHB.

Outcome Expected:				
The Competency Required: What this means:				
The Contribution required from each stakeholder				
Person	**Family/Whanau**	**Assessment and Coordination Service**	**Package of Care Providers**	**DHB**

FIGURE 8. A template for discussing stakeholder contribution.

A Process for Measurement

Outcome	Capabilities	Indicators How would you know that the outcome is being achieved or the capabilities are in place?	Measurement How would you measure the outcome or capabilities?	Target What level of achievement means that success is being achieved?

FIGURE 9. A template for discussing measurement.

It became clear that all stakeholders, but especially providers, need to be invited into an experience that they value, rather than being faced with an experience in which they felt out of their depth and threatened.

A new approach was developed for a second DHB that was also working to develop a community-based integrated continuum of care. At this stage, learning theory as applied in computer-game development was accessed. In this approach success lies in weaving the learning experience into the playing experience, so that "the user as hero" finds it easy to begin and fulfilling to continue, with an emphasis on "Giving your users an "I rule" experience."[19] The V-Model was redesigned to be more user-friendly to a non-academic audience; this is shown in FIGURE 5.

A new way of presenting the vision or "direction of travel" was developed. The idea was that if a one-page pictorial "map" showing the destination that people were working to reach was agreed upon, that would provide a starting place for discussing what success would look like and what needed to be done to achieve that success. In discussing the one-page map, stakeholders would also enhance their understanding of the concepts plus identify any real disagreements or issues. This approach builds on the V-Model theory that if people cannot agree on what success will look like, they are unlikely to agree on what has to be done to achieve it. The one-page map that the second DHB agreed on is shown in FIGURE 6.

This map was then turned into a "target" diagram (FIGURE 7). This pictorial format was developed to assist people to understand the process. A Community Consultation Group (consisting of representatives of older people, community providers, hospital providers, and funders) was shown the three diagrams—the V-Model, the "map" and the "target."

After discussing the presentation, the Group was invited to form small teams to identify what each stakeholder would have to do to ensure that a particular capability or set of capabilities was present in their district, and how they would identify success. The templates used to assist them to do this are shown in FIGURES 8 and 9.

The process and templates used with the second DHB were much more successful than the initial process and templates had been with the CCDHB. The Community Consultation Group willingly participated and produced a set of necessary activities and indicators of success for the DHB to act on.

The model is being further refined and tried with a third DHB, with ideas from other theories being included, particularly theories about contracting. This management-control system provides a mechanism for funders to develop shared understandings and trust relationships with service providers, while also providing for the development of agreed-upon indicators of success, thus giving the funders an accountability mechanism. This provides an organized approach to contract management in an area in which neither agency theory nor stewardship theory is particularly robust when explaining the reality funders and managers of community health service for older people face today.

ACKNOWLEDGMENTS

Thanks go to members of the DHB staff that have been so helpful, particularly to Shereen Moloney from CCDHB, who assisted with a presentation of this project at the International Conference on Healthy Ageing and Longevity in Melbourne.

REFERENCES

1. MINISTER OF HEALTH. 2001. The Health of Older People Strategy. Ministry of Health. Wellington, NZ. www.moh.govt.nz.
2. MINISTER OF SOCIAL DEVELOPMENT. 2001. New Zealand Positive Ageing Strategy. Ministry of Social Development. Wellington, NZ.
3. MINISTER OF HEALTH. 2001. The New Zealand Disability Strategy: making a World of Difference. Ministry of Health. Wellington, NZ.
4. MINISTER OF HEALTH. 2001. The Primary Health Care Strategy. Ministry of Health. Wellington, NZ.
5. MINISTRY OF HEALTH. 2004. Guideline for Specialist Health Services for Older People. Ministry of Health. Wellington, NZ.
6. NEW ZEALAND GUIDELINES GROUP. 2003. Assessment Processes for Older People: best Practice Evidence-Based Guideline. Guidelines Group. Wellington, NZ.
7. NEELY, A. 2003. Why measurement initiatives fail. Quality Focus **4:** 3–6.
8. DUIGNAN, P. 2001. Mainstreaming evaluation or building evaluation capacity. Journal of Alcohol and Public Health Research Unit Archive Site: Presentation to American Evaluation Association Conference. Available at www.aphru.ac.nz/services/services/pdaea2001.htm. This paper won the American Evaluation Association (AEA) 2001 President's Prize.
9. VAN SLYKE, D.M. 2005. Agents or stewards: how government manages its contracting relationship with nonprofit social service providers. Academy of Management Best Conference Paper. New York. p. PNP:D16. Available at www.pmranet.org/conferences/georgetownpapers/VanSlyke.pdf.
10. FLAMHOLTZ, E.G. 1983. Accounting, budgeting and control systems in their organizational context: theoretical and empirical perspectives. Acct. Org. Soc. **8:** 153–169.
11. SHEFFIELD, J. 2005. Systemic knowledge and the V-Model. Int. J. Bus. Inf. Syst. **1:** 83–99.
12. KAPLAN, R.S. & D.P. NORTON. 1997. Why does business need a balanced scorecard? J. Cost Mgmt. **111:** 5–11.
13. KAPLAN, R.S. & D.P. NORTON. 2001. Building a strategy focused organization. Ivey Bus. J. **May/June:** 12–19.
14. PINENEO, C.J. 2002. The balanced scorecard: an incremental approach model to health care management. J. Healthcare Finance **28:** 69–80.
15. PINK, G. *et al.* 2001. Creating a balanced scorecard for a hospital system. J. Healthcare Finance **27:** 1–20.
16. LEGGAT, S.G. & P. LEATT. 1997. A framework for assessing the performance of integrated delivery systems. Healthcare Mgmt. Forum **10:** 11–18.
17. NEELY, A., C. ADAMS & P. CROWE. 2001. The performance prism in practice. Measuring Bus. Excel. **5:** 6–12.
18. TREASURY BOARD OF CANADA. 2001. Guide to the Development of Results-Based Management and Accountability Frameworks [monograph on the internet]. Treasury Board of Canada Secretariat. Ottawa, Ontario, Canada. Available from http:/www.tbs-sct.gc.ca/eval/pubs/RMAF-CGRR/rmafcgrr_e.asp.
19. SIERRA, K. 2006. Creating passionate users. www.headrush.typepad.com.

The New Caring

Financial Asset Management and Older People

CHERYL TILSE,[a] JILL WILSON,[a] DEBORAH SETTERLUND,[a] AND LINDA ROSENMAN[b]

[a]*The University of Queensland, St. Lucia, Queensland, Australia*

[b]*Victoria University, Melbourne, Victoria, Australia*

ABSTRACT: Increasing longevity and the growing proportion of the aged in the population in most countries have served to focus on the question of how governments and older people can finance living, health, and care options in retirement. Prudent management of income and assets is an increasingly complex and important aspect of aging as assets and expectations of self-financing increase. Although many informal caregivers act as asset managers and/or substitute decision-makers for older people, little attention has been paid to this increasingly important aspect of care. This paper summaries key findings of a broad research program exploring family involvement in the management of older people's assets and the practices that constitute good practice as well as financial mismanagement and abuse. It identifies multi-level and multi-strategy responses needed to address the issues raised by the research and outlines an innovative community demonstration project aimed at improving financial management practices in relation to older people's assets.

KEYWORDS: **aging; assets; caregiving; financial abuse**

INTRODUCTION

The increasing numbers and proportion of older people living to late old age in countries with aging populations means that many older people have long periods of post-retirement living to provide for. Financial assets will thus be managed over a period of time when opportunity to generate or regenerate an asset base can be reducing and the likelihood of impairment that affects capacity to self-manage assets is increasing. Healthy aging associated with this longevity will require conserving income and prudent management of assets

Address for correspondence: Dr. Cheryl Tilse, School of Social Work and Applied Human Sciences, The University of Queensland, St. Lucia, 4012, Queensland, Australia. Voice: +65 7 3365 3341; fax +61 7 3365 1788.
c.tilse@social.uq.edu.au

Ann. N.Y. Acad. Sci. 1114: 355–361 (2007).
doi: 10.1196/annals.1396.015

to ensure that choice in lifestyle, health care, and care and accommodation in later life is protected.

Older people's financial assets are currently being managed in an environment of increasing complexity, competing interests, and concerns about financial abuse.[1] For some older people, retirement incomes in Australia have become increasingly complex, with income support comprising combinations of state-based, occupational and overseas pensions, superannuation payouts, and/or incomes from investment and property. The intersection of systems of income support, taxation, and user fees and charges for care further complicate this environment. In addition, families are also becoming more complex in their structure and responsibilities as a result of changes in divorce rates, blended families, and dispersion of the population. Older people's possession of substantial assets (in particular, home ownership in Australia) raises expectations of inheritance and potential intergenerational transfers, particularly in exchange for family care of older members. The interests of the state, service providers, the market, the family, and the older person can thus be in conflict over how the assets are managed, what and who they provide for, and who is entitled to them in the short and longer term.[2]

The form of older people's assets, the complexity of the environment in which they are managed, and the likelihood of increased levels of disability and dementia as part of longevity means that families are frequently involved in assisting older people with managing their assets. Such assistance is most often provided when sensory, physical, and/or cognitive impairment disrupt the capacity of an older person to either make or implement financial decisions. Paying bills, doing the paperwork and managing money, property, and investments are care tasks commonly undertaken by predominately by family (but also by friends and neighbors).[2] Although these tasks have generally been recognized as part of family caregiving, they have been given limited attention in the research literature and in the development of support services for caregivers. There is also a limited recognition that assistance with asset management is qualitatively different from other informal caregiving. The major differences are that assistance (and abuse) can take place from a distance: it only requires access to assets[3]; it can require specialized knowledge and skills; it takes place in the context of cultural values around money in families; and it operates within a legal and financial framework rather than a more usual social and health framework.

The Assets and Ageing Research Team at the University of Queensland have an extensive program of research (1999–2007) exploring issues relating to the management of the assets of older people. Within the research program, managing assets is defined as having some control over organizing, decision-making, and/or use of an older person's assets. Both income and capital assets are included. The program focuses on nonprofessional asset management understood as the involvement of family, friends, and neighbors in managing assets. This paper brings together key findings from several projects on asset management and financial abuse to identify a range of policy and practice issues. It then briefly describes a community demonstration project developed to address the issues identified.

KEY FINDINGS FROM THE RESEARCH

The research findings reported here are based on four research projects completed in Australia between 2002 and 2006. These comprise:

- a national prevalence survey;
- in-depth interviews with 81 nonprofessional asset managers (e.g., family friends and neighbors),
- in-depth interviews with 34 older people receiving assistance with asset management; and
- an analysis of the files of the Guardianship and Administration Tribunal in Queensland to establish the factors associated with the financial abuse of older people with impaired capacity.

The methodology and sampling strategies for the projects are reported elsewhere.[1–4] The focus of this paper is on key findings and their implications for policy and practice responses. The key findings are: (*a*) managing assets is a common task undertaken by informal carers; (*b*) older people receive help for a range of reasons and express a range of views on the desired level of participation in managing their assets; (*c*) while many carers have good practices in place, a significant group of carers have a range of risky practices and attitudes that enhance opportunities for financial abuse; and (*d*) current legal mechanisms for substitute decision-making, such as enduring or durable powers of attorney, do not necessarily protect older people from financial abuse.

Managing assets is a common task of informal carers. In the national prevalence study, one in four Australians reported having assisted an older person with managing their assets in the previous year. Although most commonly assistance was provided with the day-to-day tasks of paying bills, doing the paperwork, and accessing money, more complex tasks of managing pensions, investments, and property were also reported. Asset managers are most commonly adult children assisting parents and/or parents-in-law, although friends and neighbors were involved in 25% of asset management cases reported. Families were more likely than friends and neighbors to be involved in the more complex tasks and the management of larger assets, perhaps reflecting the cultural view that decision-making about money and property is primarily a family concern rather than the responsibility of friends and neighbors.[5]

Older people receive asset management assistance from family members for a range of reasons and have varying perspectives on this experience. Major reasons for family involvement in this area of their lives are: a lack of confidence in managing assets or inexperience in dealing with financial institutions or complex papers as well as physical and sensory impairment, poor health, mobility problems and cognitive incapacity.[1,2] This suggests there are environmental barriers to self-management as well as individual impairments. In the study exploring older people's perspectives on having their assets managed, older people reported three types of experiences ranging from full, partial, or no participation in asset management.[4] The outcomes of this assistance as assessed by older people ranged from reflecting their wishes to not having the desired level of participation.

A number of scenarios from this study illustrate levels of assistance in conjunction with the desirability of the outcome from the older person's perspective. In the first scenario, the older adult has full or partial control/participation in keeping with her/his wishes and expectations. This group was primarily older people receiving assistance with assets because of physical or sensory impairment. Their asset managers included them in decision-making or monitoring asset management by compensating for the gaps arising from impairments. There was some recognition that this, at times, places considerable stress on the carer. Carers who reported asset management as a difficult task were more likely to be helping with a range of tasks, to report complexity in the tasks, to be inconvenienced by the time and effort involved, and to be assisting an older person with dementia.

In the second scenario, the older adult has very limited participation, the asset manager has assumed full responsibility for management, and the older person is satisfied with their low level of participation.

> Mrs. M expresses relief at being cared for and recalls that her son *"took over"* her finances. Mrs. M is extremely happy with this arrangement. However, her son does not keep any records of his withdrawals from his mother's account and he provides irregular payments to his brother.

In the most worrying scenario the older adult does not have the level of control or participation he/she seeks.

> Mrs. L's daughter manages all of her finances. Mrs. L would like to be more involved and informed, but fears her daughter's reaction. *I wish that I was not frightened to ask.*

Good practice was also identified in careful record-keeping, maximizing the involvement of older person where possible in monitoring activities, and keeping finances separate and making decisions transparent by including others. These practices were identified by some carers as time-consuming, inconvenient, and onerous. Informal asset managers use a range of risky practices to asset-manage for and with older relatives for reasons of convenience, family values based on trust, ignorance of legal responsibilities, or intentional abuse. Asset managers in the in-depth study and in the national prevalence survey reported a range of informal mechanisms to manage assets. Informal mechanisms included having the older person sign checks and blank withdrawal forms and using the personal identification number of the older person to withdraw money electronically as well as more formal arrangements with banks and/or an enduring power of attorney or guardianship arrangement. The research also shows undeveloped accountability processes reflected in poor or no record-keeping, mingling of the finances of the older person and those of the asset manager, and limited knowledge of responsibilities as a substitute decision-maker under an enduring power of attorney arrangement. Underpinning attitudes of expecting that families can or should be trusted with this task may be barriers to setting up or monitoring accountability processes. When this set of beliefs is combined with a sense of entitlement to older people's assets, based on ageist beliefs and expectations that future inheritances could be accessed early for personal and family use, an environment is created where abusive practices can develop.

In summary, the research suggests that knowledge, attitudinal, systemic, and environmental barriers exist to older people's self-managing assets and/or carers' managing effectively on behalf of older people. Interventions will need to take account of sociocultural attitudes, issues of information, support, and convenience for carers, and concerns about protection and empowerment of older people.

A COMMUNITY RESPONSE TO ASSET MANAGEMENT

The identified issues of risk, lack of knowledge, and barriers to protective management practices suggest that a whole-of-community approach is needed to build asset management capacity. A whole-of-community approach is multi-level (addresses individual, family community, systems, and sociocultural issues); multi-strategy (education, support, advocacy, service delivery responses and policy changes), proactive as well as reactive, and seeks to enhance monitoring and involvement of carers as well as facilitating autonomy and protection of older people.

A whole-of-community approach was used by the research team as the guiding framework for implementation of a one-year community development demonstration project aimed at improving asset management capacity in Redcliffe, a coastal city with a high proportion of older people in Queensland, Australia. Multi-level, inter-sectoral strategies were developed to focus on (*a*) raising awareness of elder financial abuse and the need for protective asset management practices at the service provider and community levels; (*b*) developing inter-organizational protocols for responding to elder financial abuse; and (*c*) delivery of training programs and resource materials for older people, carers, and staff working with older people and their families. A community development worker was located in Redcliffe. All project development ideas and decisions were developed collaboratively with key public and private sector interest groups, including representatives from older people's and carer organizations, the local city council, local banks, local police, financial advisors, indigenous (Aboriginal and Torres Strait Islander) and multi-cultural organizations, government protective services, and community and residential aged-care service providers.

At the sociocultural level, strategies, such as radio interviews, presentations at relevant organizations, newspaper articles, and an information booth at a health/aging exposition, aimed to raise awareness of asset management issues, risky practices, and financial abuse. The strategy also sought to improve financial literacy more generally as well as addressing ageism, attitudes toward older people's assets, and planning for later life. Educational workshops for older people aimed to build confidence, skills, and feelings of self-efficacy for older people who wish to retain involvement and encourage planning and arrangements for future asset management. Educational workshops for carers aimed to provide support and improved asset management skills by increasing knowledge of responsibilities as well as addressing attitudes and risky practices.

The evaluation of the community project is still in progress. Different workshop formats and approaches and a range of educational materials (fact and information sheets) have been put to trial and evaluated. A workbook for indigenous people has been produced and is currently being further developed by involving young indigenous people in contributing to the artwork. Inter-sectoral focus groups and a survey were used to develop and put to trial a referral pathway that identifies asset management problems and potential or actual financial abuse and sets out an appropriate, early, and coordinated response. The project findings suggest that education in this area is difficult, requiring creative ways of engaging with the topic, particularly with older people who are considering involving family members in managing their assets.

CONCLUSIONS

Asset management for and with older people must be accorded an increased role in healthy aging policy development and practice. As the patterns of assistance, the emerging tasks for carers, the complexity of the changing environments in which such tasks are undertaken, and the risks of actual or alleged financial abuse for older people and their carers are increasingly understood, the challenge is to develop innovative and whole-of-community responses that operate at the social, cultural, legal, service system, and financial level. At a systems level strategies are needed that address barriers to self-management. Enhancing good asset management practice will include ensuring access to advocacy, information, and advice; professional education that raises awareness of issues; improvement of referral networks, agency protocols, and cross-sectoral responses to concerns; and review of current safeguards in legal mechanisms, such as enduring powers of attorney. This paper seeks to make a contribution to healthy aging over the life span by raising awareness of the issues relating to asset management and suggesting some ways forward in developing a range of responses.

ACKNOWLEDGMENTS

This research was funded by Australian Research Council grants and industry partners of the Queensland Government: Office of the Adult Guardian, The Public Trustee, The Public Advocate, the Guardianship and Administration Tribunal, and the Office for Seniors. The community demonstration project was supported by the Wicking Trust, the Office for Seniors and Carers, Queensland. We wish to acknowledge the contributions of community and service providers in Redcliffe and Jennie Peut and Leona Berrie, the community development workers.

REFERENCES

1. TILSE, C., J. WILSON, D. SETTERLUND & L. ROSENMAN. 2005. Older people's assets: a contested site. Australas. J. Ageing. Special Issue on the Future of Elder Care **4:** S51–S6.
2. TILSE, C., D. SETTERLUND, J. WILSON & L. ROSENMAN. 2005. Minding the money: a growing responsibility for informal carers. Ageing & Society **25:** 215–227.
3. MCCAWLEY, A., C. TILSE, J. WILSON, *et al.* 2006. Older people with impaired capacity and financial abuse. J. Adult Protect. **8:** 20–33.
4. TILSE, C., J. WILSON, D. SETTERLUND & L. ROSENMAN. 2007. Managing the financial assets of older people: balancing independence and protection. Br. J. Soc. Work. **37:** S65–S72.
5. BARKER, J. 2002. Neighbours, friends and other non-kin caregivers of community living dependent elders. J. Gerontol. Soc. Sci. **59B:** 158–167.

Involving Service Users and their Carers as Equal Partners in a Project Using Electronic Communication

SUSAN M. ASHBY, SIAN E. MASLIN-PROTHERO, AND AMELIA C. ROUT

Research Institute of Life Course Studies, Keele University School of Nursing and Midwifery, Stoke-on-Trent, Staffordshire, United Kingdom

ABSTRACT: This paper presents a research team's experience using a virtual steering group to manage a research project. The main research project was concerned with the evaluation and development of interprofessional working among health and social care staff across primary and secondary care in the United Kingdom, with a focus on older people receiving intermediate care. Intermediate care is a service aimed at preventing unnecessary hospital admission; supporting early discharge from hospital; and reducing or delaying the need for residential care in the community. Intermediate care covers multi-agency and interprofessional working across the continuum of care. We examine one aspect—the use of a virtual steering group, which is an innovative approach aimed at facilitating user and carer participation in the research in a supportive and less intimidating way. The research team includes academics, practitioners, older people and carers, and representatives from the voluntary sector—all are equal members. The team believes it is important to actively involve clients and their carers in the development and direction of research because they can provide a unique perspective ensuring that the research addresses issues of importance to them as service users. We examine and discuss the importance of preparation, support, and training to meaningfully engage in the virtual steering group and the direction of this research. The paper concludes by making recommendations about this model of involving service users in research.

KEYWORDS: intermediate care; interprofessional; multi-agency; older people; carer; service users; virtual steering group

INTRODUCTION

The experience of using a virtual steering group to manage a research project is presented in this paper. The research team used an active learning approach to

Address for correspondence: S. M. Ashby, Keele University, School of Nursing and Midwifery, Clinical Education Centre, University Hospital of North Staffordshire, National Health Service Trust, Newcastle Road, Stoke-on-Trent, Staffordshire, England, UK. Voice: 00 44 (0) 1782 556550; fax: 00 44 (0) 1782 556576.
s.m.ashby@nur.keele.ac.uk

Ann. N.Y. Acad. Sci. 1114: 362–368 (2007).
doi: 10.1196/annals.1396.018

evaluate and develop interprofessional work among health and social care staff in both statutory and voluntary sectors, at a local and national level within the United Kingdom (UK), with a focus on older people receiving intermediate care. Intermediate care is the focus of the research because it covers multi-agency and interprofessional working, across both primary and secondary care, and includes older people, who are the main but not exclusive beneficiaries of these services. Intermediate care is the layer of care that sits between primary care and specialist services to help prevent unnecessary hospital admission, support early discharge, and reduce or delay the need for long-term residential care.[1]

The research embraces collaborative working with service users/carers where they are equal members of the research team. The research was managed virtually to allow all members to participate equally at a place of convenience. The research has received ethical approval from the Local Research Ethics Committee and the Association of Directors of Social Services.

THE RESEARCH PROJECT

Examination of current health and social policy in the UK demonstrates a clear commitment by government to make users/carers the focus of services.[2] Interprofessional collaboration in health care has been identified as an important area for development within the National Health Service (NHS) plan[1] in the UK; any changes in role boundaries necessitate a greater understanding of one another.[3] It has been suggested that increased collaboration should improve clinical competencies[4] as well as improve patient outcomes.[5]

Communication has been identified as an area where practitioners fail to meet the needs of their users. The Health Service Ombudsman's report of 2004 stated "Lack of communication is evident between healthcare professionals themselves, between professionals and their GP colleagues, and between professionals and carers" (p. iv).[6] It has also been argued that interprofessional communication affects healthcare outcomes, and failures have contributed to inquiries in the UK.[7]

The National Service Framework for Older People states that, "An essential component of intermediate care services is that they should be integrated within a whole system of care including primary and secondary health care, health and social care, the statutory and independent sectors" (p. 43).[1] There is anecdotal evidence that intermediate care is built around professional boundaries rather than being service user–focused. Through improving communication and teamwork between professions in intermediate care teams, the delivery and experience for older people will be improved.

SERVICE USER INVOLVEMENT

UK policy advises that intermediate care is a crucial element for improving services for older people,[8] and there is a clear commitment to make service users the focus of services.[9] It has been acknowledged that service users have had little involvement in the development of intermediate care.[10] The researchers acknowledged the significance of facilitating the engagement of service users/carers for full and active engagement and contribution to the steering group. Building on previous experience,[11] our aim was to ensure effective communication throughout this diverse group, respecting individual circumstances, that is, current commitments, time, and geographical location, ultimately ensuring equity of all members. Electronic communication was the approach adopted, resulting in a virtual steering group.

ELECTRONIC COMMUNICATION: PREPARATION AND SUPPORT

Regular steering group meetings were held at six-weekly intervals throughout the duration of the research. This was initially by electronic mail (e-mail); clear instructions were given to all steering group members with a set agenda and timespan of 14 days from circulation of steering group e-mail to closure date of responses for each meeting. Special consideration was given to the needs of older service users/carers, offering access to extra support. This was facilitated by face-to-face meetings 1 day prior to closure of the scheduled electronic meetings. Hard copies of the meeting agenda were mailed out to the users/carers at the same time the e-mail was sent to allow time to read, question, and prepare for the face-to-face meeting. The steering group was reminded to respond to a group e-mail address, enabling all members to view comments and have dialogue. This approach directed the research, drawing on the expertise and experiences of the steering group; for example, the service users made comments, which changed participant documentation, making it more "user-friendly."

This has progressed to the development of a web forum, which provides a public space as well as an area where the researchers store and display all the scheduled meetings and responses from the members, post out latest agendas, and provide a calendar and reminders for scheduled meetings and relevant dates related to progression of the research. All documentation relating to the research was accessible via this electronic forum. Useful links were also posted on this forum. This area also had an integrated e-mail system within it to enable all e-mail messages to be stored within one place, enabling easy access to all responses. As the research progressed, further information was added.

The electronic area was username- and password-protected to steering group members for the sole purpose of communication between this membership;

no participant information was held in this domain. The users/carers had the opportunity to access this web forum within their scheduled meetings or at their own leisure using their username and password. This web forum has been developed further to facilitate an "online chat area," enabling all members to communicate simultaneously at a given time to allow "live discussion." There is also a designated space for members to post topics they wish to explore and to revisit and add comments when they visit the forum area. All members are reminded to visit the web forum by e-mail (letter to service users/carers). A timed-release facility is also in use within the web space, enabling the researchers to prepare information in advance to be released by date and time throughout stages of the research.

The web site was designed with the support and expertise of a masters' student of computer science at Keele University. This resulted in the authoring of a web page accessible to the public domain which gives information regarding the research. This can be accessed at www.keeleuniversity.org/intermediatecare.

SERVICE USER/CARER TRAINING

Training was staged in line with progression of the research and followed three elements—issues around the topic area, the research process, and computer support. The service users/carers face-to-face meetings were assisted by an independent facilitator and a researcher. This approach allowed service users/carers to clarify terms and develop an understanding of the research methodology in order to apply their knowledge and experiences in a timely manner to inform and manage the research. E-mail exchange was encouraged and training given in the use of the web forum. While some may hold the view that for many older adults using a computer and looking for information on the web is a new experience, there is a move toward greater involvement and internet use. The British Broadcasting Company (BBC) in 2002 advised that "going online" had become the number one hobby for British pensioners.[12] A UK survey undertaken the following year reported 50% of people aged 60- to 64-years-old owned a computer, of whom 37% had internet access at home.[13] A more recent study[14] stated that around 83% of older people in the UK go online regularly.

EXPERIENCES OF SERVICE USERS/CARERS USING ELECTRONIC COMMUNICATION

The users/carers had differing levels of experience in using computers, as had the other members of the steering group. All were given a simple step-by-step guide in the use of the web forum, and the user/carer group supported each other and progressed with their development. The users/carers could access the internet at their regular meetings, but not all had this facility at home. Electronic communication allowed the team to communicate in a timely manner, but we had to ensure that everyone could access internet facilities at some time.

The users/carers preferred to read the hard copies posted to them, thus having the opportunity to understand the information at their own pace, share information within their own group, and collectively respond via electronic means at their scheduled meetings. Toward the later stages of the research confidence had grown and members were exploring their individual access. The problem posed by the use of technological terms and professional jargon was overcome by the development of a glossary specific to their needs and compiled throughout the stages of the research process.

There was one face-to-face steering group between all members. On reflection, the group felt that an earlier meeting would have been beneficial, to reinforce interaction between members and responsibilities of the group. It was expressed that this would have broken down the barrier of "titles" relating to the people involved rather than just "the professor" or "the consultant."

OPPORTUNITIES

The users/carers have identified gaps in knowledge of older people regarding intermediate care services and the need for further information. Currently, they are exploring ways of pulling together information in a format this is easily understood and targets appropriate groups and individuals.

In addition, the research team had an abstract accepted at the INVOLVE Conference held in 2006 in the UK. INVOLVE is specifically focused on getting service users actively involved in research from the initial bid through to analysis of research findings.[15] A service user, accompanied by the independent facilitator and researcher, made a presentation based on his or her involvement in this research. Networking with fellow service users resulted in a sharing of experiences. Confidence was gained and interest was generated in future involvement in research projects concerning the issues raised and the development of possible solutions and ways forward.

CONCLUSION

The use of electronic communication facilitates communication across a wide audience if preparation and support are given appropriate to need. This means of communication is less intimidating to service users once the initial anxieties of using computers has been overcome. The service users/carers gained confidence in debating with professionals. The staged learning allowed time for this group to examine documentation at their own pace before putting it in the wider domain. Timely exchange of information was also made easier with group access and central storage of information which could be disseminated quickly. This also provided an audit trail. It is recommended that all information stored electronically have a back-up system. The electronic format also enabled insertion of comments to documentation, assisting the process of tracking changes. This format can also be adapted to meet the needs of the individual user (such as font size, colored backgrounds, and adjustment to individual username access on the home page). Navigation can also be simplified via web design.

The research team has specifically endeavored to ensure the users/carers were actively involved in this research as equal members of the steering group and the use of virtual communication has facilitated involvement and made changes valuable to the direction of the research. The service user involvement has also informed the direction of doctoral study being undertaken by the lead author.

ACKNOWLEDGMENTS

This project was funded by the Burdett Trust for Nursing. Thanks are also extended to Jane Snape from the Beth Johnson Foundation and service users Margaret Brandreth, Mary Peake, Ben Mountford, Margaret Williams, and Joan Walker for their energy, enthusiasm, and time, and to Zhang Ying, the website creator.

REFERENCES

1. DEPARTMENT OF HEALTH. 2001. National Service Framework for Older People. Department of Health. London.
2. DEPARTMENT OF HEALTH. 2000. The NHS Plan: a Plan for Investment, a Plan for Reform. HMSO. London.
3. HUMPHRIS, D. & A. MASTERSON, Eds. 2000. Developing New Clinical Roles: a Guide for Health Professionals. Churchill Livingstone. London.
4. BARR, H., D. FREETH, M. HAMMICK, *et al.* 2000. Evaluations of Interprofessional Education. A United Kingdom Review for Health and Social Care. CAIPE/British Educational Research Association. London.
5. ZWARENSTEIN, M., S. REEVES, H. BARR, *et al.* 2000. Interprofessional education: effects on Professional Practice and Health Care Outcomes (Cochrane Review). *In* The Cochrane Library, Issue 3. Update Software. Oxford.
6. HEALTH SERVICE OMBUDSMAN. 2004. Selected Investigations Completed October 2003–March 2004. 2nd Report – Session 2003–2004. The Stationery Office. London.
7. DEPARTMENT OF HEALTH. 2001b. Working Together, Learning Together: a Framework for Lifelong Learning in the NHS. Department of Health. London.
8. DEPARTMENT OF HEALTH. 2006. A New Ambition for Old Age: next Steps in Implementing the National Service Framework for Older People. Department of Health. London.

9. DEPARTMENT OF HEALTH. 2005. Creating a Patient-Led NHS: delivering the NHS Improvement Plan. Department of Health. London.
10. NATIONAL HEALTH SERVICE EXECUTIVE NORTH WEST REGIONAL TASK FORCE FOR OLDER PEOPLE. 2002. Intermediate Care: First Impressions; Key Messages from Older People. Report of a meeting for older members of Local members of Local Implementation Teams (LITS) conducted in Liverpool, September, pp. 1–8. National Health Service Executive North West. Manchester.
11. MASLIN-PROTHERO, S. 2003. Developing user involvement in research. J. Clin. Nurs. **12:** 412–421.
12. BRITISH BROADCASTING COMPANY NEWS. 2002. Elderly take up new surfing, May 10 2002. Accessed from http://news.bbc.co.uk/1/hi/sci/tech/1977823.stm.
13. JEFFREY, S. 2003. Over 60s reach for the mouse. Guardian Unlimited, July 8. Accessed from http://www.guardian.co.uk/guesteditor/story/0,994114,00.html.
14. KURNIAWAN, S.H., A. KING, D.G. EVANS & P.L. BLENKHORN. 2006. Personalising web page presentation for older people. Interact. Comput. **18:** 457–477.
15. INVOLVE. 2007. Promoting public involvement in NHS, public health and social care research. Accessed from http://www.invo.org.uk/.

Nutritional Interventions in Aging and Age-Associated Diseases

GEORGE S. ROTH,[a] DONALD K. INGRAM,[b] AND JAMES A. JOSEPH[c]

[a]*GeroScience, Inc. Pylesville, Maryland 21132, USA*

[b]*Nutritional Neuroscience and Aging Laboratory, Pennington Biomedical Research Center, Louisiana State University System, Baton Rouge, Louisiana 70808, USA*

[c]*USDA/Tufts University Human Nutrition Research Center on Aging, Boston, Massachusetts 02111, USA*

ABSTRACT: Dietary interventions have proven to be the most robust and consistent means of ameliorating the diseases and dysfunctions of aging. A large literature includes both quantitative (caloric restriction [CR]) and qualitative (micronutrients, antioxidants, etc.) alterations of nutrient and caloric intake and the resulting positive outcomes for age-related disease protection, maintenance of vitality, and quantitative and qualitative life-span extension. In recognition of the importance of this research area, two sessions concerning nutritional interventions in aging and age-associated diseases were organized for this symposium, one dealing with diet and the other with dietary supplements. This brief report covers both, providing selected examples of the most recent research into both these types of interventions and their mechanisms of action.

KEYWORDS: aging; nutrition; interventions;age-associated diseases

Calorie restriction (CR) has been employed for more than 70 years in well-controlled long-term studies in species ranging from single-celled organisms to primates. Ingram and Roth initiated the first large-scale long-term CR study in monkeys in 1987, which continues to this day. In the first session (on diet), one of the pioneers in CR research, Dr. Arthur Everitt from the Centre on Education and Research on Aging at the University of Sydney, reviewed key aspects of this work with particular emphasis on protection against cardiovascular disease, cancer, and diabetes, in addition to extending maximal life span in animals. Citing several past reviews of the literature, Dr. Everitt challenged the notion of CR in humans producing major life-span extension. Rather, he cited numerous

Address for correspondence: Dr. James A. Joseph, Chief, Neuroscience Laboratory, USDA/Tufts University, Human Nutrition Research Center on Aging, 711 Washington Street, Boston, MA 02111. Voice: 617-556-3178; fax: 617-556-3222.
james.joseph@tufts.edu

**Ann. N.Y. Acad. Sci. 1114: 369–371 (2007). © 2007 New York Academy of Sciences.
doi: 10.1196/annals.1396.048**

other epidemiologic studies emphasizing the health benefits of diets rich in fruits and vegetables, especially when combined with exercise.

This introduction was very much "food for thought," since there are very few of us who don't remember Mom's familiar refrain: "Eat your fruits and vegetables. They're good for you"! Little did we realize at the time how much Mom knew about what we should eat to stay healthy. Now, it seems as though Mom is finally being listened to. We are told that a high intake of fruits and vegetables will reduce your risk of encountering such ubiquitous killers as cardiovascular diseases and cancer. However, questions arise as to the nature of their beneficial properties in relation to these and possibly neurodegenerative diseases, such as Alzheimer's disease. It turns out that all of these conditions may have a common initiator—aging—where, in the case of the brain, there are decrements in neuronal communication that, even in normal aging, can lead to motor and cognitive behavioral deficits. A great deal of data suggest that these deficits arise as a result of an increasing inability of the aging organism to protect itself against inflammation and oxidative stress (INF/OS), providing "fertile ground" for the development of a host of age-related diseases. The good news reported at these sessions is that it appears that compounds (e.g., polyphenolics) found in fruits and vegetables that enhance the survivability of the plant, presumably through their antioxidant and anti-inflammatory properties, may provide the necessary protection to prevent the demise of a number of aging parameters.

In this respect Dr. Mary Ann Lila from the University of Illinois brought her considerable expertise in botanicals to bear and discussed the wide array of bioflavonoids found in fruits and vegetables and their biological effects. The demonstrated actions of flavonoids now extends beyond their actions as antioxidants to their roles in cell signaling, particularly in enhancing cell-protective mechanisms. Dr. Lila presented particularly interesting data on the search for botanicals that act as effective "performance enhancers," that is, increase function and stamina. The major points in Dr. Lila's presentation have been emphasized in the research by Dr. James Joseph. Dr. Joseph and colleagues have documented the beneficial effects of high-antioxidant fruits (bluberries, [BBs]) and a vegetable (spinach) on brain aging and behavior in aged rats fed diets containing strawberry, BB, or spinach extracts. He has also reported experiments in both cell and animal models indicating that these enriched diets have both antioxidant and anti-inflammatory properties, but also may have direct effects on neuronal communication by enhancing signaling and increasing the growth of new neurons (neurogenesis). He has also shown that there is some localization of anthocyanins in the brains of the BB-supplemented animals. Subsequent cell studies showed that BBs can decrease several oxidative stress/inflammatory signals (e.g, cyclic AMP response element binding protein [CREB]).

Finally, in this session, Dr. Carol Greenwood from the Rotman Institute in Toronto, Ontario, Canada, brought attention to the negative impact of diet on performance. Dr. Greenwood discussed her findings regarding impaired memory performance of rodents on diets high in saturated fats or in glycemic responses. The effect of diets high in glycemic value was most evident at the time immediately post ingestion. A possible relationship to inflammatory responses mediated through tumor necrosis factor-alpha was illustrated.

In the second session (on dietary supplements), Dr. Elizabeth Head from the Institute for Brain Aging and Dementia at the University of California, Irvine, made cogent arguments for the value of dog models of aging, particularly related to brain aging and the effects of diet thereon. She reviewed studies conducted in her lab that examined the effects of a diet enriched in antioxidants on brain aging in dogs. Certain aspects of memory performances were enhanced in older dogs on the enriched diet compared to a control diet, and this improvement appeared correlated to reduced levels of amyloid-beta in the brains of the aged dogs.

A similar approach has been used by Dr. Giovanni Scapagnini from the Institute of Neurological Sciences in Catania, Italy, who investigated the beneficial effects of curcumin (a key ingredient of curry powder) on brain aging in a variety of models. The actions of curcumin appeared to be mediated by heme oxygenase.

The other participant in this session was Dr. Salvatore Pepe from the Baker Heart Institute in Melbourne, Australia. He reviewed his work on stress protection in the cardiovascular system using omega-3 fatty acids. These lipids alter the membrane composition of the heart and serve as a key defense against oxidative damage, both by attenuating deleterious age changes in gene expression and impaired signal transduction.

In summary, the scientific evidence provided by these presentations underscore that diet and lifestyle could be major interventions that will attenuate age-related decrements and age-associated diseases. These findings have important implications for human health and nutrition and provide a scientific basis for the numerous citations in the popular media concerning the value of diets containing fish oils and high-antioxidant fruits and vegetables, as well as those which advocate paying particluar attention to weight control and maintaining a low glycemic index. Interestingly, the research discussed here also attests to the oft-repeated statement that lifestyle actions that increase heart health also are important for brain and eye health. Clearly, such behaviors will probably contribute a great deal to increasing the healthy life span of individuals and societies as a whole and should provide a significant "longevity dividend" in terms of healthcare costs later in life.

From Beans to Berries and Beyond

Teamwork between Plant Chemicals for Protection of Optimal Human Health

MARY ANN LILA

Division of Nutritional Sciences, University of Illinois, Urbana, Illinois, USA

ABSTRACT: It is now well known to consumers around the world that certain fruits and vegetables can help prevent or treat chronic human diseases. But, what many people don't fully appreciate is that it is not a single component in these plant-derived foods, but rather complex mixtures of interacting natural chemicals, that produce such powerful health-protective effects. These natural components accumulate simultaneously together in a plant, and provide a multifaceted defensive strategy for both the plant, and the human consumer. In order to investigate the strength of natural chemical cooperation in highly-pigmented, flavonoid-rich functional foods, our lab has relied on analysis of both whole fruits, and continuous, reliable plant cell culture production systems which accumulate anthocyanins and proanthocyanidins in high concentrations. Successive rounds of relatively gentle, rapid, and large-volume fractionations are linked to bioassay of complex to simple mixtures and semi-purified compounds. By means of this strategy, additive interactions or synergies between related compounds in health maintenance can be sorted out. Interestingly, phytochemical interactions between the same classes of compounds intensify the efficacy of flavonoid-rich fruits against multiple, not necessarily discrete, human disease conditions including CVD, cancer, metabolic syndrome, and others.

KEYWORDS: **bioflavonoids; phytoecdysteroids; potentiating interactions**

OVERVIEW

It is now well known to consumers throughout the world that certain fruits and vegetables can help prevent or treat chronic human diseases. By itself, this health-protective capacity available through the foods we eat is a tremendous proactive advantage for savvy consumers, because the multi-component, diversified strategies that bioactive food constituents use to combat

Address for correspondence: Mary Ann Lila, University of Illinois, 211 Mumford Hall MC 710, 1301 W. Gregory Drive, Urbana, IL 61801. Voice: 217-333-5154; fax: 217-333-5816. imagemal@uiuc.edu

Ann. N.Y. Acad. Sci. 1114: 372–380 (2007).
doi: 10.1196/annals.1396.047

disease progression can be more comprehensive and broad-spectrum than the effects of any single-component synthetic drugs.[1–3] In addition, because such a plethora of overlapping and complementary strategies are employed, the natural plant-based disease-combative strategy is robust and not easily overcome by pathogen resistance.

Still, healthy aging and maintenance of an optimal healthy life span involves more than just circumventing disease. In addition, a vigorous metabolism must be encouraged and sustained, and attention must be paid to such factors as strength, endurance, stamina, body weight, bone density, and muscle tone. Fortuitously, some key health-protective edible plant foods have demonstrated capacity to *both* combat disease and optimize metabolism and performance through diverse mechanisms of action.[4,5] Many people don't fully appreciate that it is not a single component in these plant-derived foods, but rather the complex mixtures of interacting natural chemicals, that produce such powerful health-protective effects. These natural components accumulate simultaneously together in a plant, and provide a multifaceted defensive strategy for both the plant, and the human consumer. Specific phytochemical mixtures are capable of generally enhancing human metabolism, beyond any nutritional benefits, and again in ways that are more robust than any single-component drug could accomplish.

In some cases, plants may utilize the same classes of natural phytochemical constituents and use similar mechanisms of action to inhibit disease *and* enhance human metabolism. In other cases, the phytochemicals and the *modus operandi* may be unique to particular health effects, but invariably in all these events, it is a complex mixture of interacting phytochemicals that is ultimately responsible for providing healthy benefits. The daunting complexity of interacting natural chemical complexes responsible for disease-resisting and health-protecting benefits to the human consumer have accelerated the momentum of scientific investigations which aim to uncover the agents and their mechanisms. Research advances in analytical techniques have led to compilation of vast, expanding databases of biologically active phytochemicals, and their roles have been further elucidated in part through emerging fields of bioinformatics and metabolomics.

MULTIPLE MECHANISMS FOR DEFENSE, PROTECTION, AND METABOLIC ENHANCEMENT

It is a basic precept of traditional Chinese medicine that complex disease conditions are best treated with complex, multifaceted compounds which mix extracts from multiple, different plant species, in order to capture the different defense and protective mechanisms of action unique to each one, and simultaneously aim at multiple human therapeutic targets.[3,6] However, even within a single plant, interacting arrays of accumulated phytochemicals complement and augment the protective or defensive effects by using slightly divergent strategies for intervention. Reliance on multiple phytochemical complexes has evolved as the most efficacious strategy for protection in the plant, and protection for the humans who consume the plant.

Probably the most widely recognized attribute of foods that are integral to maintenance of health and well-being is antioxidant capacity—the ability of a food constituent to enhance the endogenous antioxidant defense system. Although we all have built-in internal antioxidant mechanisms for protecting membranes and organs from damage, the efficiency of our internal defenses seems to fail with age, leading to increased vulnerability to many chronic diseases and fraility. In fact, oxidative stress is a precursor to so many chronic diseases that the simple antioxidant capacity of a plant extract has been used as a marker in screening to find potentially useful plant extracts for natural pharmaceutical preparations. Antioxidants from food sources scavenge free radicals in the human body that would otherwise lead to oxidative stress. Oxidative stress is also a key factor in aging, and contributes to skin wrinkling and loss of resiliency, which is why antioxidants are frequently incorporated into cosmetic or cosmeceutical products designed to inhibit the visible signs of human aging. Yet antioxidant capacity by itself may not be the most biologically relevant attribute to assess; there are questions as to whether plant phytochemicals act directly through antioxidant mechanisms or whether these processes indirectly affect cellular signaling or oxidation signal amplification pathways.[7]

Another recognized diversified mechanism employed by plant-derived components is direct or indirect interaction with enzymes residing within the human body. Phytochemicals from plant-derived foods can inhibit enzymes that would otherwise accelerate disease progression. Alternatively, the phytochemicals can help to catalyze the activities of enzymes which detoxify or otherwise counteract disease symptoms.

Natural plant chemicals can have powerful anti-inflammatory influences. Much like antioxidant capacity, anti-inflammatory phytochemicals are widely sought to provide broad-ranging health protection and bolster the human immune system, because inflammation is such a prominent early symptom of so many chronic human illnesses and injuries. Antimicrobial properties, antiallergic effects, and anticarcinogenic or antihypertensive properties are also linked to specific classes of phytochemicals in edible plant sources.

An excellent example of the multi-tasking accomplished by plant constituents in health maintenance for the human consumer can be illustrated by just one particular class of phytochemicals—the flavonoids—which are particularly rich in fruits, such as berries, blue corn and beans, and many fruits and vegetables. The flavonoids, which include anthocyanin (red–blue) pigments in many fruits and vegetables, proanthocyanins (tannins), phenolic acids, and the estrogen-mimicking isoflavones (phytoestrogens) have been called "bioflavonoids" because of the diverse health benefits attributed to this

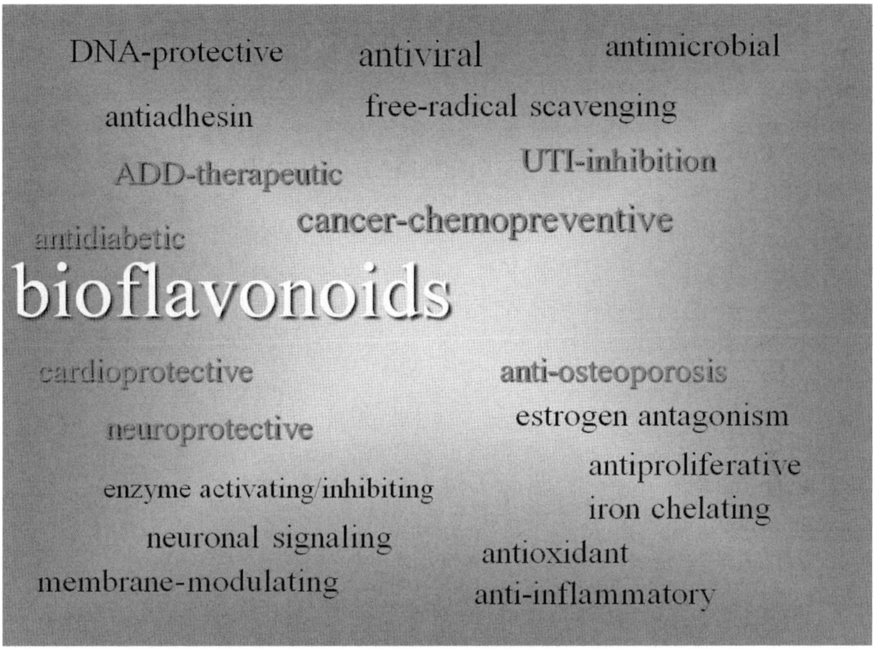

FIGURE 1. The flavonoid class of phytochemicals are frequently referred to as *bioflavonoids* on account of the wide-ranging roles they play in human health protection. When a flavonoid molecule or complex from an ingested plant food source interfaces with a human therapeutic target (enzyme, receptor, etc.) in the body, a series of distinct mechanisms of action can be exhibited: protection of cellular DNA structure; antiviral, antimicrobial, or antiadhesin activity; antioxidant capacity and the ability to scavenge free radicals; interaction with estrogen receptors; ability to curtail cancer cell growth or cause premature cancer cell death; intercellular signaling; membrane fortification; interaction with enzymes resident in the body; and inhibition of the inflammatory response. These various *modi operandi* translate into metabolism-enhancing, disease-inhibiting outcomes, making the functional foods containing flavonoids players in control of specific diseases and maintenance of optimal performance.

class (FIG. 1). These compounds have demonstrated antioxidant/free radical scavenging abilities; they can chelate minerals, modulate mammalian enzyme activity, enhance intracellular signaling, strengthen membranes, and bind to receptor sites, among other things. Each of these mechanisms translates into the more clearly recognized attributes of cancer chemoprevention, CVD treatment, prevention of obesity and type II diabetes, wound healing, enhanced stamina and cognitive clarity, enhanced bone density, selective inhibition of human intestinal pathogens and support of lactic acid bacteria, and more. These diverse mechanisms of intervention which have an impact on human health are accomplished by different classes of chemicals under the broad "flavonoid" umbrella, and quite frequently interactions between two or more of the distinct classes have potentiated larger benefits than would have been achieved when only one class of compound is given as a supplement to a patient.

EXAMPLES OF PHYTOCHEMICAL TEAMWORK

Historical connections between plants, food, and medicine have always relied on the amazing multi-tasking capacity of phytochemicals; the standard subsistence diets in tribal and traditional cultures consisted of foods rich in bioactive phytochemicals that guaranteed ability to withstand adverse conditions over the long term and extended human life. Modern science has more recently begun to elucidate the complementary strategies plant phytochemicals utilize to enhance human metabolism and optimize performance parameters. A few key examples can be put forward to illustrate these points.

Routine consumption of the tomato, a common dietary staple, has been associated with decreased risk of CVDs and cancer (particularly prostate cancer), and the carotenoid lycopene has most frequently been cited as the active ingredient. However, very recent studies have clearly illustrated that it is the complex cocktail of chemicals in the tomato, as well as the potentiating interactions between tomato fruits and other foods (such as oils) in the diet that are actually responsible for the health-protective benefits. Recently, the roles of other tomato carotenoids and compounds in prevention of carcinogenesis have been demonstrated,[8–10] and the synergistic interactions of polyphenolic compounds, not just carotenoids, in the tomato fruit toward inhibition of carcinogenesis were clearly demonstrated.[11] Not only did the research establish that multiple components from the tomato worked interactively to provide the most potent anticancer protection, but also the same team recently showed that combinations of broccoli and tomato in the diet were more effective at reducing cancer cell proliferation and causing cancer cell death than either bioactive food source alone. These results further demonstrate the importance of phytochemical interactions for optimal protection of human health, and support the public health recommendation to increase intake of a variety of fruits and vegetables.[12] Similar proof that multiple flavonoid components including the pigments and tannins in the common dietary black bean work together to inhibit carcinogenesis,[13] and that grapes contribute to inhibition of multiple stages of carcinogenesis using synergistically interacting chemical constituents resident in the fruit and seed[14] emphasize the relevance and frequency of the use of complementary strategies in natural systems.

Throughout history, more than 1000 plant species have been consumed to combat the symptoms of diabetes mellitus (type II adult onset diabetes) and metabolic syndrome, including various herbs like goat's rue (*Galega officinalis*, the original source of the drug metformin used clinically today to treat hyperglycemia), spices, vegetable sources, and fruits, and typically, more than one plant resource was consumed concurrently to maintain health and curtail obesity.[15,16] The wild berries integral to many tribal diets provided multiple, interacting benefits relevant to diabetes and metabolic syndrome control, by naturally improving serum insulin levels,[17] reducing incidence of hyperglycemia,[18,19] and inhibiting aldose reductase, an enzyme whose increased activity during hyperglycemia can lead to microvascular damage.[20] These multiple effective mechanisms of action are a consequence of the phytochemical composition within the wild-fruit pulp, conferring a multi-pronged defense for the consumer. The flavonoid-rich content of the fruit is most known for antioxidant capacity, relevant to many forms of health maintenance, but it also inhibits enzymatic processes leading to diabetic complications. The co-occuring oil-soluble terpenes in the fruit are also capable of this enzyme inhibitory benefit. As another example, fruits of Cornelian cherry dogwood (*Cornus mas*), another source of medicinal foods, are high in anthocyanins and ursolic acid; both of these compounds maintained glucose tolerance in mice fed a high-fat diet that typically induces glucose intolerance.[21]

As a final example, certain plants produce mixtures of interacting biologically active phytochemicals with potent adaptogenic properties; consumption of these plants increases nonspecific resistance in humans, with both anabolic and tonic influences on the consumer.[22] Some, such as the phytoecdysteroids present in *Ajuga turkestanica* (an important Asian herbal source) are mixtures of metabolic steroid-like compounds that increase muscle mass, stamina, enhance performance, improve cardiovascular function, and have anti-inflammatory properties.[23,24] Some of our common vegetable sources, such as spinach, contain adaptogenic phytochemicals. The wildly popular herbal "goldenroot" (*Rhodiola rosea*) as well as several related species indigenous to various communities in Asia have demonstrated potent adaptogenic properties, and are specifically valued for their ability to enhance work performance, eliminate fatigue, increase resistance to anoxia, alleviate depression, and increase endurance and mental alertness.[25,26] These performance-related benefits are attributed to phenolic glycosides and specifically, the rosavins and salidroside.[27] Plants with adaptogenic phyochemicals have been traditionally valued for sport and military purposes, to enhance human performance, but they are equally beneficial to the aging population, in order to maintain optimal healthy life span.

These are only a few examples of how the typical plant strategy relies on mixtures of phytochemicals with overlapping or complementary roles, providing more broad-spectrum and robust strategies for health intervention for the consumer. Naturally, given the complex and multifaceted nature of the interactions, and the intricate changes in composition of different biologically active plant resources, it is more difficult to pinpoint efficacy and ideal dosages for natural medicinal foods, as compared to the more targeted synthetic drugs.

TRACKING THE LEVELS OF PHYTOCHEMICAL TEAMWORK

Clearly the most advantageous diets are those that incorporate a variety of healthful food sources so as to maximize the interaction potential for human health maintenance and metabolic performance. Given the multiplicity of interacting phytochemicals that are ideally incorporated into the diet, the task of trying to track the kinetics and distribution of individual bioactive compounds has been quite daunting. In order to investigate the incidence and mechanisms of natural chemical cooperation, our lab has relied on analysis of both whole fruits, and continuous, reliable plant cell culture production systems which accumulate the same phytochemicals in high concentrations.[28–30] Successive rounds of relatively gentle, rapid, and large-volume fractionations are linked to bioassay of complex to simple mixtures and semi-purified compounds. With this strategy, additive interactions or synergies between related compounds that potentiate human healthspan can be sorted out. Of interest, phytochemical interactions between the various classes of compounds intensify the efficacy of bioactive plant extracts against multiple, not necessarily discrete, human disease conditions including CVD, cancer, metabolic syndrome, and others.

REFERENCES

1. LILA, M.A. & I. RASKIN. 2005. Health-related interactions of phytochemicals. J. Food Sci. **70:** R20–R27.
2. RASKIN, I., D. RIBNICKY, S. KOMARNYTSKY, *et al.* 2002. Plants and human health in the twenty-first century. Trends Biotechnol. **20:** 522–531.
3. SCHMIDT, B., D. RIBNICKY, P. LIPSKY & I. RASKIN. 2007. Revisiting the ancient concept of botanical therapeutics. Nature Chem. Biol. **3:** 360–366.
4. CORNWELL, T., W. COHICK & I. RASKIN. 2004. Dietary phytoestrogens and health. Phytochemistry **65:** 995–1016.
5. KRAFT, T., M. DEY, R. ROGERS, *et al.* 2007. Phytochemical composition and metabolic performance enhancing activity of dietary berries traditionally used by native North Americans. J. Agric. Food Chem. In press.
6. RASKIN, I. & C. RIPOLL. 2004. Can an apple a day keep the doctor away? Curr. Pharm. Design **10:** 3419–3429.
7. WATERHOUSE, A. 2006. Polyphenolics: diverse sources and effects implicate diet. J. Sci. Food Agric. **86:** 2243–2244.
8. BOILEAU T., Z. LIAO, S. KIM, *et al.* 2003. Prostate carcinogenesis in N-methyl-N-nitrosourea (NMU)-testosterone-treated rats fed tomato powder, lycopene, or energy-restricted diets. J. Natl. Cancer Inst. **95:** 1578–1586.
9. CAMPBELL, J., M. LILA, M. NAKAMURA & J. ERDMAN, JR. 2006. Serum testosterone reduction following short-term phytofluene, lycopene, or tomato powder consumption in F344 rats. J. Nutrition **136:** 2813–2819.
10. ZARIPHEH, S., T. BOILEAU, M. LILA & J. ERDMAN, JR. 2003. [^{14}C] lycopene and [^{14}C]-polar products are differentially distributed in tissues of F344 rats prefed lycopene. J. Nutrition **133:** 4189–4195.

11. Campbell, J., J. King, M. Harmston, *et al.* 2006. Synergistic effects of flavonoids on cell proliferation in Hepa-1c1c7 and LNCaP cancer cell lines. J. Food Sci. **71:** S358–S363.
12. Canene-Adams, K., B. Lindshield, S. Wang, *et al.* 2007. Combinations of tomato and broccoli enhance antitumor activity in Dunning R3327-H prostate adenocarcinomas. Cancer Res. **67:** 836–843.
13. Aparicio-Fernandez, X., T. Garcia-Gasca, G.G. Yousef, *et al.* 2006. Chemopreventive activity of polyphenolics from black Jamapa bean (*Phaseolus vulgaris* L.) on HeLa and HaCaT cells. J Agric. Food Chem. **54:** 2116–2122.
14. Jo, J.Y., E. de Mejia & M.A. Lila. 2006. Cytotoxicity of bioactive polymeric fractions from grape cell culture on human hepatocellular carcinoma, murine leukemia and non-cancerous PK15 kidney cells. Food Chem. Toxicol. **44:** 1758–1767.
15. Baxter, A., T. Coyne & C. McClintock. 2006. Dietary patterns and metabolic-syndrome: a review of epidemiologic evidence. Asia Pacific J. Clin. Nutrition **15:** 134–142.
16. Browning, L.M. & S. Jebb. 2006. Nutritional influences on inflammation and type 2 diabetes risk. Diabetes Technol.Ther. **8:** 45–54.
17. El-Alfy, A.T., A. Ahmed & A. Fatani. 2005. Protective effect of red grape seed proanthocyanidins against induction of diabetes by alloxan in rats. Pharmacol. Res. **52:** 264–270.
18. Iwai, K., M.-Y. Kim, A. Onodera & H. Matsue. 2006. α-Glucosidase inhibitory and anthihyperglycemic effects of polyphenols in the fruit of *Viburnum dilatatum* Thunb. J. Agric. Food Chem. **54:** 4588–4592.
19. Simeonov, S.B., N. Botushanov, E. Karahanian, *et al.* 2002. Effects of *Aronia melanocarpa* juice as part of the dietary regimen in patients with diabetes mellitus. Folia Med. (Plovdiv) **44:** 20–23.
20. Ueda H. Y. Tachibana, M. Moriyasu, *et al.* 2001. Aldose reductase inhibitors from the fruits of *Caesalpinia ferrea* Mart. Phytomedicine **8:** 377–381.
21. Jakaprakasam, B., L.K. Olso, R.E. Schutzki, *et al.* 2006. Amelioration of obesity and glucose intolerance in high-fat-fed C57BL/6 mice by anthocyanins and ursolic acid in Cornelian Cherry (*Cornus mas*). J. Agric. Food Chem. **54:** 243–248.
22. Baltaev, U. 2000. Phytoecdysteroids: structure, sources, and biosynthesis in plants. Russ. J. Bioorgan. Chem. **36:** 799–831.
23. Dinan, L. 2001. Phytoecdysteroids: biological aspects. Phytochemistry **57:** 325–339.
24. Syrov, V.N., S. Nasyrova & A. Khushbaktova. 1997. The results of experimental study of phytoecdysteroids as erthropoiesis stimulators in laboratory animals. Eksp. Klin. Farmakol. **60:** 41–44.
25. Khanum, F., A. Bawa & B. Singh. 2005. *Rhodiola rosea*: a versatile adaptogen. Comp. Rev. Food Sci. Food Safety **4:** 55–62.
26. Xu, J.F., P. Ying, A. Han & Z. Su. 1999. Enhanced salidroside production in liquid-cultivated compact callus aggregates of *Rhodiola sachalinensis*: manipulation of plant growth regulators and sucrose. Plant Cell. Tiss. Cult. **55:** 53–58.
27. Brown, R.P., P. Gerbarg & Z. Ramazanov. 2002. *Rhodiola rosea*: a phytomedical overview. Herbalgram **56:** 40–52.
28. Janle, E., M.A. Lila, L. Wood, *et al.* 2007. Kinetics and tissue distribution on ^{14}C labeled grape polyphenol fractions. FASEB Journal Abstracts. Washington, DC.

29. Jo, J.Y., E. de Mejia & M.A. Lila. 2006. Cytotoxicity of bioactive polymeric fractions from grape cell culture on human hepatocellular carcinoma, murine leukemia and non-cancerous PK15 kidney cells. Food Chem. Toxicol. **44:** 1758–1767.
30. Weaver, C., S. Barnes, J. Wyss, *et al.* 2007. Botanicals for age-related diseases: From field to practice. Am. J. Clin. Nutr. In press.

Dietary Polyunsaturated Fatty Acids and Age-Related Membrane Changes in the Heart

SALVATORE PEPE

Department of Surgery, Monash University, Alfred Hospital, Melbourne, 3004, Australia

ABSTRACT: Age-associated modification of cardiac membrane composition and related metabolic processes are major factors underlying the reduced capacity for invoking intrinsic cardioprotective survival mechanisms. In this brief review we examine some of these key membrane lipid modifications, which include alterations in the type and proportion of long–carbon chain polyunsaturated fatty acids (PUFAs). These changes have profound consequences on the efficacy of membrane proteins and lipids involved with numerous processes including ion homeostasis, intracellular signal transduction, free radical metabolism, and mitochondrial energy production. The resultant postoxidative modification of lipids, proteins, and DNA leads to a reduction in capacity for repair and ultimately worsening cellular dysfunction. However, some age-related detrimental adaptations may be counteracted by selectively manipulating membrane lipid and protein composition via dietary treatment with key omega-3 PUFAs. Such pervasive treatment results in manipulation of cellular function at multiple levels from cell membrane to the genome.

KEYWORDS: omega-3 polyunsaturated fatty acids; heart; aging; dietary lipid

INTRODUCTION

Age-associated modification of cardiac membrane composition, lipid–protein interactions, and related metabolic processes are major factors underlying the reduced capacity to recover from acute injury or to adapt during chronic disease in advanced age.[1] In this brief review we examine important adaptive and maladaptive membrane phospholipid changes that occur in

Address for correspondence: Dr. Salvatore Pepe, Head, Laboratory of Cardiothoracic Surgical Research, Department of Surgery, Monash University, Alfred Hospital, Melbourne, 3004, Australia. Voice: +61 3 9903 0815; fax: +61 3 99030618.
Salvatore.Pepe@med.monash.edu.au

Ann. N.Y. Acad. Sci. 1114: 381–388 (2007).
doi: 10.1196/annals.1396.046

response to age-linked heart disease which involves omega (ω)-6 polyunsaturated fatty acids (PUFAs), ω-3 PUFAs, phosphatidylcholine, and cardiolipin [see Refs. 2 and 3 for recent detailed reviews].

OMEGA-3 POLYUNSATURATED FATTY ACID DECLINE

Evidence is mounting in support of the proposal that a deficiency of long-chain ω-3 PUFAs in myocardial membranes may underlie increased vulnerability to cardiac arrhythmias and age-related cardiovascular disease. Phospholipids, and their constituent fatty acids, are integral to cell membrane function and to biologically active membrane-derived signaling products, including fatty acid diacylglycerols, inositol triphosphate, and fatty acid metabolites. Many PUFAs are required in the diet since mammals cannot synthesize them *de novo*.[4,5] Of the two main families, ω-6 PUFAs are readily available in the diet as linoleic acid (LA). The main PUFAs in myocardial membranes are the ω-6 PUFA, LA, and arachidonic acid (AA), and the ω-3 PUFA, docosahexaenoic acid (DHA). With abundant LA in the normal Western diet, α-linolenic acid (ALA) cannot be converted to adequate levels of eicosapentaenoic acid (EPA) and DHA on account of competition for the desaturase enzymes with the vast excess of dietary ω-6 PUFA.[4–8] Delta-5 and delta-6 fatty acid desaturases are crucial for the conversion of ω-6 PUFA (LA to AA) and for the conversion of ω-3 PUFA (ALA to EPA and DHA).[4–8] The activity of microsomal delta-6 desaturase is less than that of delta-5, making it the rate-limiting step involved in two stages of DHA production.[8] Thus, in the face of dietary intake that is low in ALA intake and high in AA (e.g., meat) there is commonly a reduction in ω-3 PUFA content in membranes and it will be exacerbated by any downturn in desaturase enzyme function.[5,6]

ALTERED METABOLIC ADAPTATION AND RESPONSE TO INJURY

Notably, the consequences of these dietary effects further exaggerate low ω-3 PUFA and high ω-6 PUFA content in cardiac membranes and compound the development and impact of heart disease in advanced age. In senescent rat hearts, ischemia–reperfusion injury results in augmented Ca^{2+} accumulation in cytoplasm and mitochondria, and reduced recovery of contractile and mitochondrial function, compared to that of young counterparts.[9,10] In human atrial trabeculae, we have also shown the impact of advanced age in reducing recovery of post-ischemic contractile function.[11] Dietary treatment for only a few weeks with fish oil can circumvent the effects of a high ω-6-PUFA diet in old and young adult rats, because it increases the incorporation of cardiac membrane ω-3 PUFAs, the recovery of contractile work, Ca^{2+} homeostasis, and O_2 utilization efficiency, and it reduces the incidence of arrhythmias after acute ischemia and reperfusion.[9,12–16] The consequences of a decrease in the ratio of ω-3 to ω-6 PUFAs include changes in crucial membrane lipid–protein interactions that lead to altered intracellular signaling, reduced efficiency of oxygen utilization, failing mitochondrial energy production, augmented free radical formation, lipid and protein oxidation and modification, and decreased ion homeostasis.[1,17] These changes create a vicious spiral that is triggered by adverse events, such as ischemic injury and oxidative stress. Notably, superoxide free radicals form ω-6 PUFA-derived cytotoxic and chemotactic reactant lipid hydroperoxides (i.e., 4-hydroxy-2-nonenal, 4-HNE) that interact with the sulfhydryl and thiol groups of adjacent proteins to form protein carbonyls and other structural modifications which limit or inactivate numerous enzyme and ion exchange systems, thus creating dysfunction in energy metabolism and ion homeostasis.[1,17] Thus, high cardiac membrane ω-6 PUFAs, coupled with increased ROS in aging, magnify myocardial dysfunction after adverse events, such as ischemia, or in chronic conditions, such as hypertrophy, heart failure, or diabetes.

Prostacyclin, a vasodilator eicosanoid, is produced from the ω-6 PUFA, AA, and the ω-3 PUFA, EPA, whereas the active form of the vasoconstrictor, thromboxane, is only produced from the ω-6 PUFA, AA. It is widely accepted that a low ratio of EPA/AA predisposes to a balance of eicosanoids, favoring vasoconstriction, platelet aggregation, inflammatory mediator signaling and cardiac arrhythmias.[16,18–23] During acute stress, augmented free radical production, peroxidation of ω-6 PUFA, and production of specific prostanoids, leukotrienes, cytokines, and other inflammatory mediators promote metabolic, ionic, and electrical instability, and depress contractile dysfunction.[18–23] Recently it has been shown in CABG surgery patients that epicardial adipose tissue around the proximal coronary arteries is a rich source of inflammatory cytokines, such as IL-6, IL-1β, TNF-α, and that surgery promotes their release.[19–25] Thus the resultant rise in ω-6 PUFA-derived products, particularly those predisposed to increase in disease and surgical interventions may be targeted by dietary intervention to augment ω-3 PUFA incorporation into tissues, thus diminishing the excess of ω-6 PUFA-derived metabolites.

CORRECTING ω-3 PUFA DEFICITS AND CARDIAC DYSFUNCTION

Rodent and primate hearts are responsive to increases in dietary ω-3 PUFA intake and preferentially accumulate DHA well above the circulating levels, even after having been fed pure EPA.[15,16,26] Notably dietary supplementation with ω-3 PUFA can reverse the age-associated decline in ω-3 PUFA and accumulation of ω-6 PUFA in cardiac membranes. This has important implications in disease states that are more common in advanced age. After ischemia–reperfusion, ω-3 PUFA treatment is associated with markedly reduced vulnerability to arrhythmia, improved pumping and O_2 utilization efficiency, and maintenance of Ca^{2+} homeostasis.[6,9,12–18,26,27] Glucose intolerance and insulin resistance seen in diabetes and diabetic-like states (heart failure, aging) depresses fatty acid desaturase enzyme expression and function, increasing the selective decrease of ω-3 PUFA in membrane phospholipids.[26,27] It has been shown from animal studies that insulin resistance is increased when membrane levels of ω-3 PUFA are low, but can be restored by ω-3 PUFA treatment, thus circumventing the disturbance in fatty acid metabolism due to deficits in delta-5 and delta-6 fatty acid desaturase activities that occur in diabetes.[26,27] In acute Type I diabetes in rats, DHA levels in myocardial membranes decline compared to ω-6 PUFA levels, which remain largely unchanged.[27]

Previous studies with isolated cardiac cells and intact hearts have shown that ω-3 PUFAs specifically normalize the transient outward current, the voltage-dependant Na^+ current, the delayed rectifier current, Ca^{2+}-Mg^{2+}-ATPase, Na^+-Ca^{2+}-exchanger, Ca^{2+} uptake in the sarcoplasmic reticulum, and L-type Ca^{2+} current, and they also prevent Ca^{2+} overload in cytoplasm and mitochondria[9,13,28,29]; further, they also reduce the activity of membrane phospholipases responsible for the generation of key intracellular messengers for Ca^{2+} handling: diacylglycerol and IP_3. Although considerable more work is required to fully define these and other effects on membrane proteins that contribute to the maintenance of homeostasis, cellular electrical stability, metabolism, and sinus rhythm, we expect restoration of ω-3 PUFA to myocardial membranes to confer multiple benefits. To date the most convincing clinical benefit relates to a reduction in the vulnerability to the triggers of arrhythmogenesis.

CLINICAL POTENTIAL

Recent clinical data support the extensive animal data that have demonstrated a specific antiarrhythmic action of dietary fish oil.[14–16] A low incidence of both primary cardiac arrest and postinfarction sudden death plus improved heart rate variability in patients consuming moderate levels of fish or fish oil all strongly indicate that the antiarrhythmic effect of ω-3 PUFA treatment is very potent in humans.[30–35] Studies of post-MI patients and heart failure patients have identified reduced mortality associated with modest consumption of ω-3 PUFA ($\sim$1 g/day), which does not significantly influence classical risk factors such as plasma triacylglycerols, blood pressure or thrombogenic factors even though these factors are ameliorated by ω-3 PUFA at higher doses ($\sim$5 g/day).[30–35] Diabetic cardiomyopathy, independent of vascular disease, contributes to high patient morbidity and mortality due to heart failure. Heart failure predisposes to sudden arrhythmic death, and the incidence of heart failure is more than doubled in diabetic patients.[36] Thus, it is not surprising that a regular fish diet, providing a rich source of ω-3 PUFA, is linked to reduced mortality for all cardiovascular disease.[37] In 1997, the U.S. Food and Drug Administration extensively reviewed more than 2600 studies to conclude that dietary intake of fish oil (DHA+EPA) up to and including 3 g per day was safe and had no clinically adverse effects or contraindications.[38] In light of this and recent positive clinical trials, the American Heart Association now recommends clinical use of dietary ω-3 PUFA.[39] Notably, a recent population-based cohort study prospectively examined 4815 elderly adults ($\geq$65 years) for 12 years, and found that dietary intake of ω-3 PUFA from fish was associated with markedly lower incidence of atrial fibrillation (AF).[40] Cardiac surgery represents a predictable acute stress to the heart that precipitates a defined degree of postoperative AF that is not easily prevented or managed and, if left untreated, may lead to stroke and heart failure. Most recently a study involving preoperative randomized open-label treatment of elective cardiac surgery patients with only 2 g/day/5 days of regular ω-3 PUFA capsules ($n = 79$) or placebo ($n = 81$) demonstrated a relative reduction in postoperative AF by 54%. AF in the ω-3 PUFA group was 15.2%; in the placebo group it was 33.3%. Hospital stay was significantly reduced by 1 full day in the ω-3 group.[41]

SUMMARY AND CONCLUSION

In summary, ω-3 PUFAs are prone to be diminished in cardiac membranes with advanced age in individuals on a Western-type diet, particularly when compounded by disease. Omega-3 PUFAs are antiarrhythmic and protective against ischemia–reperfusion injury and circumvent the perturbations in cardiac membranes due to advanced age: action potential and contractility, loss of Ca^{2+} homeostasis, inefficient O_2 and energy metabolism, oxidative stress, and inflammation. Omega-3 PUFA dietary supplementation may target many of these factors, as has been reported by the numerous animals studies now required to be better demonstrated in humans. Large clinical trials have shown ω-3 PUFA therapy reduces mortality, particularly mortality related to fatal ventricular fibrillation in post-MI patients. Early clinical studies suggest that ω-3 PUFA intake via a fish-rich diet is associated with reduced incidence of AF in the general elderly population, and after cardiac surgery. However, substantially more work is required to generate the definitive clinical evidence of ω-3 PUFA's therapeutic efficacy. In particular, more detailed delineation of all the molecular mechanisms of ω-3 PUFA action in humans is required to determine the extent of its capacity to minimize, circumvent, prevent, or reverse key age-linked cardiac membrane dysfunction.

REFERENCES

1. Lakatta, E.G., S.J. Sollott & S. Pepe. 2001. The old heart: operating on the edge. Novartis Foundation Symposia **235:** 172–196.
2. Pepe, S. 2005. Effect of dietary polyunsaturated fatty acids on age-related changes in cardiac mitochondrial membranes. Exp. Gerontol. **40:** 751–758.
3. Sheeran, F.L. & S. Pepe. 2006. Energy deficiency in the failing heart: linking increased reactive oxygen species and disruption of oxidative phosphorylation rate. Biochim. Biophys. Acta **1757:** 543–552.
4. Sprecher, H., D.L. Luthria, B.S. Mohammed & S.P. Baykousheva. 1995. Reevaluation of the pathways for the biosynthesis of polyunsaturated fatty acids. J. Lipid Res. **36:** 2471–2477.
5. Clandinin, M.T., J. Jumpsen & M. Suh. 1994. Relationship between fatty acid accretion, membrane composition, and biologic functions. J. Pediat. **125:** S25–32.
6. McLennan, P.L., P.R. Howe & M.Y. Abeywardena, *et al.* 1996. The cardiovascular protective role of docosahexaenoic acid. Eur. J. Pharmacol. **300:** 83–89.
7. Cho, H.P., M. Nakamura & S.D. Clarke. 1999. Cloning, expression, and fatty acid regulation of the human delta-5 desaturase. J. Biol. Chem. **274:** 37335–37339.
8. Cho, H.P., M.T. Nakamura & S.D. Clarke. 1999. Cloning, expression, and nutritional regulation of the mammalian delta-6 desaturase. J. Biol. Chem. **274:** 471–477.
9. Pepe, S., N. Tsuchiya, E.G. Lakatta & R.G. Hansford. 1999. PUFA and aging modulate cardiac mitochondrial membrane lipid composition and Ca^{2+} activation of pyruvate dehydrogenase. Am. J. Physiol. **276:** H149–H158.
10. Hansford, R.G., N. Tsuchiya & S. Pepe. 1999. Mitochondria in heart ischemia and aging. Biochem. Soc. Trans. **66:** 141–147.
11. Mariani, J.A., R. Ou, M. Bailey, *et al.* 2000. Tolerance to ischemia and hypoxia is reduced in aged human myocardium. J. Thorac. Cardiovasc. Surg. **120:** 660–667.
12. Pepe, S. & P.L. McLennan. 2002. Cardiac membrane fatty acid composition modulates myocardial oxygen consumption and post-ischemic recovery of contractile function. Circulation **105:** 2303–2308.
13. Pepe, S., K. Bogdanov, H. Hallaq, *et al.* 1994. Omega-3 PUFA modulates dihydropyridine effects on L-type Ca^{2+} channels, cytosolic Ca^{2+}, and contraction in adult rat cardiac myocytes. Proc. Natl. Acad. Sci. USA **91:** 8832–8836.
14. Pepe, S. & P.L. McLennan. 1996. Dietary fish oil exerts direct antiarrhythmic action in myocardium of the rat. J. Nutr. **126:** 34–42.
15. McLennan, P.L., M.Y. Abeywardena, J.S. Charnock. 1988. Dietary fish oil prevents ventricular fibrillation following coronary artery occlusion and reperfusion. Am. Heart J. **116:** 709–717.
16. McLennan, P.L., T.M. Bridle, M.Y. Abeywardena & J.S. Charnock. 1993. Comparative efficacy of n-3 and n-6 PUFA in modulating VF threshold in marmoset monkeys. Am. J. Clin. Nutr. **58:** 666–669.
17. Pepe, S. 2000. Mitochondrial function in ischemia and reperfusion of the aging heart. Clin. Exp. Physiol. Pharmacol. **27:** 745–750.
18. Abeywardena, M.Y., P.L. McLennan & J.S. Charnock. 1991. Differential effects of dietary fish oil on myocardial prostaglandin I2 and thromboxane A2 production. Am. J. Physiol. **260:** H379–H385.

19. SCHAUDER, P., U. ROHN, G. SCHAFER, *et al.* 2002. Impact of fish oil enriched total parenteral nutrition on DNA synthesis, cytokine release and receptor expression by lymphocytes in the postoperative period. Br. J. Nutr. **87**(Suppl 1): S103–110.
20. WACHTLER, P., W. KONIG, M. SENKAL, *et al.* 1997. Influence of a total parenteral nutrition enriched with omega-3 fatty acids on leukotriene synthesis of peripheral leukocytes and systemic cytokine levels in patients with major surgery. J. Trauma **42:** 191–198.
21. NILSEN, D.W., K. DALAKER, A. NORDOY, *et al.* 1991. Influence of a concentrated ethylester of n-3 fatty acids on lipids, platelets and coagulation in patients undergoing coronary bypass surgery. Thromb Haemost. **66:** 195–201.
22. DECATERINA, R., D. GIANNESSI, A. MAZZONE, *et al.* 1990. Vascular prostacyclin is increased in patients ingesting omega-3 PUFA before coronary artery bypass graft surgery. Circulation **82:** 428–438.
23. RUPP, H., D. WAGNER, T. RUPP, *et al.* 2004. Risk stratification by EPA+DHA level and EPA/AA ratio: anti-inflammatory & antiarrhythmogenic effects of PUFA. Herz **29:** 673–685.
24. GAUDINO, M., F. ANDREOTTI, R. ZAMPARELLI, *et al.* 2003. The -174G/C interleukin-6 polymorphism influences postoperative interleukin-6 levels and postoperative atrial fibrillation. Is atrial fibrillation an inflammatory complication? Circulation **108**(Suppl 1): II195–II199.
25. MAZUREK, T., L. ZHANG, A. ZALEWSKI, *et al.* 2003. Human epicardial adipose tissue is a source of inflammatory mediators. Circulation **108:** 2460–2466.
26. FAAS, F.H. & W.J. CARTER. 1980. Altered fatty acid desaturation and microsomal fatty acid composition in the streptozotocin diabetic rat. Lipids **15:** 953–961.
27. MCLENNAN, P.L. & D. RAEDERSTORFF. 1999. Diabetes puts myocardial omega-3 fatty acid status at risk in the absence of supplementation in the rat. Lipids **34:** S91–S92.
28. BOGDANOV, K.Y., H.A. SPURGEON, T.M. VINOGRADOVA & E.G. LAKATTA. 1998. Modulation of the transient outward current in adult rat ventricular myocytes by PUFA. Am. J. Physiol. 274: H571–H579.
29. LEAF, A., J.X. KANG, Y.F. XIAO & G.E. BILLMAN. 2003. Clinical prevention of sudden cardiac death by n-3 polyunsaturated fatty acids and mechanism of prevention of arrhythmias by n-3 fish oils. Circulation **107:** 2646–2652.
30. CHRISTENSEN, J.H., H.A. SKOU, L. FOG, *et al.* 2001. Marine n-3 fatty acids, wine intake, & heart rate variability in patients referred for coronary angiography. Circulation **103:** 651–657.
31. SISCOVICK, D.S., T.E. RAGUNATHAN, I. KING, *et al.* 1995. Dietary intake and cell membrane levels of long-chain n-3 PUFA and the risk of primary cardiac arrest. JAMA **274:** 1363–1367.
32. GISSI-PREVENZIONE INVESTIGATORS. 1999. Dietary supplementation with n-3 polyunsaturated fatty acids and vitamin E after myocardial infarction: the GISSI-Prevenzione trial. Lancet **354:** 447–455.
33. DAVIGLUS, M.L., J. STAMLER, A.J. ORENCIA, *et al.* 1997. Fish consumption and the 30-year risk of fatal myocardial infarction. N. Engl. J. Med. **336:** 1046–1053.
34. BURR, M.L., A.M. FEHILY, J.F. GILBERT, *et al.* 1989. Effects of changes in fat, fish, and fibre intakes on death and myocardial reinfarction: diet and reinfarction trial. Lancet **2:** 757–761.
35. ALBERT, C.M., H. CAMPOS, M.J. STAMPFER, *et al.* 2002. Blood levels of long-chain n-3 fatty acids and the risk of sudden death. N. Engl. J. Med. **346:** 1113–1118.

36. FEIN, F.S. 1996. Clinical manifestations of diabetic cardiomyopathy. Develop. Card. Med. **185:** 1–22.
37. CESARI, M., B.W. PENNINX, A.B. NEWMAN, *et al.* 2003. Inflammatory markers and onset of cardio-vascular events: the Health ABC study. Circulation **108:** 2317–2322.
38. FOOD & DRUG ADMINISTRATION, HHS. 1997. Substances Affirmed as Generally Recognized as Safe: Menhaden Oil. 30751–30757. Federal Register, USA, June 5, 1997.
39. KRIS-ETHERTON, P.M., W.S. HARRIS & L.J. APPEL. 2002. American Heart Association. Fish consumption, fish oil, omega-3 fatty acids, & cardiovascular disease. Circulation **106:** 2747–2757.
40. MOZAFFARIAN, D., B.M. PSATY, E.B. RIMM, *et al.* 2004. Fish intake and risk of incident atrial fibrillation. Circulation **110:** 368–373.
41. CALO, L., *et al.* 2005. N-3 Fatty acids for the prevention of atrial fibrillation after coronary artery bypass surgery: a randomized, controlled trial. J. Am. Coll. Cardiol. **45:** 1723–1728.

Dietary Influences on Cognitive Function with Aging

From High-Fat Diets to Healthful Eating

MATTHEW D. PARROTT[a,b] AND CAROL E. GREENWOOD[a,b,c]

[a]*Department of Nutritional Sciences, Faculty of Medicine, University of Toronto, Toronto, Ontario, Canada*

[b]*Kunin–Lunenfeld Applied Research Unit, Baycrest, Toronto, Ontario, Canada*

[c]*Food and Nutrition Services, Baycrest, Toronto, Ontario, Canada*

ABSTRACT: Human epidemiologic studies provide convincing evidence that dietary patterns practiced during adulthood are important contributors to age-related cognitive decline and dementia risk. Diets high in fat, especially trans and saturated fats, adversely affect cognition, while those high in fruits, vegetables, cereals, and fish are associated with better cognitive function and lower risk of dementia. While the precise physiologic mechanisms underlying these dietary influences are not completely understood, modulation of brain insulin activity and neuroinflammation likely contribute. Not surprisingly, deficits in cognitive functions, especially those dependent on the medial temporal lobes, are apparent in type 2 diabetes mellitus (T2DM). Special care in food selection at meals should be exercised by those with T2DM since ingestion of rapidly absorbed, high–glycemic index carbohydrate foods further impairs medial temporal lobe function, with food-induced increases in oxidative stress and cytokine release likely explaining the association between food ingestion and reduction in cognitive function in those with T2DM.

KEYWORDS: diet; dietary fat; cognition; memory; aging; insulin resistance; type 2 diabetes mellitus; inflammation; neuroinflammation; Alzheimer's disease

INTRODUCTION

The maintenance of cognitive function with aging is a current and growing concern. While pharmaceutical and therapeutic approaches continue to be actively researched, increasing evidence indicates a role for lifestyle factors in

Address for correspondence: Carol Greenwood, Department of Nutritional Sciences, Faculty of Medicine, University of Toronto, Toronto, ON, Canada M5S 3E2. Voice: 416-978-4261; fax: 416-978-5882.

carol.greenwood@utoronto.ca

Ann. N.Y. Acad. Sci. 1114: 389–397 (2007).
doi: 10.1196/annals.1396.028

successful brain aging. One such factor is diet, or a person's usual pattern of eating and drinking, which is proposed to exert both beneficial and detrimental influences. Our early studies in rats were among the first to demonstrate adverse cognitive effects of chronic consumption of high-fat diets, especially diets high in saturated fat, in young adult animals.[1] In the intervening years, numerous animal and human epidemiologic studies have both confirmed the detrimental role of consumption of high saturated fats and identified positive attributes of more healthful eating patterns. Of interest, dietary profiles thought to benefit or impair cognitive function correspond to those profiles exerting similar effects on risk for many chronic diseases which are not thought to have primary cognitive involvements (e.g., diabetes, coronary heart disease). As will be briefly discussed, this relationship may result from shared dietary influences on the overlapping physiologic mechanisms responsible for disease risk and altered cognition with aging.

EPIDEMIOLOGIC EVIDENCE LINKING DIETARY PATTERNS WITH MODULATED COGNITIVE FUNCTION

A number of large, prospective and cross-sectional observational studies find that the dietary profile benefiting cognitive function with aging contains weekly servings of fish[2–5] and multiple daily servings of cereals, darkly or brightly colored fruits, and leafy vegetables.[6–9] Alternatively, these and other studies find that decreased cognitive function associates with diets relatively high in total, trans, and saturated fat.[10] How each dietary profile is contributing to metabolism and cognitive function is debatable. For instance, leafy vegetables are good sources of folate and other B-vitamins whose serum concentrations are associated with better cognitive function,[11,12] and, perhaps more importantly, strongly modulate homocysteine levels—another cognitive risk factor.[13–15] Similarly, although darkly or brightly colored fruits and vegetables are sources of antioxidant vitamins, increasing evidence points to the importance of the polyphenolic compounds found in plants, which may have direct effects on cell signaling, growth, and differentiation in addition to being the most abundant and powerful dietary antioxidants.[16–18] Although not well understood, plant polyphenols (PPs) as a group may play a large role in findings of better cognitive outcomes with consumption of plant matter (fruits/vegetables), which is a view supported by findings that food sources of antioxidant vitamins are associated with cognitive benefits while pill-forms are not.[19,20] Certainly, studies in animal models of aging and Alzheimer's disease (AD) have found that dietary PPs reduce neuroinflammation and stimulate the activity of phosphotidylinositol-3 kinase (PI3K)—a ubiquitous enzyme involved in many cellular responses with particular importance to insulin signaling—resulting in improved neuronal survival and memory.[21–23]

Similarly, omega-3 (n-3) fatty acids, in addition to their acknowledged role in supporting membrane-bound protein functions and neurotransmisson, may have anti-inflammatory and pro-survival capabilities by modulating cytokine activity, neurotrophin expression, and anti-apoptotic pathways, including those influenced by PI3K.[24–26] Animal studies find that diets low in n-3 fat compromise cognitive function, promote neuroinflammation, and increase neuropathology in animal models.[27–29] These roles are complemented by epidemiologic data showing that intake of n-3 fatty acids, particularly those from fish and fish oil, reduces risk for dementia and cognitive decline.[3–5] Alternatively, feeding animals diets high in saturated fat leads to cognitive impairment, reduced neurotrophin levels, and increased AD-related pathology—mirroring epidemiologic findings.[30–32] Human food consumption patterns indicate that diets high in saturated fat usually contain lower amounts of plant matter, and place individuals at increased risk for obesity.[33,34] Observational studies indicating positive or negative effects of dietary fat on cognitive decline also find that such influences are modified by the relative abundance of different fatty acids so that negative effects of saturated fat are most evident in the context of low n-3 intake and vice versa. These results could indicate that foods contained in "bad" and "good" dietary profiles are somewhat exclusive.

DIETARY MODULATION OF BRAIN INSULIN ACTIVITY AND NEUROINFLAMMATION

Given the contrary effects of these alternative diets on cognitive function, there is potential that these diets are modulating the same overall mechanisms albeit in opposite directions. While many systems may be involved, existing evidence points to the overall importance of two critical mechanisms: (1) sustainability of insulin and insulin-related cell signaling (especially through PI3K) and (2) limiting inflammation. The dietary components discussed above all share the ability to influence these processes, which themselves can beneficially and adversely influence each other. It is increasingly recognized that disruptions in brain insulin–responsive cell signaling, apparent in those with reduced tissue sensitivity to insulin despite high blood concentration—termed *insulin resistance* (IR)—may result in brain insults leading to compromised cognitive function and in some cases neuropathologic progression of AD.[35,36] IR and/or hyperinsulinemia are also well known to increase systemic inflammatory responses and oxidative stress,[37] which can translate into increased markers of central nervous system (CNS) inflammation.[38] Increased pro-inflammatory cytokine levels can downregulate PI3K/Akt in aged rat brains, leading to subsequent caspase activation,[39] suggesting that insulin-related signaling is negatively affected by inflammation, ultimately leading to increased potential for neuronal death.

Interestingly, increased CNS inflammation has been positively correlated with amyloid-beta (Aβ) levels and insulin-resistant individuals with greatest inflammation exhibit greater cognitive deficits,[40] suggesting that synchronous hyperinsulinemia-induced increases in Aβ and inflammation may represent an important pathway through which IR promotes both cognitive deterioration and AD pathology.[35] Generalized age-related increases in brain oxidative stress/inflammation[41] and findings that half of all adults over 60 years of age have some degree of IR[42] highlight the importance of understanding the potential adverse consequences of IR and its most obvious pathologic result, type 2 diabetes mellitus (T2DM), in compromising cognitive function by impairing appropriate cellular responses and exacerbating inflammation.

We have linked memory impairments in obese and insulin-resistant Zucker rats with evidence of compromised brain insulin signaling compared to that of lean littermates,[43] and found that inflammatory cytokine expression modulates cognitive decline in the extreme IR of human T2DM.[44] Highly prevalent chronic conditions like diabetes[45] and cardiovascular disease[46,47] are themselves linked to cognitive dysfunction, highlighting a relationship between peripheral and brain health. Of importance, both conditions are associated with IR, which is well known to be promoted by saturated fat intake,[48] whereas lower fat diets relatively high in foods providing PPs and n-3 fat (fish, fruits and vegetables) are thought to prevent or reduce it.[49,50] Therefore, IR becomes an important underlying mechanism common to each risk factor that unites experimental and observational examinations of diet and disease on age-related cognitive changes.

Given modification of single nutrient/food effects by other dietary components and the interrelationship between insulin activity and inflammation, any future, and much needed, randomized controlled trials would be wise to intervene with global dietary changes allowing for protection of these two critical points by simultaneously recruiting multiple systems associated with their protection and minimizing adverse consequences of a poor-quality diet that confound supplemental or single-nutrient approaches. Supporting such a view are the results of one study indicating that a low-fat diet enriched with both fish oils and polyphenol extract improved learning in old rats compared to those receiving a diet high in saturated fat. Importantly, benefits were only seen in old rats with the combined use of fish oil and polyphenols, but not when either component was separately provided.[51]

ADVERSE EFFECTS OF MEALS IN ADULTS WITH TYPE 2 DIABETES MELLITUS

While chronic dietary profiles that lead to development of IR, oxidative stress, and inflammation are associated with poorer cognitive function and risk of dementia, even circadian glycemic fluctuations experienced in those with T2DM may be of sufficient magnitude to further compromise cognitive function.[52] For example, our own studies of meal/food ingestion in adults with T2DM demonstrate further cognitive decrements, especially on measures of delayed verbal memory, following ingestion of high–glycemic index carbohydrate foods, such as white bread, in those with T2DM.[53,54] Most notable is the fact that many studies in healthy older adults or those with IR, but not T2DM, repeatedly demonstrate that ingestion of 50 g of glucose, either as a glucose drink or in the form of high-carbohydrate foods,[55] enhances cognitive functions, especially those served by the hippocampus and surrounding structures.[56–58] Animal-based studies demonstrate that transient decreases in hippocampal, extraneuronal glucose levels occur during memory-demanding cognitive tasks in older animals and that these levels can be replenished with glucose administration.[59,60] Since glucose supplies energy and also serves as a substrate for synthesis of neurotransmitters, such as acetylcholine, glutamate, and GABA,[58] local glucose depletion may be detrimental to actively firing neurons.

These studies, conducted predominantly in healthy older adults, suggest that raising blood glucose levels into an optimal range enhances cognition in those with poor memories[61–63]; yet provide indications that excessively high blood glucose levels no longer produce this benefit and may be associated with cognitive dysfunction. This adverse effect of hyperglycemia presumably parallels the cognitive decrements apparent in those with T2DM after food ingestion since impaired glucose disposal in those with T2DM results in higher and prolonged elevations in blood glucose levels. Consequently, it is not surprising that food-induced cognitive decrements in those with T2DM are highly associated with postprandial elevations in blood glucose levels and are minimized by consumption of lower–glycemic index carbohydrate foods, such as pasta, which result in lower blood glucose levels.[54]

While the precise mechanism linking acute hyperglycemia to declines in cognitive functions is unknown, enhancement of oxidative stress and release of proinflammatory cytokines likely contribute. This potential mechanism is supported by results from our recent study in which a deficit following ingestion of a meal high in fat and simple carbohydrates was prevented by the co-consumption of the antioxidant vitamins C and E (Chui and Greenwood, submitted for publication).

CONCLUSIONS

Long-term dietary exposures appear to positively and negatively influence age-related cognitive changes. Both "good" and "bad" dietary profiles may do so by influencing the interrelated mechanisms of insulin signaling and inflammation. Therefore, a diet that maximizes consumption of plant matter and includes fish, while limiting total and saturated fats, presents the greatest potential for successful brain aging. Special attention to dietary practices is needed in those with T2DM as even meal-induced elevations in glucose levels can further compromise cognitive function.

ACKNOWLEDGEMENTS

Results reported from our studies were supported by grants from the Natural Sciences and Engineering Research Council of Canada to C.E.G. M.D.P. is the recipient of a Banting & Best Diabetes Centre-Novo Nordisk Fellowship.

REFERENCES

1. WINOCUR, G. & C.E. GREENWOOD. 2005. Studies of the effects of high fat diets on cognitive function in a rat model. Neurobiol. Aging **26:** 46–49.
2. BARBERGER-GATEAU, P. *et al.* 2002. Fish, meat, and risk of dementia: cohort study. Br. Med. J. **325:** 932–933.
3. KALMIJN, S. *et al.* 1997. Dietary fat intake and the risk of incident dementia in the Rotterdam Study. Ann. Neurol. **42:** 776–782.
4. MORRIS, M.C. *et al.* 2003. Consumption of fish and n-3 fatty acids and risk of incident Alzheimer disease. Arch. Neurol. **60:** 940–946.
5. MORRIS, M.C. *et al.* 2005. Fish consumption and cognitive decline with age in a large community study. Arch. Neurol. **62:** 1849–1853.
6. KANG, J.H., A. ASCHERIO & F. GRODSTEIN. 2005. Fruit and vegetable consumption and cognitive decline in aging women. Ann. Neurol. **57:** 713–720.
7. MORRIS, M.C. *et al.* 2006. Associations of vegetable and fruit consumption with age-related cognitive change. Neurology **67:** 1370–1376.
8. ORTEGA, R.M. *et al.* 1997. Dietary intake and cognitive function in a group of elderly people. Am. J. Clin. Nutr. **66:** 803–809.
9. PANZA, F. *et al.* 2004. Mediterranean diet and cognitive decline. Public Health Nutr. **7:** 959–963.
10. MORRIS, M.C. *et al.* 2003. Dietary fats and the risk of incident Alzheimer disease. Arch. Neurol. **60:** 194–200.
11. RAMOS, M.I. *et al.* 2005. Low folate status is associated with impaired cognitive function and dementia in the Sacramento Area Latino Study on Aging. Am. J. Clin. Nutr. **82:** 1346–1352.
12. WANG, H.X. *et al.* 2001. Vitamin B(12) and folate in relation to the development of Alzheimer's disease. Neurology **56:** 1188–1194.
13. ELIAS, M.F. *et al.* 2005. Homocysteine and cognitive performance in the Framingham offspring study: age is important. Am. J. Epidemiol. **162:** 644–653.
14. HAAN, M.N. *et al.* 2007. Homocysteine, B vitamins, and the incidence of dementia and cognitive impairment: results from the Sacramento Area Latino Study on Aging. Am. J. Clin. Nutr. **85:** 511–517.
15. SESHADRI, S. *et al.* 2002. Plasma homocysteine as a risk factor for dementia and Alzheimer's disease. N. Engl. J. Med. **346:** 476–483.
16. JOSEPH, J.A., B. SHUKITT-HALE & G. CASADESUS. 2005. Reversing the deleterious effects of aging on neuronal communication and behavior: beneficial properties of fruit polyphenolic compounds. Am. J. Clin. Nutr. **81:** 313S–316S.

17. MANDEL, S.A. *et al.* 2005. Multifunctional activities of green tea catechins in neuroprotection: modulation of cell survival genes, iron-dependent oxidative stress and PKC signaling pathway. Neurosignals **14:** 46–60.
18. SCALBERT, A., I.T. JOHNSON & M. SALTMARSH. 2005. Polyphenols: antioxidants and beyond. Am. J. Clin. Nutr. **81:** 215S–217S.
19. ENGELHART, M.J. *et al.* 2002. Dietary intake of antioxidants and risk of Alzheimer disease. JAMA **287:** 3223–3229.
20. MORRIS, M.C. *et al.* 2002. Dietary intake of antioxidant nutrients and the risk of incident Alzheimer disease in a biracial community study. JAMA **287:** 3230–3237.
21. CASADESUS, G. *et al.* 2004. Modulation of hippocampal plasticity and cognitive behavior by short-term blueberry supplementation in aged rats. Nutr. Neurosci. **7:** 309–316.
22. JOSEPH, J.A. *et al.* 2003. Blueberry supplementation enhances signaling and prevents behavioral deficits in an Alzheimer disease model. Nutr. Neurosci. **6:** 153–162.
23. LIM, G.P. *et al.* 2001. The curry spice curcumin reduces oxidative damage and amyloid pathology in an Alzheimer transgenic mouse. J. Neurosci. **21:** 8370–8377.
24. HORROCKS, L.A. & A.A. FAROOQUI. 2004. Docosahexaenoic acid in the diet: its importance in maintenance and restoration of neural membrane function. Prostaglandins Leukot. Essent. Fatty Acids **70:** 361–372.
25. LUKIW, W.J. *et al.* 2005. A role for docosahexaenoic acid-derived neuroprotectin D1 in neural cell survival and Alzheimer disease. J. Clin. Invest. **115:** 2774–2783.
26. MARCHESELLI, V.L. *et al.* 2003. Novel docosanoids inhibit brain ischemia-reperfusion-mediated leukocyte infiltration and pro-inflammatory gene expression. J. Biol. Chem. **278:** 43807–43817.
27. BARCELO-COBLIJN, G. *et al.* 2003. Modification by docosahexaenoic acid of age-induced alterations in gene expression and molecular composition of rat brain phospholipids. Proc. Natl. Acad. Sci. USA **100:** 11321–11326.
28. CALON, F. *et al.* 2004. Docosahexaenoic acid protects from dendritic pathology in an Alzheimer's disease mouse model. Neuron **43:** 633–645.
29. CATALAN, J. *et al.* 2002. Cognitive deficits in docosahexaenoic acid-deficient rats. Behav. Neurosci. **116:** 1022–1031.
30. GREENWOOD, C.E. & G. WINOCUR. 2005. High-fat diets, insulin resistance and declining cognitive function. Neurobiol. Aging **26:** 42–45.
31. HO, L. *et al.* 2004. Diet-induced insulin resistance promotes amyloidosis in a transgenic mouse model of Alzheimer's disease. FASEB J. **18:** 902–904.
32. MOLTENI, R. *et al.* 2004. Exercise reverses the harmful effects of consumption of a high-fat diet on synaptic and behavioral plasticity associated to the action of brain-derived neurotrophic factor. Neuroscience **123:** 429–440.
33. DAVIS, J.N., V.A. HODGES & M.B. GILLHAM. 2006. Normal-weight adults consume more fiber and fruit than their age- and height-matched overweight/obese counterparts. J. Am. Diet. Assoc. **106:** 833–840.
34. NELSON, L.H. & L.A. TUCKER. 1996. Diet composition related to body fat in a multivariate study of 203 men. J. Am. Diet. Assoc. **96:** 771–777.
35. CRAFT, S. 2005. Insulin resistance syndrome and Alzheimer's disease: age- and obesity-related effects on memory, amyloid, and inflammation. Neurobiol. Aging **26:** 65–69.

36. MESSIER, C. & K. TEUTENBERG. 2005. The role of insulin, insulin growth factor, and insulin-degrading enzyme in brain aging and Alzheimer's disease. Neural Plast. **12:** 311–328.
37. CABALLERO, A.E. 2003. Endothelial dysfunction in obesity and insulin resistance: a road to diabetes and heart disease. Obes. Res. **11:** 1278–1289.
38. FISHEL, M.A. *et al.* 2005. Hyperinsulinemia provokes synchronous increases in central inflammation and beta-amyloid in normal adults. Arch. Neurol. **62:** 1539–1544.
39. MAHER, F.O., D.S. MARTIN & M.A. LYNCH. 2004. Increased IL-1beta in cortex of aged rats is accompanied by downregulation of ERK and PI-3 kinase. Neurobiol. Aging **25:** 795–806.
40. YAFFE, K. *et al.* 2004. The metabolic syndrome, inflammation, and risk of cognitive decline. JAMA **292:** 2237–2242.
41. JOSEPH, J.A. *et al.* 2005. Oxidative stress and inflammation in brain aging: nutritional considerations. Neurochem. Res. **30:** 927–935.
42. KAHN, R. *et al.* 2005. The metabolic syndrome: time for a critical appraisal: joint statement from the American Diabetes Association and the European Association for the Study of Diabetes. Diabetes Care **28:** 2289–2304.
43. WINOCUR, G. *et al.* 2005. Memory impairment in obese Zucker rats: an investigation of cognitive function in an animal model of insulin resistance and obesity. Behav. Neurosci. **119:** 1389–1395.
44. CHUI, M.H. *et al.* 2007. The TNF-α-238G>A single nucleotide polymorphism protects against memory decline in older adults with type 2 diabetes. Behav. Neurosci. **121**(3)**:** 619–624.
45. ARVANITAKIS, Z. *et al.* 2004. Diabetes mellitus and risk of Alzheimer disease and decline in cognitive function. Arch. Neurol. **61:** 661–666.
46. HONIG, L.S. *et al.* 2003. Stroke and the risk of Alzheimer disease. Arch. Neurol. **60:** 1707–1712.
47. NEWMAN, A.B. *et al.* 2005. Dementia and Alzheimer's disease incidence in relationship to cardiovascular disease in the Cardiovascular Health Study cohort. J. Am. Geriatr. Soc. **53:** 1101–1107.
48. RICCARDI, G., R. GIACCO & A.A. RIVELLESE. 2004. Dietary fat, insulin sensitivity and the metabolic syndrome. Clin. Nutr. **23:** 447–456.
49. FELDEISEN, S.E. & K.L. TUCKER. 2007. Nutritional strategies in the prevention and treatment of metabolic syndrome. Appl. Physiol. Nutr. Metab. **32:** 46–60.
50. NETTLETON, J.A. & R. KATZ. 2005. n-3 long-chain polyunsaturated fatty acids in type 2 diabetes: a review. J. Am. Diet. Assoc. **105:** 428–440.
51. SHIRAI, N. & H. SUZUKI. 2004. Effect of dietary docosahexaenoic acid and catechins on maze behavior in mice. Ann. Nutr. Metab. **48:** 51–58.
52. COX, D.J. *et al.* 2005. Relationships between hyperglycemia and cognitive performance among adults with type 1 and type 2 diabetes. Diabetes Care **28:** 71–77.
53. GREENWOOD, C.E. *et al.* 2003. Carbohydrate-induced memory impairment in adults with type 2 diabetes. Diabetes Care **26:** 1961–1966.
54. PAPANIKOLAOU, Y. *et al.* 2006. Better cognitive performance following a low-glycaemic-index compared with a high-glycaemic-index carbohydrate meal in adults with type 2 diabetes. Diabetologia **49:** 855–862.
55. KAPLAN, R.J. *et al.* 2000. Cognitive performance is associated with glucose regulation in healthy elderly persons and can be enhanced with glucose and dietary carbohydrates. Am. J. Clin. Nutr. **72:** 825–836.

56. Gold, P.E. 2005. Glucose and age-related changes in memory. Neurobiol. Aging **26**(Suppl 1): 60–64.
57. Korol, D.L. & P.E. Gold. 1998. Glucose, memory, and aging. Am. J. Clin. Nutr. **67:** 764S–771S.
58. Messier, C. 2004. Glucose improvement of memory: a review. Eur. J. Pharmacol. **490:** 33–57.
59. McNay, E.C., T.M. Fries & P.E. Gold. 2000. Decreases in rat extracellular hippocampal glucose concentration associated with cognitive demand during a spatial task. Proc. Natl. Acad. Sci. USA **97:** 2881–2885.
60. McNay, E.C. & P.E. Gold. 2001. Age-related differences in hippocampal extracellular fluid glucose concentration during behavioral testing and following systemic glucose administration. J. Gerontol. A Biol. Sci. Med. Sci. **56:** B66–71.
61. Benton, D., P.Y. Parker & R.T. Donohoe. 1996. The supply of glucose to the brain and cognitive functioning. J. Biosoc. Sci. **28:** 463–479.
62. Manning, C.A., M.E. Ragozzino & P.E. Gold. 1993. Glucose enhancement of memory in patients with probable senile dementia of the Alzheimer's type. Neurobiol. Aging **14:** 523–528.
63. Parsons, M.W. & P.E. Gold. 1992. Glucose enhancement of memory in elderly humans: an inverted-U dose-response curve. Neurobiol. Aging **13:** 401–404.

Combining an Antioxidant-Fortified Diet with Behavioral Enrichment Leads to Cognitive Improvement and Reduced Brain Pathology in Aging Canines

Strategies for Healthy Aging

ELIZABETH HEAD

Institute for Brain Aging & Dementia, Department of Neurology, University of California, Irvine, Irvine, California 92697, USA

ABSTRACT: The number of elderly individuals in our population is rapidly rising and age-associated neurodegenerative disease is becoming more prevalent. Thus, identifying ways by which we can promote healthy aging are becoming more critical. Lifestyle factors, such as engaging in physical, intellectual, and social activities, are protective against dementia in aged individuals. Similarly, there is some evidence to suggest that antioxidants are beneficial. Observational studies in humans have been confirmed and extended in rodent model systems. We present additional evidence that, in a canine model of aging, combining an antioxidant-enriched diet and behavioral enrichment (including social, physical, and cognitive components) can lead to substantial improvements in cognition and reduced brain pathology. These results suggest that modifying lifestyle factors can have a beneficial impact on the aging process, even in aged individuals with existing cognitive decline and brain pathology.

KEYWORDS: beta-amyloid; brain; dog; fruits; mitochondrial co-factors; neuron number; vegetables; vitamin E; vitamin C

INTRODUCTION

The preservation of cognitive function throughout life is becoming an increasingly important goal of aging research as our population ages. Current estimates suggest that the number of individuals over the age of 60 years is rapidly rising both nationally and internationally. With increasing age, individuals become more vulnerable to age-associated neurodegenerative diseases, such as Alzheimer's disease (AD). Recently, the United States Alzheimer's Association

Address for correspondence: Elizabeth Head, Ph.D., Institute for Brain Aging & Dementia, Department of Neurology, University of California at Irvine, 1259 Gillespie NRF, Irvine, CA 92697-4540. Voice: +949-824-8700; fax: +949-824-2071.
ehead@uci.edu

Ann. N.Y. Acad. Sci. 1114: 398–406 (2007). © **2007 New York Academy of Sciences.**
doi: 10.1196/annals.1396.004

reported that the 5 million individuals in the United States in 2007 who have AD is higher than the previous prevalence nationwide estimate of 4.5 million (www.alz.org). Promoting healthy aging through beneficial lifestyle choices is one of the most promising approaches to reducing risk for developing neurodegenerative diseases. Interventions that can be incorporated into a healthy lifestyle include maintaining or increasing physical, social and cognitively engaging activities, and adequate levels of dietary antioxidant (AOX) intake.

There are a number of human epidemiologic studies that suggest that maintaining an active lifestyle can protect against pathologic aging.[1] For example, participation in leisure,[2] cognitively stimulating,[3,4] or physical activities that lead to improved function[5,6] reduces the risk for developing AD. Further, reduced, passive intellectual or physical activity in middle age can lead to a higher risk for AD in later life.[7] Additional lifestyle factors, such as a higher level of education or more intellectually challenging occupations, may provide similar life-long behavioral enrichment (BEH) (see Kramer *et al.* for an excellent review[1]). Consistent with these observational studies, in systematic and controlled experiments using cognitive training interventions, significant long-lasting cognitive improvements are observed.[8,9] The possible neurobiological mechanisms underlying the beneficial effects of BEH (social, physical, and intellectual) are actively being studied in animal models. In rodents, improved behavioral function as a consequence of BEH or physical exercise may be mediated by increased neurogenesis,[10,11] increased availability of brain growth factors,[12,13] and possibly by reductions in brain pathology, such as beta-amyloid (Aβ) accumulations typically seen in patients with AD.[14,15]

In contrast to the consistent effects of BEH and lifestyle parameters that engage social, physical, and cognitive activities, results from studies on the use of AOXs in human observational studies are less uniform. While some studies have shown a positive effect of AOX supplementation on cognition and risk reduction for developing AD,[16,17] other studies do not report significant effects.[18,19] Difficulties in interpreting human observational studies stem from inconsistencies in the amount of supplements taken, their form and source, their duration and regularity of use, and from the challenges of determining the exact dietary intake of AOXs. For example, in one study, combinations of AOXs were superior to single supplementation,[20] and dietary intake of AOXs has been shown to be superior to supplements in human studies (i.e., tablets).[17] Overall, recent reviews of the human epidemiologic literature have emphasized that AOXs may be one of the most promising treatments for preventing AD.[21] There have been few systematic studies evaluating AOXs in clinical trials. Vitamin E, however, has been shown to delay institutionalization in AD patients, suggesting some beneficial effects.[22] However, vitamin E alone did not improve cognition in patients with mild cognitive impairment, which is thought to be a precursor to AD.[23] In normally aging individuals, vitamin E alone has little benefit in elderly women[24]; however, one study showed that supplementation with a combination of vitamins E and C led to improved cognition.[25] In contrast, research in animal models consistently supports the hypothesis that AOXs can improve function and behavior. In aged rodents, AOXs in the form of supplements and fruits and vegetables can improve learning, memory, and motor function and reduce oxidative damage.[26–29]

The major advantage of using animal models is the ability to control many of the experimental variables that can contribute to differences in human observational studies. Animal studies further allow the researcher to understand the brain mechanisms and molecular cascades that might be critical for beneficial treatment effects. However, there are few studies in animal models that develop similar types of brain pathology as that found in the human and that can be studied for extended periods of time. Further, studies in animal models rarely evaluate the effects of combinatorial interventions. For example, can the benefits of an AOX diet be enhanced by inclusion of BEH? For these reasons, we have been using a canine model of human aging.

THE CANINE MODEL OF HUMAN BRAIN AGING

Aged canines naturally develop cognitive dysfunction on tests of spatial memory that may begin as early as middle age.[30] Spatial attention also declines with age,[31] as do measures of executive function.[32,33]

In parallel, there are numerous reports that the aging canine brain develops diffuse plaques that contain the Aβ peptide similar to those observed in aging humans.[34–37] Canine Aβ has an amino acid sequence identical to that of human Aβ.[38,39] And, like humans, some cortical regions of the canine brain are more vulnerable to Aβ accumulation than others. For example, disease accumulates in the prefrontal cortex in beagles as young as 8 years of age.[35] Correlation studies provide some support for a link between extent of Aβ pathology and level of cognitive dysfunction.[40–43] Specifically, animals with significant prefrontal Aβ disease perform more poorly on a visual reversal learning task, which is sensitive to frontal lobe function.[42] Another key feature of the aged canine brain is the progressive accumulation of oxidative damage.[44–48]

Dog brains accumulate oxidative damage to protein carbonyl groups with age.[44,48] Further, reduced endogenous AOX enzymes, such as glutamine synthetase activity and superoxide dismutase, occur in parallel with increased protein damage.[44,45] In several studies, a relation between age and increased oxidative damage has been inferred by measuring the amount of end products of lipid peroxidation (oxidative damage to lipids), including the extent of 4-hydroxynonenal (4HNE),[43,46,47] lipofuscin,[43] lipofuscin-like pigments,[46,47] and malondialdehyde.[44] Last, there is evidence of increased oxidative damage to DNA or RNA (8OHdG) in the aged dog brain.[43]

There is also evidence for a link between Aβ accumulation and oxidative damage. Several recent studies suggest that oxidative damage precedes Aβ production and that the two enter into a cyclic cascade in which oxidative damage stimulates Aβ production and accumulation and Aβ accumulation leads to more oxidative damage. If unsuccessfully opposed, both oxidative damage and Aβ may progressively accumulate and impair neuronal function. Oxidative stress increases the production of both Aβ precursor protein and Aβ.[49] *In vivo* imaging in transgenic mice shows increased reactive oxygen species (ROS) production associated with Aβ deposits.[50] In transgenic mice, lipid peroxidation rises prior to Aβ deposition, further suggesting that oxidative damage may lead to Aβ accumulation.[51] Aβ may subsequently enter mitochondria, where it can bind Aβ-binding alcohol dehydrogenase, leading to increased ROS production.[52] Aβ itself can lead to the generation of ROS, superoxide radicals, HNE, and membrane lipid peroxidation.[53–56] Thus, the mitochondria may be a key therapeutic target for AOXs.

A STUDY OF ANTIOXIDANTS AND BEHAVIORAL ENRICHMENT IN AGED CANINES

To test the hypothesis that dietary AOX supplementation might improve age-related cognitive function and brain pathology, aged beagles received a long-term treatment (>2.5 yr) with a diet rich in AOXs, either alone or in combination with BEH. Twenty-four aged beagles (mean age = 10.69 yr) were placed into one of four treatment groups: (1) control/control, (2) BEH/control diet, (3) control enrichment/AOX diet, or (4) BEH/AOX diet. The AOX diet contained α-tocopherol (800 IU), vitamin C (10 mg/kg), and fruit and vegetable extracts (1%). In addition, the diet contained two mitochondrial co-factors thought to improve mitochondrial function, DL-lipoic acid (2.7 mg/kg) and L-carnitine (6 mg/kg). BEH consisted of social (housing with a kennel-mate) and cognitive enrichment (provided with learning problems 5 days/week), and physical exercise (2–20-min walks outdoors).

Treatment with the AOX diet led to rapid improvements in learning and within 2 weeks of beginning the diet, aged animals began showing significant improvements in spatial attention.[31] Subsequent testing of animals with a more difficult complex learning task, oddity discrimination, also revealed positive AOX effects.[57] Improved learning ability was maintained over time with the AOX treatment, while untreated animals showed a progressive decline.[58] We observed similar improvements in cognition in animals provided with BEH. In addition, for each cognitive task that allowed a comparison between all four treatment conditions (control, AOX alone, BEH alone, or AOX+BEH), the combination treatment group showed consistently more benefits than either treatment alone.[58] For example, spatial memory showed a trend toward improvement in singly treated animals, but only after long-term treatment with the combined interventions did these differences reach statistical significance. To determine the neurobiological mechanisms underlying improved cognition in response to AOX and BEH treatments, we measured the extent of Aβ plaque accumulation.[59] Aβ deposition in the AOX-treated animals was reduced by 27–84% in the parietal, entorhinal, and occipital cortices, but not in the prefrontal cortex. Aβ-lowering effects were not observed in animals receiving the BEH intervention alone. In contrast, animals provided with BEH treatment showed less neuronal loss in the hippocampus compared to results in untreated dogs and those treated with the AOX diet alone.[60]

DISCUSSION

The study of lifestyle factor effects on brain aging in humans is challenging. Most observational studies support a link between engaging in intellectual, social, and/or physical activities and successful brain aging, but inconsistent results have been reported with AOX intake. Systematic clinical trials using AOX interventions have been somewhat disappointing in humans. There may be several reasons for this. First, single AOX interventions (e.g., vitamin E alone) may not produce consistent beneficial effects because AOX combinations (e.g., combining vitamin E with vitamin C) might be necessary to have an impact on brain function. Second, AOX intake through food may be more potent than using supplements, which are typically employed in clinical trials. Third, controlling or quantifying additional AOX intake through food that may be both substantial and/or variable across individual subjects may also reduce the observed extent of cognitive benefits. Many of these issues can be controlled in animal studies.

Our study in canines was unique in that two interventions were compared and combined. A diet rich in a broad spectrum of AOXs, either alone or when combined with BEH, improved cognition and reduced human-type brain pathology in aged beagles. The AOX dose levels used in the current study are consistent with human clinical trials and indicate that dietary supplementation with a broad spectrum of AOXs may shift Aβ precursor protein processing in favor of nonamyloidogenic pathways. BEH appeared to improve neuronal function independently of Aβ, but may have served to maintain neuronal number and/or plasticity. Thus, the combination of an AOX diet with BEH may improve brain health through separate yet complementary molecular pathways. The results of this study also suggest that modifying environmental factors and diet can have a substantial effect on the aging process even in older animals with existing cognitive impairments and brain pathology.

ACKNOWLEDGMENTS

Funding for this study was provided by NIH/NIA Grant AG12694 and by U.S. Department of the Army Contract DAMD17-98-1-8622. I appreciate the helpful discussions with Dr. Lori-Ann Christie during the preparation of this manuscript.

REFERENCES

1. KRAMER, A.F. *et al.* 2004. Environmental influences on cognitive and brain plasticity during aging. J. Gerontol. A Biol. Sci. Med. Sci. **59:** M940–M957.
2. ROVIO, S. *et al.* 2005. Leisure-time physical activity at midlife and the risk of dementia and Alzheimer's disease. Lancet Neurol. **4:** 705–711.
3. WILSON, R.S., C.F. MENDES DE LEON, L.L. BARNES, *et al.* 2002. Participation in cognitively stimulating activities and risk of incident Alzheimer disease. JAMA **287:** 742–748.
4. SCARMEAS, N. *et al.* 2001. Influence of leisure activity on the incidence of Alzheimer's disease. Neurology **57:** 2236–2242.
5. WANG, L. *et al.* 2006. Performance-based physical function and future dementia in older people. Arch. Intern. Med. **166:** 1115–1120.
6. LARSON, E.B. *et al.* 2006. Exercise is associated with reduced risk for incident dementia among persons 65 years of age and older. Ann. Intern. Med. **144:** 73–81.
7. FRIEDLAND, R.P., T. FRITSCH, K.A. SMYTH, *et al.* 2001. Patients with Alzheimer's disease have reduced activities in midlife compared with healthy control-group members. PNAS **98:** 3440–3445.
8. BALL, K., D.B. BERCH, K.F. HELMERS, *et al.* FOR THE ACTIVE STUDY GROUP. 2002. Effects of cognitive training interventions with older adults: a randomized controlled trial. JAMA **288:** 2271–2281.
9. YESAVAGE, J.A. 1985. Nonpharmacologic treatments for memory losses with normal aging. Am. J. Psychiatry **142:** 600–605.
10. VAN PRAAG, H., G. KEMPERMANN & F.H. GAGE. 2000. Neural consequences of environmental enrichment. Nat. Rev. Neurosci. **1:** 191–198.
11. VAN PRAAG, H. *et al.* 2005. Exercise enhances learning and hippocampal neurogenesis in aged mice. J. Neurosci. **25:** 8680–8685.
12. COTMAN, C.W. & N.C. BERCHTOLD. 2002. Exercise: a behavioral intervention to enhance brain health and plasticity. TINS **25:** 295–230.
13. COTMAN, C.W. & S. NEEPER. 1996. Activity-dependent plasticity and the aging brain. *In* Handbook of the Biology of Aging, 4th ed. E.L. Schneider *et al.*, Eds.: 283–299. Academic Press Inc. San Diego, CA.
14. LAZAROV, O. *et al.* 2005. Environmental enrichment reduces Abeta levels and amyloid deposition in transgenic mice. Cell **120:** 701–713.
15. ADLARD, P.A. *et al.* 2005. Voluntary exercise decreases amyloid load in a transgenic model of Alzheimer's disease. J. Neurosci. **25:** 4217–4221.
16. ENGELHART, M.J., M.I. GEERLINGS, A. RUITENBERG, *et al.* 2002. Dietary intake of AOXs and risk of Alzheimer disease. JAMA **287:** 3223–3229.
17. MORRIS, M.C., D.A. EVANS, J.L. BIENIAS, *et al.* 2002. Vitamin E and cognitive decline in older persons. Arch. Neurol. **59:** 1125–1132.
18. LUCHSINGER, J.A., M.X. TANG, S. SHEA & R. MAYEUX. 2003. Antioxidant vitamin intake and risk of Alzheimer disease. Arch. Neurol. **60:** 203–208.
19. MASAKI, K.H., K.G. LOSONCZY, G. IZMIRLIAN, *et al.* 2000. Association of vitamin E and C supplement use with cognitive function and dementia in elderly men. Neurology **54:** 1265–1272.
20. ZANDI, P.P., J.C. ANTHONY, A.S. KHACHATURIAN, *et al.*, CACHE COUNTY STUDY GROUP. 2004. Reduced risk of Alzheimer disease in users of AOX vitamin supplements: the Cache County Study. Arch. Neurol. **61:** 82–88.

21. RUTTEN, B.P., H.W. STEINBUSCH, H. KORR & C. SCHMITZ. 2002. Antioxidants and Alzheimer's disease: from bench to bedside (and back again). Curr. Opin. Clin. Nutr. Metab. Care **5:** 645–651.
22. SANO, M., C. ERNESTO, R.G. THOMAS, *et al.* 1997. A controlled trial of selegiline, alpha-tocopherol, or both as treatment for Alzheimer's disease. N. Engl. J. Med. **336:** 1216–1222.
23. PETERSEN, R.C. *et al.* 2005. Vitamin E and donepezil for the treatment of mild cognitive impairment. N. Engl. J. Med. **352:** 2379–2388.
24. KANG, J.H. *et al.* 2006. A randomized trial of vitamin E supplementation and cognitive function in women. Arch. Intern. Med. **166:** 2462–2468.
25. CHANDRA, R.K. 2001. Effect of vitamin and trace-element supplementation on cognitive function in elderly subjects. Nutrition **17:** 709–712.
26. BICKFORD, P.C. *et al.* 2000. Antioxidant-rich diets improve cerebellar physiology and motor learning in aged rats. Brain Res. **866:** 211–217.
27. SOCCI, D.J., B.M. CRANDALL & G.W. ARENDASH. 1995. Chronic AOX treatment improves the cognitive performance of aged rats. Brain Res. **693:** 88–94.
28. JOSEPH, J.A., B. SHUKKITT-HALE, N.A. DENISOVA, *et al.* 1998. Long-term dietary strawberry, spinach or vitamin E supplementation retards the onset of age-related neuronal signal-transduction and cognitive behavioral deficits. J. Neurosci. **18:** 8047–8055.
29. JOSEPH, J.A., B. SHUKITT-HALE, N.A. DENISOVA, *et al.* 1999. Reversals of age-related declines in neuronal signal transduction, cognitive and motor behavioral deficits with blueberry, spinach, or strawberry dietary supplementation. J. Neurosci. **19:** 8114–8121.
30. STUDZINSKI, C.M. *et al.* 2006. Visuospatial function in the beagle dog: an early marker of cognitive decline in a model of human aging and dementia. Neurobiol. Learn. Mem. **86:** 197–204.
31. MILGRAM, N.W. *et al.* 2002. Landmark discrimination learning in the dog: effects of age, an AOX fortified diet, and cognitive strategy. Neurosci. Biobehav. Rev. **26:** 679–695.
32. MILGRAM, N.W. *et al.* 1994. Cognitive functions and aging in the dog: acquisition of nonspatial visual tasks. Behav Neurosci. **108:** 57–68.
33. TAPP, P.D. *et al.* 2003. Size and reversal learning in the beagle dog as a measure of executive function and inhibitory control in aging. Learn. Mem. **10:** 64–73.
34. CUMMINGS, B.J., J.H. SU, C.W. COTMAN, *et al.* 1993. Beta-amyloid accumulation in aged canine brain: a model of plaque formation in Alzheimer's disease. Neurobiol. Aging **14:** 547–560.
35. HEAD, E. *et al.* 2000. Region-specific age at onset of beta-amyloid in dogs. Neurobiol. Aging **21:** 89–96.
36. HOU, Y., R.G. WHITE, M. BOBIK, *et al.* 1997. Distribution of beta-amyloid in the canine brain. NeuroReport **8:** 1009–1012.
37. WISNIEWSKI, H.M. *et al.* 1990. Aged dogs: an animal model to study beta-protein amyloidogenesis. *In* Alzheimer's disease. Epidemiology, Neuropathology, Neurochemistry and Clinics. K. MAURER *et al.*, Eds.: 151–167. Springer-Verlag. New York.
38. JOHNSTONE, E.M. *et al.* 1991. Conservation of the sequence of the Alzheimer's disease amyloid peptide in dog, polar bear and five other mammals by cross-species polymerase chain reaction analysis. Brain Res. Mol. Brain Res. **10:** 299–305.

39. SELKOE, D.J., D.S. BELL, M.B. PODLISNY, *et al.* 1987. Conservation of brain amyloid proteins in aged mammals and humans with Alzheimer's disease. Science **235:** 873–877.
40. COLLE, M.-A., J.-J. HAUW, F. CRESPEAU, *et al.* 2000. Vascular and parenchymal Ab deposition in the aging dog: correlation with behavior. Neurobiol. Aging **21:** 695–704.
41. CUMMINGS, B.J. *et al.* 1996. Beta-amyloid accumulation correlates with cognitive dysfunction in the aged canine. Neurobiol. Learn Mem. **66:** 11–23.
42. HEAD, E. *et al.* 1998. Visual-discrimination learning ability and beta-amyloid accumulation in the dog. Neurobiol. Aging **19:** 415–425.
43. ROFINA, J.E. *et al.* 2006. Cognitive disturbances in old dogs suffering from the canine counterpart of Alzheimer's disease. Brain Res. **1069:** 216–226.
44. HEAD, E. *et al.* 2002. Oxidative damage increases with age in a canine model of human brain aging. J. Neurochem. **82:** 375–381.
45. KIATIPATTANASAKUL, W. *et al.* 1997. Immunohistochemical detection of anti-oxidative stress enzymes in the dog brain. Neuropathology **17:** 307–312.
46. PAPAIOANNOU, N. *et al.* 2001. Immunohistochemical investigation of the brain of aged dogs. I. Detection of neurofibrillary tangles and of 4-hydroxynonenal protein, an oxidative damage product, in senile plaques. Amyloid **8:** 11–21.
47. ROFINA, J.E. *et al.* 2004. Histochemical accumulation of oxidative damage products is associated with Alzheimer-like pathology in the canine. Amyloid **11:** 90–100.
48. SKOUMALOVA, A. *et al.* 2003. The role of free radicals in canine counterpart of senile dementia of the Alzheimer type. Exp. Gerontol. **38:** 711–719.
49. FREDERIKSE, P.H. *et al.* 1996. Oxidative stress increases production of beta-amyloid precursor protein and beta-amyloid (Abeta) in mammalian lenses, and Abeta has toxic effects on lens epithelial cells. J. Biol. Chem. **271:** 10169–10174.
50. MCLELLAN, M.E., S.T. KAJDASZ, B.T. HYMAN & B.J. BACSKAI. 2003. *In vivo* imaging of reactive oxygen species specifically associated with thioflavine S-positive amyloid plaques by multiphoton microscopy. J. Neurosci. **23:** 2212–2217.
51. PRATICO, D. *et al.* 2001. Increased lipid peroxidation precedes amyloid plaque formation in an animal model of Alzheimer amyloidosis. J. Neurosci. **21:** 4183–4187.
52. LUSTBADER, J.W. *et al.* 2004. ABAD directly links Abeta to mitochondrial toxicity in Alzheimer's disease. Science **304:** 448–452.
53. MARK, R.J., Z. PANG, J.W. GEDDES, *et al.* 1997. Amyloid β-peptide impairs glucose transport in hippocampal and cortical neurons: involvement of membrane lipid peroxidation. J. Neurosci. **17:** 1046–1054.
54. BEHL, C., J. DAVIS, G.M. COLE & D. SCHUBERT. 1992. Vitamin E protects nerve cells from amyloid-beta protein toxicity. Biochem. Biophys. Res. Commun. **186:** 944–950.
55. PEREIRA, C., M.S. SANTOS & C. OLIVEIRA. 1999. Involvement of oxidative stress on the impairment of energy metabolism induced by Ab peptides on PC12 cells: protection by AOXs. Neurobiol. Dis. **6:** 209–219.
56. MCDONALD, D.R., K.R. BRUNDEN & G.E. LANDRETH. 1997. Amyloid fibrils activate tyrosine kinase-dependent signaling and superoxide production in microglia. J. Neurosci. **17:** 2284–2294.
57. COTMAN, C.W. *et al.* 2002. Brain aging in the canine: a diet enriched in AOXs reduces cognitive dysfunction. Neurobiol. Aging **23:** 809–818.

58. MILGRAM, N.W. *et al.* 2005. Learning ability in aged beagle dogs is preserved by BEH and dietary fortification: a two-year longitudinal study. Neurobiol. Aging **26:** 77–90.
59. POP, V. *et al.* 2003. Reduced Ab deposition with long-term AOX diet treatment in aged canines. Program No. 525.4.2003 *Abstract Viewer/Itinerary Planner*. Washington, DC: Society for Neuroscience, 2003. Online.
60. SIWAK-TAPP, C.T. *et al.* 2006. Region specific neuron loss in the aged canine hippocampus is reduced by enrichment. Neurobiol. Aging. Nov. 6 [Epub ahead of print].

The Effect of Resveratrol on a Cell Model of Human Aging

MAURIZIO STEFANI,[a] M. ANDREA MARKUS,[a] RUBY C. Y. LIN,[b] MARK PINESE,[b] IAN W. DAWES,[b] AND BRIAN J. MORRIS[a]

[a]*Basic & Clinical Genomics Laboratory, School of Medical Sciences and Bosch Institute, The University of Sydney, Sydney, New South Wales, Australia*

[b]*Ramaciotti Centre for Gene Function Analysis, University of New South Wales, Sydney, New South Wales, Australia*

ABSTRACT: The natural polyphenol resveratrol stimulates sirtuins and extends lifespan. Here resveratrol inhibited expression of replicative senescence marker *INK4a* in human dermal fibroblasts, and 47 of 19,000 genes from microarray experiments were differentially expressed. These included genes for growth, cell division, cell signaling, apoptosis, and transcription. Genes involved in Ras and ubiquitin pathways, *Ras-GRF1*, *RAC3*, and *UBE2D3*, were downregulated. The changes suggest resveratrol might alter sirtuin-regulated downstream pathways, rather than sirtuin activity. Serum deprivation and high confluency caused nuclear translocation of the SIRT1-regulated transcription factor FOXO3a. Our data indicate resveratrol's actions might cause FOXO recruitment to the nucleus.

KEYWORDS: resveratrol; nicotinamide; senescence; cell culture; microarray; fibroblasts

INTRODUCTION

Calorie restriction (CR) increases the mean and maximum lifespan of all species tested.[1] This involves an increase in the activity and/or abundance of sirtuins, a conserved family of NAD^+-dependent deacetylases.[1] Resveratrol (trans-3,5,4′-trihydroxystilbene), a polyphenol found in grape skins and red wine, has also been shown to increase the median and maximum lifespan of a variety of organisms,[2,3] and may do so by activation of sirtuin-mediated deacetylation.[2] This mechanism, has, however, been disputed.[4,5] Resveratrol affects other pathways, including those involving activation of phosphatidylinositol 3-kinase (PI3K),[6] inhibition of protein kinase B,[7] and inhibition of the

Address for correspondence: Brian J. Morris, D.Sc., Basic & Clinical Genomics Laboratory, School of Medical Sciences and Bosch Institute, Bldg. F13, The University of Sydney, Sydney, New South Wales 2006, Australia. Voice: +61-2-93513688; fax: +61-2-93512227.
brianm@medsci.usyd.edu.au

Ann. N.Y. Acad. Sci. 1114: 407–418 (2007).
doi: 10.1196/annals.1396.001

effects on transcription of AP-1 and NF-κB.[8] Importantly, studies of the effect of resveratrol in *human* models of aging are lacking.

The aim of the present study was first to test the effect of resveratrol on the number of cumulative population doublings (CPDs) required to reach replicative senescence in mortal human foreskin fibroblasts, and then to similarly test nicotinamide, an inhibitor of the human sirtuin homolog SIRT1,[9] to see whether SIRT1 inhibition might have an effect on senescence opposite to that of resveratrol. The expression of *INK4a* has been shown to increase in many organs in aging mice, and this increase is attenuated by CR.[10] In addition, *INK4a* deficiency prevents an aging-related phenotype in mouse hematopoietic stem cells[11] and other cells, pointing to its involvement in the aging process. We therefore measured *INK4a* mRNA in order to monitor the effects of resveratrol and nicotinamide on replicative senescence of these cells. To gain a clearer picture of the mechanism of resveratrol's action on mortal human fibroblasts, we used oligonucleotide microarrays to determine global gene expression changes induced by acute treatment with this natural polyphenol. We also determined the localization of the mammalian forkhead box O transcription factor homolog, FOXO3a, in our cell culture model. Overexpression or increased transcriptional activity of FOXO homologs increases lifespan of *Drosophila*[12] and IGF-KO mice.[13] In nematodes, at least, increased lifespan conferred by Sir-2.1 overexpression is dependent on the presence of functional Daf-16 (a nematode FOXO homolog).[14] In mammalian cells, deacetylation of FOXO proteins by SIRT1 stimulates transcription of genes involved in resistance to cellular stress and cell survival.[15–17] Thus, by mapping FOXO3a localization, we hoped to gain further insight into the interaction between resveratrol, SIRT1, and FOXO transcription factors in our cell model.

MATERIALS AND METHODS

Cell Culture

Mortal human diploid foreskin fibroblasts (FB0603 cells) at passage 1 were cultured at 37°C under 5% CO_2 in Dulbecco's Modified Eagle Medium (DMEM) (Invitrogen, Carlsbad, CA) containing 5% fetal bovine serum (FBS), 80 units/mL penicillin, and 80 μg/mL streptomycin (Sigma, St. Louis, MO).

Real-Time RT-PCR

RNA used for real-time quantitative (q) reverse transcription (RT)-PCR analysis as well as for the microarray analysis were extracted using the SV Total RNA Isolation System (Promega, Madison, WI). An SYBR green-based qRT-PCR kit (Qiagen, Hilden, Germany) was used according to the manufacturer's

TABLE 1. Primer sequences used for qRT-PCR of mRNAs of genes shown

Gene	Forward primer (5′ → 3′)	Reverse primer (5′ → 3′)
INK4a	caacgcaccgaatagttacg	atgtctgagggaccttccg
A2M	gcatggccttgtacagttctc	ctgagtatggccacagggtagt
GAPDH	catgagaagtatgacaacagcct	agtccttccacgataccaaagt
ACTB	attgccgacaggatgcagaag	tgctgatccacatctgctgga
CDKN2A	caaattggcagggagcaat	atccttgccatccttgagc
UBE2D3	aggtccagttggggatgata	gggctgtcattaggtcccata

DESIGN CRITERIA: The amplicon must cross an intron–exon boundary; the T_m of the primers must be approximately 64°C; and the primers must have weak propensity for secondary structure and primer–dimer formation.

ABBREVIATIONS: *INK4a*, cyclin-dependent kinase inhibitor 2A gene; *A2M*, α_2-macroglobulin gene; *GAPDH*, glyceraldehyde-3-phosphate dehydrogenase gene; ACTB, β-actin gene; *CDKN2A*, cyclin-dependent kinase inhibitor 2A gene; *UBE2D3*, ubiquitin-conjugating enzyme E2D 3 gene.

instructions, except for minor changes in the 25-μL final volume, with 2.5 μL of gene-specific forward primer (0.5 μM) (TABLE 1) being added after completion of the RT step. qRT-PCR was performed using a RotorGene 3000 (Corbett Research, Sydney, Australia) instrument. Bestkeeper[18] and REST[19] freeware were used to calculate amplification efficiency. A $P < 0.05$ was accepted as statistically significant. Results for nicotinamide treatment were normalized to values for mRNA encoding β-actin and GAPDH. Interestingly, we found that β-actin and GAPDH expression were regulated by resveratrol (data not shown). Thus the qRT-PCR results for resveratrol treatment were normalized to total RNA input.

Intracellular Localization Studies

FB0603 cells were seeded onto 8-well Labtek® Permanox chamber slides (Nunc, Rochester, NY) and allowed to grow to various levels of confluency. The cells were incubated with medium containing different concentrations of FBS (1–5%). After 4-h incubation, the cells were fixed with ice-cold absolute methanol for 2 min and blocked with 10% (w/v) BSA for 1 h. Cells were then incubated with goat polyclonal anti-FOXO3a antibody (1:500 in PBS) (Abcam, Cambridge, UK) for 1 h, followed by incubation with secondary donkey anti-goat IgG, Alexa 495–conjugated antibody (1:500 in PBS) (Invitrogen) for 1 h. Nuclei were stained by incubating cells with 4′-6-diamidino-2-phenylindole (DAPI) (0.1 μg/mL in PBS) for 1 min. Cells were visualized with an Axioplan microscope and photographed with an AxioCam camera (both Zeiss, Oberkochen, Germany).

Microarray Analysis

Normal postmitotic, quiescent foreskin fibroblasts of low (<8) passage were treated with 1 μM resveratrol in 0.01% dimethylsulfoxide (DMSO; Sigma) or

0.01% DMSO control for 24 h and then harvested for RNA extraction. RNA integrity was analyzed using an RNA 6000 LabChip on a Bioanalyzer 2100 (Agilent Technologies, Santa Clara, CA). Dye assignments were swapped for each biological replicate to eliminate gene-specific dye bias ($n = 2$). Compugen 19K human oligonucleotide arrays from the Adelaide Microarray Facility (University of Adelaide, Adelaide, South Australia) were used. Each slide consists of 20,000 spots (18,889 oligonucleotides and 1111 control spots), with no duplicate oligonucleotide spots except for GAPDH. RNA (20 μg) was reverse-transcribed to cDNA according to the manufacturer's instructions (Invitrogen) with minor amendments of addition of 1.35 μL of 10 mM amino allyl-dUTP (Sigma) for subsequent dye incorporation. The protocol involved incubation at 42°C for 2.5 h after 2 μL of Superscript III (Invitrogen) was added. The reaction was inactivated by addition of 5 μL of 0.5 M EDTA. RNA–DNA heteroduplexes were hydrolyzed at 65°C with 10 μL of 0.25 M NaOH, and then neutralized with 15 μL of 0.2 M acetic acid. This was then purified using a column-based PCR clean-up kit (Qiagen), except that nuclease-free water was used instead of elution buffer. The eluents were then dried in a Speed-Vac (Savant, Waltham, MA) set at 45°C for approximately 20 min or until they were reduced to 1–2 μL. Incorporation of 2 μL of Cy3™ or 2 μL Cy5™ dyes (GE Healthcare Life Sciences, Chalfont St. Giles, Bucks, UK) was assigned to respective eluents with 9 μL of 0.1 M $NaHCO_3$ added, and then incubated at room temperature in the dark for 1 h. The dyes were resuspended in 20 μL of DMSO. Cy-dye-labeled cDNAs were purified to remove unincorporated dye (Qiagen), and then concentrated again to 1–2 μL using a Speed-Vac at 45°C. Prior to this, the slides were treated as follows: post-printing treatment of the Compugen 19K arrays was performed by baking at 80°C for 80 min. The pre-hybridization treatment included 0.1% sodium dodecyl sulfate (SDS) at 95°C for 1 min, 5% ethanol for 1 min at room temperature, double-distilled water for 1 min, and drying by centrifugation at 100*g*. The blocking step was performed immediately before hybridization. The purified Cy-dye-labeled cDNAs were then added to 50 μL DIG Easy Hyb buffer (Roche, Basel, Switzerland). In addition, 4 μL yeast tRNA (25 mg/mL) and 10 μL herring sperm DNA (10 mg/mL) were added to reduce nonspecific hybridization. For each biological replicate, DMSO vehicle and resveratrol-treated labeled cDNA solutions were added accordingly. This cocktail was then incubated at 65°C for 5 min and centrifuged at 13,000*g* for 5 min. The solution was then applied to a microarray slide covered with a 25 mm × 60 mm Lifterslip (Erie Scientific, Portsmouth, NH). Bubbles were reduced by applying this from the top right corner by capillary action. Then the slides were sealed in a humidified chamber and incubated in a 37°C oven overnight (13–18 h). Post-hybridization wash was done 3 times in 1× standard saline citrate (SSC), 0.1% SDS for 15 min at 50°C, 3 times in 1× SSC for 1 min at 50°C, and once in 0.2× SSC at 50°C for 1 min, all with continuous agitation. After drying by centrifugation at 800*g* at room temperature, the slides were scanned using a GenePix® 4000B Microarray Scanner

(Axon Instruments, Sunnyvale, CA). Manual flagging was used to eliminate misprinted spots from subsequent analysis. Intensities were calculated from the red and green channels and background was corrected using the median measure of pixel intensities. Ratios of corrected intensities for each gene were calculated and the output was saved in a tab-delimited file. The ratio data were analyzed using the LIMMA package in "R".[20] The LIMMA GUI interface was used to interact with the LIMMA package. Print-tip LOESS normalization was used for within-array normalization, and scale-type normalization was used for between-array normalization. A criterion of twofold or greater change in expression was applied to identify differentially expressed genes. These were annotated with the web-based software tool Database for Annotation, Visualization, and Integrated Discovery (DAVID).[21]

RESULTS

Effect of Long-Term Treatment of Cells with Resveratrol and Nicotinamide

In vehicle (DSMO)-treated cells, *INK4a* mRNA changed little after 10 CPDs (46 days) (FIG. 1A). In contrast, the low dose of resveratrol caused a significant ($P < 0.001$) decrease in *INK4a* mRNA by 24 h, and this level was maintained through to 10 CPDs. The higher dose of resveratrol also decreased *INK4a* mRNA. At 10 CPDs another marker of replicative senescence, α_2-macroglobulin mRNA, was reduced by 0.2 μM and 1 μM resveratrol to 44% and 46%, respectively.

In contrast, with nicotinamide treatment, a significant ($P < 0.001$) dose-dependent increase in *INK4a* mRNA by 10 CPD (57 days) was observed (FIG. 1B).

Effect of Nicotinamide and Resveratrol on Cell Division and Replicative Senescence

Cells were passaged at a 1:4 ratio, so it was assumed that for each passage the cells had undergone 2 CPDs. The cells were considered replicatively senescent when they had not divided after 4 weeks. Resveratrol treatment did not alter the rate of cell division compared to vehicle-treated controls, and both groups reached replicative senescence after 23 CPDs (FIG. 2A). Cells treated with 0.5 and 5 mM nicotinamide, however, reached replicative senescence after 25 CPDs (FIG. 2B).

Global Gene Expression Changes by Acute Resveratrol Treatment

The expression of 47 genes (21 with known function) changed by $\geq$ twofold in response to 24-h resveratrol. Those genes whose function is known are listed

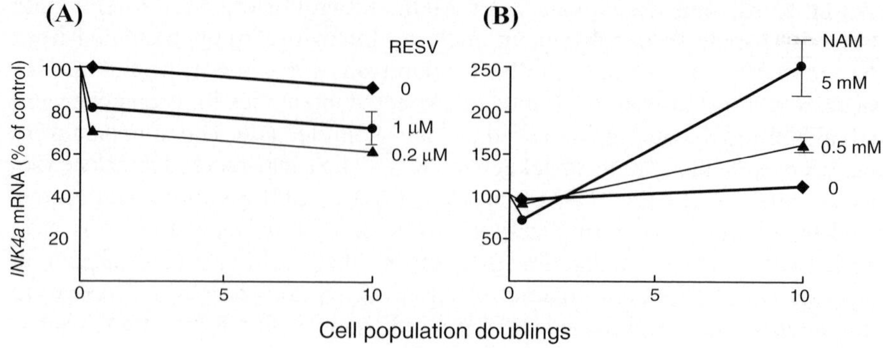

FIGURE 1. Response in *INK4a* mRNA of human fibroblast F0603 cells treated with (**A**) resveratrol or vehicle-treated control and (**B**) nicotinamide. RESV, resveratrol; NAM, nicotinamide.

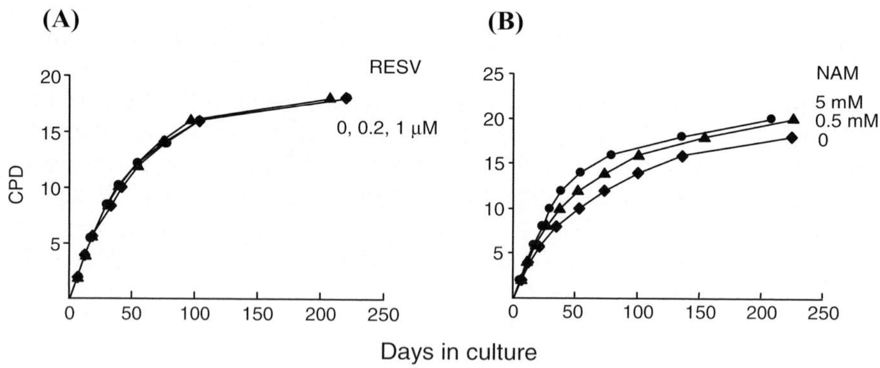

FIGURE 2. Cell division curves for F0603 fibroblasts treated with (**A**) resveratrol and (**B**) nicotinamide, and controls. RESV, resveratrol; NAM, nicotinamide.

in TABLE 2. They included genes involved in the Ras pathway, apoptosis, cell growth and division, G-protein signaling, cell–cell adhesion, immune system regulation, neuroendocrine signaling, muscle development, transcription, and proteolysis.

Validation of Microarray Results by qRT-PCR

One gene (*UBE2D3*) was selected for RT-PCR to confirm array reliability. Its expression was normalized to a reference gene for cyclophilin D. Consistent with the microarray result in TABLE 2, qRT-PCR, performed in triplicate, showed that *UBE2D3* was significantly downregulated ($P < 0.001$). This provides preliminary support for the veracity of the data in TABLE 2. Confirmation of each gene change by mRNA quantification will require considerable further work.

TABLE 2. Genes whose expression changed by $\geq$ twofold in response to acute 1-μM resveratrol treatment vs. control[a]

Gene name	Accession no.	Fold change in mRNA vs control	Annotated function
CCR6	NM_004367	+3.1	C-C chemokine receptor activity
RAC3	NM_005052	−2.6	Small GTPase-mediated signal transduction
KCNC4	NM_004978	+2.5	Voltage-gated potassium channel complex
GPR133	AL162032	−2.4	Neuropeptide signaling pathway
COL2A1	NM_001844	−2.3	Fibrillar collagen
Spliced *IL-6R* β	U58146	−2.3	Cell surface receptor–linked signal transduction; apoptosis
DNAH3	AK026793	+2.3	ATPase activity; microtubule motor activity
RASGRF1	AL109674	−2.2	Ras guanyl–nucleotide exchange factor activity
FGF23	AF263537	−2.2	Growth factor activity
PRB3	NM_006249	−2.2	G-protein–coupled receptor protein signaling pathway
SEMA6A1	AF279656	−2.1	Cell surface receptor–linked signal transduction; apotosis
EFNA1	NM_004428	+2.1	Cell–cell signaling
Ankyrin repeat domain 2 (stretch-responsive muscle)	AJ249975	+2.1	Muscle development
CD84	NM_003874	+2.1	Defense response; homophilic cell adhesion
ASF1B	NM_018154	−2.1	Histone chaperone
CDH23	AY010111	+2.0	Calcium-dependent cell–cell adhesion
P4HB	NM_000918	+2.0	Electron transport; isomerase activity
UBE2D3	NM_003340	−2.0	Ubiquitin–protein conjugating activity
MLL4	NM_014727	−2.0	Transcription factor activity; proto-oncogene
GTF2I	NM_001518	+2.0	General RNA polymerase II transcription factor activity
KIAA0769	NM_014824	−2.0	Oxidoreductase activity

[a]Annotated with DAVID web-based software tool.

FOXO3a Localization

With 5% serum in the culture medium FOXO3a was entirely cytoplasmic, whereas with 1% serum its localization was nuclear (FIG. 3). Similarly, FOXO3a was predominantly localized in the cytoplasm at low confluency,

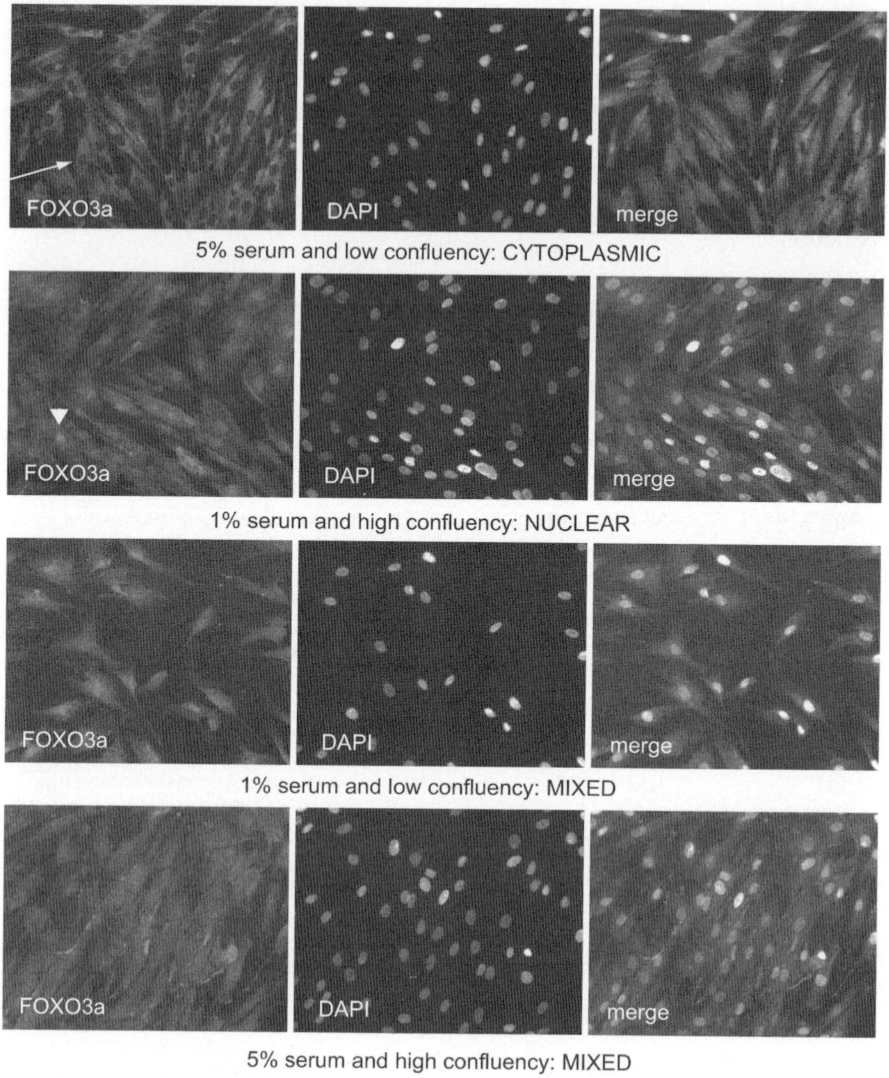

FIGURE 3. Immunolocalization of FOXO3a in F0603 fibroblasts under the conditions of cell culture indicated for each set of panels. *Arrow* indicates localization in cytoplasm. *Arrowhead* points to localization in nucleus. "DAPI" refers to nuclear staining and "merge" shows results of merging FOXO3a and DAPI images. (In color in Annals online)

whereas at high confluency FOXO3a was nuclear. There was, moreover, an additive effect of confluency and serum concentration on FOXO3a localization.

DISCUSSION

We found that resveratrol downregulates expression of *Ras-GRF1*, which encodes an AP-1-inducible Ras guanine nucleotide exchange factor that upregulates Ras activity.[22] This adds to resveratrol's known ability to reduce AP-1 transcriptional activity.[8] We also found that *RAC3* mRNA was similarly downregulated twofold. RAC3 stimulates c-Jun to activate AP-1.[23] Therefore, the reduction in Ras-GRF1 expression may be secondary to the downregulation by resveratrol of *RAC3* expression. Thus we find resveratrol likely downregulates the Ras pathway. Active Ras binds to and activates PI3K, a downstream effector of the insulin/IGF-1 growth factor pathway.[24] PI3K then activates protein kinase B, which phosphorylates FOXO3a and causes it to translocate to the cytoplasm.[24] Inhibition of Ras signaling by resveratrol would counter FOXO3a phosphorylation, so increasing its activity.

Like others[25] we found that FOXO3a localization was affected by serum concentration. Serum activates the insulin/IGF-1 growth factor pathway, leading to phosphorylation of FOXO3a and its sequestering in the cytoplasm, thus terminating its transcriptional influence. We now demonstrate that confluency also affected FOXO3a localization. At higher confluencies, fibroblasts undergo density-dependent reversible G1 cell cycle arrest,[26] and changes in FOXO3a localization could contribute to this. Indeed, FOXO3a can induce G1 cell cycle arrest.[15] Of interest, $p27^{kip1}$ and GADD45 have been implicated in density-dependent cell cycle arrest[27] and FOXO3a upregulates $p27^{kip1}$ expression,[15] leading to increased GADD45.[27] Therefore FOXO3a could directly contribute to density-dependent cell cycle arrest.

Resveratrol also downregulated UBE2DE3, an E2 enzyme involved in protein ubiquitination. E2 enzyme activity is reduced in the liver of CR mice, possibly because of reduction in oxidative stress in CR animals.[28] This might involve a reduction in oxidative damage to proteins, thereby reducing the rate of protein turnover, indicated by reduced E2 activity. E2 activity increases, moreover, with age in mouse liver, and this increase is prevented by CR.[28] The reduction in UBE2DE3 seen may indicate that resveratrol is mimicking CR by the reduction of cellular stress. This could be mediated by increased FOXO3a transcriptional activity. IκBα is part of a three-protein complex that is bound to NF-κB and that blocks its nuclear import and subsequent transcriptional activity.[29] UBE2D3 mediates conjugation of ubiquitin to IκBα, and the degradation of IκBα allows NF-κB to enter the nucleus and activate transcription of target genes.[29] Thus reduced UB2D3 after acute resveratrol would decrease NF-κB activity. Resveratrol can decrease NF-κB activity,[8] and this then leads to an increase[30] or decrease[31] in stress-induced apoptosis, depending on cell type. The reduction of ubiquitin conjugation by resveratrol as a result of its reduction of UBE2D3 may be the mechanism behind reduced NF-κB signaling. NF-κB activation, along with stress-induced apoptosis, is increased by oxidative stress.[31] It is thus possible that resveratrol decreases oxidative stress, and that this manifests itself as a decrease in UBE2D3, which would lead to a reduction in NF-κB activation, and consequently, a reduction in stress-induced apoptosis. Last, SIRT1 has been shown to inhibit NF-κB by a separate mechanism.[30] Thus the effects of SIRT1 and resveratrol overlap, and may help reconcile the confusion in the literature regarding the relationship of resveratrol with SIRT1 signaling.

Our study has also shown that chronic, but not acute, nicotinamide treatment increased *INK4a* expression. In contrast, acute low-dose resveratrol treatment decreased *INK4a* expression, and it is probable that chronic low-dose resveratrol has a similar effect. Nicotinamide inhibits SIRT1 activity with an $IC_{50} < 50\ \mu M$.[9] Therefore, assuming adequate uptake of nicotinamide, SIRT1 should have been completely inhibited in our study. SIRT1 increases stress resistance and repair in a variety of cell types via modulation of FOXO activity and effects on other pathways.[1] Thus, SIRT1 inhibition should increase susceptibility to damage, and thus increase *INK4a* expression. Resveratrol, on the other hand, decreased *INK4a* acutely, perhaps indicating a decrease in cellular damage, and this occurred on a much shorter timescale than that seen for the effect of nicotinamide. FOXO3a was predominantly cytoplasmic in the presence of growth factors. Thus in the control cells, FOXO3a would be unable to activate target genes. SIRT1 is a nuclear protein, and FOXO3a must enter the nucleus in order to interact with SIRT1 and induce stress resistance and repair genes.[15] We propose that inhibition of SIRT1 by nicotinamide has only a minor effect on repair and stress resistance, since only a small fraction of the total cellular FOXO3a pool was in the nucleus under the cell culture conditions used. Thus, it would take a long time for damage to accumulate to a significantly greater amount in the nicotinamide-treated than in the control cells. Resveratrol downregulates the Ras pathway. We therefore propose that resveratrol does not allosterically activate SIRT1, but rather stimulates interaction of SIRT1 with FOXO3a by inducing FOXO3a nuclear localization through the downregulation of Ras activity. Therefore, an increase in FOXO3a transcriptional activity by resveratrol might quickly decrease cellular damage, thereby leading to a decrease in *INK4a* expression, and consequently attenuate replicative senescence. Given that *INK4a* is induced by Ras overexpression as a tumor-suppressor mechanism,[32] the decrease in *INK4a* expression would be due directly to a reduction in Ras pathway activity.

Chronic resveratrol treatment did not increase the time required to reach replicative senescence, despite likely beneficial effects of resveratrol in reducing cellular damage, as judged by the reduction in *INK4a* it induced. On the other hand, reduced cellular $p16^{ink4a}$ should have a favorable effect on physiological function at the organismal level, unrelated to its effect on lifespan of fibroblasts. Effects on other cell types could, moreover, be quite different.

Chronic nicotinamide increased the time our fibroblasts required to reach replicative senescence. Although nicotinamide inhibits sirtuin activity, it also influences mitochondrial function.[33] Complex I of the respiratory chain is a key site of ROS generation during oxidative phosphorylation. Curiously, mitochondria that perform their major physiological function of ATP synthesis impose less oxidative danger to cells than do resting mitochondria, despite the multiple sites in mitochondria with the capacity to generate ROS.[33] Despite nicotinamide being the precursor of NADH, and ROS formation being increased by elevation in NADH:NAD$^+$ ratio, nicotinamide has been reported to protect against mitochondial damage and oxidative stress, as least in neuronal cells.[33] Further work is, however, required to explain the effect we observed of nicotinamide on time to replicative senescence.

In conclusion, resveratrol may decrease replicative senescence marker expression by inducing the localization of FOXO3a (and possibly other FOXO proteins) to the nucleus, thus promoting FOXO-dependent gene expression. Consequently, by regulating the subcellular localization of FOXO transcription factors, possibly through Ras pathway modulation, resveratrol can mimic the effects of CR.

REFERENCES

1. Sinclair, D.A. 2005. Toward a unified theory of caloric restriction and longevity regulation. Mech. Ageing Devel. **126:** 987–1002.
2. Howitz, K.T. *et al.* 2003. Small molecule activators of sirtuins extend *Saccharomyces cerevisiae* lifespan. Nature **425:** 191–196.
3. Wood, J.G. *et al.* 2004. Sirtuin activators mimic caloric restriction and delay ageing in metazoans. Nature **430:** 686–689.
4. Kaeberlein, M. *et al.* 2005. Substrate-specific activation of sirtuins by resveratrol. J. Biol. Chem. **280:** 17038–17045.
5. Borra, M.T., B.C. Smith & J.M. Denu. 2005. Mechanism of human SIRT1 activation by resveratrol. J. Biol. Chem. **280:** 17187–17195.
6. Pozo-Guisado, E., M.J. Lorenzo-Benayas & P.M. Fernandez-Salguero. 2004. Resveratrol modulates the phosphoinositide 3-kinase pathway through an estrogen receptor alpha-dependent mechanism: relevance in cell proliferation. Int. J. Cancer **109:** 167–173.
7. Poolman, T.M. *et al.* 2005. Inhibition of the respiratory burst by resveratrol in human monocytes: correlation with inhibition of PI3K signaling. Free Radical Biol. Med. **39:** 118–132.
8. Manna, S.K., A. Mukhopadhyay & B.B. Aggarwal. 2000. Resveratrol suppresses TNF-induced activation of nuclear transcription factors NF-κB, activator protein-1, and apoptosis: potential role of reactive oxygen intermediates and lipid peroxidation. J. Immunol. Meth. **164:** 6509–6519.
9. Bitterman, K.J. *et al.* 2002. Inhibition of silencing and accelerated aging by nicotinamide, a putative negative regulator of yeast sir2 and human SIRT1. J. Biol. Chem. **277:** 45099–45107.
10. Krishnamurthy, J. *et al.* 2004. *Ink4a/Arf* expression is a biomarker of aging. J. Clin. Invest. **114:** 1299–1307.
11. Janzen, V. *et al.* 2006. Stem-cell ageing modified by the cyclin-dependent kinase inhibitor p16^{INK4a}. Nature **443:** 421–426.

12. HWANGBO, D.S. *et al.* 2004. *Drosophila* dFOXO controls lifespan and regulates insulin signalling in brain and fat body. Nature **429:** 562–566.
13. NEMOTO, S. & T. FINKEL. 2002. Redox regulation of forkhead proteins through a p66shc-dependent signaling pathway. Science **295:** 2450–2452.
14. TISSENBAUM, H.A. & L. GUARENTE. 2001. Increased dosage of a sir-2 gene extends lifespan in *Caenorhabditis elegans*. Nature **410:** 227–230.
15. BRUNET, A. *et al.* 2004. Stress-dependent regulation of FOXO transcription factors by the SIRT1 deacetylase. Science **303:** 2011–2015.
16. MOTTA, M.C. *et al.* 2004. Mammalian SIRT1 represses forkhead transcription factors. Cell **116:** 551–563.
17. VAN DER HORST, A. *et al.* 2004. FOXO4 is acetylated upon peroxide stress and deacetylated by the longevity protein $hSir2^{SIRT1}$. J. Biol. Chem. **279:** 28873–28879.
18. PFAFFL, M.W. *et al.* 2004. Determination of stable housekeeping genes, differentially regulated target genes and sample integrity: BestKeeper–Excel-based tool using pair-wise correlations. Biotech. Lett. **26:** 509–515.
19. PFAFFL, M.W., G.W. HORGAN & L. DEMPFLE. 2002. Relative expression software tool (REST) for group-wise comparison and statistical analysis of relative expression results in real-time PCR. Nucleic Acids Res. **30:** e36.
20. IHAKA, R. & R. GENTLEMAN. 1996. R: a language for data analysis and graphics. J. Comput. Graphic Stats. **5:** 299–314.
21. HOSACK, D.A. *et al.* 2003. DAVID: Database for Annotation, Visualization, and Integrated Discovery. Genome Biol. **4:** R60.
22. LEANER, V.D. *et al.* 2005. p75-Ras-GRF1 is a c-Jun/AP-1 target protein: its up regulation results in increased Ras activity and is necessary for c-Jun-induced nonadherent growth of Rat1a cells. Mol. Cell. Biol. **25:** 3324–3337.
23. KARIN, M. & E. GALLAGHER. 2005. From JNK to pay dirt: jun kinases, their biochemistry, physiology and clinical importance. IUBMB Life **57:** 283–295.
24. BURGERING, B.M.T. & G.J.P.L. KOPS. 2002. Cell cycle and death control: long live Forkheads. Trends Biochem. Sci. **27:** 352–360.
25. BRUNET, A. *et al.* 1999. Akt promotes cell survival by phosphorylating and inhibiting a Forkhead transcription factor. Cell **96:** 857–868.
26. MEERSON, A., M. MILYAVSKY & V. ROTTER. 2004. p53 mediates density-dependent growth arrest. FEBS Lett. **559:** 152–158.
27. ZHANG, X. *et al.* 2003. Role of Gadd45a in the density-dependent G1 arrest induced by p27Kip1. Oncogene **22:** 4166–4174.
28. SCROFANO, M.M. *et al.* 1998. Calorie restriction, stress and the ubiquitin-dependent pathway in mouse livers. Mech. Ageing Dev. **105:** 273–290.
29. CHEN, Z.J. 2005. Ubiquitin signalling in the NF-kappaB pathway. Nat. Cell Biol. **7:** 758–765.
30. YEUNG, F. *et al.* 2004. Modulation of NF-κB-dependent transcription and cell survival by the SIRT1 deacetylase. EMBO J. **23:** 2369–2380.
31. MANNA, S.K. *et al.* 1998. Overexpression of manganese superoxide dismutase suppresses tumor necrosis factor-induced apoptosis and activation of nuclear transcription factor-κB and activated protein-1. J. Biol. Chem. **273:** 13245–13254.
32. SERRANO, M. *et al.* 1997. Oncogenic ras provokes premature cell senescence associated with accumulation of p53 and $p16^{INK4a}$. Cell **88:** 593–602.
33. ADAM-VIZI, V. & C. CHINOPOULOS. 2006. Bioenergetics and the formation of mitochondrial reactive oxygen species. Trends Pharmacol. Sci. **27:** 639–645.

Intermittent Dietary Restriction as a Practical Intervention in Aging

SANDEEP SHARMA AND GURCHARAN KAUR

Department of Biotechnology, Guru Nanak Dev University, Amritsar, India

ABSTRACT: Recent studies on the effects of dietary restriction (DR) in rodents and primates have shown that even late-onset short-term regimens can bring about comparable beneficial changes seen in animals subjected to life-long DR. We examined the effect of 3 months of DR on the expression of antioxidant enzymes and antioxidants, heat-shock protein 70 (HSP-70), neural cell adhesion molecule (NCAM) and its polysialylated form, PSA-NCAM, as well as astrocytic marker glial fibrillary acidic protein (GFAP) in 18-month-old middle-aged rats. The present study shows that DR initiated in late adulthood confers beneficial effects, such as attenuation of oxidative stress, enhanced expression of HSP-70, neural plasticity markers NCAM, and PSA-NCAM, and reduced levels of GFAP.

KEYWORDS: dietary restriction; aging; oxidative stress; neuronal plasticity; hormesis

INTRODUCTION

Experts currently have contradictory opinions on "whether dietary restriction (DR) can increase longevity in all species, particularly, in human beings."[1] However, even if DR fails to increase longevity in humans, still it could prove advantageous by preventing or delaying the onset of age-related diseases or increasing "functional longevity." In the majority of reported cases, DR is initiated early in life, that is, soon after weaning or at early growth stages. In view of recent reports on DNA array of gene expression for the effect of aging and DR in rodents and primates, even late-onset short-term regimens have been shown to produce a large part of changes seen in animals subjected to life-long DR.[2,3] Studies on late-life DR have been limited, perhaps because the lifespan extension by late-life regimen is not as remarkable as that by life-long DR. Practicing significant DR from childhood or adolescence in humans is very difficult. Studies on late-life DR in rodents provide the basis for possible

Address for correspondence: Dr. Gurcharan Kaur, Professor, Department of Biotechnology, Guru Nanak Dev University, Amritsar 143005 India. Voice: +91-183-2258431; fax: +91-183-2258272. gurcharank@yahoo.com

Ann. N.Y. Acad. Sci. 1114: 419–427 (2007).
doi: 10.1196/annals.1396.031

human application of DR for better quality of life.[4] The present study elucidates the beneficial effects of adult-onset short-term rather than life-long DR in animals for its neuroprotective role.

To test whether short-term DR of 3 months' duration initiated in middle-aged rats at the age of 15 months can attenuate the oxidative stress and induce a mild stress response, we studied the expression of antioxidant enzymes and antioxidants, such as catalase, Cu-Zn SOD, glutathione peroxidase, glutathione, and lipid peroxidation. The expression of the stress-response protein heat-shock protein 70 (HSP-70) was also studied. The expression of the neural cell adhesion molecule (NCAM) and its polysialylated form, PSA-NCAM (which shape the formation of neuronal networks during development and adult synaptic plasticity), was studied in middle-aged rats since age-dependent memory impairments have been related to a decline in synaptic plasticity. The expression of astrocytic marker glial fibrillary acidic protein (GFAP) was also studied to test whether DR can reduce age-associated reactive gliosis.

EXPERIMENTAL METHODS

The paradigm of DR involved periodic fasting, in which 15-month-old Wistar strain male albino rats were deprived of food for a full day, every other day, and were fed *ad libitum* (AL) on the intervening day for 3 months. Another group of rats of similar age was fed AL and used as control. The body weights were recorded every 15th day in DR and age-matched AL rats. Blood glucose determination was done every month. Animal care procedures were followed in accordance with the guidelines of the Institutional Animal Ethical Committee. A brief description of the experimental methods is given here; detailed procedures of biochemical assays, Western blotting and immunohistochemical studies are given elsewhere.[5] The activity of catalase,[6] Cu-Zn SOD,[7] and glutathione peroxidase[8] was determined from different brain regions of AL and DR rats. Total glutathione[9] and lipid peroxidation[10] was also estimated.

Western Blot Analysis of HSP-70, GFAP, and Total NCAM

Thirty micrograms of protein sample were electrophoresed on 10% polyacrylamide gel for HSP-70 and GFAP and 7.5% for total NCAM from the regions of the cerebral hemisphere (CH), diencephalon (DC), cerebellum (CB), and brain stem (BS) from AL and DR rats.[11] The separated proteins were then blot-transferred onto a PVDF membrane (Hybond-P) using the semidry Novablot system (Amersham Pharmacia Biotech UK Ltd., Buckinghamshire, England) at 25 V for 3 hours. After blocking, the membranes were hybridized with primary antibodies against these proteins followed by incubation with (anti-mouse IgG–peroxidase conjugated) secondary antibody. Blots were developed

TABLE 1. Effect of DR Cu-Zn SOD, catalase, and GPx activity and GSH and lipid peroxidation levels

Group/Area	CH	DC	CB	BS
Cu-Zn SOD (U · g tissue^{-1})				
AL	28.1 ± 0.76	35.75 ± 1.03	40.37 ± 1.23	39.22 ± 0.97
DR	34.1 ± 1.15*	44.87 ± 1.56*	39.22 ± 1.32	33.56 ± 1.34
Catalase (U · g tissue^{-1})				
AL	0.45 ± 0.07	0.57 ± 0.04	0.91 ± 0.08	0.31 ± 0.02
DR	0.76 ± 0.08*	0.91 ± 0.02*	1.21 ± 0.12*	0.41 ± 0.03
GPx (U · g tissue^{-1})				
AL	4.72 ± 0.23	4.96 ± 0.34	3.60 ± 0.15	2.37 ± 0.11
DR	3.10 ± 0.45*	2.41 ± 0.12*	3.23 ± 0.17	2.21 ± 0.13
GSH (mg · g tissue^{-1})				
AL	2.31 ± 0.14	2.11 ± 0.11	1.93 ± 0.21	1.46 ± 0.28
DR	3.12 ± 0.17*	3.33 ± 0.18*	3.06 ± 0.30*	1.41 ± 0.33
LPx (mg · dL^{-1})				
AL	23.4 ± 1.71	23.81 ± 1.83	24.25 ± 2.12	17.7 ± 1.14
DR	16.1 ± 1.68*	15.86 ± 1.45*	19.87 ± 1.66	14.8 ± 1.78

NOTE: Values are mean ± SEM of five experiments. *$P < 0.05$

ABBREVIATIONS: Cu-Zn SOD = copper zinc superoxide dismutase; GPx = glutathione peroxidase; GSH = glutathione (reduced); LPx = lipid peroxidation; DR = dietary restriction; CH = cerebral hemisphere; CB = cerebellum; DC = diencephalon; BS = brain stem; and AL = *ad libitum*–fed.

using enhanced chemiluminescence (Amersham Biosciences [GE Healthcare UK Ltd., Buckinghamshire, England]).

Immunofluorescent Localization of Total NCAM, PSA-NCAM, and GFAP

Brains were perfused transcardially with 4% paraformaldehyde in 0.1 M PBS and 20-micron-thick coronal sections were cut using a cryostat microtome. After the sections were blocked, they were incubated with primary antibodies (1:500) in 0.1% Triton X-100 and 1% BSA-PBS for 48 h at 4°C, followed by secondary antibodies (anti-mouse IgM μ-chain-specific FITC diluted 1:128 for PSA-NCAM and anti-mouse IgG–FITC diluted 1:200 for total NCAM and GFAP) in 0.3% Triton-PBS for 2 hours. After washing, the tissue sections were cover-slipped using the appropriate anti-fading mounting medium for fluorescence detection.

A one-way analysis of variance (ANOVA) was used to compare the results in different groups of rats. When ANOVA detected a difference, these sets of AL and DR rats were subjected to *post hoc* comparisons using Sigma Stat 2.03 (Systat Software Inc., San Jose, CA) to determine the statistical significance, which was assumed to be different when the comparison showed a significance level of $P < 0.05$.

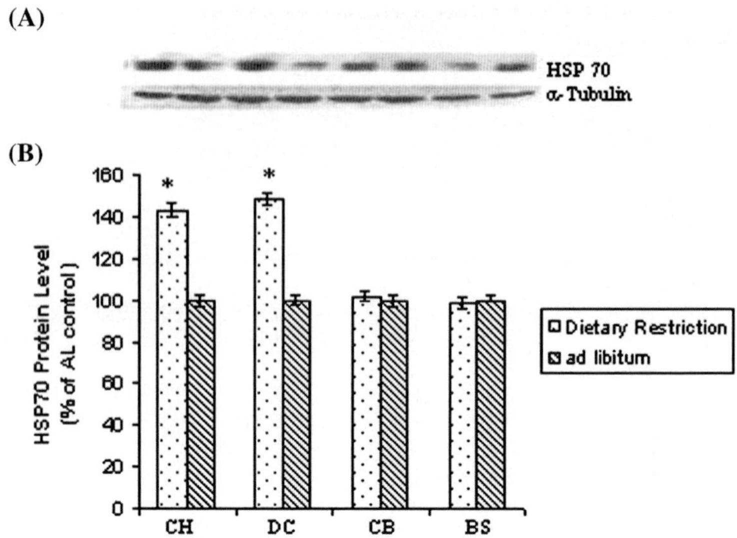

FIGURE 1. (**A**) Representative Western blot hybridization signals for HSP-70 and α-tubulin from cerebral hemisphere (CH), diencephalon (DC), cerebellum (CB), and brain stem (BS) regions from diet-restricted (DR) and *ad libitum*–fed (AL) rats. (**B**) Mean values of HSP-70 levels expressed as percentage of α-tubulin labeling. $^{*}P < 0.05$ versus AL rats; Bonferroni t-test after ANOVA.

RESULTS AND DISCUSSION

Rats subjected to DR in this study consumed around 20% less food in a 48-h period, as compared to AL rats. At the end of this study, the body weights of DR rats were about 16% less than those of the AL rats. Fasting serum glucose concentrations in AL rats averaged 140 mg/dL and in DR rats at 100 mg/dL.

Effect of DR on Oxidative Stress and HSP-70

CuZn-SOD, catalase, GPx activity, and GSH content showed a significant increase in the DR group from different regions of brain as compared to AL rats, whereas lipid peroxidation was significantly lowered in DR rats (TABLE 1). These observations suggest that DR in middle-aged rats may attenuate oxidative stress and these results are supported by recent studies on late-life DR which have been shown to reverse age-related alterations in protein metabolism.[12] Reduced GSH is a key component of the cellular defense cascade against injury caused by ROS and decreased levels of reduced GSH and increased lipid peroxidation are used as markers of oxidative stress. HSP-70 labeling in the representative samples from AL and DR groups is shown

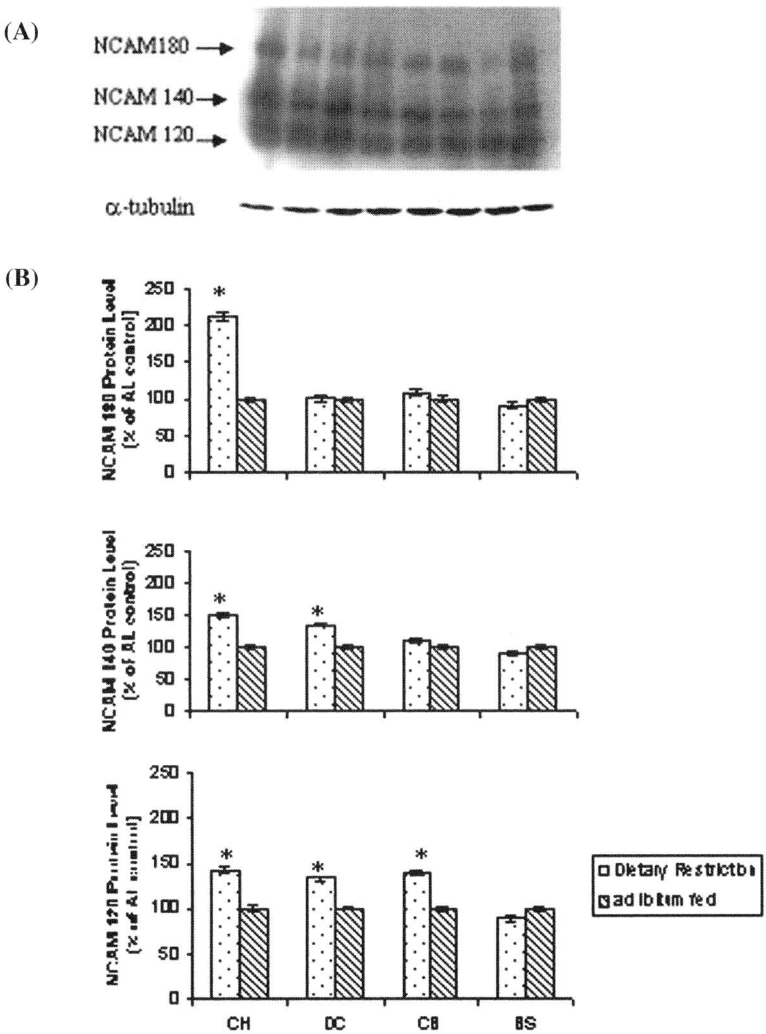

FIGURE 2. (**A**) Representative Western blot hybridization signals for total NCAM and α-tubulin from cerebral hemisphere (CH), diencephalon (DC), cerebellum (CB), and brain stem (BS) regions from diet-restricted (DR) and *ad libitum*–fed (AL) rats. (**B**) Mean values of total NCAM levels expressed as percentage of α-tubulin labeling. *$P < 0.05$ versus AL rats; Bonferroni t-test after ANOVA.

in FIGURE 1A and B. The quantitative analysis of immunoblots from the two groups revealed a significant increase in HSP-70 expression in DR rats as compared to AL rats from CH and DC ($P < 0.05$) regions of the brain. The enhanced expression of HSP-70 in DR rats may be the result of a mild stress

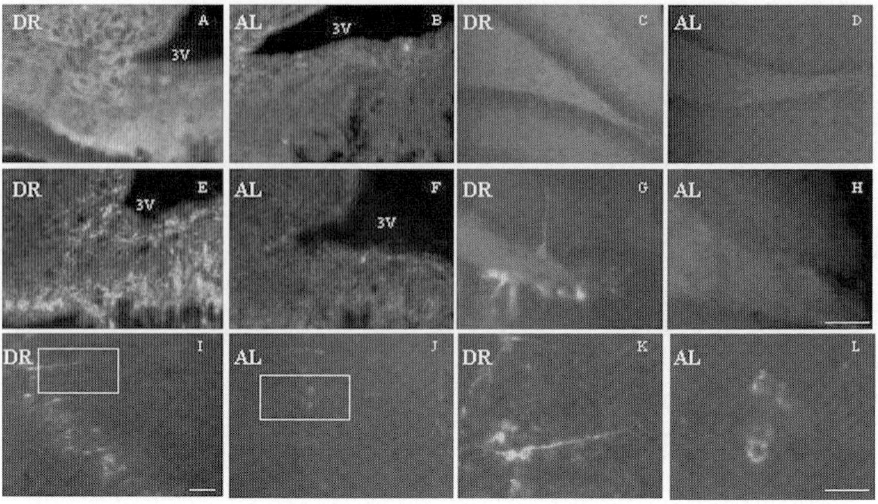

FIGURE 3. Representative immunofluorescent images of NCAM (**A–D**) and PSA-NCAM (**E–L**). Note the marked increase in NCAM immunoreactivity in DR rats as compared to AL rats in hypothalamus (**A** and **B**) and dentate gyrus (**C** and **D**) regions of brain. PSA-NCAM immunoreactivity is enhanced in DR rats as compared to AL rats in hypothalamus (**E** and **F**), dentate gyrus (**G** and **H**), and piriform cortex (**I–L**). Scale bars = 80 μm (**A–H**), 100 μm (**I–J**), and 40 μm (**K–L**). (In color in Annals online.)

response due to the decreased glucose availability in the alternate-day feeding regimen of DR rats. Mild stress is known to protect neurons against subsequent severe stress conditions.[13] In a previous study from our lab, DR was reported to reduce oxidative stress and induce HSP-70 expression under normal and excitotoxic conditions in young adult rats,[5] and the present study suggests that short-term DR has a neuroprotective effect even in middle-aged rats.

Effect of DR on NCAM and PSA-NCAM Expression

The quantitative analysis of immunoblots from the two groups revealed a significant increase in NCAM-120, -140 and 180-kDa isoform expression in DR rats as compared to AL rats (FIG. 2A and B). Immunohistochemical study in DR rats also showed enhanced expression of total NCAM (FIG. 3A–D) and PSA-NCAM (FIG. 3E–K) in the dentate gyrus region of hippocampus, piriform cortex, and hypothalamus as compared to AL rats. These observations suggest that short-term DR in middle-aged rats may reverse memory and learning functions since NCAM plays an important role in regeneration and learning in the adult nervous system[14,15] and NCAM-deficient mice are impaired in spatial learning.[16,17] PSA-NCAM is associated with morphogenetic changes during developmental processes such as cell migration, synaptogenesis, and

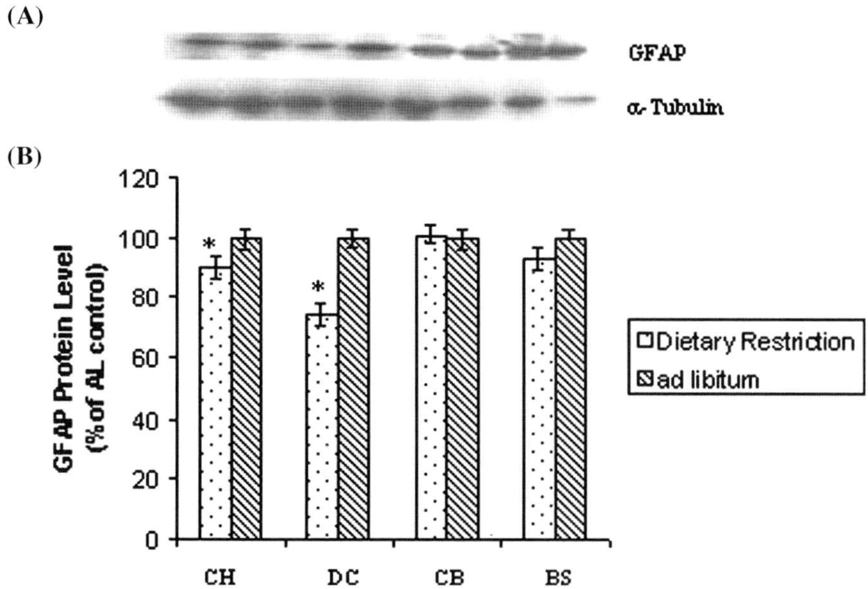

FIGURE 4. (**A**) Western blot hybridization signals for GFAP and α-tubulin from cerebral hemisphere (CH), diencephalon (DC), cerebellum (CB), and brain stem (BS) regions from diet-restricted (DR) and *ad libitum*–fed (AL) rats. (**B**) Mean values of GFAP levels expressed as percentage of α-tubulin labeling. $^{*}P < 0.05$ versus AL rats; Bonferroni t-test test after ANOVA.

axonal growth and branching and it persists in adult brain only in structures that display a high degree of functional plasticity.[18] Cell surface expression of PSA-NCAM is considered a permissive factor to allow activity-dependent neuronal and glial remodeling to occur whenever the proper inductive stimulus intervenes.[18]

Effect of DR on the GFAP Expression

The quantitative analysis of immunoblots from the two groups revealed a significant decrease in GFAP expression in DR control rats as compared to AL rats (FIG. 4A–B). Immunohistochemical study in DR rats showed reduced expression of GFAP in the dentate gyrus of hippocampus and hypothalamus as compared to AL rats (FIG. 5A–F). Beginning soon after maturation, GFAP expression increases progressively during aging in human and rodent models.[19] Recent studies have correlated an increase in GFAP during aging with age-related deficits in learning and memory and impaired synaptic plasticity.[20] To date changes of astrocytes during aging and their effects on age-related functional decline have not been well understood.

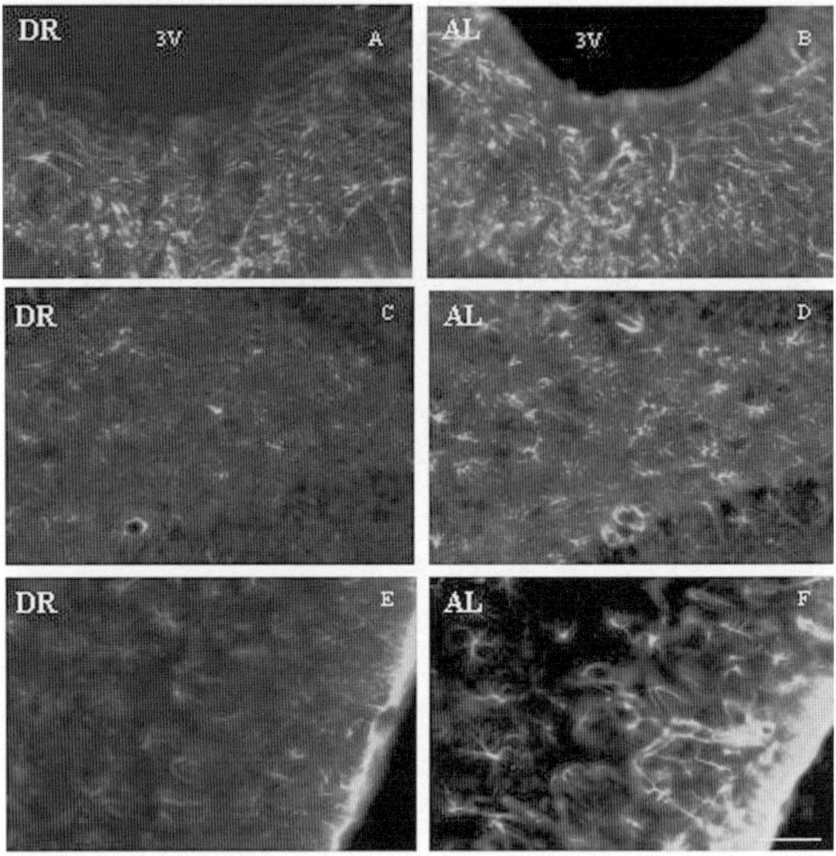

FIGURE 5. Representative immunofluorescent images of GFAP (**A–F**). Note the marked decrease in GFAP immunoreactivity in DR rats as compared to AL rats in hypothalamus (**A** and **B**), dentate gyrus (**C** and **D**), and piriform cortex (**E** and **F**) regions of brain. Scale bars = 80 μm (**A–F**). (In color in Annals online.)

The present results suggest that short-term DR in middle-aged rats confers potentially beneficial effects and support the idea of hormetic intervention in the aging process. This may be an interesting model system to study the potential of DR initiated in middle and old age in reversing age-related functional impairment and looking for targets to design drugs that mimic these effects.

REFERENCES

1. Le Bourg, E. & S.I.S. Rattan. 2006. Can dietary restriction increase longevity in all species, particularly in human beings? Introduction to a debate among experts. Biogerontology **7:** 123–125.

2. LEE, C.K., R.G. KLOPP, R. WEINDRUCH, *et al.* 1999. Gene expression profile of aging its retardation by caloric restriction. Science **285:** 1390–1393.
3. CAO, S.X., J.M. DHAHBI, P.L. MOTE, *et al.* 2001. Genomic profiling of short- and long-term caloric restriction effects in the liver of aging mice. Proc. Natl. Acad. Sci. USA **98:** 10630–10635.
4. GOTO, S. 2006. Health span extension by later-life caloric or dietary restriction: a view based on rodent studies. Biogerontology **7:** 135–138.
5. SHARMA, S. & G. KAUR. 2005. Neuroprotective potential of dietary restriction against kainate-induced excitotoxicity in adult male wistar rats. Brain Res. Bull. **67**: 482–493.
6. AEBI, H. 1984. Catalase in vitro. *In* Methods in Enzymology, Vol. **105**. L. Packer, Ed.: 121–126. Academic Press. New York.
7. KONO, Y. 1978. Generation of superoxide radical during autoxidation of hydroxylamine and an assay for superoxide dismutases. Arch. Biochem. Biophys. **186:** 189–195.
8. FLOHE, L. & W.A. GUNZLER. 1984. Assays of glutathione peroxidase. *In* Methods in Enzymology, Vol. **105**. L. Packer, Ed.: 114–121. Academic Press. New York.
9. SEDLAK, J. & R.H. LINDSAY. 1968. Estimation of total, protein bound and non-protein sulfhydryl groups in tissue with Ellman's reagent. Anal. Biochem. **25:** 192–205.
10. BEUGE, J.A. & A.D. AUST. 1978. Microsomal lipid peroxidation. *In* Methods in Enzymology, Vol **52**. L. Packer, Ed.: 302–310. Academic Press. New York.
11. LAEMMLI, U.K. 1970. Cleavage of structural proteins during the assembly of the head of bacteriophage T4. Nature **227:** 680–685.
12. GOTO, S., R. TAKAHASHI, S. ARAKI, *et al.* 2002. Dietary restriction initiated in late adulthood can reverse age-related alterations of protein and protein metabolism. Ann. N. Y. Acad. Sci. **959:** 50–56.
13. HEYDARI, A.R., S. YOU, R. TAKAHASHI, *et al.* 1996. Effect of caloric restriction on the expression of heat shock protein 70 and the activation of heat shock transcription factor 1. Dev. Genet. **18:** 114–124.
14. RONN, L.C., V. BEREZIN, E. BOCK. 2000. The neural cell adhesion molecule in synaptic plasticity and ageing. Int. J. Dev. Neurosci. **18:** 193–199.
15. VENERO, C., T. TILING, I. HERMANS-BORGMEYER, *et al.* 2006. Chronic stress induces opposite changes in the mRNA expression of the cell adhesion molecules NCAM and L1. Neuroscience **115:** 1211–1219.
16. CAMBON, K., S.M. HANSON, C. VENERO, *et al.* 2004. A synthetic neural cell adhesion molecule mimetic peptide promotes synaptogenesis, enhances presynaptic function, and facilitates memory consolidation. J. Neurosci. **24:** 4197–4204.
17. SANDI, C. 2004. Stress, cognitive impairment and cell adhesion molecules. Nat. Rev. Neurosci. **5:** 917–932.
18. THEODOSIS, D.T., R. BONHOMME, S. VITIELLO, *et al.* 1999. Cell surface expression of polysialic acid on NCAM is a prerequisite for activity-dependent morphological neuronal and glial plasticity. J. Neurosci. **19:** 10228–10236.
19. PROLLA, T.A. & M.P. MATTSON. 2001. Molecular mechanism of brain aging and neurodegenerative disorders, lessons from dietary restriction. Trends Neurosci. **24:** 521–531.
20. ROZOVSKY, I., M. WEI, T.E. MORGAN, *et al.* 2005. Reversible age impairments in neurite outgrowth by manipulations of astrocytic GFAP. Neurobiol. Aging **26:** 705–715.

Life Extension by Calorie Restriction in Humans

ARTHUR V. EVERITT AND DAVID G. Le COUTEUR

Centre for Education and Research on Ageing, University of Sydney, and Concord RG Hospital, Sydney, NSW, Australia

ABSTRACT: Long-term reduction in energy intake in the diet (calorie restriction [CR]) extends the life of the laboratory rat by about 25%. However, in humans there are no life-long studies of CR, but only short-term trials which indicate that 20% CR acting over periods of 2–6 years is associated with reduced body weight, blood pressure, blood cholesterol, and blood glucose—risk factors for the major killer diseases of cardiovascular disease and diabetes. In addition, recent research has shown that CR for 6 months is able to improve biomarkers for longevity (deep body temperature and plasma insulin) and thus should increase life expectancy. The magnitude of the life-extension effect of CR in humans can only be estimated. The Okinawans, the longest-lived people on earth, consume 40% fewer calories than the Americans and live only 4 years longer. Similarly, women in United States consume 25% fewer calories than men and live 5 years longer. From the survival studies of overweight and obese people, it is estimated that long-term CR to prevent excessive weight gain could add only 3–13 years to life expectancy. Thus the effects of CR on human life extension are probably much smaller than those achieved by medical and public health interventions, which have extended life by about 30 years in developed countries in the 20th century, by greatly reducing deaths from infections, accidents, and cardiovascular disease.

KEYWORDS: calorie restriction; longevity; life extension; cardiovascular disease

INTRODUCTION

In 1935 McCay and co-workers clearly demonstrated that long-term calorie restriction (CR) without malnutrition started soon after weaning in the laboratory rat extended maximum life span.[1] This life-prolonging effect of CR in the rat has been repeatedly confirmed and discussed extensively by later investigators.[2] Ten life-long studies of specific pathogen-free male rats on 40% CR

Address for correspondence: Professor Arthur Everitt, Centre for Education and Research on Ageing, University of Sydney, and Concord RG Hospital, Sydney, NSW 2139, Australia. Voice: +612-9767-7212; fax: +612-9767-5419.
arthureveritt@yahoo.com.au

Ann. N.Y. Acad. Sci. 1114: 428–433 (2007).
doi: 10.1196/annals.1396.005

showed mean life span to be extended by 24.7%.[3] CR over a lifetime retards the age-related decline in most physiological processes in the rat and, conversely, overeating accelerates these age changes.[4] Long-term CR also slows the development of age-related disease, which is a major determinant of life span.[5] The National Institute on Aging of the National Institutes of Health (NIH) started a long-term study in 1987 of the effects of CR on the rhesus monkey, a primate with a shorter life span than that of humans, and found that CR attenuated many age-related changes that had been found in rodents, but no life-span data are yet available.[6]

HUMAN LIFE EXPECTANCY AT BIRTH

There are no controlled life-long studies of calorie restriction in humans. However, the longest-lived people on earth are the inhabitants of the Japanese islands of Okinawa, who consume 40% fewer calories than Americans do and live 4 years longer, only about 5% of U.S. life expectancy.[7] This is much less than that found in male rats on 40% CR, whose life is extended by almost 25%.[3] The Okinawans differ from Americans in genetics, diet, and lifestyle, and thus any direct comparison of the causes of the differences in life expectancy is complex. Okinawans have a lower mortality from stroke and cancer.[7] They are lean and physically active and their diet is rich in unrefined carbohydrates, including fruits, vegetables and whole grains; sweet potatoes are their main source of energy.

Women in the United States in the year 2000 had a life expectancy 5.5 years longer than that of men (79.4 versus 73.9 years).[8] Caloric intake in one study of American women was 25% less than that in men.[9] Other factors contributing to differences in life expectancy of men and women are genes on the Y chromosome, sex hormones, and gender-related disease.

In 18 developed countries human life expectancy at birth has increased by 28 years between 1900 and 2000, a gain of almost 60%.[8] This increase in life expectancy at birth in United States is still continuing in the first decade of the 21st century at the rate of about 2 months per year.[10] During the 20th century there was a shift from infectious diseases to chronic degenerative diseases as major causes of death.[11] In the first half of the 20th century there were large reductions in infant and maternal mortalities, largely secondary to fewer deaths from infectious diseases as a result of better sanitation and nutrition. Antibiotics and vaccines were later developed to control these diseases. Also, during the 20th century, deaths from suicides and accidents were significantly reduced. Improved design of cars and roads associated with legislation to restrict speed have reduced motor accident deaths. In the first half of the 20th century there was a great loss of life in two world wars. In the second half of the 20th century there were increased deaths from chronic degenerative diseases, principally cardiovascular diseases. Since 1980 deaths from heart disease have fallen on account of treatment with drugs and diet to lower blood pressure and blood cholesterol and by a significant decrease in cigarette smoking, which also reduced deaths from lung cancer.[12]

From 1961 to 2000 the intake of calories increased by 30% in the United States and by 15% in Japan,[8] and this has led to an increased incidence of obesity and overweight in these countries. Despite this increase in body weight, life expectancy has continued to increase over this period in the United States and Japan.

CALORIE RESTRICTION IN HUMANS

The first human CR studies had their origin in World War II in the 1940s. As a result of the war, Scandinavians were living on a semi-starvation diet of about 20% CR, which was associated with a reduced incidence of heart disease.[13] CR in Norway and Estonia in the 1940s has been linked to a lower incidence of colorectal cancer in those surviving through the 1980s to 2000s.[14] Childhood overweight in 1937–1939 has been related to cardiovascular mortality 50 years later.[15] Thus, some of the diseases of old age may have their origin in the diet of childhood.

In the 1990s, two 20% CR studies on healthy adult humans showed that CR over 2–6 years reduced many physiological parameters by 20% or more.[16,17] In both studies CR produced significant reductions in body weight, blood pressure, blood cholesterol, and blood sugar, which are risk factors for cardiovascular disease and diabetes. Thus, long-term CR over 2–6 years in healthy adult humans was able to reduce the risk of developing these two major killer diseases of later life. The study of Fontana *et al.*[17] clearly showed that 6 years of CR reduced the development of atherosclerosis, a leading cause of mortality in Western countries. In a recent study, Heilbronn and co-workers showed that in overweight adults 6 months of 25% CR was not only able to lower their body weight, but was also likely to simultaneously increase life expectancy by reducing deep body temperature and plasma insulin, which are biomarkers of longevity.[18]

OVERWEIGHT AND OBESITY IN HUMANS

In most countries of the world the incidence of overweight and obesity in humans has been rising steadily over the last 20 years.[19] Overweight is a risk factor for many diseases of old age and has a life-shortening action. Data from the Framingham study show that at age 40 years being overweight (body mass index [BMI] of 25–30 kilograms per meter squared) shortens life by 3 years, and being obese (BMI > 30) reduces life expectancy by 7 years.[20] A study of very obese subjects (BMI > 45) aged 20–30 years revealed life shortening of 13 years for men and 7 years for women.[21] Excess body weight may be reduced by decreasing the intake of energy in food and by increasing physical exercise to consume excess energy.[22] Reduction of body weight decreases the risk of developing type 2 diabetes mellitus, hypertension, coronary heart disease, cancer of the uterus, breast, prostate and colon, and dementia.[19] The epidemic of overweight and obesity is predicted to become a serious threat to our survival[23] and as a consequence the continuing rise in life expectancy through the years is expected to stop by the mid 21st century.[23] Public health measures are being developed to minimize the epidemic of overweight,[24] but it is very difficult to reduce body weight for long periods.

HUMAN LIFE EXTENSION BY LONG-TERM CALORIE RESTRICTION

How many extra years of life can human beings expect to gain by life-long 20% CR started in childhood? There are no life-long CR studies to answer this question and so the answer can only be predicted. If overweight in middle age takes 3 years off life expectancy, obesity takes 7 years, and extreme obesity in young adult men takes 13 years, then CR started in early life to prevent overweight and obesity could possibly add only 3–13 years to life.

CONCLUSIONS

Increasing energy intake in humans leads to overweight, which results in premature onset of age-related diseases and shortening of life. Conversely, reducing energy intake by about 20% decreases the risk factors for cardiovascular disease and type 2 diabetes mellitus, thus increasing life expectancy. On balance, the probable increase in life expectancy by long-term CR would be about 3–13 years, which is of major importance, but less than that predicted on the basis of animal experiments.

ACKNOWLEDGMENTS

We are grateful for comments and ideas from Dr. Fiona Blyth, Dr. Hamish Jamieson, and Dr. Leonie Heilbronn. We especially thank Professor Donald K. Ingram for his input into this manuscript.

REFERENCES

1. McCay, C.M., M.F. Crowell & L.A. Maynard. 1935. The effect of retarded growth upon the length of the lifespan and ultimate body size. J. Nutr. **10:** 63–79.

2. MASORO, E.J. 2002. Caloric Restriction: a Key to Understanding and Modulating Aging. Elsevier. Amsterdam.
3. EVERITT, A.V., G.S. ROTH, D.G. LE COUTEUR, *et al.* 2005. Caloric restriction versus drug therapy to delay the onset of aging diseases and extend life. Age **27:** 39–48.
4. YU, B.P. 1994. How diet influences the aging processes of the rat. Proc. Soc. Exp. Biol. Med. **205:** 97–105.
5. SIMMS, H.S. & B.N. BERG. 1962. Longevity in relation to lesion onset. Geriatrics **17:** 235–242.
6. MATTISON, J.A., M.A. LANE, G.S. ROTH, *et al.* 2003. Calorie restriction in rhesus monkeys. Exp. Gerontol. **38:** 35–46.
7. WILLCOX B., C. WILLCOX & M. SUZUKI. 2000. The Okinawa Way. Penguin Books. London.
8. TIFFEN, R. & R. GITTINS. 2005. How Australia Compares. Cambridge University Press. Cambridge, UK.
9. STATEN, M.A. 1991. The effect of exercise on food intake in men and women. Am. J. Clin. Nutr. **53:** 27–31.
10. MINIÑO, A.M., M.P. HERON, B.L. SMITH. 2006. Deaths: preliminary data for 2004. National vital statistics reports; Vol. 54 no. 19. Hyattsville, MD: National Center for Health Statistics.
11. DE FLORA, S., A. QUAGLIA, C. BENNICELLI, *et al.* 2005. The epidemiological revolution of the 20th century. FASEB **19:** 892–897.
12. SYTKOWSKI, P.A., W.B. KANNEL & R.B. AGOSTINO. 1990. Changes in risk factors and the decline in mortality from cardiovascular disease: the Framingham Heart Study. N. Engl. J. Med. **322:** 1635–1641.
13. STROM, A. & R.A. JENSEN. 1951. Mortality from cardiovascular disease in Norway 1940–1945. Lancet **1:** 126–129.
14. SVENSSON, E., B. MOLLOR, S. TRETLI, *et al.* 2005. Early life events and later risk of colorectal cancer: age-period-cohort modelling in the Nordic countries and Estonia. Cancer Causes Control **16:** 215–223.
15. GUNNEL, D.J., S.J. FRANKEL, K. NANCHAHAL, *et al.* 1998. Childhood obesity and adult cardiovascular mortality: a 57 year follow-up study based on the Boyd-Orr cohort. Am. J. Clin. Nutr. **67:** 1111–1118.
16. WALFORD, R.L., D. MOCK, R. VERDERY, *et al.* 2002. Caloric restriction in Biosphere2: alterations in physiologic, hematologic, and biochemical parameters in humans restricted for a 2 year period. J. Gerontol. **57A:** B211–B222.
17. FONTANA, L., T.E. MEYER, S. KLEIN, *et al.* 2004. Long-term caloric restriction is highly effective in reducing the risk for atherosclerosis in humans. Proc. Natl. Acad. Sci. USA **101:** 6659–6663.
18. HEILBRONN, L.K., L. DE JONGE, M.I. FRISARD, *et al.* 2006. Effect of 6-month caloric restriction on biomarkers of longevity, metabolic adaptation, and oxidative stress in overweight individuals: a randomized controlled trial. JAMA **295:** 1539–1548.
19. BRAY, G.A. 2004. Medical consequences of obesity. J. Clin. Endocrinol. Metab. **89:** 2583–2594.
20. PEETERS, A., J.J. BARENDREGT, F. WILLEKENS, *et al.* 2003. Obesity in adulthood and its consequences for life expectancy: a life table analysis. Ann. Intern. Med. **138:** 24–32.
21. FONTAINE, K.R., D.T. REDDEN, D. WANG, *et al.* 2003. Years of life lost due to obesity. JAMA **289:** 187–193.
22. STUBBS, C.O. & A.L. LEE. 2004. The obesity epidemic: both energy intake and physical activity contribute. Med. J. Aust. **181:** 489–491.

23. OLSHANSKY, S.J., D.J. PASSORO, R.C. HERSHOY, *et al.* 2005. A potential decline in life expectancy in the United States in the 21st century. N. Engl. J. Med. **352:** 1138–1145.
24. SAVVA, S.C., M. CHADJIOANNOU & M.J. TORNARITIS. 2007. Policy options for responding to the growing challenge from obesity: Cyprus national findings. Obes. Rev. **8**(Suppl 2): 37–45.

Caloric Restriction, the Traditional Okinawan Diet, and Healthy Aging

The Diet of the World's Longest-Lived People and Its Potential Impact on Morbidity and Life Span

BRADLEY J. WILLCOX,[a,b] D. CRAIG WILLCOX,[a,c] HIDEMI TODORIKI,[d] AKIRA FUJIYOSHI,[b] KATSUHIKO YANO,[a] QIMEI HE,[a] J. DAVID CURB,[a,b] AND MAKOTO SUZUKI[e]

[a]*Pacific Health Research Institute, Honolulu, Hawaii 96813, USA*

[b]*Departments of Geriatric Medicine and Medicine, John A. Burns School of Medicine, 1356 Lusitana Street, 7F, Honolulu, Hawaii 96813*

[c]*College of Nursing, Okinawa Prefectural University, 1-24-1 Yogi, Naha, Okinawa, Japan 902-0076*

[d]*Department of Environmental and Preventive Medicine, Faculty of Medicine, University of the Ryukyus, 207 Uehara, Nishihara-cho, Okinawa, Japan 903-0215*

[e]*Faculty of Medicine, University of the Ryukyus, 207 Uehara, Nishihara, Okinawa, Japan 903-0215 and Okinawa Research Center for Longevity Science, 1-27-8 Ahacha, Urasoe, Okinawa, Japan 901–2114*

ABSTRACT: Long-term caloric restriction (CR) is a robust means of reducing age-related diseases and extending life span in multiple species, but the effects in humans are unknown. The low caloric intake, long life expectancy, and the high prevalence of centenarians in Okinawa have been used as an argument to support the CR hypothesis in humans. However, no long-term, epidemiologic analysis has been conducted on traditional dietary patterns, energy balance, and potential CR phenotypes for the specific cohort of Okinawans who are purported to have had a calorically restricted diet. Nor has this cohort's subsequent mortality experience been rigorously studied. Therefore, we investigated six decades of archived population data on the elderly cohort of Okinawans (aged 65-plus) for evidence of CR. Analyses included traditional diet composition, energy intake, energy expenditure, anthropometry, plasma DHEA, mortality from age-related diseases, and current survival patterns. Findings include low caloric intake and negative energy balance at younger ages, little weight gain with age, life-long low BMI, relatively

Address for correspondence: Bradley J. Willcox, M.D., Pacific Health Research Institute, 846 South Hotel Street, Suite 301, Honolulu, HI 96813. Voice: 808-524-4411; fax: 808-524-5559. bjwillcox@phrihawaii.org

Ann. N.Y. Acad. Sci. 1114: 434–455 (2007).
doi: 10.1196/annals.1396.037

high plasma DHEA levels at older ages, low risk for mortality from age-related diseases, and survival patterns consistent with extended mean and maximum life span. This study lends epidemiologic support for phenotypic benefits of CR in humans and is consistent with the well-known literature on animals with regard to CR phenotypes and healthy aging.

KEYWORDS: **caloric restriction; longevity; Okinawa; DHEA; biomarker; human; healthy aging; life span**

INTRODUCTION

Energy restriction, also known as caloric restriction (CR), is the most robust and reproducible means of reducing age-related diseases and extending life span in short-lived animals, but the effects in humans are unknown.[1–7] Preliminary data from ongoing research with CR on long-lived nonhuman primates (rhesus and squirrel monkeys) also suggest potential health benefits—although it will take some years before final results become available and firmer conclusions can be reached regarding aging and life span.[8,9]

Some epidemiologic and short-term human studies support CR-related health benefits.[9–13] However, a role for CR in human aging is difficult to ascertain since human life span makes long-term investigations impractical and there are no universally accepted biomarkers to measure the rate of human aging.[14] Whether CR (without malnutrition) affects human aging may be among the most significant unanswered questions in modern biogerontology.[9] Yet we have few available human populations or studies that can address this question. Most human populations who have experienced low calorie intake have suffered from high mortality due to infectious diseases and malnutrition.

Studies of human CR volunteers are currently under way with promising early results,[12,15,16] but will be unable to address longevity as an outcome. Only one long-term (>30 years) epidemiologic study has linked CR to human longevity.[17] This 36-year follow-up study reported a weak trend for lower all-cause mortality in healthy never-smoking Japanese–American men whose caloric intake was 15% lower than the cohort average, suggesting that those who maintained a modestly low energy intake (mean kcal/day 1882, range 1705–2061) in mid-life had the lowest late-life mortality risk. There was higher risk for mortality when caloric intake dropped below 50% of the group mean. These data are consistent with previous findings in animals, but much more study is required of healthy human populations with low caloric intake in order to understand the effects of CR on human aging and life span.

Therefore, it is of significant interest that Kagawa[18] reported low caloric intake in the Okinawan population, relative to other Japanese, and hypothesized that this may have been, at least in part, responsible for their healthy longevity. Kagawa[18] reported dietary information from the 1972 Japan National Nutrition survey that suggested Okinawan adults consumed only 83%

of the Japan average caloric intake. This report followed an earlier report by Hokama *et al.*[19] that Okinawan school children consumed only 62% of the calories of other Japanese school children in the early 1960s. Kagawa[18] also presented anthropometric and morbidity data from selected small samples of older Okinawans that were consistent with CR and reported markedly lower risk for age-associated diseases in middle-aged Okinawans (60–64 year olds) than in other Japanese. A later study by Chan *et al.* also reported dietary and phenotypic data in Okinawan septuagenarians and centenarians consistent with CR.[20]

There are several weaknesses with these previous reports. One, to our knowledge no population-based dietary information has been reported in a peer-reviewed journal on Okinawan adults before the 1972 National Nutrition Survey. Since the Japanese lifestyle underwent radical changes from the 1950s,[18] including changes in food choices, caloric intake, and energy expenditure, it is unlikely that the 1972 Japan National Nutrition Survey reflects the traditional CR diet that may be implicated in Okinawan longevity. Two, Okinawans are smaller than other Japanese (and Americans) and likely require fewer calories. Therefore, accounting for their unique energy requirements is necessary before concluding that Okinawans were calorically restricted. For example, Okinawans currently consume 8% fewer calories than other Japanese,[21] yet they now have the largest body mass index (BMI) in Japan.[22] Three, since the Okinawan mortality advantage has all but disappeared except in older cohorts (aged 65-plus),[22,23] it would be informative to have a more detailed, population-based epidemiologic analysis of the traditional diet, energy intake, energy expenditure, phenotype, and the subsequent mortality experience of this older cohort. These data might help answer the question as to whether Okinawans were truly calorically restricted and to what degree, the phenotypic consequences, and the current consequences for age-related mortality and life span. Importantly, an in-depth epidemiologic analysis of this older cohort using longer-term population data might also provide significant new information on the potential human effects of CR.

Fortunately, detailed population surveys of the traditional Okinawan diet and anthropometry were conducted in 1949.[24] Smaller, more limited population surveys were conducted at irregular intervals thereafter by U.S. and Okinawan postwar administrations.[25] These surveys were archived and have not been systematically analyzed. Therefore, we investigated these data in detail and asked three questions: (1) were older Okinawans truly calorically restricted across the life course? (2) is there longer-term anthropometric evidence to support this hypothesis? and (3) is there evidence for slower aging in this cohort?

To address question (1), we assessed population food-consumption patterns, energy intake, and energy expenditure for older Okinawans (aged 70s) at two different times—when they were young adults (aged in their 30s) and when they were middle-aged (aged 50s). To answer question (2), we analyzed archived data on height and weight, and calculated BMI for multiple time points in the Okinawan adult population coincident with whole-population energy balance. To answer question (3), whether delayed aging has occurred in this cohort, we took three approaches. First, since the adrenal steroid dehydroepiandrosterone (DHEA) has been purported to satisfy the main criteria for a candidate biomarker of aging in nonhuman primates,[26] and is of hypothetical utility as a biomarker of the rate of human aging,[27] we measured plasma levels in a sample of older Okinawans (aged 70s). Second, we analyzed late-life survival patterns for potential evidence of extended average and maximum life span. Third, we calculated age-adjusted mortality patterns standardized to the world standard population for age-related diseases. Finally, we compared selected findings to published data from non-CR Americans and Japanese of similar chronological age, in whom differences in diet, energy balance, anthropometry, DHEA, age-related mortality and late-life survival should be robust, if older Okinawans truly experienced long-term CR.

METHODS

Study Materials and Measurements

Population Dietary Intake and Energy Expenditure

Archived dietary data are used to estimate energy balance and potential CR status in Okinawan septuagenarians at younger ages. Okinawan data were derived from the Office of the Civil Administrator of the Ryukyu Islands (Okinawa) for the year 1949, when the current septuagenarians were aged approximately 30 years,[24] and approximately every 5–10 years thereafter by the Okinawa prefectural government.[25] Studies relied on 3-day food records of usual dietary intake to estimate caloric intake in conjunction with dietitian interviews.

For the U.S. population, data collected with comparable methodology for the same birth cohort of Americans (circa 1915–1925 birth cohort) is extremely limited. The closest data set appears to be the NHANES I data set, which used 24-hour dietary recall methods to estimate dietary intake in adult subjects (aged 20–74 years) in 1971–1974.[28] We could find no comparative data set before the 1970s for the U.S. population.

Archived data of anthropometric measurements (height and weight) and demographic data (age, occupation) in the Okinawan population and in the NHANES population were used for estimation of energy expenditure using the Harris–Benedict equation.[29] This equation estimates basal or resting metabolic rate (BMR). An activity factor was then used to calculate additional calories burned according to the following levels of activity:

Sedentary = BMR $\times$ 1.2 (little or no exercise, desk job)

Lightly active = BMR × 1.375 (light exercise/sports 1–3 days/wk)
Moderately active = BMR × 1.55 (moderate exercise/sports 3–5 days/wk)
Very active = BMR × 1.725 (hard exercise/sports 6–7 days/wk or physical job)
Extremely active = BMR × 1.9 (hard daily exercise/sports and physical job).

Energy balance was estimated in the Okinawans and Americans by subtracting estimated energy expenditure from caloric intake measured in the dietary surveys. BMI was also calculated by body weight (kg)/height $(m)^2$ and used as an additional, longer-term marker of energy balance in the Okinawan and U.S. populations.

DHEA Measurement in Study Subjects

All subjects from Okinawa were healthy, community-dwelling men and women aged approximately 75 years ($n = 54$ septuagenarians; 29 males, mean age 74.5 ± 0.7 years and 25 females, mean age 74.7 ± 0.6 years) in 1988, the time of DHEA measurement,[30] and were born into the cohort from which energy balance data were derived. Septuagenarians were selected from subjects who were attending their annual physical exam and were recruited as part of the annual Okinawa Centenarian Study, a population-based study of Okinawans over the age of 100 and selected other elderly controls that begun in 1976.[31]

The reference population for DHEA levels in Americans consisted of healthy, community-dwelling American septuagenarians from the Rancho Bernardo Study ($n = 991$ septuagenarians; 534 men, mean age 68.6 ± 9.0 years; 457 women, mean age 72.1 ± 8.0 years) who had plasma DHEA measured in 1984–1987.[32]

DHEA Measurement Protocol

Similar protocols were used in both the U.S. reference population from the Rancho Bernardo Study and the septuagenarian subjects from the Okinawa Centenarian Study. Specifically, nonfasting venous blood samples were drawn between 8 AM and 4 PM, separated and stored at −20 to −80 °C for up to 8 months. Plasma was assayed for DHEA using a solid-phase 125I RIA (radioimmunoassay) (Okinawa samples: SRL Laboratories, Tokyo, Japan; Rancho Bernardo samples: Rancho Bernardo Study Laboratory, San Diego, CA). The inter- and intra-assay coefficients of variations averaged 5.2% and 10.0% in the Okinawan samples and 6.7% and 6.1% in the Rancho Bernardo samples, respectively.

Statistical Analysis of DHEA Levels

Septuagenarian DHEA levels were compared between the study population in Okinawa and the U.S. reference population in Rancho Bernardo. Two sample unpaired Student *t*-tests with unequal variances were used to compare the Okinawans to the U.S. reference population. A two-tailed *P*-value of 0.05 was considered a statistically significant difference.

Mortality and Survival Comparisons

The late-life survival experience (mortality rates) for Okinawan, Japanese, and American populations was modeled according to the SAS procedure LIFEREG, which fits parametric models to failure-time data.[33] Period life-table data from the U.S. National Center for Health Statistics[34] and the Japan Ministry of Health and Welfare[35] for the calendar year 1995 were used for constructing the survival models. Cause-specific mortality for particular age-related diseases was also calculated for the calendar year 1995 from data obtained from the World Health Statistics Annual[36] and Okinawa Prefectural Government, Division of Statistics[37] and age-standardized to the World Standard Population.[36]

RESULTS

To assess whether there was evidence for CR in elderly Okinawans and if so, for what period of time, we analyzed long-term trends in whole-population caloric intake and energy balance for the years 1949–1998. These data demonstrate that the Okinawan population appeared to be in a relative "energy deficit" consistent with CR until the 1960s, eating approximately 10.9% fewer calories than would normally be recommended for maintenance of body weight, according to the Harris–Benedict equation (FIG. 1A).[29] Consistent with adaptation to a long-term energy deficit, the BMI of adult Okinawans remained stable at a very lean level of approximately 21 kg/m^2 until the 1960s. During the 1960s the Okinawan adult BMI began to rise (FIG. 1A). This was coincident with a shift to consistently positive energy balance.

To further clarify whether these population data indeed support a CR state for Okinawans before 1960, analysis of body weight by age strata was performed for the year 1949, when the Okinawan population appeared to be under CR conditions. Studies in nonhuman primates show that there is consistently lower body weight at all ages and relatively small weight gain beyond adulthood in male CR monkeys in contrast to those with *ad libitum* access to food.[8] Comparisons between the effects of CR on body weight in nonhuman primates (FIG. 1B) and Okinawan men in 1949 (FIG. 1C) show several similarities. There

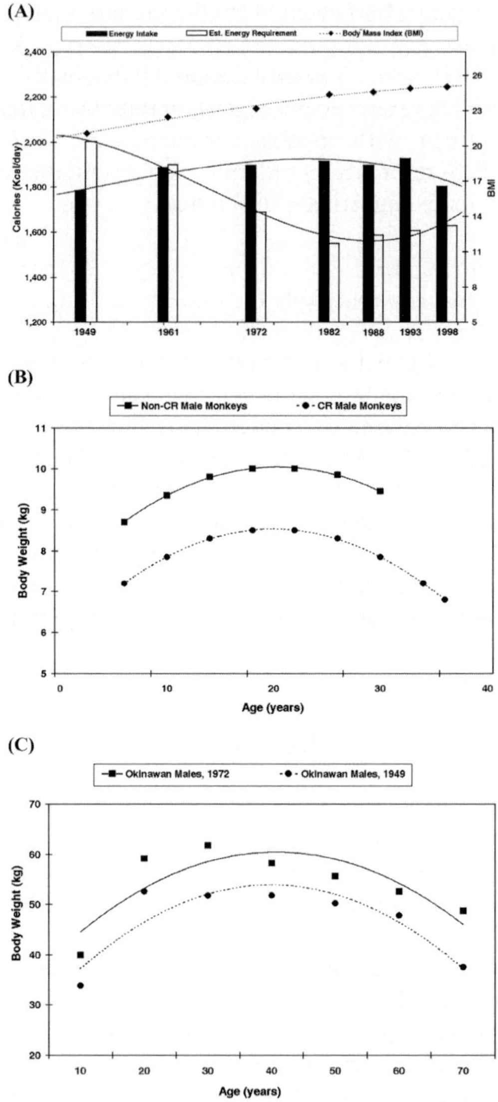

FIGURE 1. Population energy intake, energy expenditure, and anthropometric data for adult Okinawans for the years 1949–1998. (**A**) Data standardized for 50-year-old Okinawans from 1949–1998 show that Okinawans were in negative energy balance (CR) of approximately 11% until the 1960s. As Okinawans transitioned to positive energy balance, BMI began to increase. This was due to a combination of increased energy intake and decreased energy expenditure. BMI peaked at the height of positive energy balance in the mid 1990s. (**B**) Body weight in CR nonhuman primates is markedly lower across all age strata versus non-CR controls.[8] (**C**) Anthropometric data from the year 1949 in Okinawa while under CR conditions demonstrate that body weight is markedly lower across all age strata versus 1972, when energy balance had shifted to non-CR conditions.

was a marked difference in body weight at all ages and relatively small change in body weight across age strata in both groups after peak body weight is achieved at adult ages. In Okinawans during 1949, peak body weight was reached at approximately age 30 years and appeared relatively stable until elderly ages, when it began to decline.

To quantify the degree of CR in the Okinawan population, energy balance calculations are presented in FIGURE 2A. Okinawan whole-population data for the year 1949, when the septuagenarian cohort was aged approximately 30, show an energy intake of 1785 kcal/day and energy requirement estimated at 2003 kcal/day. This energy deficit is quantified in FIGURE 2B as a negative energy balance of −218 kcal/day (10.9%) and is consistent with the very lean adult population BMI of 21.2, thus supporting a CR phenotype. Population data collected in 1972 demonstrate a positive energy balance of 212 calories per day and a corresponding BMI of 23.3, suggesting that the CR phenomenon was largely over for the Okinawan population sometime in the 1960s. For comparative purposes, caloric intake and energy expenditure calculations for the U.S. population using data collected during the years 1971–1974 in the NHANES I study,[28] demonstrate a positive energy balance of 239 calories per day, consistent with a higher population BMI of 25.6 (FIG. 2B).

Since the CR paradigm rests upon "undernutrition" without malnutrition we analyzed whether the traditional Okinawan diet was of sufficient nutritional quality to ensure that widespread nutritional deficiency did not occur and to generate hypotheses as to the impact that other food components might have had on health status (e.g., protein intake, antioxidant vitamins). TABLE 1 presents whole-population dietary intake for the year 1949.[24] Notable is the high intake of vegetables, particularly sweet potatoes and soy, as a principal protein. TABLE 2 presents micronutrient data that suggest that the diet was adequate in most micronutrients. Notable is that intake of the antioxidant vitamins C and E, as well as folate and vitamin B_6 were very high at 289%, 190%, 295% and 221% of recommended intake, respectively, whereas vitamins D, B_2, and B_{12} were quite low at 2%, 45%, and 27% of recommended intake, respectively.

TABLE 3 presents physical examination data from the same subjects from whom dietary information was collected in 1949. Notable is the relatively high prevalence of cheilosis (dry, cracked lips and mouth) at 10.7% of the population. This is consistent with the low consumption of vitamin B_2 (riboflavin) reported in TABLE 2. Notable as well is the relatively high prevalence of delayed menstruation and deficient lactation, consistent with low caloric intake and/or low body fat levels in women.[38,39]

If indeed a CR phenomenon occurred in elderly Okinawans at younger ages, it is theoretically possible that there might be biomarker evidence of CR-linked delayed physiological aging, as suggested by animal data.[8] Therefore, we measured DHEA in 54 Okinawan septuagenarians, who would have undergone CR until at least middle age, according to the previous population data. As

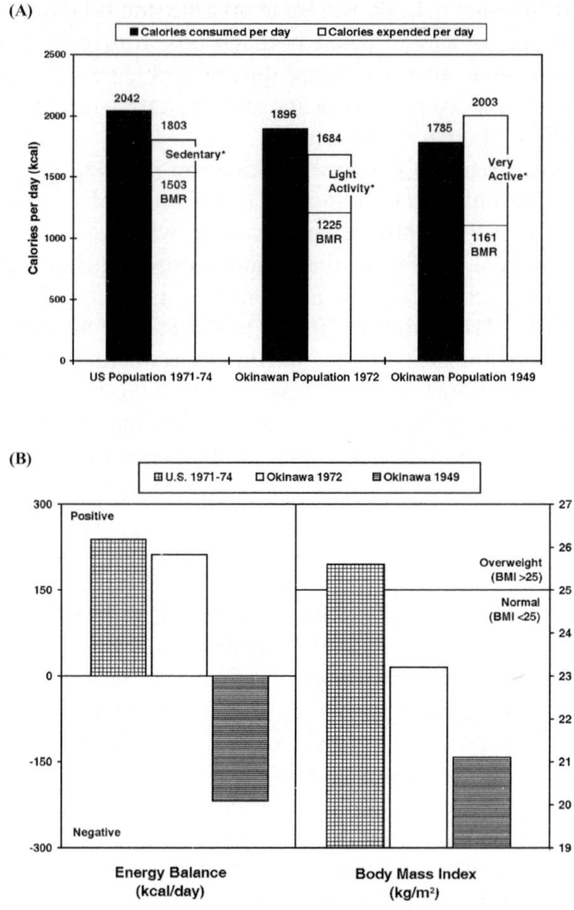

FIGURE 2. Daily energy balance and BMI in Okinawans and Americans (kcal/day). (**A**) Caloric expenditure for various levels of activity is based on reported occupation and activity levels from NHANES I (U.S. National Center for Health Statistics 1978), the Office of the Civil Administrator of the Ryukyu Islands (1949), the U.S. Department of the Office of the Civil Administrator of the Ryukyu Islands (1949), and the National Nutrition Survey, Japan Ministry of Health, Labor and Welfare (1972). BMR = basal metabolic rate (based on sex, height, body weight at age 50, Harris–Benedict equation).[29] Energy balance shifted from negative to positive in Okinawa from 1949 to 1972, supporting an early-life CR phenotype for older Okinawans. By the 1970s, population data from the U.S. and Okinawa standardized to 50 year olds show that both Americans and Okinawans were in positive energy balance. (**B**) Population data from the U.S. and Okinawa show that Okinawans were in negative energy balance of approximately −218 kcal/day in 1949. Both Okinawans and Americans were in positive energy balance in the 1970s. Americans had a positive energy balance of approximately 239 kcal/day in the 1970s, while Okinawans were in positive energy balance of approximately 212 kcal/day. This supports an energy balance shift of approximately 400 kcal/day for Okinawans during these years.

TABLE 1. Traditional dietary intake of Okinawans and other Japanese circa 1950

	Okinawa, 1949[a]	Japan, 1950[b]
Total calories	1785[c]	2068
Total weight (grams)	1262	1057
Caloric density (calories/gram)	1.4	2.0
Total protein in grams (% total calories)	39 (9)	68 (13)
Total carbohydrate in grams (% total calories)	382 (85)	409 (79)
Total fat in grams (% total calories)	12 (6)	18 (8)
Saturated fatty acid	3.7	4.7
Monounsaturated fatty acid	3.6	5.3
Polyunsaturated fatty acid	4.8	8.0
Total fiber (grams)	23	23
Food group	**Weight in grams (% total calories)**	
Grains		
Rice	154 (12)	328 (54)
Wheat, barley, and other grains	38 (7)	153 (24)
Nuts, seeds	<1 (<1)	<1 (<1)
Sugars	3 (<1)	8 (1)
Oils	3 (2)	3 (1)
Legumes (e.g., soy and other beans)	71 (6)	55 (3)
Fish	15 (1)	62 (4)
Meat (including poultry)	3 (<1)	11 (<1)
Eggs	1 (<1)	7 (<1)
Dairy	<1 (<1)	8 (<1)
Vegetables		
Sweet potatoes	849 (69)	66 (3)
Other potatoes	2 (<1)	47 (2)
Other vegetables	114 (3)	188 (1)
Fruit[d]	<1 (<1)	44 (1)
Seaweed	1 (<1)	3 (<1)
Pickled vegetables	0 (0)	42 (<1)
Foods: flavors & alcohol	7 (<1)	31 (2)

[a]Data derived from analysis of U.S. National Archives, archived food records, 1949 and based on survey of 2279 persons.

[b]Japan National Nutrition Survey, 1950.

[c]Total daily caloric intake was originally reported as 1785 kcal/day in 1949. This was estimated to be 17% less than government-recommended daily intake. Differences in assumptions regarding particular foods, cooking methods, and choice of nutritional analysis programs result in a range of 1605 to 2012 kcal/day.

[d]Papaya and tomatoes were classified as vegetables.

predicted by the CR hypothesis, the mean value of DHEA for male septuagenarians in Okinawa was significantly higher at 2.59 ng/mL (95% CI: 2.24–2.94) versus 2.00 ng/mL (95% CI: 1.91–2.10) in non-CR American men ($P < 0.001$) (FIG. 3). Larger differences were seen in similarly aged women from the two populations with levels of 3.03 ng/mL (95% CI: 2.48–3.58) found in Okinawans versus 1.13 ng/mL (95% CI: 1.06–1.20) in Americans ($P < 0.001$) (FIG. 3).

TABLE 2. Micronutrient sufficiency of the traditional Okinawan diet and Japanese diet

	Okinawa, 1949[b]		Japan, 1950[c]	
Micronutrient	Amount	% RDAJ[a]	Amount	% RDAJ[a]
Vitamin A (RE)	602	110	337	62
Vitamin D (mcg)[d]	0.4	2	7.9	31
Vitamin E (mg)	16.6	190	6.3	72
Vitamin K (mcg)	87.6	160	65.8	120
Vitamin B_1: thiamin (mg)	1.4	137	1.1	113
Vitamin B_2: riboflavin (mg)	0.5	45	0.5	47
Niacin (mg)	13.2	93	18.1	127
Vitamin B_6: pyridoxine (mg)	3.0	221	1.6	118
Folate (mcg)	557.4	295	267.2	141
Vitamin B_{12}: cobalamin (mcg)[e]	0.6	27	4.0	176
Vitamin C (mg)	273.4	289	94.9	100
Calcium (mg)[f]	505.3	82	325.5	53
Iron (mg)	11.6	109	11.0	103
Phosphorus (mg)	864.1	115	1191.2	159
Magnesium (mg)	396.1	151	327.4	125
Potassium (mg)	5199.6	272	2712.3	142
Zinc (mg)	6.2	62	10.6	107
Sodium (mg)	1133.0	113	2450.8	245

[a]Micronutrient requirements are from 6th Recommended Dietary Allowances for the Japanese (RDAJ), 1996.

[b]Calculated from U.S. National Archive, archived food records, 1949.

[c]Calculated from the Japan National Nutrition Survey and the Statistics Record of the Ministry of Agriculture, Forestry and Fisheries, Government of Japan, 1950.

[d]Okinawa is located at a subtropical (26.4°) latitude and subjects would likely have manufactured enough *in vivo* vitamin D from sunlight to meet RDAJ.[70]

[e]Periodic festivals (approximately monthly) in which pork and other meats were consumed are not accounted for in this analysis.

[f]Okinawan drinking water is high in calcium and other minerals, which are not accounted for in this analysis.

In addition, if delayed aging occurred in Okinawans, there may be a rightward shift in the survival curve as seen in animal studies,[3] with increases in both average life span (defined here as age at death for 50th percentile of population) and maximum life span (defined here as age at death of 99th percentile of survival). FIGURE 4 displays survival curves for Okinawan, Japanese, and U.S. populations for the year 1995. These data show increases in both average and maximum life span in the Okinawan population compared to Japanese and American populations, consistent with CR. Average life span and maximum life span in the Okinawan, Japanese, and U.S. populations was 83.8 and 104.9 years, 82.3 and 101.1 years, and 78.9 and 101.3 years, respectively. These data are based on a conservative survival model (LIFEREG Procedure) and differences are even greater in maximum life span between Okinawans and Japanese and/or Americans using a Weibull survival model (data not shown).

TABLE 3. Nutritional deficiency symptoms in Okinawa, 1949

	Urban	Rural	All Okinawa
Number of persons examined	797	1029	1826
Persons with no symptoms (%)	74.3	81.6	80.5
Persons with one or more symptoms (%)	25.7	18.4	19.5
Anemia	3.1	1.1	1.4
Hyperkeratosis	0.4	0.7	0.7
Xerophthalmia	–	0.7	0.6
Cheilosis	14.1	10.1	10.7
Glossitis	2.0	0.7	0.9
Loss of knee jerk	5.5	2.7	3.1
Edema	0.5	1.6	1.5
Chronic diarrhea	0.9	0.9	0.9
Bradycardia	0.8	1.8	1.6
Delayed menarche[a]	4.7	10.2	9.4
Deficient lactation[a]	5.6	20.0	17.8

[a]In women only.

Finally, since CR induces profound reductions in risk for chronic disease and increases the age at onset of chronic diseases,[2] we summarize age-adjusted mortality data for specific age-related diseases in FIGURE 5. Coronary heart disease, and forms of cancer, such as lymphoma, and cancer of the prostate, breast, and colon are remarkably low in age-matched Okinawans versus other Japanese and Americans. Since the forces of mortality act most strongly at older ages, these mortality differences reflect mainly the mortality experience of the older Okinawan cohort, which appears to have been subjected to mild CR.

DISCUSSION

The older cohort of Okinawans (aged 65-plus) is remarkable in many ways. Of particular interest is that they possess among the highest functional capacity and the longest survival in Japan, the country with world's longest-lived population. Life expectancy at birth for the year 2000 was 86.0 years for Okinawan women and 77.6 years for Okinawan men, respectively. Life expectancy for the septuagenarian cohort (life expectancy at age 65) is the highest in Japan, and possibly the world, at 24.1 years for females and 18.5 years for males, respectively.[40] This compares to 22.5 years and 17.6 years for the same birth cohort in mainland Japan and 19.3 years and 16.2 years for the corresponding U.S. birth cohort of females and males, respectively.[41]

The question addressed by this article is whether or not CR is linked to this phenomenon. Since this is an epidemiologic study, reporting mainly population data, the results must be viewed as hypothesis-generating and not conclusive. The question of whether CR or some other unknown factor, such as genetics,

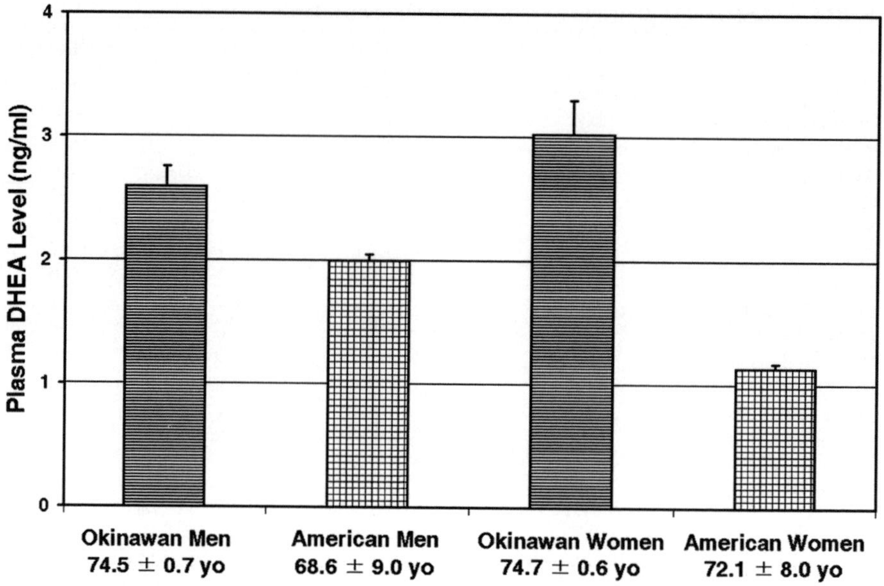

FIGURE 3. Plasma DHEA levels (ng/mL) in elderly Okinawans and Americans. DHEA was significantly higher in older Okinawan men ($P < 0.001$) and Okinawan women ($P < 0.001$) versus older American men and women, respectively. yo = years old.

is responsible for the healthy longevity of the Okinawans are hypotheses that require further testing. Nevertheless, this study is supportive of at least a partial role for CR in Okinawan longevity and is significant for several reasons.

One, this is the first in-depth epidemiologic study of long-term CR in the Okinawan population, which has been cited as a potential human example of the effects of CR on aging and age-related disease.[18] Two, this study is the only study to report population-based dietary and phenotypic data over the adult life span for the older cohort of Okinawans. Previous studies reported on small selected samples of Okinawans, which are more subject to bias, and reported data at a single point in time.[18,20] Three, this study supports and extends earlier observations from the 1970s regarding the potential impact of CR on Okinawan health and longevity.[18] Okinawans indeed appear to be one of the few populations in the world that may have experienced mild long-term CR without significant malnutrition, and this may be linked to their exceptionally healthy survival. That this phenomenon appears to have occurred in their natural environment also makes them a population of special interest.

Four, there is a dearth of research on factors that may affect the rate of human aging and controversy on how exactly to measure this. While there are no universally accepted biomarkers of aging,[14] this study adds human data that are consistent with the literature in animals with regard to the effects of CR on DHEA levels.[8,27] Furthermore, this is the first study that has shown extended

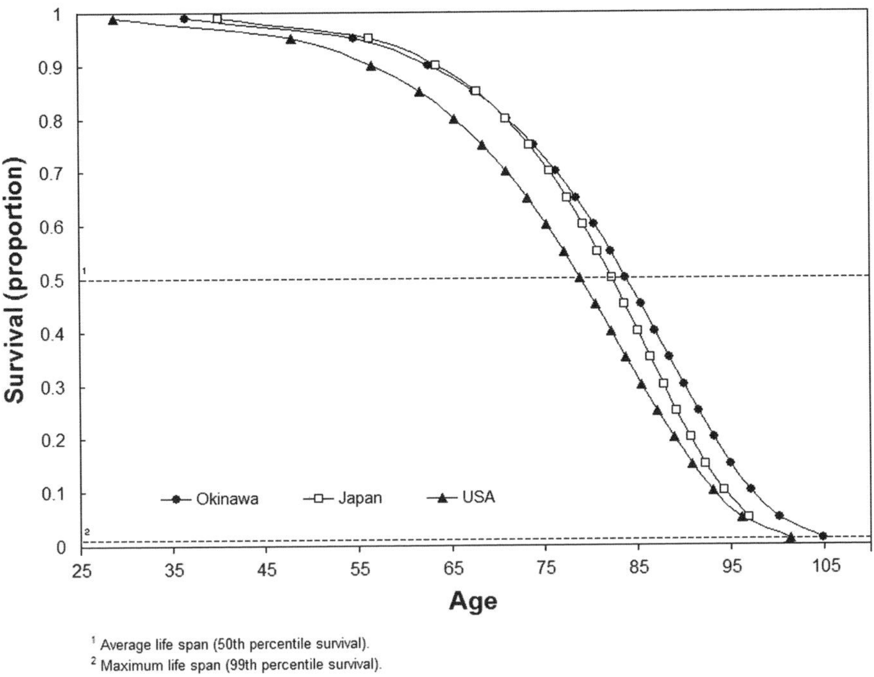

FIGURE 4. Survival model for Okinawans, Japanese, and American populations. Data show extended life span (average, maximum) for Okinawans (83.8 years, 104.9 years) compared to other Japanese (82.3 years, 101.1 years), and Americans (78.9 years, 101.3 years).

average *and* maximum life span in a human population that is potentially due to CR, as indicated by a rightward shift of the survival curve for Okinawans. While mortality due to age-related disease was much lower in the Okinawans and helped lead to a higher average life span, the apparent increase in maximum life span is suggestive of slower aging in the Okinawans. Finally, this study population has among the longest life expectancy in the world, and includes what may be the world's highest prevalence of exceptionally aged individuals, such as centenarians, and therefore is an important population for the study of environmental and genetic factors that may predict or lead to exceptional survival.[42]

There are several strengths to this study. One, this is among the most comprehensive studies of dietary and phenotypic data related to CR in Okinawa to date and the first to report population-based energy balance and phenotypic data in older Okinawan adults before the 1970s, when they were young adults. This is important for establishing a link between CR and longevity in Okinawa since it appears that only older Okinawans have a survival advantage compared to other Japanese.[22] Two, a sample of fairly well-defined subjects from the older cohort of Okinawans was available for biomarker

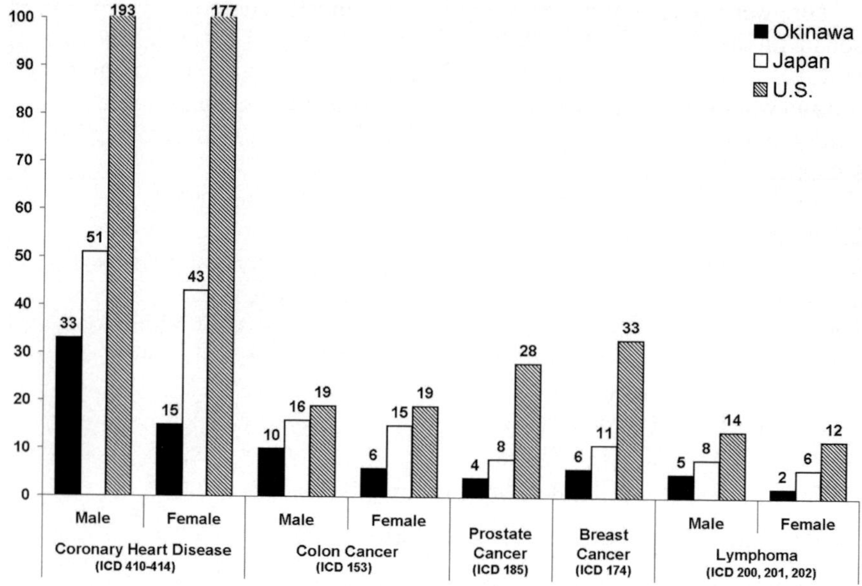

FIGURE 5. Mortality from age-associated diseases in Okinawans versus Americans. Numbers represent age-adjusted mortality rate in deaths per hundred thousand persons per year for 1995. Coding was according to ICD-9 codes; populations were age-adjusted to World Standard Population. These data show markedly lower mortality risk from age-related diseases in Okinawans versus other Japanese and Americans.

(DHEA) study, and both the Okinawan and American samples were selected from specific populations of healthy septuagenarians in geographically defined communities living in their natural environments. Three, while exact dietary data are not known for these individuals, population data are otherwise known for their birth cohorts, collected with good dietary methodology. While these are only crude measurements of energy balance, the population data support an energy deficit in Okinawans (but not Americans) that would require physiological adaptations consistent with a CR phenotype.

Four, anthropometric data, such as BMI, are consistent with a large difference in energy balance between the study populations and support a CR status for Okinawan septuagenarians at younger ages. The study cohort to which these Okinawans belong was relatively thin and appeared to have been calorically restricted at least until middle age. Support for a CR phenotype for members of this older cohort can also be derived from later-life measurements of their BMI, which was 23.5 for males and 24.1 for females when aged approximately 75 years.[20] When corrected for potential loss of height with aging[43] the true BMI for these subjects may be closer to 22.4 for males and 22.0 for females. This low BMI, compared with an average BMI of 21.2 in 1949 at approximately age 30, suggests minimal weight gain with age that is also consistent with CR.

There were several limitations to this study, mostly centered on the comparisons with the U.S. study population. However, this study is mainly concerned with the cumulative evidence that CR occurred in the older Okinawan population and whether there is evidence that CR-linked outcomes may have occurred, including phenotypic outcomes, lower mortality from age-associated diseases, and extended survival.

Limitations include the fact that population data are used to estimate whether this older generation of Okinawans were calorically restricted and these data are subject to measurement error. Clearly it is not possible to maintain an energy deficit in the long term and still be healthy. Nevertheless, the low BMI from the Okinawan population until the 1960s is consistent with a long-term adaptive response to limited energy availability and periodic energy deficits. These energy deficits are supported by historical reports of periodic crop failures that occurred in Okinawa in the early 20th century and a long history of marginal food supply.[44] A CR status appears to have remained until at least middle age for Okinawan septuagenarians since dietary and physical activity patterns remained relatively consistent in Okinawa until the 1960s (FIG. 1A). After the 1960s standards of living improved, physical activity declined, and Westernization (increased fat, meat and bread) and Japanization (increased polished Japanese or "Japonica" white rice) of the diet occurred, with large shifts in energy balance.

Another weakness of this study centers on interpretation of the DHEA levels. Since the populations were of different ethnicity and the Okinawans come from an island population with restricted gene flow, there may be genetic or other unique features that account, at least in part, for the differences in DHEA levels and health status of the populations.[45] However, we could find no published data on DHEA in other Japanese populations, and the only prior reported comparisons of Japanese and Americans involved dehyroepiandrosterone sulfate (DHEAS). These data demonstrate that Japanese men actually have lower DHEAS levels than American men when younger.[46] Therefore, persistently higher life-long DHEA levels in Okinawans compared with Americans is unlikely to account for the older-age differences in DHEA we found in this study. More likely is a slower age-related decline in DHEA in the Okinawans, consistent with CR.

In support of this, Nafziger *et al.*[47] reported DHEA levels from a study of community-dwelling white Americans that were not markedly different from the values for the comparison group in this study (Rancho Bernardo cohort). Plasma levels were reported as 1.6 ng/mL for septuagenarians of both sexes—lower than Okinawans of similar age. In addition, in a separate study of plasma estrogen levels in these same subjects, similar differences were observed between Okinawan and U.S. populations, with significantly higher levels of estrogen seen in Okinawans at older ages.[48] Estrogen, as a downstream byproduct of DHEA, usually reflects DHEA's plasma level (i.e., high DHEA means relatively high estrogen).

Finally, other factors might be responsible for the mortality advantages and phenotypic differences between Okinawans, other Japanese, and Americans. These include differences in dietary macronutrients, such as differences in protein intake or amino acid composition of the diet, which has been linked to longevity in rodents[49,50] and fruit flies[51] and to stroke risk in animals and humans,[52,53] lower glycemic index carbohydrates in the Okinawans, which may lower cardiovascular and cancer risk,[54] higher consumption of flavonoids (e.g., soy foods and Okinawan sweet potatoes),[55] which may stimulate sirtuin proteins and have been implicated as CR mimetics,[56] higher intake of antioxidant-rich vegetables, lower intake of sodium and higher K/Na ratio (both implicated in lower blood pressure),[57] or multiple other dietary differences.

Chronic low caloric intake and chronic low protein and/or low intake of particular amino acids (e.g., methionine) may also be chronic low-intensity stressors and may have contributed to a phenomenon known as "hormesis," where low levels of otherwise damaging agents have positive, potentially life-extending effects.[6,58]

Other factors, such as genetic differences[41,59,60] and psychosocial factors,[61] are important survival factors that have not been accounted for in this study. In particular, recent work suggests that Okinawans may possess familial survival advantages that account, at least in part, for their remarkable longevity.[42] On the other hand, fuller expression of the potential health and longevity benefits of CR in Okinawans would have been limited by the lack of a good public health infrastructure until the 1960s, and resultant high death rates due to tuberculosis and other infectious diseases.[62] Furthermore, of the nutritional factors, only CR has been found in animal studies capable of consistently increasing both average *and* maximum life span, indicated by a rightward shift of the survival curve.[63]

Thus, the septuagenarian cohort of Okinawans appears to have experienced CR of approximately 11% with respect to their estimated energy requirements, until middle age. Older cohorts in Okinawa may have experienced CR for even longer or at greater levels, consistent with their increased survival advantage compared with that of other Japanese, although estimates of the degree of CR should be viewed with caution since they were obtained from energy balance estimates for the whole Okinawan population and were not validated by precise metabolic methods, such as indirect calorimetry or doubly labeled water.

The CR status of the older Okinawans appears to be due, in part, to a high level of occupational energy demand from their main vocation as farmers. This was coupled to low caloric intake from an energy-poor but nutrient-dense diet rich in Okinawan sweet potatoes, other vegetables, legumes, and other foods low in energy density (FIG. 1).[24] The Okinawan sweet potato, with a caloric density of 1.0 kcal/gram, has been the main carbohydrate of the Okinawan diet from the 1600s until approximately 1960, accounting for more than 50% of calories.[64] The higher caloric density of the traditional Japanese diet is partly due to their higher consumption (75% of calories) of high caloric density Japonica (white) rice, with a caloric density of 1.5 kcal/gram.

The CR phenotype appears to be disappearing in Okinawa except among older birth cohorts (aged 65-plus years). Calorie intake has been increasing and activity levels decreasing since the 1960s with concomitant population-wide increases in BMI, although this trend appears to have stabilized after reversion of Okinawa from U.S. to Japanese governance in the 1970s (FIG. 1A). Loss of the CR phenotype among younger Okinawans and increased weight gain in this cohort is also associated with higher mortality from obesity-linked diseases. This can be inferred from a mortality crossover, which has followed a BMI crossover in Okinawans versus other Japanese.[22] Concomitantly, there has been a significant slowing of life expectancy gains over the last two decades. This may, as is hypothesized in Americans,[65] have an unfavorable impact upon the health and longevity of subsequent generations.

Early-life energy deficits experienced by older Okinawans would have required compensatory responses in terms of more efficient use of energy[66] since long-term energy deficits are unsustainable and inconsistent with health. These adaptations may have included less body fat, lower blood sugar and insulin levels, hormetic responses, and multiple changes in gene expression, among other adaptations potentially linked to longer life span.[3–6,8,67–69] Animal studies suggest that CR at any stage of the life cycle (early or later life) may potentially result in mortality advantages versus non-CR controls.[4]

In conclusion, we observed low calorie intake coupled with high physical activity levels that appear to have contributed to a CR phenotype in older Okinawans. This phenotype includes a life-long low BMI, relatively high plasma levels of DHEA at older ages, reduced mortality from age-associated diseases, and extended average and maximum survival. While these conclusions are tentative in nature, an adaptive response to early and mid-life energy restriction in the older cohort of Okinawans may be implicated in their low morbidity and exceptionally long survival. This is consistent with the well-known animal literature that supports a beneficial effect of CR on BMI, age-related biomarkers, morbidity/mortality, and life span. More studies using the CR paradigm (energy restriction without malnutrition) are required to validate the effects of low energy intake and/or energy balance in human populations.

ACKNOWLEDGMENTS

This work was supported by the National Institute on Aging (Grants RO3 AG021293-01, K08 AG22788-02 and R01 AG027060-01 to B.W.) and the Japan Ministry of Health, Labor and Welfare. All authors contributed to the design of the experiment, collection of data, analysis of data, and writing of the manuscript. We thank Sayaka Mitsuhashi and Joshua Spiegelman for research assistance and manuscript preparation.

REFERENCES

1. McCay, C.M., M.F. Crowell & L.A. Maynard. 1935. The effect of retarded growth upon the length of life span and upon the ultimate body size. Nutrition **5:** 155–171.
2. Maeda, H., C.A. Gleiser, E.J. Masoro, *et al.* 1985. Nutritional influences on aging of Fischer 344 rats: II. Pathology. J. Gerontol. **40:** 671–688.
3. Weindruch, R., R.L. Walford, S. Fligiel, *et al.* 1986. The retardation of aging in mice by dietary restriction: longevity, cancer, immunity and lifetime energy intake. J. Nutr. **116:** 641–654.
4. Dhahbi, J.M., H.J. Kim, P.L. Mote, *et al.* 2004. Temporal linkage between the phenotypic and genomic responses to caloric restriction. Proc. Natl. Acad. Sci. USA **101:** 5524–5529.
5. Masoro, E.J. 2005. Overview of caloric restriction and ageing. Mech. Ageing Dev. **126:** 913–922.
6. Masoro, E.J. 2006. Caloric restriction and aging: controversial issues. J. Gerontol. A. Biol. Sci. Med. Sci. **61:** 14–19.
7. Dirks, A.J. & C. Leeuwenburgh. 2006. Caloric restriction in humans: potential pitfalls and health concerns. Mech. Ageing Dev. **127:** 1–7.
8. Mattison, J.A., M.A. Lane, G.S. Roth, *et al.* 2003. Caloric restriction in rhesus monkeys. Exp. Gerontol. **38:** 35–46.
9. Lane, M.A., J.A. Mattison, G.S. Roth, *et al.* 2004. Effects of long-term diet restriction on aging and longevity in primates remain uncertain. J. Gerontol. A. Biol. Sci. Med. Sci. **59:** 405–407.
10. Lee, I.M., S.N. Blair, D.B. Allison, *et al.* 2001. Epidemiological data on the relationships of caloric intake, energy balance, and weight gain over the life span with longevity and morbidity. J. Gerontol. A. Biol. Sci. Med. Sci. **56:** 7–19.
11. Walford, R.L., D. Mock, R. Verdery, *et al.* 2002. Caloric restriction in biosphere 2: alterations in physiologic, hematologic, hormonal, and biochemical parameters in humans restricted for a 2-year period. J. Gerontol. A. Biol. Sci. Med. Sci. **57:** B211–B224.
12. Heilbronn, L.K. & E. Ravussin. 2003. Calorie restriction and aging: review of the literature and implications for studies in humans. Am. J. Clin. Nutr. **78:** 361–369.
13. Fontana, L., T.E. Meyer, S. Klein, *et al.* 2004. Long-term calorie restriction is highly effective in reducing the risk for atherosclerosis in humans. Proc. Natl. Acad. Sci. USA **101:** 6659–6663.
14. Butler, R.N., R. Sprott, H. Warner, *et al.* 2004. Biomarkers of aging: from primitive organisms to humans. J. Gerontol. A. Biol. Sci. Med. Sci. **59:** B560–B567.
15. Meyer, T.E., S.J. Kovacs, A.A. Ehsani, *et al.* 2006. Long-term caloric restriction ameliorates the decline in diastolic function in humans. J. Am. Coll. Cardiol. **17:** 398–402.
16. Fontana, L. & S. Klein. 2007. Aging, adiposity, and calorie restriction. JAMA **297:** 986–994.
17. Willcox, B.J., K. Yano, R. Chen, *et al.* 2004. How much should we eat? The association between energy intake and mortality in a 36-year follow-up study of Japanese American men. J. Gerontol. A. Biol. Sci. Med. Sci. **59:** 789–795.
18. Kagawa, Y. 1978. Impact of Westernization on the nutrition of Japanese: changes in physique, cancer, longevity and centenarians. Prev. Med. **7:** 205–217.

19. HOKAMA, T., H. ARAGAKI, H. SHO, *et al.* 1967. Nutrition survey of school children in Okinawa. Sci. Bull. Coll. Agr. Univ. Ryukyus **14:** 1–15.
20. CHAN, Y.C., M. SUZUKI & S. YAMAMOTO. 1997. Dietary, anthropometric, hematological and biochemical assessment of the nutritional status of centenarians and elderly people in Okinawa, Japan. J. Am. Coll. Nutr. **16:** 229–235.
21. JAPAN MINISTRY OF HEALTH, LABOR AND WELFARE. 2002. Results from Ministry of Health, Labor and Welfare Nutrition Survey, 2000. Daiichi Publishers. Tokyo, Japan.
22. TODORIKI, H., D.C. WILLCOX & B.J. WILLCOX. 2004. The effects of post-war dietary change on longevity and health in Okinawa. Okinawa J. Amer. Studies **1:** 52–61.
23. WILLCOX, D.C. 2005. Okinawan longevity: where do we go from here? Nutr. Dietetics **8:** 9–17.
24. U.S. DEPARTMENT OF THE OFFICE OF THE CIVIL ADMINISTRATOR OF THE RYUKYU ISLANDS: RECORDS OF HEALTH, EDUCATION AND WELFARE. 1949. U.S. Occupation Headquarters, World War II. Record Group 260.12.5. National Archives at College Park, 8601 Adelphi Road, College Park, MD 20740-6001.
25. DIVISION OF HEALTH, LABOR AND WELFARE, OKINAWA PREFECTURAL GOVERNMENT. 1976. Health and Welfare Dataset 1880–1976. Okinawa Prefectural Government. Okinawa.
26. LANE, M.A., D.K. INGRAM, S.S. BALL, *et al.* 1997. Dehydroepiandrosterone sulfate: a biomarker of primate aging slowed by caloric restriction. J. Clin. Endocrinol. Metab. **82:** 2093–2096.
27. ROTH, G.S., M.A. LANE, D.K. INGRAM, *et al.* 2002. Biomarkers of caloric restriction may predict longevity in humans. Science **297:** 811.
28. NATIONAL CENTER FOR HEALTH STATISTICS. 1978. National Health and Nutrition Examination Survey I Dataset. 1971–74. Vital and Health Statistics. U.S. Government Printing Office. Washington, DC.
29. FRANKENFIELD, D.C., E.R. MUTH & W.A. ROWE. 1998. The Harris-Benedict studies of human basal metabolism: history and limitations. J. Am. Diet Assoc. **98:** 439–445.
30. SUZUKI, M. & N. HIROSE. 1999. Endocrine function of centenarians. *In* Japanese Centenarians. Medical Research for the Final Stages of Human Aging. H. Tauchi, T. Sato & T. Watanabe, Eds.: 101–110. Aichi Medical University Press. Aichi, Japan.
31. SANABE, E., I. ASHITOMI & M. SUZUKI. 1977. Social and medical survey of centenarians. Okinawa J. Pub. Health **9:** 98–106.
32. GREENDALE, G.A., S. EDELSTEIN & E. BARRETT-CONNOR. 1997. Endogenous sex steroids and bone mineral density in older women and men: the Rancho Bernardo Study. J. Bone. Miner. Res. **12:** 1833–1843.
33. SAS INSTITUTE INC. 1999. The LIFEREG Procedure, SAS/STAT® User's Guide, Version 8. SAS Institute Inc. Cary, NC.
34. NATIONAL CENTER FOR HEALTH STATISTICS. 1998. Vital Statistics of the United States, 1995, preprint of Vol. II, Mortality, part A, section 6, life tables. National Center for Health Statistics. Hyattsville, MD.
35. JAPAN HEALTH AND WELFARE STATISTICS ASSOCIATION. 1996. 1995 Annual Statistical Report of National Health Conditions. Tokyo, Japan.
36. WORLD HEALTH ORGANIZATION. 1996. 1995 World Health Statistics Annual. WHO. Geneva.
37. OKINAWA PREFECTURAL GOVERNMENT, DIVISION OF STATISTICS. 1996. 39th Okinawa Statistics Annual, 1995. Okinawa, Japan.

38. MEYER, F., J. MOISAN, D. MARCOUX, *et al.* 1990. Dietary and physical determinants of menarche. Epidemiology **1:** 377–381.
39. DEWEY, K.G. 1998. Effects of maternal caloric restriction and exercise during lactation. J. Nutr. **128**(2 Suppl): 3865–3895.
40. JAPAN MINISTRY OF HEALTH, LABOR AND WELFARE. 2000. Prefectural Life Tables. Statistics and Information Department, Health and Welfare Statistics Association. Tokyo, Japan.
41. CENTERS FOR DISEASE CONTROL AND PREVENTION, NATIONAL CENTER FOR HEALTH STATISTICS. 2003. National Vital Statistics System. NCHS. Hyattsville, MD. Available at http://www.cdc.gov/nchs/.
42. WILLCOX, B.J., D.C. WILLCOX, Q. HE, *et al.* 2006. Siblings of Okinawan centenarians exhibit lifelong mortality advantages. J. Gerontol. A. Biol. Med. Sci. **61:** 345–354.
43. SORKIN, J.D., D.C. MULLER & R. ANDRES. 1999. Longitudinal change in height of men and women: implications for interpretation of the body mass index: the Baltimore Longitudinal Study of Aging. Am. J. Epidemiol. **150:** 969–977.
44. KERR, G. 2000. Okinawa: the History of an Island People. Tuttle Publishing. Boston, MA.
45. ROTTER, J.I., F.L. WONG, E.T. LIFRAK, *et al.* 1985. A genetic component to the variation of dehydroepiandrosterone sulfate. Metabolism **34:** 731–736.
46. LACROIX, A.Z., K. YANO & D.M. REED. 1992. Dehydroepiandrosterone sulfate, incidence of myocardial infarction, and extent of atherosclerosis in men. Circulation **86:** 1529–1535.
47. NAFZIGER, A.N., S.J. BOWLIN, P.L. JENKINS, *et al.* 1998. Longitudinal changes in dehydroepiandrosterone concentrations in men and women. J. Lab. Clin. Med. **131:** 316–323.
48. WILLCOX, B.J., D.C. WILLCOX, M. SUZUKI, *et al.* 2000. Serum estrogen and long term survival in Okinawan-Japanese men and women. J. Am. Geriatr. Soc. **48:** P348.
49. ZIMMERMAN, J.A., V. MALLOY, R. KRAJCIK, *et al.* 2003. Nutritional control of aging. Exp. Gerontol. **38:** 47–52.
50. SANZ, A., P. CARO & G. BARJA. 2004. Protein restriction without strong caloric restriction decreases mitochondrial oxygen radical production and oxidative DNA damage in rat liver. J. Bioenerg. Biomembr. **36:** 545–552.
51. MAIR, W., M.D. PIPER & L. PARTRIDGE. 2005. Calories do not explain extension of life span by dietary restriction in *Drosophila*. PloS. Biol. **3:** e223.
52. ABBOTT, R.D., J.D. CURB, B.L. RODRIGUEZ, *et al.* 1996. Effect of dietary calcium and milk consumption on risk of thromboembolic stroke in older middle-aged men: the Honolulu Heart Program. Stroke **27:** 813–818.
53. YAMORI, Y., S. MURAKAMI, K. IKEDA, *et al.* 2004. Fish and lifestyle-related disease prevention: experimental and epidemiological evidence for anti-atherogenic potential of taurine. Clin. Exp. Pharmacol. Physiol. **31:** S20–S23.
54. JENKINS, D.J., C.W. KENDALL, A. MARCHIE, *et al.* 2004. Too much sugar, too much carbohydrate, or just too much? Am. J. Clin. Nutr. **79:** 711–712.
55. WILLCOX, B.J., K. FUCHIGAMI, D.C. WILLCOX, *et al.* 1995. Isoflavone intake in Japanese and Japanese-Canadians. Am. J. Clin. Nutr. **61:** 901.
56. WOOD, J.G., B. ROGINA, S. LAVU, *et al.* 2004. Sirtuin activators mimic caloric restriction and delay ageing in metazoans. Nature **430:** 686–689.
57. BRAY, G.A., W.M. VOLLMER, F.M. SACKS, *et al.* 2004. A further subgroup analysis of the effects of the DASH diet and three dietary sodium levels on blood pressure: results of the DASH-Sodium Trial. Am. J. Cardiol. **94:** 222–227.

58. MASORO, E.J. 1998. Hormesis and the antiaging action of dietary restriction. Exp. Gerontol. **33:** 61–66.
59. TAKATA, H., M. SUZUKI, T. ISHII, *et al.* 1987. Influence of major histocompatibility complex region genes on human longevity among Okinawan-Japanese centenarians and nonagenarians. Lancet **2:** 824–826.
60. AKISAKA, M., M. SUZUKI & H. INOKO. 1997. Molecular genetic studies on DNA polymorphism of the HLA class II genes associated with human longevity. Tissue Antigens **50:** 489–493.
61. GOTO, A., S. YASUMURA, Y. NISHISE, *et al.* 2003. Association of health behavior and social role with total mortality among Japanese elders in Okinawa, Japan. Aging Clin. Exp. Res. **15:** 443–450.
62. WILLCOX, B.J., D.C. WILLCOX & M. SUZUKI. 2006. Exceptional human longevity. *In* Aging and Age-Related Diseases. The Basics. M. Karasek, Ed.: 459–509. Nova Science Publishers. New York, NY.
63. MASORO, E.J. 1990. Assessment of nutritional components in prolongation of life and health by diet. Proc. Soc. Exp. Biol. Med. **193:** 31–34.
64. SHO, H. 2001. History and characteristics of Okinawan longevity food. Asia Pacific J. Clin. Nutr. **10:** 159–164.
65. OLSHANSKY, S.J., D.J. PASSARO, R.C. HERSHOW, *et al.* 2005. A potential decline in life expectancy in the United States in the 21st century. N. Engl. J. Med. **352:** 1135–1137.
66. SHRAUWEN, P., K. WALDER & E. RAVUSSIN. 1999. Human uncoupling proteins and obesity. Obes. Res. **7:** 97–105.
67. LEE, C.K., R.G. KLOPP, R. WEINDRUCH, *et al.* 1999. Gene expression profile of aging and its retardation by caloric restriction. Science **285:** 1390–1393.
68. BLUHER, M., B.B. KAHN & C.R. KAHN. 2003. Extended longevity in mice lacking the insulin receptor in adipose tissue. Science **299:** 572–574.
69. SPEAKMAN, J.R., D.A. TALBOT, C. SELMAN, *et al.* 2004. Uncoupled and surviving: individual mice with high metabolism have greater mitochondrial uncoupling and live longer. Aging Cell **3:** 87–95.
70. HOLICK *et al.* 1996. Vitamin D and bone health. J. Nutr. **126:** 1159S–1164S.

Evidence-Based Practice

Limitations and Successful Implementation

RHONDA NAY AND DEIRDRE FETHERSTONHAUGH

ACEBAC, La Trobe University, Melbourne, Victoria 3083, Australia

ABSTRACT: Evidence-based practice (EBP) has been heralded as the most appropriate way of ensuring that patients receive the most effective care possible. It does, however, have several limitations. It is also not enough just to locate and evaluate the evidence without implementing it. This paper discusses EBP and its limitations and potential barriers; it also suggests strategies for changing current practices to more evidence-based ones according to an accurate understanding of the concept of EBP.

KEYWORDS: evidence-based practice (EBP); barriers, limitations; implementation; strategies for change

INTRODUCTION TO EVIDENCE-BASED PRACTICE

Evidence-based practice (EBP) is "the conscientious, explicit and judicious use of current best evidence in making decisions about the care of individual patients" (p. 71).[7] It involves the best available external clinical evidence from systematic research, clinical judgment, and expertise,[1] and individual patients' preferences.[2–4] "Optimal decision making is based on clinical data, clinical acumen, and individual patients' characteristics" (p. 356).[5] EBP should begin and end with the patient. Therefore, the goal of EBP is to translate clinical research into the best possible patient care. It is highly unlikely that patient-desired outcomes will be achieved if research evidence is applied "without due regard for clinical expertise, the patient or the family's wants or needs, and the situational context" (p. 211).[6] Potentially, the appropriate use of EBP can reduce inconsistency in care and support clinical effectiveness and therefore can assist clinicians in justifying and improving the quality of care they provide.[7] In nursing, best research evidence refers to "methodologically sound, clinically relevant research about the effectiveness and safety of nursing

Address for correspondence: Rhonda Nay, ACEBAC, La Trobe University, 1231 Plenty Road, Bundoora, Melbourne, Victoria 3083, Australia. Voice: +61 3 495 3141; fax: +61 3 495 3154. r.nay@latrobe.edu.au

Ann. N.Y. Acad. Sci. 1114: 456–463 (2007).
doi: 10.1196/annals.1396.038

interventions, the accuracy and precision of nursing assessment measures, the power of prognostic markers, the strength of causal relationships, the cost-effectiveness of nursing interventions, and the meaning of illness or patient experiences" (p. 4).[8] According to Thompson,[9] EBP has five key processes: converting healthcare problems into focused healthcare questions; searching for the best available research evidence; critically appraising the evidence retrieved; implementing the evidence; and auditing the implementation of the evidence.

LIMITATIONS AND BARRIERS TO EBP

The ideal of EBP at first glance may seem unproblematic. However, it does have several limitations. These are due fundamentally to the way in which standards of "evidence" are prioritized and the types of research that are funded and subsequently undertaken from which this evidence is derived.[10] There is a reliance on the randomized controlled trial (RCT) as the "gold standard" and therefore an emphasis on meta-analyses and systematic reviews of multiple studies with effectiveness as the ultimate outcome. There is usually not a focus on the feasibility and applicability of the evidence to the specific clinical and individual context. There are several fundamental issues that arise from such a bias. With such a reliance on the RCT, the evidence from which decisions are based has often come from studies with highly homogenous samples. Findings from a study of antihypertensive medication that only enrolled males aged between 40 and 60 years of age, and whose only diagnosed disease was high blood pressure, cannot be applied to frail elderly patients with multiple co-morbidities.[5] Older people are especially under-represented in clinical trials. Of the clinical trials reported in the English-language articles cited in Medline in 2000, only 3% involved people over 65 years,[11] and yet in most developed countries older people make up a much larger percentage of the population (in 2005 13.1% of the Australian population were aged 65 years or older[12]) and are high users of health care. Patients themselves also have little influence over the priorities of research.[13]

There is also the issue that for many clinical areas, and this is most pertinent in the area of nursing practice, evidence, especially RCTs, is lacking.[4] It would be unethical to undertake RCTs in some areas as patients cannot be deprived of care and interventions that have proven to have some efficacy; therefore the evidence that would be designated as "best" can never be gathered. RCTs in many situations are not the most appropriate way in which to obtain evidence. For example, a RCT will never provide the appropriate evidence in relation to a person's experience of living on dialysis or of coping with rheumatoid arthritis.

There is also an assumption that by relying on evidence derived from quantitative data that this information is "objective" and therefore true. Such an assumption does not take into account the social, political, and economic contexts in which research is undertaken. Funding bias, whereby studies are sponsored by, for example, pharmaceutical companies or large private corporations, means that certain interventions are never researched.[14] Research has found that industry-supported research is not always transparent, may not highlight methodological limitations, and often reports more favorable results.[15–17] "Publication bias, in which certain types of data are knowingly kept from publication, can also distort the dissemination of research data" (p. 296).[14]

The limitations discussed above are related to the type of research that is funded and undertaken on which the evidence is based. There are, however, other impediments to EBP. Clinical staff may be unable to interpret or access the evidence or to implement the practice that is derived from it. Several barriers to the implementation of evidence into nursing practice have been identified: lack of time; lack of availability of pertinent research; limited access to journals; little experience and low comfort levels with libraries and search techniques; limited ability to understand and interpret reports; limited exposure to research-use techniques during education; negative staff attitudes toward research; lack of administrative support; cost constraints; concerns about reallocation of staff time; and lack of support and leadership from nurse–managers to implement evidence-based findings.[7,18–21] EBP may also lack support where potential users do not fully understand the integral nature of patient choice and clinician experience; in such situations EBP is criticized for ignoring these and taking a "recipe book" approach.

Another potential criticism of EBP is that while patient preferences are a fundamental component, patients may still feel excluded. This is because EBP is often interpreted and implemented as disease-oriented and the protocols, guidelines, and clinical recommendations that are produced are not necessarily developed to answer the question about "what is the best treatment or procedure for this patient at this particular time" (p. 357).[5] Evidence should not be put ahead of patient choice. Clinicians do not necessarily know what is best for the individual patient. This is especially true in the case of chronic illness, where the person who has to live with the disease and its treatment must make choices about what he or she values and the goals that are important to them.

Yet another possible criticism of EBP is the potential that it may be put up as the cure-all when there are other factors that are also having an impact on the standard of care and/or care outcomes, such as poor management, sociodemographic factors, staff skills, and the effects of the patient's environment.[22]

IMPLEMENTING EBP

Establishing what is the best evidence is not the same as implementing the best evidence.[8,23,24] "Changing practice requires energy, motivation, and support from others and needs to be sustained over a period of time" (p. 269).[25]

Numerous barriers have been identified in relation to implementation and the following section will highlight some of these along with strategies for addressing them.

LEARNING TO SELL CHANGE

Change is frequently feared and resisted, and volumes have been written on how to lead change. Changing practice to implement evidence faces similar challenges. Change experts note the only change we can expect is that there will be more change and it will happen more rapidly! It is a fundamental part of life. Clinicians need to see therefore that change can be used by them for something positive, an opportunity for improvement, rather than something that happens or is imposed by others. Clinicians need to be convinced that the change is not about someone else's investment. They also need to view the activity as everyday clinical practice rather than as something different or special.[25]

If the problem or area targeted for practice change has been identified by clinicians "working at the coal face," then they are more likely to "own" and participate in that practice change.[7] If change is approached negatively, then it turns into a negative experience. Alternatively, we can learn from skilled marketers how to sell new ideas and achieve "take up" of products that the consumer didn't ever consider as important to him or her!

Change strategies need to be implemented at both an institutional and an individual level. Dobbins *et al.*[23] (p. 180) note that organizations that "have high functional differentiation, a culture that values the use of research evidence in practice, effective communication systems, decentralized decision making, managerial support for change, and adequate resources are more likely to adopt innovations."

Changing practices through implementation and utilization of research evidence will only occur, however, if there is an awareness of the impact of culture in the clinical workplace. Providing information through such avenues as in-service lectures and handouts with an expectation that clinicians will then adopt this into their practice is generally considered to be ineffective and is referred to as the passive-diffusion model.[24] The development of a shared philosophy within the organization, however, can guide change. If those involved do not have a common understanding of the problem, the stakeholders, and the outcomes to be achieved, then positive change will not take place. An example that highlights a lack of a shared philosophy can be seen in the different understandings of how to address alcoholism. If management assumes alcoholism is an individual problem, whereas the staff see it as a social problem, it is likely both parties will also have quite different views on what an appropriate intervention might be. If staff can see no rationale for the intervention, they are more inclined to resist. When trying to implement change it is also important to use the same language and have shared definitions.

Leadership (as opposed to management) and modeling are integral to implementation of change. Having key staff who know the system and players, are good facilitators, and recognized opinion leaders to be the everyday champions at care level has been associated with more success.[24,25,27] The appointment of champions and rewards for involvement in changing practice can be very effective. According to Perry,[25] there are obvious advantages in having clinical leaders/champions from within the organization as they know both the system and who does what and they therefore have access to data, personnel, and clinical areas. Their credibility within the facility has already been established.

IMPLEMENTATION OF CHANGE

There can be many challenges to the introduction of evidence into practice. It is therefore necessary to ascertain as much information as possible about potential challenges in a particular context so that they can be addressed as part of the planning process.[26] Information can be gathered from many sources, including formal meetings and focus groups, ward meetings, corridor conversations, nursing handovers, case conferences, and grand rounds. Any opportunity for discussion and feedback should be exploited. The following steps can be important in diagnosing likely challenges. First, it is essential to identify all stakeholder groups who might be involved in, influenced by, or able to influence the change proposed.[7,9,23] An analysis of stakeholders can assess the values, interests, and beliefs of key persons who may have an impact on change proposed.[23] Such an analysis also assesses stakeholders' level of support and identifies potential ways to engage their interest.[23] Second, there needs to be an assessment of the characteristics of the proposed change that might influence its acceptance.[9] Third, there needs to be an assessment of the readiness of clinicians to make the change and the factors within this group of clinicians that may have an impact on their capacity to do so.[9] Finding out what change strategies worked previously may provide a good starting place.[25] Fourth, possible external barriers to proposed change should be canvassed—for example, professional colleges, the government, industrial unions, or community leaders may have vested interests that could interfere with the change process. This may be avoided by ensuring they are informed and included in any planning they have the potential to scuttle!

CONTEXTUALIZE

Regardless of the strength of evidence, implementation will be enhanced by attention to the applicability and feasibility for each context. Current practice needs to be audited and benchmarked against evidence-based guidelines.[7,25] Just because an intervention has been found to be effective in one context does not mean that it can implemented in all situations. An audit of current practice also provides an indication of what can be done in the specific environment and what needs to be adapted to the particular context so that the best available evidence can be integrated. Current practice cannot be audited, however, without appropriate tools. For example, an audit of unmet-need behaviors (such as wandering, aggression, or restlessness) and the ways in which they are addressed, cannot be undertaken if current practice does not document these behaviors or measure their incidence and severity. It is also important however, to ensure that current organizational documentation is not duplicated as clinicians do not need to increase their workload. An audit may actually identify areas in which the workload could be decreased. Using audit tools that require identification of structure, process, and outcome criteria[28] facilitates attention to the full context rather than, as so often happens, focusing solely on education, which is known to be insufficient to bring about sustainable change. For example, it may be that the assessment tools are inappropriate, the documentation is tedious and irrelevant, or the staff who are supposed to implement the change do not have the authority to do so. Focusing only on outcome will not identify these issues and in tight economic and time-poor environments it is important that energy and resources are directed efficiently and effectively to gain maximum results.

CONCLUSION

EBP involves much more than locating, analyzing, and appraising the best evidence available about the effectiveness of an intervention. While there are certainly limitations to EBP, criticisms should be grounded in an understanding of the broader definition of EBP and not on a narrow misinterpretation of practice based only on RCTs.

Clinical judgment and expertise together with patient preferences are fundamental to successful EBP. Criteria for judging the level of evidence should be related to the question at hand, not just on effectiveness. The implementation of EBP involves an understanding of the context in which the change is to take place and the use of strategies that empower all stakeholders to participate.

REFERENCES

1. Sackett, D.L., W. Rosenberg, J.A.M. Gray, *et al.* 1996. Evidence based medicine: what it is and what it isn't. Br. Med. J. **312:** 71–72.
2. Waldman, M.H. 2006. Evidence-based medicine: how to translate research into patient care. J. Am. Podiatric Med. Assoc. **96:** 374–377.

3. CRAIG, J.V. & R.L. SMYTH, Eds. 2007. The Evidence-Based Practice Manual for Nurses. Elsevier. London.
4. GERRISH, K. & J. CLAYTON. 1998. Improving clinical effectiveness through an evidence-based approach: meeting the challenge for nursing in the United Kingdom. Nurs. Admin. Q. **22:** 55–65.
5. DRACUP, K., C.W. BRYAN-BROWN. 2006. Evidence-based practice is wonderful. . .sort of. Am. J. Crit. Care **15:** 356–358.
6. LIVESLEY, J. & M. HOWARTH. 2007. Integrating research evidence into clinical decisions. *In* The Evidence-Based Practice Manual for Nurses. J.V. Craig & R.L. Smyth, Eds.: 209–233. Elsevier. London.
7. NAY, R. 2003. Evidence-based practice: does it benefit older people and gerontic nursing? Geriatr. Nurs. **24:** 338–342.
8. DICENSO, A., G. GUYATT & D. CILISKA. 2005. Evidence-Based Nursing: a Guide to Clinical Practice. Elsevier Mosby. St. Louis, MO.
9. THOMPSON, C. 2007. How can we develop an evidence-based culture? *In* The Evidence-Based Practice Manual for Nurses. J.V. Craig & R.L. Smyth, Eds.: 305–337. Elsevier. London.
10. DICKENSON, D. & P. VINEIS. 2002. Evidence-based medicine and quality of care. Health Care Anal. **10:** 243–259.
11. NAIR, B.R. 2002. Evidence based medicine for older people: available, accessible, acceptable, adaptable. Australas. J. Ageing **21:** 58–60.
12. AUSTRALIAN INSTITUTE OF HEALTH AND WELFARE. 2006. Australia's Health 2006 Cat. No. AUS73. AIHW. Canberra.
13. KERRIDGE, I.H., M. LOWE & D. HENRY. 1998. Ethics and evidence based medicine. Br. Med. J. **316:** 1151–1153.
14. GUPTA, M. 2004. Evidence-based medicine: ethically obligatory or ethically suspect. Evid. Based Ment. Health **7:** 296–298.
15. MOYNIHAN, R. 2003. Who pays for the pizza? Redefining the relationships between doctors and drug companies. 1: Entanglement. Br. Med. J. **326:** 1189–1192.
16. JORGENSEN, A.W., J. HILDEN & P.C. GOTZSCHE. 2006. Cochrane reviews compared with industry supported meta-analyses and other meta-analyses of the same drugs: systematic review. Br. Med. J. **333:** 782–786.
17. SHAH, R.V., T.J. ALBERT, V. BRUEGEL-SANCHEZ, *et al.* 2005. Industry support and correlation to study outcome for papers published in *Spine*. Spine **30:** 1099–1104.
18. COLLINS, S., T. VOTH, A. DICENSO, *et al.* 2005. Finding the evidence. *In* Evidence-Based Nursing: a Guide to Clinical Practice. A. DiCenso, G. Guyatt & D. Ciliska, Eds.: 20–43. Elsevier Mosby. St. Louis, MO.
19. CRANSTON, M. 2002. Clinical effectiveness and evidence-based practice. Nurs. Stand. **16:** 39–43.
20. SIMPSON, B. 1996. Evidence-based nursing practice: the state of the art. Can. Nurse **92:** 22–25.
21. MCCURREN, C.D. 1995. Research utilization: meeting the challenge. Geriatric Nursing **16:** 132–135.
22. MUIR GRAY, J. 1997. Evidence-Based Healthcare. Churchill Livingstone. London.
23. DOBBINS M., D. CILISKA, C. ESTABROOKS, *et al.* 2005. Changing nursing practice in an organization. *In* Evidence-Based Nursing: a Guide to Clinical Practice. A. DiCenso, G. Guyatt & D. Ciliska, Eds.: 172–200. Elsevier Mosby. St. Louis, MO.

24. DAVIES, P. 2003. Introducing change. *In* Evidence-based Practice. M. Dawes, P. Davies, A. Gray, J. Mant, K. Seers & R. Snowball, Eds.: 223–241. Elsevier Churchill Livingston. London.
25. PERRY, L. 2007. Implementing best evidence in clinical practice. *In* The Evidence-Based Practice Manual for Nurses. J.V. Craig & R.L. Smyth, Eds.: 267–304. Elsevier. London.
26. THOMPSON, C. & M. LEARMONTH. 2002. How can we develop an evidence-based culture? *In* The Evidence-Based Practice Manual for Nurses. J.V. Craig & R.L. Smyth, Eds.: 211–239. Elsevier Science. London.
27. TITLER, M.G., J.C. MENTES, B.A.A. RAKEL, *et al.* 1999. From book to bedside: putting evidence to use in the care of the elderly. Jt. Comm. J. Qual. Improv. **25:** 545–556.
28. MORRELL, C. & G. HARVEY. 2003. The Clinical Audit Handbook. Elsevier Science Limited. London.

Erratum

In Ref. 1, an author's name was inadvertently omitted from the list of authors on page 24; the list read as follows:

Amir Tanay,[a] Eyal Leibovitz,[b] Angela Frayman,[b] Reuven Zimlichman,[c] and Dov Gavish[b]

Below is the correct list of authors:

Amir Tanay,[a] Eyal Leibovitz,[b] Angela Frayman,[c] Reuven Zimlichman,[c] Marina Shargorodsky,[c] and Dov Gavish[b]

REFERENCE

1. TANAY, A., E. LEIBOVITZ, A. FRAYMAN, *et al.* 2007. Vascular Elasticity of Systemic Lupus Erythematosis Patients Is Associated with Steroids and Hydroxychloroquine Treatment. Ann. N.Y. Acad. Sci. **1108:** 24–34.

Ann. N.Y. Acad. Sci. 1114: 464 (2007).
doi: 10.1196/annals.1396.052

Index of Contributors